EUTROPHICATION PROCESSES

IN COASTAL SYSTEMS

Origin and Succession of Plankton Blooms and Effects on Secondary Production in Gulf Coast Estuaries

Frontispiece

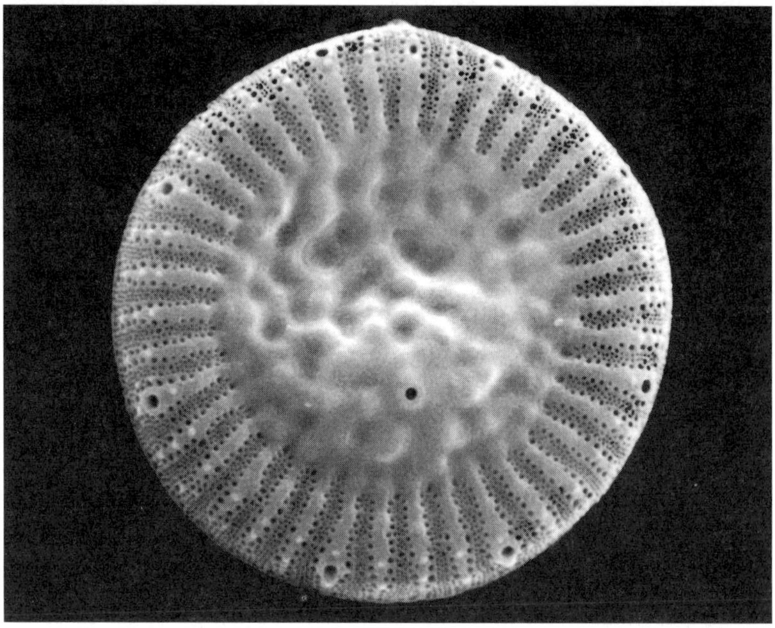

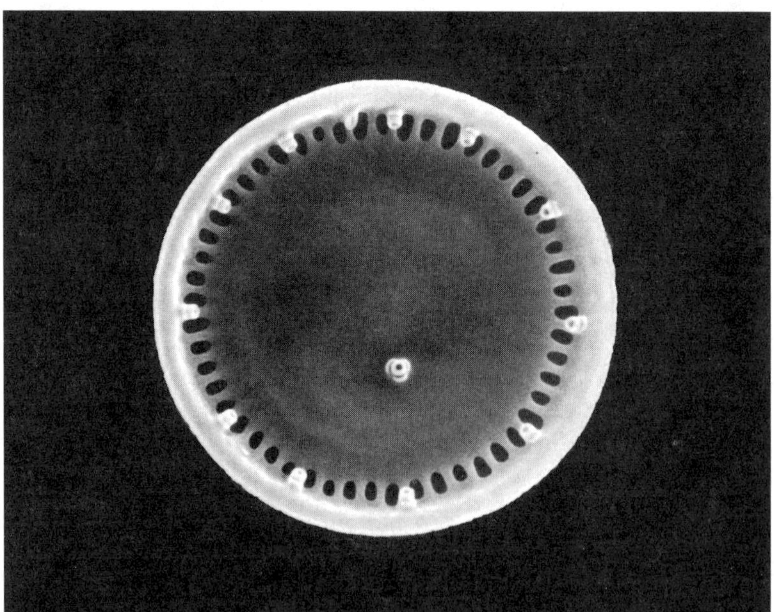

Cyclotella choctawhatcheeana Prasad, a bloom-forming, planktonic diatom from Perdido Bay system. Top: External view of the valve. Diameter of the valve = 6 μm. Acid-cleaned material. Bottom: Internal view of the valve. Diameter of the valve = 7 μm. Acid-cleaned material. (Scanning electron micrographs courtesy of A. K. S. K. Prasad, Florida State University.)

EUTROPHICATION PROCESSES IN COASTAL SYSTEMS

Origin and Succession of Plankton Blooms and Effects on Secondary Production in Gulf Coast Estuaries

ROBERT J. LIVINGSTON
Director
Center for Aquatic Research and Resource Management
Florida State University
Tallahassee

CRC Press
Boca Raton London New York Washington, D.C.

COVER ART

The cover represents a pictorial metaphor for the chief findings of the book. Anthropogenous alterations of nutrient loading to coastal systems initiates complex changes of associated phytoplankton communities. Temporal progressions of bloom species lead to reduced phytoplankton species richness and altered dominance relationships. These changes are passed along to associated food webs which, through basic alterations in the response of key trophic groups, lead to major reductions of secondary production.

Senior Editor: John B. Sulzycki
Project Editor: Christine Andreasen
Production Manager: Suzanne Lassandro
Marketing Manager: Carolyn Spence

Library of Congress Cataloging-in-Publication Data

Livingston, Robert J.
Eutrophication processes in coastal systems : origin and succession of plankton blooms and effects on secondary production in Gulf Coast estuaries / Robert J. Livingston.
p. cm.
Includes bibliographical references (p.).
ISBN 0-8493-9062-1
1. Marine eutrophication--Mexico, Gulf of. 2. Coastal ecology--Mexico, Gulf of. 3. Estuarine ecology--Mexico, Gulf of. I. Title. II. Series.

QH91.8.E87 L58 2000
577.7′86364--dc21 00-058572

Direct all inquiries to CRC Press LLC, 2000 N.W. Corporate Blvd., Boca Raton, Florida 33431.

International Standard Book Number 0-8493-9062-1
Library of Congress Card Number 00-058572
Printed in the United States of America 1 2 3 4 5 6 7 8 9 0
Printed on acid-free paper

Marine Science Series

The CRC Marine Science Series is dedicated to providing state-of-the-art coverage of important topics in marine biology, marine chemistry, marine geology, and physical oceanography. The Series includes volumes that focus on the synthesis of recent advances in marine science.

CRC MARINE SCIENCE SERIES

SERIES EDITOR

Michael J. Kennish, Ph.D.

PUBLISHED TITLES

Artificial Reef Evaluation with Application to Natural Marine Habitats, William Seaman, Jr.
Chemical Oceanography, Second Edition, Frank J. Millero
Coastal Ecosystem Processes, Daniel M. Alongi
Ecology of Estuaries: Anthropogenic Effects, Michael J. Kennish
Ecology of Marine Bivalves: An Ecosystem Approach, Richard F. Dame
Ecology of Marine Invertebrate Larvae, Larry McEdward
Environmental Oceanography, Second Edition, Tom Beer
Estuary Restoration and Maintenance: The National Estuary Program, Michael J. Kennish
Handbook of Marine Mineral Deposits, David S. Cronan
Handbook for Restoring Tidal Wetlands, Joy B. Zedler
Intertidal Deposits: River Mouths, Tidal Flats, and Coastal Lagoons, Doeke Eisma
Morphodynamics of Inner Continental Shelves, L. Donelson Wright
Ocean Pollution: Effects on Living Resources and Humans, Carl J. Sindermann
Physical Oceanographic Processes of the Great Barrier Reef, Eric Wolanski
The Physiology of Fishes, Second Edition, David H. Evans
Pollution Impacts on Marine Biotic Communities, Michael J. Kennish
Practical Handbook of Estuarine and Marine Pollution, Michael J. Kennish
Practical Handbook of Marine Science, Second Edition, Michael J. Kennish
Seagrasses: Monitoring, Ecology, Physiology, and Management, Stephen A. Bortone

Preface

The research effort on which this book is based has involved continuous analyses of various river–estuarine and coastal systems in the northeastern Gulf of Mexico since 1970. Multiyear studies (over 60 field-years of data in coastal areas) have been carried out using a combination of field descriptive and experimental (laboratory and field) approaches. The research team for this work has included field collection personnel, chemists, taxonomists, experimental biologists, physical oceanographers, hydrological engineers, statisticians, computer programmers, and modelers. A long-term, interdisciplinary, comparative database has been created that is currently being published in a series of books and peer-reviewed scientific journals. This field program has been used to evaluate system-level processes that determine the relationships of nutrient loading/nutrient dynamics, primary productivity and eutrophication processes, and associated food web responses. Efforts have been made to determine how human activities affect these processes. The emphasis of the research has been on seasonal and interannual trophic response to habitat changes and an evaluation of the interrelationships of natural and anthropogenous nutrient loading, primary and secondary production, and the effects of toxic substances. This research is currently being used to develop realistic indices of ecosystem condition.

This book includes several features that have been integrated into an ecosystem evaluation of the role of eutrophication in shaping primary and secondary production in coastal systems. Detailed, long-term, species-specific analyses of column phytoplankton and benthic microalgae were analyzed relative to long-term changes of nutrient loading and the response of the food webs in Perdido Bay. The food web data were based on a long-term stomach content program whereby data concerning invertebrates and fishes taken in a series of coastal systems along the northeastern Gulf of Mexico were organized into ontological feeding units. These data were then translated into a trophic organization. By reviewing detailed changes of the phytoplankton communities over different nutrient loading regimes, it was possible to determine the response of the food webs to such changes. The Perdido Bay data were then compared to similar information taken in a series of other coastal systems in the northeastern Gulf.

Author

Robert J. Livingston is currently Professor of Biological Science and Director of the Center for Aquatic Research and Resource Management at Florida State University (Tallahassee, Florida, U.S.A.). His interests include aquatic ecology, pollution biology, field and laboratory experimentation, and long-term ecosystem-level research in freshwater, estuarine, and marine systems. Past research includes multidisciplinary studies of lakes and a series of drainage systems on the East and Gulf Coasts of the United States. Over the past 31 years, Dr. Livingston's research group has amassed over 70 field-years of data in various freshwater and marine systems. Dr. Livingston has directed the programs of 49 graduate students who have carried out research in behavioral and physiological ecology with individual aquatic populations and communities in lakes, rivers, and coastal systems. Dr. Livingston is the author of more than 140 scientific papers and 4 books on the subject of aquatic ecology, and he has been the principal investigator for over 100 projects since 1970.

The primary research program of the author has been carried out largely in the southeastern United States. The areas studied include the Apalachicola drainage system, the Choctawhatchee drainage system, the Perdido drainage system, the Escambia Bay system, the Blackwater River–estuary, the Escatawpa–Pascagoula drainage system, the Mobile River–estuary, the Winyah Bay system (including the Sampit River), Apalachee Bay (the Econfina and Fenholloway River–estuaries), and a series of north Florida lake systems. This work has included determinations of the impact of various forms of anthropogenous activities on a range of physicochemical and biological processes. A sophisticated computer system has been developed to aid in the analysis of the established long-term databases.

Acknowledgments

There were many people who contributed to the research effort over the past 31 years. It is impossible to acknowledge all of these people. Literally hundreds of undergraduate students, graduate students, technicians, and staff personnel have taken part in the long-term program.

The author is grateful for the astute reviews and editing of this book by Dr. D. M. Anderson, Dr. A. K. S. K. Prasad, Mr. G. C. Woodsum, Dr. M. J. Kennish, and Ms. V. Tschinkel. Dr. A. K. S. K. Prasad carried out phytoplankton taxonomy and systematics. Dr. K. Rhew has provided support in the microalgal analyses. Mr. Robert L. Howell IV ran the field collections. Mr. G. C. Woodsum designed the overall database organization and, together with Mr. P. Homann, ran database management and day-to-day data analysis and presentation. Dr. F. Graham Lewis III was instrumental in the development of analytical procedures and statistical analyses. Dr. D. A. Flemer designed the overall approach for analytical chemistry.

The early Perdido Bay work was carried out with the services of the following individuals: Dr. D. A. Birkholz (toxicology, residue analyses), Dr. G. S. Brush (long-core analysis), Dr. L. A. Cifuentes (nutrient studies, isotope analyses), Dr. R. A. Coffin (nutrient studies, isotope analyses), Dr. W. P. Davis (biology of fishes), Dr. J. H. Epler (aquatic insects), Dr. M. Franklin (riverine hydrology), Dr. T. Gallagher (hydrological modeling), Dr. C. R. Gilbert (fish systematics), Dr. W. C. Isphording (marine geology), Dr. R. D. Kalke (estuarine zooplankton), Dr. C. J. Klein III (aquatic engineering, estuarine modeling), Dr. C. C. Koenig (biology of fishes), and Dr. A. W. Niedoroda (physical oceanography).

The author would like to thank the following phycologists for their helpful comments and advice on taxonomy and nomenclature of phytoplankton species over the years: for diatom taxonomy, Dr. C. W. Reimer (Academy of Natural Sciences of Philadelphia), Dr. G. A. Fryxell (University of Texas at Austin), Dr. G. R. Hasle (University of Oslo, Norway), Dr. P. Hargraves (University of Rhode Island), Professor F. Round (University of Bristol, England), Mr. R. Ross, Ms. P. A. Sims, Dr. E. J. Cox, and Dr. D. M. Williams (British Museum of Natural History, London, U.K.), Dr. L. K. Medlin and Dr. R. M. Crawford (Alfred-Wegener Institute, Bremerhaven, Germany), Professor T. V. Desikachary (University of Madras, India), Dr. J. A. Nienow (Valdosta State University, Georgia), Dr. P. Silva (University of California at Berkeley), Dr. M. A. Faust (Smithsonian Institution), Dr. R. A. Andersen (Bigelow Laboratory for Ocean Sciences, Maine), Dr. D. Wujeck (Michigan State University), Dr. M. Melkonian (University of Köln, Germany), Dr. C. J. Tomas (Florida Department of Environmental Regulation, St. Petersburg), Professor T. H. Ibaraki (Japan), Dr. Y. Hara (University of Tsukuba, Japan), and Dr. J. Throndsen (University of Oslo, Norway). Technical assistance with transmission and scanning electron microscopes was provided by Ms. K. A. Riddle (Department of Biological Science, Florida State University). Mrs. A. Black and Mr. D. Watson (Histology Division, Department of Biological Science, Florida State University) aided in specimen preparation and serial thin sectioning.

I am also grateful to the curators of the following International Diatom Herbaria for arranging the loan of type collections and other authentic materials: Academy of Natural Sciences of Philadelphia (ANSP), California Academy of Sciences (CAS, San Francisco), National Museum of Natural History, Smithsonian Institution (Washington, D.C.), Harvard University Herbaria (Cambridge, Massachusetts), The Natural History Museum (British Museum, London, U.K.), and the F. Hustedt Collections, Alfred-Wegener Institute (Bremerhaven, Germany).

Dr. D. A. Meeter, Dr. X. Niu, and Ms. L. E. Wolfe performed statistical analyses. Dr. D. C. White performed microbiological analyses. Dr. S. E. McGlynn ran the chemistry laboratories. Mr. C. R. Cripe, Mr. W. Meeks and Mr. P. Moreton provided laboratory assistance. Mr. D. Fiore, Mr. O. Salcedo, Ms. I. Salcedo, Mr. R. Wilt, Mr. S. Holm, Mr. B. Bookman, Ms. L. Bird, Ms. L. Doepp, Ms. C. Watts, and Ms. M. Guerrero-Diaz provided other forms of laboratory assistance. Field support for the various projects was provided by Mr. H. Hendry, Mr. A. Reese,

Mr. T. Bevis, Mr. C. J. Boschen, Mr. J. G. Kobylinski, Ms. J. Scheffman, Ms. H. S. Greening, Mr. B. McLane, Mr. S. Holm, Mr. M. Hollingsworth, Dr. H. Jelks, Ms. K. Burton, Ms. E. Meeter, Ms. C. Meeter, Ms. S. Solomon, Mr. S. S. Vardaman, Ms. S. van Beck, Mr. M. Goldman, Mr. S. Cole, Mr. K. Miller, Mr. C. Felton, Ms. B. Litchfield, Mr. J. Montgomery, Ms. S. Mattson, Ms. C. Phillips, Mr. J. D. Ryan, Mr. P. Rygiel, Mr. J. Duncan, Ms. S. Roberts, Mr. T. Shipp, Ms. A. Fink, Ms. J. Livingston, Ms. J. Huff, Mr. R. M. Livingston, Mr. A. S. Livingston, Mr. T. Space, Ms. R. A. Reese, Mr. M. Wiley, Mr. D. Dickson, Ms. L. Tamburello, and Mr. C. Burbank.

Field collections, taxonomic identification, and associated laboratory analyses were made by the following individuals: Dr. F. Graham Lewis III (aquatic macroinvertebrates), Mr. R. L. Howell III (infaunal and epibenthic invertebrates and fishes), Mr. W. R. Karsteter (aquatic macroinvertebrates), Dr. P. Sheridan (trophic analyses), Dr. J. Schmidt (aquatic macroinvertebrates), Dr. A. W. Stoner (trophic analyses), Dr. R. A. Laughlin (trophic analyses), Dr. B. M. S. Mahoney (experimental biology), Dr. K. Main (experimental biology), Dr. K. Leber (experimental biology), Dr. W. H. Clements (trophic analyses), Dr. G. L. Ray (infaunal macroinvertebrates), Dr. B. MacFarlane (fishes), Dr. K. R. Smith (oligochaetes), Dr. J. L. Luczkovich (experimental biology), Dr. J. Holmquist, (experimental biology), Dr. D. Bone (experimental biology), Dr. E. L. Bousfield (amphipods), Dr. F. Jordan (trophic analyses), Ms. J. Schmidt-Gengenbach (experimental biology), Dr. C. J. Dawes (submerged aquatic vegetation), Dr. R. W. Yerger (fishes), Mr. M. Zimmerman (submerged aquatic vegetation), Mr. M. Kuperberg (submerged aquatic vegetation), and Ms. J. J. Reardon (zooplankton, submerged aquatic vegetation).

Project administrators included the following individuals: Dr. T. W. Duke, Dr. R. Schwartz, Dr. M. E. Monaco, Ms. J. Price, Dr. N. P. Thompson, Mr. J. H. Millican, Mr. D. Arceneaux, Mr. W. Tims, Jr., Dr. C. A. Pittinger, Mr. M. Stellencamp, Ms. S. A. Dowdell, Mr. C. Thompson, Mr. K. Moore, Dr. E. Tokar, Dr. D. Tudor, Dr. E. Fernald, Ms. D. Giblon, Ms. S. Dillon, Ms. A. Thistle, Ms. M. W. Livingston, and Ms. C. Wallace.

The book cover was designed by Julia B. Livingston and Robert J. Livingston.

To every thing there is a season,
And a time to every purpose under the heaven.

—Ecclesiastes 3:1

"I weep for you," the Walrus said:
"I deeply sympathize."
With sobs and tears he sorted out
Those of the largest size
Holding his pocket-handkerchief
Before his streaming eyes.
But answer came there none —
And this was scarcely odd, because
They'd eaten every one.

—*Through the Looking-Glass* by Lewis Carroll

A man and what he loves and builds have but a day and then disappear; nature cares not — and renews the annual round untired. It is the old law, sad but not bitter. Only when man destroys the life and beauty of nature, there is the outrage.

—*Grey of Fallodon* by George Macaulay Trevelyan

Contents

Chapter 1 Introduction
I. Eutrophication Processes 1
II. Aspects of Excessive Nutrient Loading to Estuaries 3
III. Plankton Response to Nutrient Loading 4
IV. Phytoplankton Blooms and the HAB Concept 5
V. Natural History of HABs 6
VI. Phytoplankton as Indicators of Water Quality 7
VII. Food Web Response to Nutrient Loading 9

Chapter 2 Northeast Gulf of Mexico
I. Physiography 11
II. Regional Geology 14
III. Climate 17
A. Temperature 17
B. Precipitation 18
C. Wind 20
IV. River Flow 21
V. Tides 22
VI. Aquatic Habitats 22

Chapter 3 Primary Study Sites
I. Apalachee Bay (Econfina and Fenholloway River–Estuaries) 26
II. Apalachicola River and Bay System 27
III. The Choctawhatchee River and Bay System 29
IV. The Pensacola River and Bay System 31
V. The Perdido River and Bay System 33

Chapter 4 The Perdido Bay System: Eutrophication Processes and Plankton Blooms
I. Introduction 37
A. The Perdido Bay Study 38
B. Phytoplankton Bloom Status 39
II. River Flow Trends 40
A. Seasonal and Interannual River Flow Patterns 41
B. Riverine Influence on Bay Stratification and Flushing 42
III. Nutrient Dynamics 44
A. Nutrient Loading 44
B. Nutrient Concentration Gradients 49
C. Nutrient Limitation 51
IV. Sediment Quality 56
A. Introduction 56
B. Physical Characteristics and Liquid Mud 57
C. Sediment Nutrients 58
D. Stable Isotope Analyses of Carbon Cycling 63
E. Toxic Agents 64
V. Water Quality: Spatial/Temporal Trends 65
A. Introduction 65
B. Temperature and Salinity 68
C. Color, Turbidity, and Secchi Depths 69

D. Light Distribution and Primary Productivity 70
E. Dissolved Oxygen, Biochemical Oxygen Demand, and pH 72
F. Nutrients and Chlorophyll *a* 76
VI. Long-Term Phytoplankton Trends in Perdido Bay 81
A. Introduction 81
B. Chronology of Phytoplankton Trends in Perdido Bay 85
1. 1988 to 1991 85
2. 1992 to 1994 86
3. 1995 to 1997 86
4. 1998 to 1999 86
C. Bloom Species 91
D. Long-Term Changes at the Division/Class Level 100
E. Silica and Diatom Abundance 104
F. Net Phytoplankton and Zooplankton 106
G. Whole-Water Phytoplankton 110
1. Long-Term Changes in Phytoplankton Distribution 110
2. Plankton Response to Nutrient Loading 118
H. Phytoplankton Blooms and the Benthic Microalgae 121
I. Phytoplankton Blooms and Water Quality 137
1. Introduction 137
2. Long-Term Water Quality Trends 137
3. Phytoplankton Blooms and Chlorophyll *a* 140
J. Long-Term Sediment Quality Trends in Perdido Bay 143
K. Statistical Analyses of Phytoplankton Distributions 146
1. The 1993–1994 Bloom Period 146
2. Comparison of Long-Term Trends 148

Chapter 5 Effects of Blooms on Secondary Production
I. Introduction 153
II. Baywide Trends of Invertebrates and Fishes 154
A. Introduction 154
B. Community Interactions 154
C. Trophic Organization 162
1. Research Approach 162
2. Long-Term Trophodynamic Trends 166

Chapter 6 Comparison of Gulf Coastal Systems
I. Habitat Conditions 175
II. Salinity Stratification and Habitat Deterioration 176
A. Stratification, Dissolved Oxygen, and Sediment Condition 176
III. Nutrient Loading and Nutrient Concentrations 182
A. Introduction 182
B. Flow Rates and Nutrient Loading/Concentrations 182
IV. Water Quality 188
V. Sediment Comparisons 191
VI. Phytoplankton Organization 196
A. Introduction 196
B. Phytoplankton Communities in the Study Areas 197
1. Wolf Bay 197
2. Escambia Bay 203
3. Apalachee Bay 210

4. Choctawhatchee Bay 212
5. Perdido Bay 217
C. Phytoplankton Indices 217
VII. Food Web Response to Plankton Organization 224
A. Introduction 224
B. Existing Conditions 228
C. Food Web Comparisons 228

Chapter 7 Restoration, Research, and Regulation
I. Restoration of the Perdido System 237
II. "Ecosystem" Research 240
A. Gulf of Mexico 241
B. Chesapeake Bay 241
C. Florida Bay 244
D. Scientific Overview of the Eutrophication Question 246
III. Information, Regulatory "Action," and Political Control 251

Chapter 8 Conclusions 259

Chapter 9 Summary of Results 269

References 281

Appendix I 301

Appendix II 315

Index 319

CHAPTER 1

Introduction

I. EUTROPHICATION PROCESSES

Eutrophication represents and defines one of the central themes of aquatic ecology: the production of organic matter that forms the basis of aquatic food webs. There is a rich scientific literature concerning many aspects of the eutrophication process whereby nutrients are taken up by aquatic plants (phytoplankton, benthic microalgae, epiphytes, emergent vegetation, and submergent macrophytes) to form organic carbon. The basic process is generally the same for most aquatic systems. However, the ultimate end products exemplified by the resultant food webs vary from system to system depending on the multifold interactions of the nutrient loading process, nutrient limitation characteristics of the receiving system, habitat distribution, and existing biological conditions.

River–estuarine and associated coastal systems are among the most productive areas on earth due to nutrient enrichment from land runoff, the usually shallow nature of the receiving system, and natural energy supplements in the form of wind, tidal currents, and thermohaline circulation. Phytoplankton and benthic microalgae represent one of the primary sources of organic carbon in river-dominated estuaries. Because of high diversity, taxonomic complexity, and rapid reproduction rates, community interactions of coastal microalgae are poorly understood even though such organisms form the basis of many coastal food webs. The food web structure of natural aquatic systems is molded by the eutrophication process, with complex interactions that depend largely on multiple biological processes of the primary producers and highly evolved feeding strategies of the consumers (Livingston et al., 1997). In addition to being highly productive, many temperate, river-dominated estuarine and coastal systems are known for rapid changes in the physical habitat that contribute to stress-induced changes in food web response. The myriad juxtapositions of the eutrophication process in coastal systems reflect varying combinations of nutrient loading and response mechanisms in habitats that change both seasonally and interannually. Variation of the trophic organization of estuarine systems is an important part of the eutrophication process.

Periodic loading of carbon, nitrogen, phosphorus, and other nutrients and organic compounds from allochthonous (land runoff) and autochthonous sources affect river-dominated estuaries. The often shallow, coastal receiving areas are also influenced by mixing characteristics and current patterns that dominate the physical environment. The resulting food webs are based on the combination of high primary production and limitations on species richness imposed by the extremely variable estuarine habitat. It can be argued that this paradigm has been overused since there is a considerable range of variability in overall primary and secondary production both within and between various coastal systems. The processes that determine primary productivity and the resulting food webs vary considerably from one system to the next. This variation depends on factors such as the physiographic structure of the receiving system, temperature, salinity, wind, tides, currents, stratification characteristics, light transmission, nutrient loading events, nutrient concentrations, sediment characteristics, primary and secondary production, and biological processes such as competition and predation. Thus, different combinations of common variables lead to highly variable food web responses from one system to the next that are then translated into major differences in population dynamics, community structure, and overall secondary production. Despite such variation, the basic eutrophication process remains essentially the same, whereby light energy is captured and redistributed into myriad forms of coastal plants and animals.

There is growing evidence that human activities increasingly affect both the quantitative and qualitative aspects of nutrient loading and phytoplankton response (i.e., bloom formation) in coastal systems around the world. Bricker et al. (1999), in a comprehensive review of the extent of the eutrophication problem in the conterminous United States, found that 44 estuaries, representing 40% of the total estuarine surface area studied, suffer "high expressions of eutrophic conditions" with an additional 40 estuaries having "moderate conditions." The primary symptoms on which the survey was based included decreased light availability, high chlorophyll *a* concentrations, high epiphytic/macroalgal growth rates, changes in algal dominance (diatoms to flagellates, benthic to pelagic dominance), and increased decomposition of organic matter due to the high chlorophyll *a* concentrations and macroalgal growth. These effects were most pronounced along the coasts of the Gulf of Mexico and the Middle Atlantic states. Commercial and shellfish fishery use was impaired in 43 and 46 estuaries, respectively. Most of the estuaries studied were affected by human influence although only 14% of the affected estuaries had correspondingly high nitrogen inputs. More than half of these estuaries had a high capacity to retain nutrients. It was projected that eutrophic conditions would worsen in 86 estuaries by the year 2020. The authors emphasized that better data are needed in terms of estimators of anthropogenous forms of nutrient pressure and nutrient inputs by source.

According to Bricker et al. (1999), there is currently little in the way of available "process-oriented data relative to the mechanisms involved in the progressive development of eutrophication." Currently, thresholds of nutrient loading leading to toxic blooms remain largely undetermined and the relationships of nitrogen and phosphorus to such blooms need "further clarification." The influence of climate changes on blooms remains undefined and the combination of nutrient inputs and other anthropogenous stressors remains largely unknown. There has been virtually no information generated concerning the species-specific qualitative and quantitative changes of the phytoplankton communities relative to the generation of the blooms. The authors found little information with regard to the impacts of plankton blooms on coastal food webs. In other words, the causative factors involved in the generation of phytoplankton blooms, the effects of such blooms on plankton community organization, and the impacts of altered plankton assemblages on secondary production through effects on coastal food webs all remain largely undefined (Bricker et al., 1999).

A comprehensive understanding of the effects of cultural eutrophication on diverse coastal systems is complicated by the extreme complexity of the response of dynamic phytoplankton assemblages to background variation. Despite their importance in terms of overall estuarine primary production, water column phytoplankton and benthic microalgae represent one of the most complex and least well understood parts of any given coastal system. The daunting taxonomic problems associated with microalgal associations usually preclude detailed and quantitative determinations of what is basically an ephemeral series of rapidly cycling bursts of production. The cycles of these microalgal populations can be measured in periods ranging from seconds to years. The assimilative capacity of any given coastal system relative to nutrient loading remains a major question in the evaluation of the transformation from natural eutrophication processes to hypereutrophication and postulated deterioration of associated food webs. It is likely that the responses of the microalgae to nutrient loading determine the structure of the estuarine food webs. In turn, food web processes are central to the nature and extent of population distribution. Although these aspects of eutrophication have been subject to extensive research for decades, there is still a considerable lack of understanding concerning how nutrient loading and habitat characteristics affect the quantitative and qualitative aspects of primary production, and how the resulting microalgal associations are related to associated food webs. This fundamental inconsistency of our knowledge of coastal systems is largely the result of the almost complete absence of long-term, interdisciplinary studies that integrate the various aspects of nutrient loading, microalgal response, and food web structure.

Human population increases in coastal areas throughout the world have become an increasingly important source of nutrient loading to estuarine systems. Recent studies indicate that phytoplankton are the single most important factor in eutrophication processes of river-dominated estuaries. It is therefore necessary to understand phytoplankton population and community dynamics if the postulated effects of plankton blooms on the trophic organization of coastal systems are to be understood. Unfortunately, scientific inquiry in this phase of aquatic ecology is still in the early stages of descriptive research. The taxonomy and natural history of coastal phytoplankton populations remain among the least-understood aspects of coastal ecology. The recent increase of destructive plankton blooms due to anthropogenous nutrient loading has led to renewed interest in this discipline. However, information at the ecosystem level relating to the role of microalgal community structure on the eutrophication of estuarine and coastal systems remains poorly developed even though such processes are central to a basic understanding of the origin and succession of plankton blooms.

II. ASPECTS OF EXCESSIVE NUTRIENT LOADING TO ESTUARIES

Kennish (1997) has reviewed the problems associated with anthropogenous loading of nutrients to coastal systems. Resulting hypereutrophication has long been correlated with increased biochemical oxygen demand (BOD), hypoxia and anoxia in receiving estuaries. Chronic hypoxia has been a problem in coastal areas such as Chesapeake Bay, the Florida Keys, the Pamlico River–estuary, Long Island Sound, and broad areas of the northern Gulf of Mexico associated with runoff from the Mississippi and Atchafalaya Rivers. Hypoxia is also considered a problem in estuaries and bays throughout Europe, the Far East, and Australia (Kennish, 1997). Other postulated water quality changes due to excessive nutrient loading include increased turbidity, reduced light penetration, and deterioration of sediment quality. Proliferation of nuisance algal macrophytes and plankton blooms, often associated with losses of productive seagrass beds, are considered major factors in the response of coastal areas to anthropogenous nutrient loading. Toxic agents associated with plankton blooms have been responsible for fish kills and the loss of commercially important species in addition to creating public health problems following consumption of affected sea life.

The primary effects of nutrient loading on shallow estuaries and lagoons have been outlined by McComb (1995). A common aspect in the review of cultural eutrophication in diverse estuaries includes association of nutrient loading with the development of phytoplankton blooms, associated fish kills, and the general decline of fisheries (Hodgkiss and Yim, 1995). Nutrient enrichment also adversely affects seagrass beds through changes associated with stimulation of micro- and macroalgae (Duarte, 1995; Hein et al., 1995; Valiela et al., 1997). The emphasis on shallow systems highlights the importance of sediments in the nutrient loading and accumulation process (McComb and Lukatelich, 1995). Sediment resuspension as a product of water column dynamics can be a factor in nutrient storage and loading processes (de Jonge and Raaphorst, 1995; de Jonge, 1995), although such effects are not always associated with hypereutrophication (Marcomini et al., 1995). The timed release and bioavailability of sediment nutrients, as modified by resuspension, remobilization, and regeneration through biotic activity, remain poorly understood even though such processes could have an important effect on the response of benthic microalgae and macrophytes (Thornton et al., 1995a). The lack of information regarding non-point-source nutrient loading due to urban and agricultural runoff is another consideration when reviewing the causes of hypereutrophication (King and Hodgson, 1995). De Jonge and Raaphorst (1995) reviewed the positive and negative effects of increased nutrient loading. There can be initial increases of secondary production and fisheries output with increased nutrient loading (Streckis et al., 1995; Thornton et al., 1995b). However, prolonged increases often end in plankton blooms and accompanying declines of coastal populations that are often not discovered until there is an advanced state of hypereutrophication. The complex processes associated with the interaction and competition of microphytes and macrophytes often complicate a uniform response of different estuaries to increased anthropogenous nutrient loading (Streckis et al., 1995). However, in most of these studies, there is a general lack of detailed data concerning how phytoplankton blooms are initiated and how changes in the phytoplankton community structure actually affect secondary production. Despite a plethora of studies concerning hypereutrophication in estuarine and coastal systems, the processes involved in nutrient loading that ultimately lead to altered phytoplankton populations and associated food web changes remain largely undefined.

Howarth et al. (2000) conducted a review of the effects of excess nutrients on coastal systems. This worldwide phenomenon has led to increased algal biomass, excessive concentrations of sometimes toxic algae in the form of harmful brown and red tides, reduced seagrass and coral reef habitat, altered marine biodiversity, hypoxia/anoxia, and the loss of commercial fisheries. The authors emphasized that a considerable number of coastal systems in the United States have some of the symptoms of excessive nutrification. The report was designed to make recommendations for the implementation of management efforts to reduce nutrient loading. These recommendations included expansion of monitoring programs, development of ways to reduce non-point sources of nitrogen and phosphorus, an increased federal role in eutrophication issues, development of a classification scheme for management of nutrient overenrichment, improvement of comprehensive assessments of environmental quality and associated modeling efforts, and expansion of our knowledge concerning eutrophication questions. The dimensions of the nutrification problem in the coastal waters of the United States have been generally well established (Bricker et al., 1999). The key to any restoration effort rests on the development of adequate scientific knowledge of the eutrophication process and the determination of how specific changes in anthropogenous nutrient loading affect individual estuaries and coastal systems.

III. PLANKTON RESPONSE TO NUTRIENT LOADING

The response of individual phytoplankton populations to nutrient loading in coastal systems has been extensively studied (Anderson and Garrison, 1997). Short-term studies of bloom species at well-defined taxonomic levels are the rule both in funding of the research and publication of the results. There are studies that go beyond autecological analyses. Squires and Sinnu (1982) addressed the complexity of the response of diatom assemblages in estuarine areas to varying habitat conditions. Currents and salinity were important factors in the determination of diatom assemblages in the study area. Marshall (1982a,b) showed that distributions of marine diatoms in the Gulf of Maine were related to the location of large bay systems and the Georges Bank. Anthropogenous effects on water quality often are evaluated relative to the distribution of marine phytoplankton. According to Reddy and Venkateswarlu (1986), pulp mill effluents caused reduced numbers of phytoplankton in the effluent channel. The Cyanophycean types were dominant in such areas, with an almost total lack of green algae. *Oscillatoria* spp. were dominant in the polluted regions, with *Rhopalodia gibberula* and *Nitzschia palea* as the primary diatom species. Premila and Rao (1977) showed that blue-green algae (*O. nigroviridis*) were indicative of waters affected by sewage relative to less-polluted areas. Low salinity along with the sewage effluents in such areas tended to encourage this species. However, blue-green algae are also abundant in marine areas under natural conditions (Potts, 1980). Chang (1988) described a *Prorocentrum micans* bloom in a New Zealand estuary that was coincident with increased nitrogen from upwelling. The author found onshore–offshore relationships of size fractions with larger cells found in higher proportions in inshore areas. Inshore fractions accounted for lower percentages of chlorophyll *a*. However, in all these studies, little consideration was given to the response of the phytoplankton community to nutrient loading. Most such studies were related to changes in single-species populations.

One constant in the evaluation of plankton blooms is the central role of nutrient loading in destabilization of plankton populations. Another commonly found result is the range of adaptations of bloom species that enhance their ability to outcompete other plankton species. Riegman (1998) related the occurrence of harmful algal blooms (HABs) to macronutrient dynamics. He noted that diatom growth during blooms associated with anthropogenous N + P loading can be inhibited by silica availability. This could be related to enhanced N:Si and P:Si supply ratios that favor nonsiliceous phytoplankton (Smayda, 1990). Shifts in N:P supply ratios can also cause changes in species composition. Nutrient enrichment with macronutrients can also be indirectly related to changes in phytoplankton community composition through induced changes in predation, resource limitation, light requirements, and biological effects on sediments. Altered nutrient speciation can also influence these changes. The ecological role of bloom-associated toxins in such interspecific interactions remains unclear (Riegman, 1998). The fact that various dinoflagellates are not usually eaten by zooplankton (compared with non-dinoflagellates) could be related to the generation of toxins. Bloom-associated toxins could thus be related to protection of populations from losses due to predation. Of course, such protection could then result in competitive advantages of bloom species leading to elimination of other species through competition for nutrients, light, or other potentially limiting factors.

The production of resting spores by various plankton species is another possible explanation for the predominance of bloom species during periods of destabilized nutrient conditions. Some dinoflagellate cysts have spines and others are toxic; the resting spores of dinoflagellates may have more toxin than the reproducing vegetative cells (Lirdwitayaprasit et al., 1990; Oshima et al., 1992; Riegman, 1998). Thus, the often-complex life history of bloom species could be associated with the mechanism of relative dominance and elimination of competing plankton species. However, the exact process of how such dominance is established and how shifts in dominance affect long-term changes of phytoplankton community structure remain largely unknown. Philippart et al. (2000) monitored a single station over a prolonged period and made identifications of various taxonomic units within the phytoplankton community of a tidal inlet in the Wadden Sea over a 21-year period. They found major changes in phytoplankton biomass and species composition that coincided with shifts of a phosphorus-controlled system to a more eutrophic nitrogen-controlled system. The authors found a strong causal relationship between TN:TP ratios and phytoplankton community structure with increases of the size of the plankton.

IV. PHYTOPLANKTON BLOOMS AND THE HAB CONCEPT

The composition and numbers of estuarine and coastal phytoplankton depend on a delicate balance of various factors, and these aspects of phytoplankton ecology need to be viewed within the twin aspects of spatial heterogeneity and seasonal/interannual variation. Of the (approximate) total number of marine phytoplankton species (5000), some 300 species are known to occur at numbers high enough to discolor seawater (Sournia et al., 1991). About 40 or 50 of these species produce toxins that can affect natural marine populations of plants and animals as well as human beings (Hallegreaff et al., 1995). These so-called harmful algal blooms or HABs have become the focus of increased research over the past decade as part of the perception that these blooms are increasing in both frequency and severity of impact on a worldwide basis. Although the HAB concept may be refined (Smayda, 1997a), there is general agreement that an understanding of the causes of such blooms and bloom dynamics with respect to impacts on secondary production has become the principal research interest in aquatic systems.

HABs take three general forms (Hallegreaff et al., 1995):

1. Nontoxic blooming populations reaching concentrations that eventually affect important environmental factors such as dissolved oxygen with the resulting hypoxia/anoxia ending in debilitation and/or extirpation of other populations.
2. Toxic bloom species that introduce toxic agents into associated food webs to the extent that upper trophic levels (including humans) are adversely affected.
3. Toxic bloom species that produce and release substances having direct and/or indirect effects on associated populations. These species are usually not harmful to humans but are known to affect other aquatic plant and animal species adversely.

Representative groups that have documented histories as HABs include blue-green algae, diatoms, pyrmnesiophytes, dinoflagellates, and raphidophytes. Unfortunately, there has been little unequivocal coupling established between microalgal concentrations and their harmful effects (Hallegreaff et al., 1995). In addition, most of the studies of individual HABs have been undertaken after the blooms are under way, thus precluding the determination of specific cause-and-effect relationships. Smayda and Shimizu (1993) have summarized what is known in this area. However, background information concerning bloom causation and the ecological effects of HABs remain largely dependent on serial autecological studies, thus precluding realistic appraisals of how HABs are created and how such populations affect phytoplankton assemblages as a whole. The implications of increased bloom activity relative to long-term changes in coastal primary and secondary production have simply not been addressed. The combination of increased specialization and a reductionist approach to coastal research has resulted in a serious lack of understanding regarding the long-term (i.e., chronic) impacts of plankton blooms on coastal systems.

One of the longest series of studies of harmful algal blooms concerns outbreaks of red tide that have been noted along the west coast of Florida since the 1840s. This phenomenon is based on the sudden appearance of high concentrations of the unarmored dinoflagellate *Gymnodinium breve.* This species has been associated with discoloration of the water, fish kills, shellfish poisoning, and human-related ailments associated with aerosols (Steidinger et al., 1966). Characteristics of stream flows in areas of the Gulf of Mexico affected by red tides have been reviewed (Martin et al., 1971). However, causation of the red tide events remains largely undocumented. Martin and Taft (1998) reviewed efforts for control and mitigation of the red tide phenomenon, noting that recent (1995–1996) red tide outbreaks have been associated with the deaths of manatees. There is a growing literature concerning possible control measures for red tide events (Anderson, 1996). Martin and Taft (1998) developed criteria for the use of pesticides in the control of red tide. However, studies that relate specific combinations of causative factors to the red tide outbreaks have not been developed, and determination of causation as a potential way of solving the growing red tide problem remains illusory.

There are many problems associated with the ingestion of seafood contaminated by toxic algal blooms. Organisms affected include fishes, seabirds, porpoises, whales, and humans (National Research Council, 1999). Documented increases of red tide incidences have been reported from Tolo Harbor, Hong Kong (Lam and Ho, 1989), and in the Inland Sea in Japan (Okaichi, 1997). Qualifications of these findings are given by Howarth et al. (2000). Increased incidence of *Chattonella* blooms has been associated with fish kills in the Seto Inland Sea (Okaichi, 1997). The recently described dinoflagellate *Pfiesteria* has been associated with extensive fish kills along the eastern coast of the United States. Burkholder and Glasgow (1997) provided evidence of the increasing prevalence of problems associated with this species relative to increased loading of nutrients to coastal areas.

V. NATURAL HISTORY OF HABS

One key to an understanding of the impact of phytoplankton blooms involves a clear understanding of the life history of bloom species. Burkholder et al. (1992) have illustrated the complexity of this issue and indicate that relatively little is known about the life history of many bloom species. The effects of toxic plankton blooms may be associated with a number of complex life history stages of the species involved. Dinoflagellates, for example, can have toxic vegetative cells and cysts that produce neurotoxins. However, relatively few species of cyst-forming dinoflagellates are toxic, and very few cells actively form cysts late in their brief lives. Dinoflagellates usually encyst during unfavorable conditions and may do so in two ways. They may form temporary cysts (without flagella) which can return to the motile stage when favorable environmental conditions occur. They can also form resting cysts that lie in bottom sediments for several months, especially during winter, or even longer before becoming resuspended in the water column and germinating (excystment) to form further motile stages. Once the cysts are present in the sediments, they may be transported elsewhere by currents, thus creating the possibility of reseeding the plankton over time. Also, phenomena such as upwelling and downwelling can have different effects on the formation of toxic blooms. Recent studies with the dinoflagellate *P. piscicida* (Steidinger et al., 1998a, b) indicate that there can be direct attacks by specific life stages with attachment to a dying fish and subsequent digestion of tissue debris. Lethal flagellated vegetative cells can move from the sediment surface to the water column in response to finfishes that have been adversely affected by the release of neurotoxins from this dinoflagellate. The organism then kills the fish and rapidly descends once again so that the entire process is relatively ephemeral and cannot be detected using routine field sampling methods.

Regardless of the mechanism of toxicity, there is strong evidence that there is a direct correlation of toxic phytoplankton blooms with increasing nutrient enrichment (i.e., anthropogenous loading of nitrogen and phosphorus compounds) that makes the estuarine habitat more favorable for algal blooms and toxic effects (Burkholder et al., 1992). In addition, there is some evidence that dinoflagellates may have symbiotic relationships with estuarine/marine bacteria that could also be involved in the ecological consequences of the blooms. These general observations are consistent with what is known of other forms of HABs.

VI. PHYTOPLANKTON AS INDICATORS OF WATER QUALITY

A central theme of phytoplankton research has been the use of microalgae as indicators of water quality. Differences in essential nutrients, the physiological state of the indigenous algal populations, and other factors that are difficult to measure all contribute to the complexity of causal relationships between phytoplankton assemblages and water quality in aquatic systems. The relationship of freshwater microalgae with water quality indices has been well developed in the scientific literature and includes both qualitative and quantitative lines of research (Patrick, 1967; 1971; Patrick and Reimer, 1966). Various water quality changes have been associated with plankton blooms (Foehrenbach, 1971; Bidagare and Ondrusek, 1996; Latasa and Bidigare, 1998). Previous information in freshwater systems indicates that clean water supports speciose assemblages of algae, whereas polluted water tends to reduce the number of species and to increase the relative dominance of some species due to the survival and well-being of a few resistant forms. Some microalgae flourish in highly eutrophic conditions or are stimulated by various types of toxic wastes. These species can then outcompete other species leading to the use of such species as indicators of pollution. In addition, the effects of such species on community organization leads to the use of indices such as species diversity in determination of the effects of pollution on a given water body. This aspect of phytoplankton as indicators of pollution has not been well defined in coastal systems.

Various authors have reviewed the use of freshwater algae as indicators of nutrient input, eutrophication, and different forms of pollution. For instance, due to gas vacuole formation, an ability to escape nitrogen limitation, and a colonial growth habit, blue-green algal species of the genera *Anabaena*, *Anacystis*, and *Aphanizomenon* are considered indicative of water pollution of various types in freshwater systems. Blue-green algae such as *Oscillatoria nigroviridis* are known indicators of sewage in freshwater areas. However, blue-green algae have also been shown to be abundant in marine areas under natural conditions (Potts, 1980). Thus, direct relationships of specific single-species indicators are not always clear-cut in marine systems, and there is considerable variation in the response of various algal species from system to system.

Diatoms can be useful in the analysis of the history of a given water body (Bonin et al., 1986). Such algae are easy to collect, are cosmopolitan, and are responsive to short- and long-term changes of environmental conditions. Diatoms are also relatively well understood ecologically and are suitable for historic surveys and the calculation of diversity analyses. Shifts of diatom assemblages can thus be used to indicate changes of ambient water quality conditions. Different species of diatoms flourish under varying concentrations of nutrients and pollutants. Often, the response of a given population may be indirect, due to the effects of increased or decreased competition from bloom species that are affected by water quality conditions. However, although the basic causes of plankton blooms (i.e., increased nutrient loading and creation of nutrient imbalances) are similar in freshwater and marine systems, there may be basic differences in the responses of freshwater and marine plankton to such stimuli (Livingston, unpublished data).

Maestrini et al. (1984a, b) reviewed the subject of microalgae as indicators of water quality in marine systems. The authors pointed out the various problems involved in attempting to make associations of nutrient loading, water quality factors, and descriptive algal data. Marshall (1982a,b, 1984) and Marshall and Ranasinghe (1989) have described the phytoplankton distribution along the eastern coast of the United States with numerical dominance by pico-nanoplankton, centric diatoms, and *Skeletonema costatum*. Overall, environmental controls of coastal microalgae are extremely complex and are not well understood (Malone, 1977; Lehman, 1992). This is particularly true of coastal waters of the Gulf of Mexico where relatively little research has been carried out concerning long-term changes of phytoplankton communities. The few studies that address control of phytoplankton assemblages in coastal systems indicate that water circulation, temperature, salinity, light, and nutrients along with predation and interspecific competition appear to be the dominant controlling factors of phytoplankton associations in any given area. However, extreme spatial/temporal variation of phytoplankton communities confounds simple explanations of the associations of specific microalgae with water quality factors. More importantly, such studies usually do not address the effects of nutrient-induced changes of phytoplankton community structure on water and sediment quality. Differences in essential nutrients, the physiological state of the indigenous algal populations, and other factors that are difficult to measure all contribute to the complexity of the interactive relationships of phytoplankton and water and sediment quality in estuarine and marine systems. The toxic effects of contaminants on algal communities in the field have been addressed only recently. This entire subject remains relatively complex and poorly understood.

The search for water quality indicators of phytoplankton activity has been the subject of extensive scientific investigation. In fact, most references to the phytoplankton components of a given estuary are made through measured factors that are considered cognates. Ease of measurement has been substituted for accuracy and taxonomic precision. It follows that such studies have been carried out with little in the way of long-term changes of population and community structure of microalgal associations. O'Connor (1981) and O'Connor et al. (1977) discussed water quality indicators of the effects of hypereutrophication. Factors such as chlorophyll *a* have been widely used to represent phytoplankton communities in descriptive determinations of trophic organization and in a wide range of modeling efforts. Dissolved oxygen has also been used as an indicator of plankton activity. Dissolved oxygen in estuarine and coastal systems depends on complex interactions of physiographic, physical/chemical, and biological factors (Odum and Wilson, 1962; Foehrenbach, 1971; Livingston, 1997a). Often, dissolved oxygen distribution is strongly influenced by microalgal activity. Large, well-circulated, unstratified systems usually have fewer problems with low dissolved oxygen than smaller, more-stratified or isolated aquatic systems. However, the impacts of nutrient loading on dissolved oxygen can be felt in relatively large areas. Justic (1987) demonstrated long-term changes in the northern Adriatic Sea due to nutrient enrichment from sewage input with resulting benthic anoxia and mass mortalities of bottom organisms.

The basic lack of understanding of phytoplankton community processes should lead to caution in the use of phytoplankton as indicators of water quality. The usefulness of phytoplankton as indicators of water quality has been reviewed in the early literature (Officer et al., 1984). The long-term adverse impacts of nutrient loading through sewage disposal in the New York Bight were well documented (Mahoney et al., 1973; Young and Barber, 1973; Botton, 1979). However, major hypoxic events in areas associated with the New York Bight were not systematically associated with phytoplankton components although it was thought that such areas were periodically affected adversely by plankton populations. Such associations with water quality were only tentative and the major losses of marine life along the entire coast of New Jersey during a particularly severe and widespread incidence of coastal hypoxia and anoxia remain largely unexplained. Thus, definitive determinations of bloom episodes, hypoxic conditions, and the massive loss of secondary production remain the subject of considerable speculation even though cause-and-effect relationships associated with increased activity of so-called brown tides along this coast remain virtually undetermined. Although hypereutrophic events have been extensively reported, relatively few studies include detailed documentation of the response of plankton communities to nutrient loading and the secondary effects of the resulting blooms on water and sediment quality. Since many of the HAB species have relatively complex and poorly understood life histories, the relationships of bloom outbreaks and responses of co-occurring species remain vague and poorly documented.

The concentration of research on the relationships of water quality changes and phytoplankton blooms is paralleled by an almost complete lack of emphasis concerning how estuarine and coastal blooms affect food web processes. This subject is fundamental to an understanding of how water quality changes related to anthropogenous nutrient loading can lead to ecosystem-level shifts and the loss of secondary production through indirect effects on phytoplankton associations due to increased dominance by a small numbers of bloom species. Changes in water quality due to hypereutrophication may only be the hands of the clock; the actual mechanism of the response of coastal ecosystems to nutrient loading may be related more to fundamental changes in food web response as a result of altered phytoplankton community structure than to associated water quality changes.

VII. FOOD WEB RESPONSE TO NUTRIENT LOADING

The question of how much nutrient loading is necessary to maintain naturally high secondary productivity of a given coastal system remains speculative at this time. The associated question concerning the point at which anthropogenous nutrient loading becomes a source of destructive plankton blooms is also undetermined. The effects of anthropogenous increases in nutrient loading on coastal systems remains one of the central questions facing aquatic ecologists today. Because of the diversity and complexity of ecological processes of estuarine/coastal systems, a simple resolution of the nutrient question is not likely. In addition, there are usually multiple nutrient sources in most coastal systems that include both natural and anthropogenous sources. Multiple point and non-point sources present a major problem in the determination of specific cause-and-effect mechanisms. These problems should be addressed in any study of the causes and effects of plankton blooms on estuarine and coastal areas.

Nutrient imbalances have been associated with destabilization of plankton populations that have then led to fish kills, postulated imbalances of food webs, and general reductions of useful productivity in coastal areas. Early reviews of predation (Frost and Grazing, 1980) and interspecific competition (Smayda, 1980) illustrate the complexity of these processes both in the generation of bloom concentrations of estuarine phytoplankton populations and in the effects of such blooms on secondary production. Nitrogen fertilization of Long Island Sound from duck farms led to proliferations of chlorophytes (*Nannochlorius atomus* and *Stichococcus cylindrus*) that replaced the natural phytoplankton populations (Ryther, 1954; Smayda, 1980). These changes led to food web disruptions that were related to reductions of important fisheries in the receiving areas. Nutrient-based disruptions can also be translated into competitive exclusions due to bloom demands for essential nutrients such as iron. The resulting blooms can then induce production and excretion of chelators that suppress the growth of other algal species.

Reduction of phytoplankton species richness during nutrient-induced algal blooms has been described in the scientific literature, although how such changes affect food web responses remains largely unknown. Long-term changes in the bloom successions complicate the process of identifying impacts on secondary production. Smayda (1980) outlined how successions in nutrient-enriched inshore waters may occur. Small-celled diatoms such as *Skeletonema costatum* and *Leptocylindricus danicus* can give way to mixed communities of large-celled diatoms and dinoflagellates such as *Chaetoceros* spp. and *Prorocentrum* spp. Dinoflagellate blooms commonly occur in highly stratified conditions with predominance of *Gymnodinium* spp. and *Prorocentrum* spp. following intermediate replacement of the smaller-celled diatoms. Nutrient availability, vertical mixing, turnover rates, and water mass stratification also contribute to this form of succession. Predation rates and the order of abundance at the time of the nutrient loading also are determinants of bloom sequences. Unfortunately, there are virtually no detailed studies of the connection of nutrient loading and long-term changes of phytoplankton community response; consequently, the general model of phytoplankton successions remains largely inferential at this time.

Roelke et al. (1997) have developed models that, based on successions of phytoplankton populations in response to increased nutrient loading, may provide guidelines for phytoplankton management. The complex interaction of nutrient loading, nutrient limitation, grazing, interspecific (phytoplankton) competition, and seasonal/interannual changes in plankton population distributions all contributed to the response of Nueces Bay (Texas Gulf coast) to sewage wastes. Edible forms of phytoplankton appeared first in a release of nutrients to the estuary. These populations were then diminished through preferential grazing leading to the ascendance of less edible phytoplankton types. Phytoplankton changes appeared to react to specific ranges of nutrient loading ratios so that management of phytoplankton through control of nutrient ratios may be one form of nutrient management (Roelke et al., 1997). However, the basic mechanism of food web response to the nutrient loading over long-term periods remained beyond model predictions.

Overall, despite considerable research on phytoplankton assemblages in estuarine and coastal systems, there is still relatively little knowledge concerning the response of both water column and sedimentary microalgal communities to long-term trends of increased anthropogenous nutrient loading. Even less well understood are the implications of food web response to postulated changes of plankton community structure. Based on 30 years of continuous research on a series of river-estuarine and coastal systems in the northeast Gulf of Mexico, this book addresses the following issues: nutrient loading and coastal eutrophication, phytoplankton response to natural and anthropogenous nutrient loading, and long-term changes of coastal food webs as a result of phytoplankton blooms and associated changes in phytoplankton communities.

CHAPTER 2

Northeast Gulf of Mexico

The estuarine and coastal areas along the northeastern Gulf Coast are characterized by a series of habitats that are controlled to a considerable degree by the physiographic features of the adjoining freshwater drainage basins and associated Gulf connections (Figure 2.1). The coastal zone and associated freshwater drainages of the north Florida area extend from the Perdido River and Bay system in eastern Alabama and the western Panhandle of Florida to the small blackwater rivers on Apalachee Bay. Nine of the 12 major rivers and 5 of the 7 major tributaries of Florida occur in this region. The upland watershed of the north Florida Gulf Coast, which is located in Alabama, Florida, and Georgia, includes about 135,000 km^2. The combined area of the inshore estuarine systems in the region ≈2150 km^2. The coastal zone in this area is characterized by saltwater marshes, sandy beaches, tidal creeks, intertidal flats, oyster reefs, seagrass beds, subtidal unvegetated soft bottoms, and a broad assemblage of transitional habitats. Most of the inshore habitats of the major bays and estuaries are dominated by freshwater runoff and groundwater contributions from associated river basins. The salinity regimes of these areas are variously affected by major river systems or, as is the case in Apalachee Bay, by a series of small rivers and groundwater flows. The combined drainage of springs and streams contributes about 1 billion gallons of fresh water per day to Apalachee Bay, which is the only study area not affected by a major alluvial river system (Livingston, 1990).

I. PHYSIOGRAPHY

Fernald and Patton (1984) and Wolfe et al. (1988) give comprehensive reviews of the physiography of the study area. The land region from Apalachee Bay to Perdido Bay (Figure 2.1) is called the Florida Panhandle. In terms of human population numbers and the degree of urbanization and industrialization, this is one of the least-developed coastal areas in the continental United States. Much of the Panhandle landscape is the result of stream and river flows and wave action that has acted on the land surface over the past 10 to 15 million years. From the Pleistocene Period to the Recent (about 2 million years), there have been major changes in sea level that have greatly affected the land types in the Panhandle region. During the periods of lowered sea levels in the Pleistocene, much of present-day shallow areas was terrestrial resulting in an increase of 1.5 to 2 times the terrestrial acreage of the Panhandle coastal regions. This has resulted in a current coastal physiography that is dominated by beach ridges (marine in origin), barrier islands, spits, cliffs, swales, sloughs, dunes, lagoons, and estuaries along a relatively flat upland configuration. Barrier islands start in the eastern end of the Apalachicola Bay system, and include a series through Apalachicola Bay and west to the Pensacola and Perdido Bay systems. Brackish water lagoons are common in the Panhandle region, and include St. George Sound (Apalachicola estuary) and Santa Rosa Sound

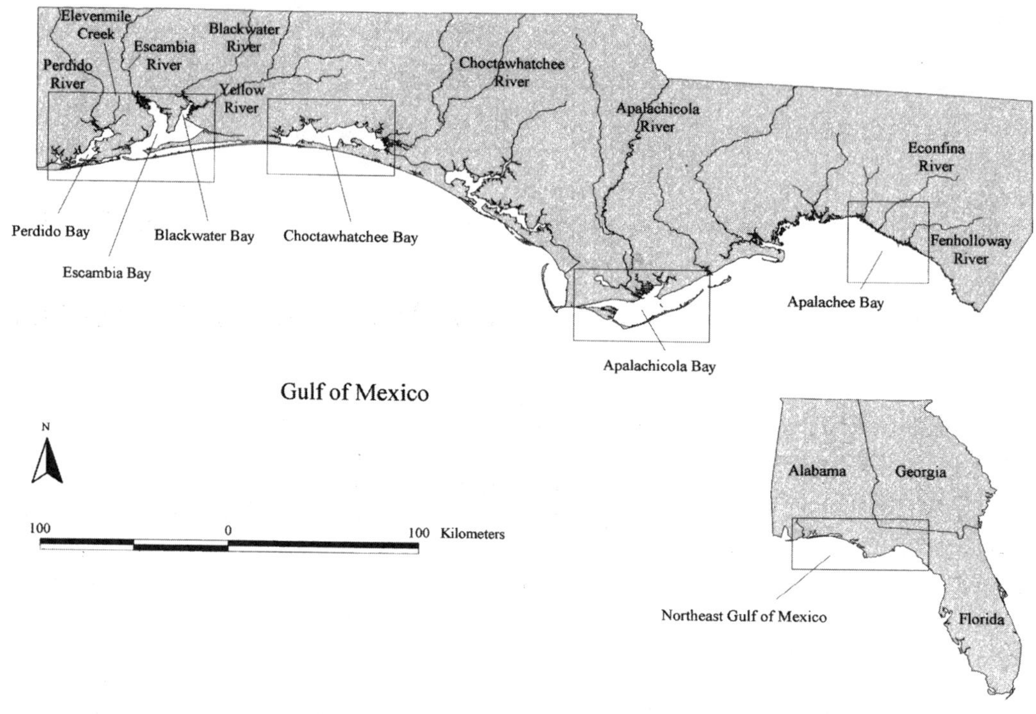

Figure 2.1 The Florida Panhandle showing the primary river basins and distribution of associated river–estuaries and coastal systems. This figure is a composite of information provided by the National Oceanic and Atmospheric Administration, the Florida Department of Environmental Protection, and the ESRI Corporation.

(Pensacola Bay system). On the eastern end of the Panhandle coast, there is a massive development of coastal swamps and marshlands that rim Apalachee Bay. This area is characterized by the absence of barrier islands as a result of the shallow, sloping margins, the lack of wave action, and an inadequate supply of sand (Tanner, 1960).

The major alluvial rivers of the northwest Florida Panhandle (Perdido, Escambia, Choctawhatchee, Apalachicola) have their headwaters in Georgia and Alabama (Figure 2.1). A series of smaller streams along the panhandle coast include the Blackwater and Yellow Rivers of the Pensacola Bay system, the Chipola River (part of the Apalachicola drainage), and the Ochlockonee River on Apalachee Bay. Farther down the coast, a series of small streams (St. Marks, Aucilla, Econfina, and Fenholloway) with drainage basins in Florida flow into Apalachee Bay.

The Perdido drainage system (Figure 2.2a) is a relatively shallow inshore embayment characterized by a dredged connection with the offshore Gulf, Perdido Pass. The bay complex lies in an area of submergence on the north flank of the active Gulf Coast geosyncline. The Pensacola Bay system (Figure 2.2a), a complex of three component bays with a restricted connection to the Gulf, is dominated by sediments provided by stream discharge (mainly the Escambia River) and wave/current action from the Gulf (Wolfe et al., 1988). Santa Rosa Sound has generally coarser sand-dominated sediments derived largely from offshore areas. Choctawhatchee Bay (Figure 2.2b) occupies a Pleistocene river valley that has been recently enclosed by a barrier system. Until the late 1920s, the bay was connected with the Gulf via a shallow pass and, as such, was periodically a fresh to brackish water lake. However, with the human creation of East Pass in 1929, Choctawhatchee Bay became a highly stratified estuarine system dominated by freshwater input from the Choctawhatchee River flowing over relatively salty Gulf water. Human modifications to the

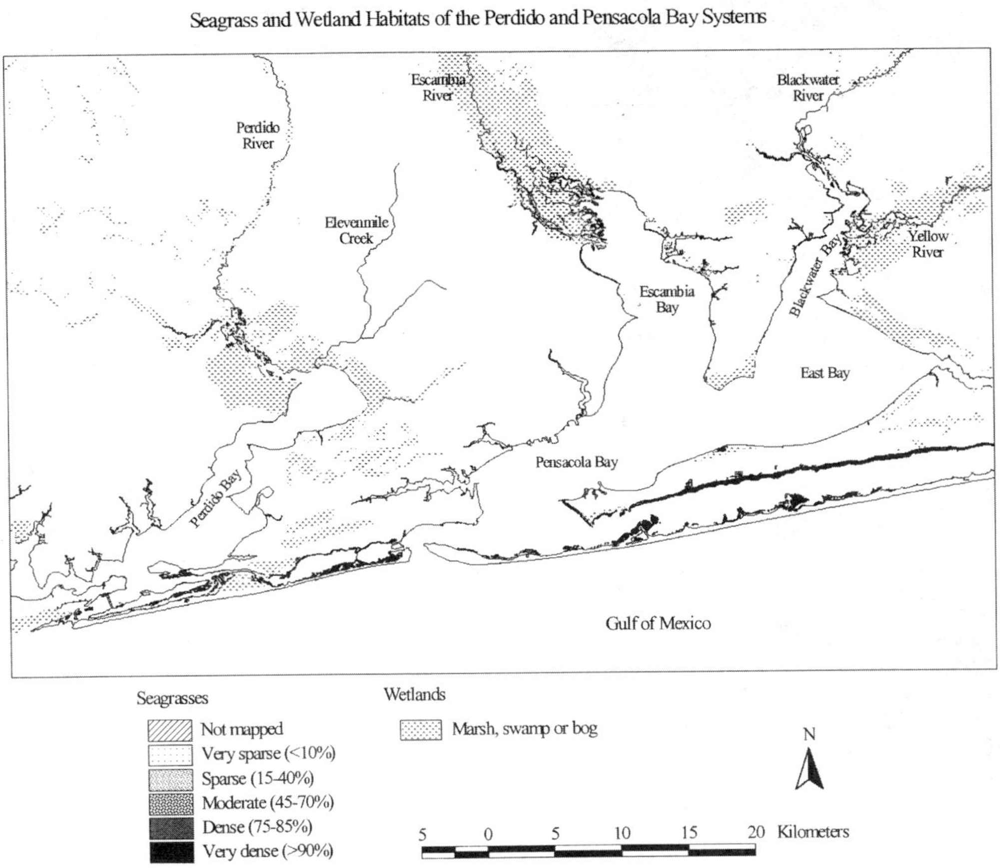

Figure 2.2 Detailed maps of bay systems in the northeast Gulf of Mexico. (a) The Perdido River–estuary showing sampling stations for long-term synoptic monitoring from October 1988 through September 1999 and the Pensacola Bay system (Escambia, Blackwater, Yellow) showing sampling stations for studies from May 1997 through October 1998.

connection of the bay with the Gulf changed the sedimentary environment that is currently dependent in large part on the entrapment of fine-grained sediments from the Choctawhatchee River. The artificial creation of the channel across Santa Rosa Island has also contributed to the erosional and depositional patterns in large portions of the Choctawhatchee Bay system.

The Apalachicola Bay system (Figure 2.2c), bounded gulfward by three barrier islands, is an extension of the Apalachicola River. This barrier system is a shallow width-dominated coastal plain estuary (Livingston, 1984a). Sand, silt, and shell components that have accumulated over tertiary limestones and marls of alluvial origin with some recent modifications due to hurricanes dominate sediments in this estuary. The region along the upper Gulf Coast of peninsular Florida from the Ochlockonee River to the Suwannee River is characterized by a series of drainage basins that include the Aucilla, Econfina, and Fenholloway Rivers. These streams drain into Apalachee Bay (Figure 2.2d). The smaller basins are wholly within the coastal plain as part of a poorly drained region that is composed of springs, lakes, ponds, freshwater swamps, and coastal marshes. The Econfina and Fenholloway river-estuaries both originate in the San Pedro Swamp. This basin has been affected, in terms of water flow characteristics, by long-term physical modifications through forestry activities. Apalachee Bay is part of a broad, shallow shelf area that occurs along the entire Gulf coast of peninsular Florida.

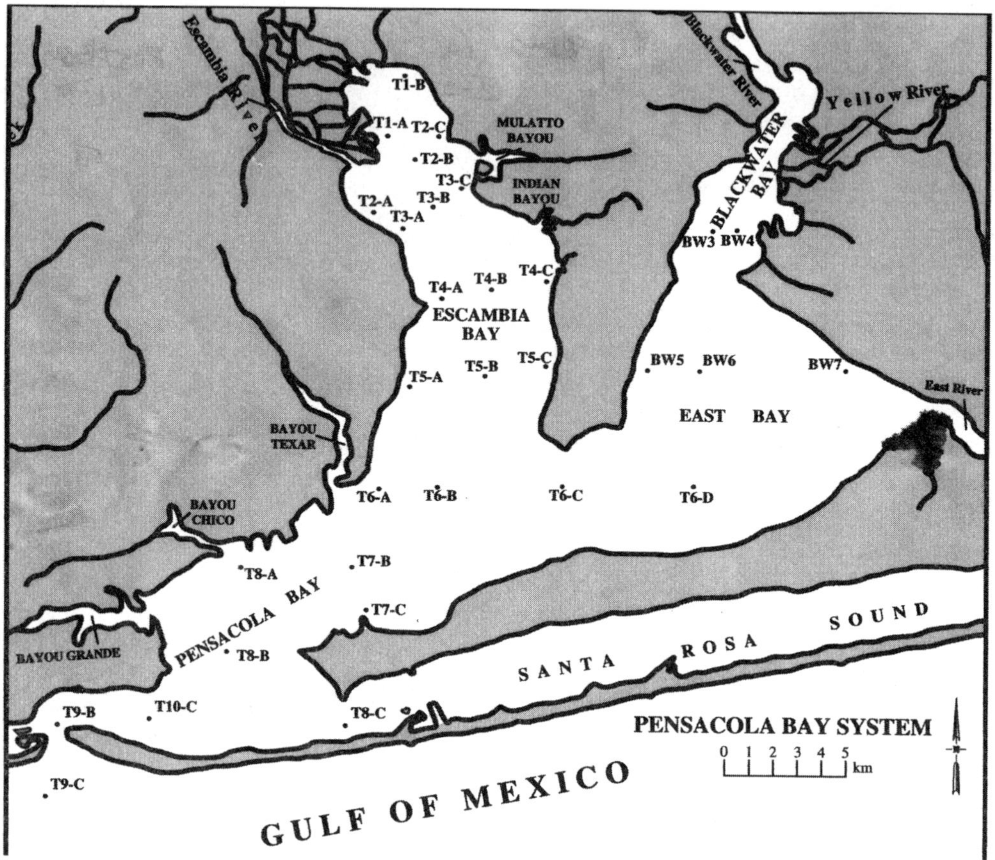

Figure 2.2(a) (continued)

II. REGIONAL GEOLOGY

The geomorphology of the Florida Panhandle is complex with strata that range in age from 450 million years ago (Paleozoic) to the present (Fernald and Patton, 1984). Wolfe et al. (1988) have reviewed the geology of the Panhandle region. The surface geology of the Florida Panhandle is sedimentary, with limestones, organics, and clastics (gravel, sand, silt, clay) forming the primary sediment types. The western Panhandle coast from the Perdido system to the Ocklochonee drainage is dominated by medium to fine sand and silt, whereas the big bend area (Apalachee Bay) is predominantly limestone or limestone/dolomite (Fernald and Patton, 1984). In the eastern half of the Panhandle, ground surfaces are affected by surface limestones and solution activity. Only those predominantly marine deposits of the past 60 million years are important for groundwater resources. Sandy clays dominate northern portions of northwest Florida or clayey sands deposited by ancient rivers, whereas sands deposited along the constantly changing shorelines characterize southern areas. The Panhandle area has been emergent since the Miocene; however, about half of the open Gulf portions of the beach areas have been eroding due to postulated sea level increases (Wolfe et al., 1988).

The geology of the northern Gulf coast of Florida is dominated by four structural features: the Gulf of Mexico Sedimentary Basin (west of Choctawhatchee Bay), the Chattahoochee Anticline (a rise at the eastern end of the Panhandle), the Apalachicola Embayment (from Alligator Point south of Tallahassee, Florida to the Choctawhatchee drainage), and the Suwannee Straits (in the

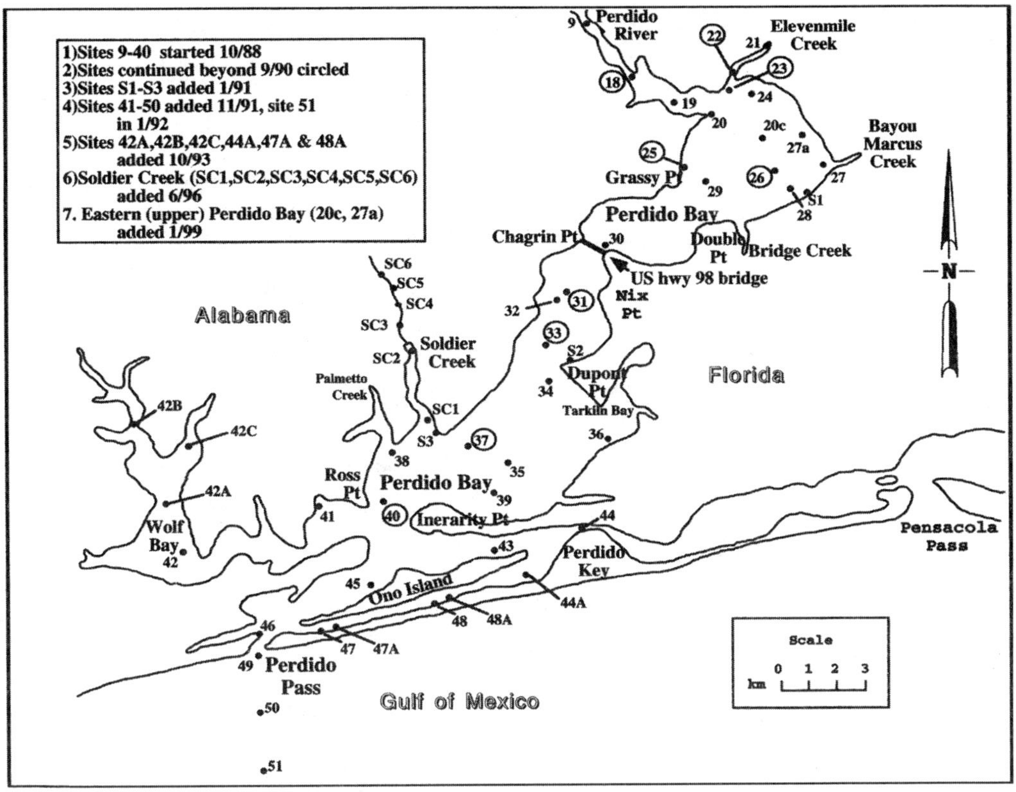

Figure 2.2(a) (continued)

Big Bend area of north central Florida). The Apalachicola Delta lies at the center of the Apalachicola Embayment, which is a relatively shallow basin that is characterized by the ground surface exposure of Oligocene and Eocene carbonate rocks. These carbonate deposits substantially affect the physical and chemical properties of water in this region. Offshore of this area lies the North Gulf Coast Sedimentary Province, whereas the Florida Peninsula Sedimentary Province underlies the Apalachee Bay region.

The present sea level along the northern Gulf coast of Florida was established approximately 6000 years ago (Wolfe et al., 1988). Due to post-Pleistocene geological changes, the coastal regions of the study area are the result of complex combinations of erosion, deposition, rainfall, waves, tides, and wind. Human influence is considerable in some of the more highly populated areas such as lower Perdido Bay, Escambia–Pensacola Bay, and western sections of Choctawhatchee Bay. The various physiographic features such as beach ridges, bars, dunes, spits, and coastal flats are the result of stranding as ancient sea levels receded. During the Pleistocene, much of the currently submerged areas along the inshore northern Gulf coast of Florida was at one time exposed to air and to colonization by terrestrial organisms.

The coastal area from the Ocklochonee drainage to the southern reaches of Apalachee Bay is characterized by the flat, Karst topography of the Gulf Coastal Lowlands. There are no barrier islands in this area. A series of small streams flow through freshwater swamps and salt marshes into a coastal region dominated by low wave activity, a paucity of sand supplements to the offshore area, and a shallow marine surface that is part of the broad, marine shelf (Tanner, 1960). Westward from Ocklochonee Bay to the Apalachicola region, the Ocklochonee and Apalachicola rivers provide alluvial input for a series of barrier islands (St. Vincent, St. George, and Dog). From western

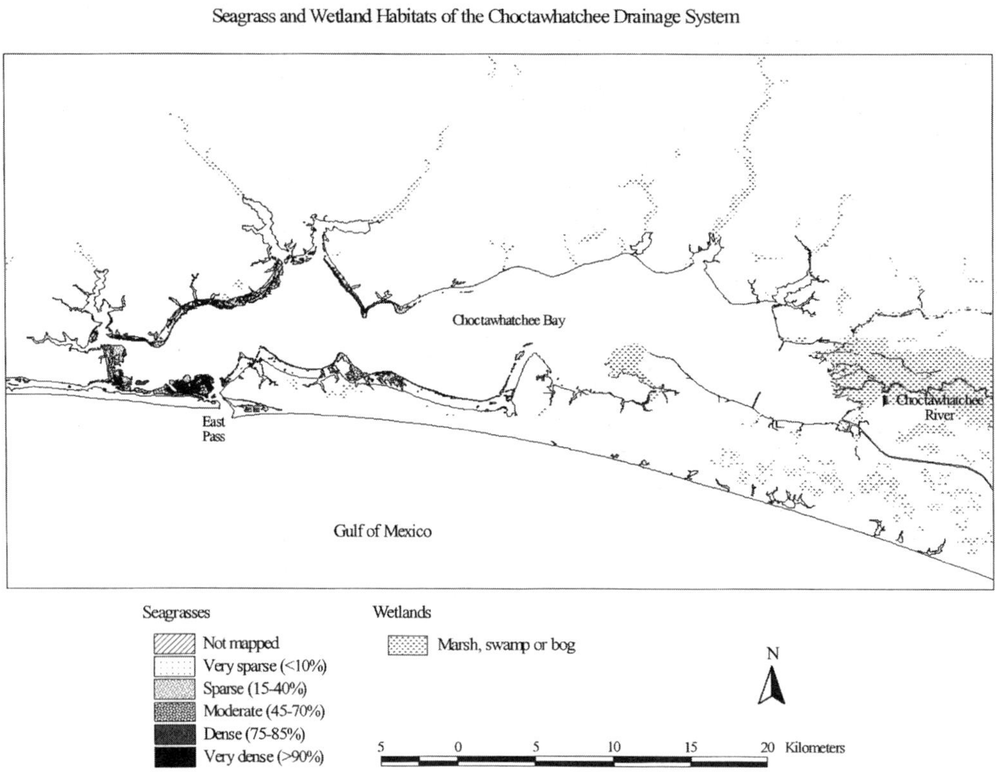

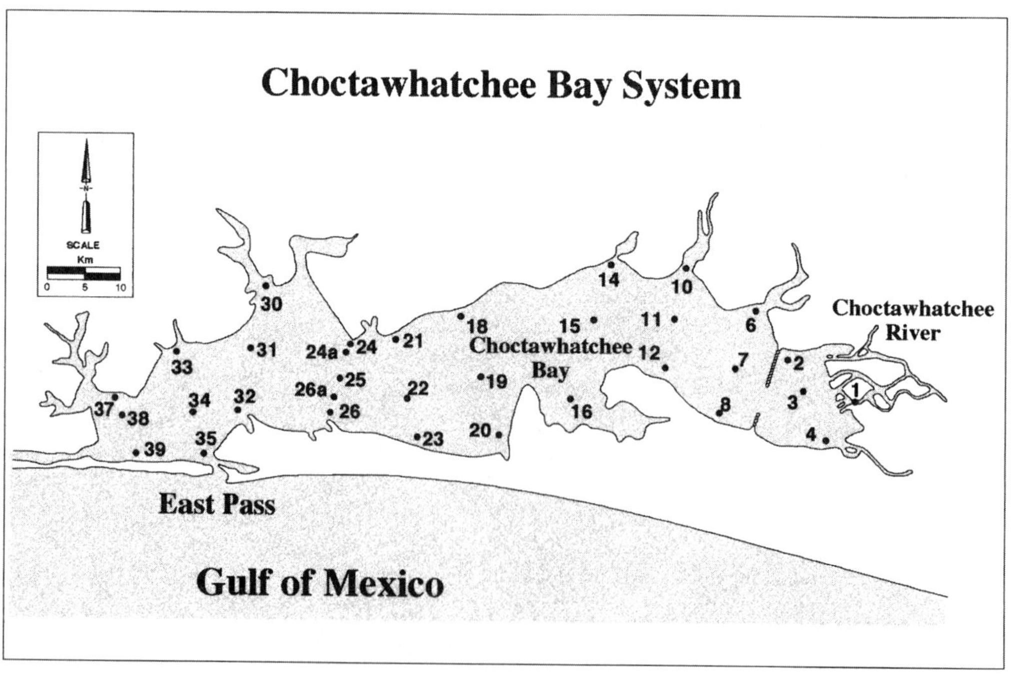

Figure 2.2(b) The Choctawhatchee River and Bay system showing sampling stations for studies from September 1985 through February 1988.

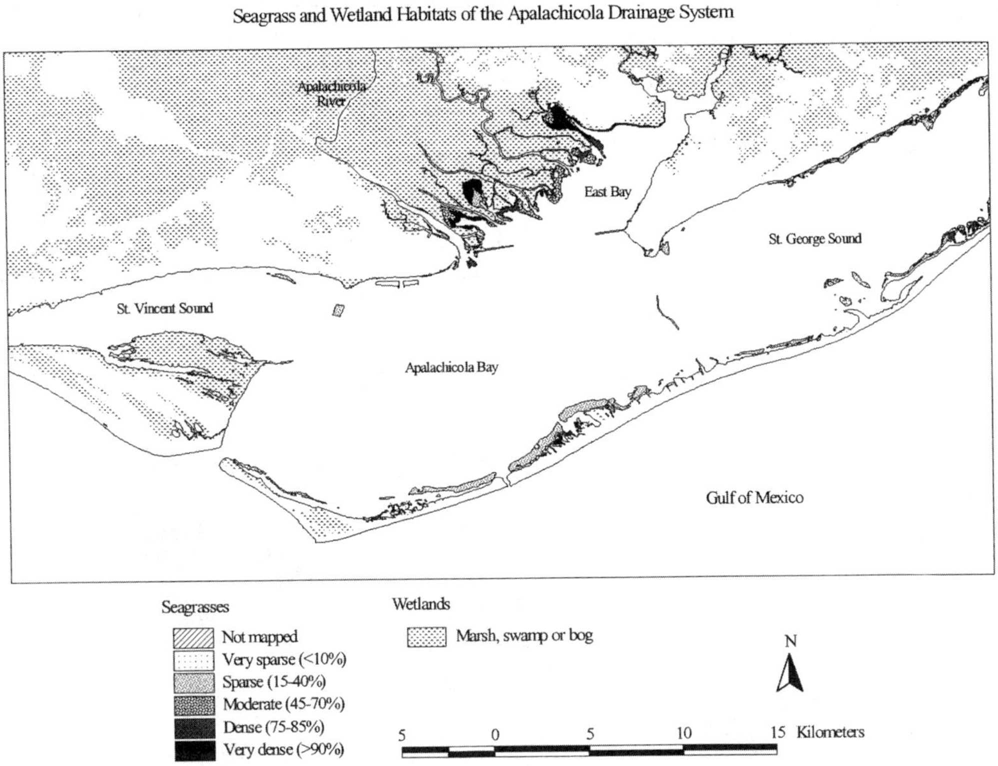

Figure 2.2(c) The Apalachicola Bay system showing sampling stations for studies from March 1971 through December 1992.

Apalachicola Bay to the Choctawhatchee Bay area, alluvial and shelf sediments are channeled into deeper water via westward drift. There are no true barrier islands in this region where moderate wave action forms a series of beaches with little in the way of estuarine development. The Choctawhatchee River provides sands for the adjoining barrier system, Santa Rosa Island, to the west. Down-drift from the bay and from the Continental Shelf supplies sediments for Santa Rosa Sound to the Pensacola Bay system. Perdido Key encloses the Perdido system at its southern terminus.

III. CLIMATE

The northwest Florida region is considered to be south temperate with mild temperatures and a humid atmosphere (Fernald and Patton, 1984; Wolfe et al., 1988).

A. Temperature

The Gulf of Mexico moderates the coastal temperatures. Major temperature changes occur because of winter cold fronts and general cooling during summer periods. Mean annual temperatures in the region approximate 20°C, with some variation depending on elevation and proximity to the coast (Livingston, 1984a). The mean number of days with temperatures at or near freezing is 20 in upland areas and 5 along the coast. Average winter temperatures in the bays are ~15 to 16°C, although there is considerable short-term variation due to the frequent passage of cold fronts with

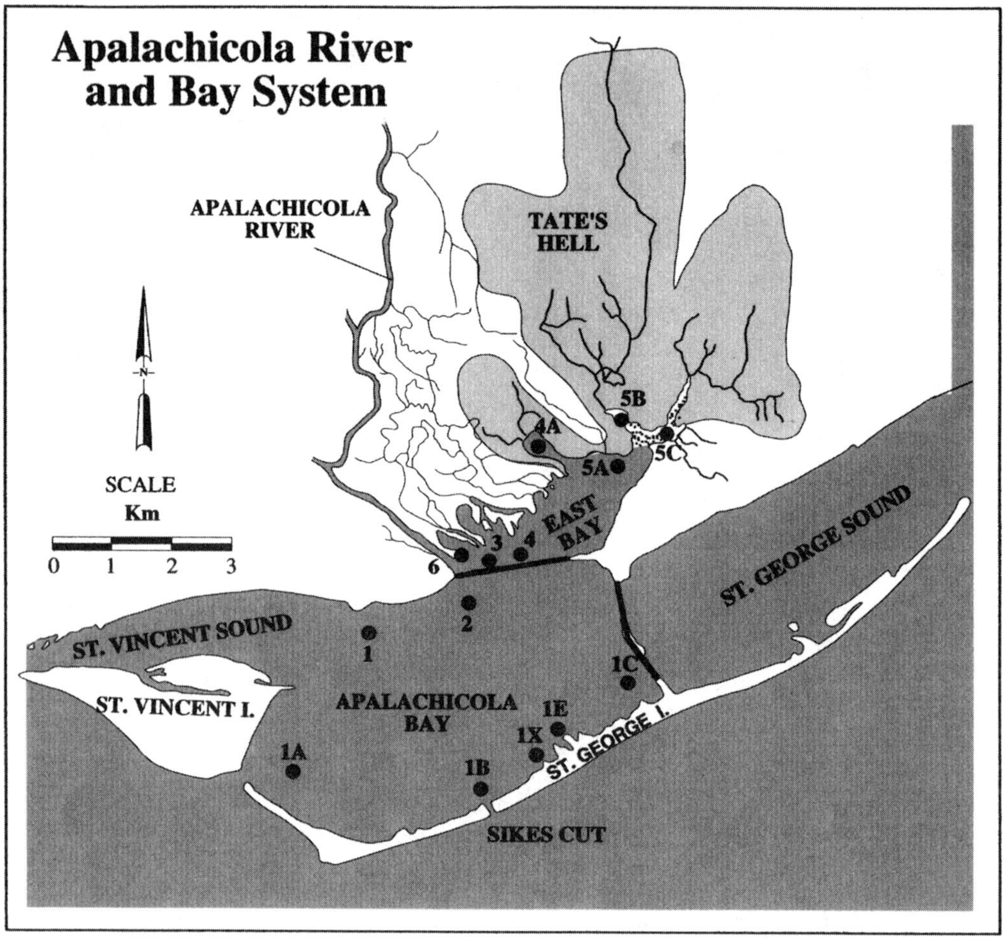

Figure 2.2(c) (continued)

low bay temperatures ranging from 4 to 7°C. The coldest months occur from December through February. Spring temperature transition occurs during March, whereas a transition from the mild, dry fall periods to winter usually occurs during November. Summer temperatures have mean values ~30°C with peaks reaching 32 to 33°C, usually in August. There are strong seasonal differences in temperature, and there is also considerable interannual variation in winter temperature conditions (Livingston, 1984a). Coastal waters closely follow changes in air temperatures due in large part to the shallowness of the inshore areas and the highly active conditions resulting from tidal currents and wind-driven water movements.

B. Precipitation

Rainfall along the northeast Gulf coast of Florida is usually bimodal on an annual basis, with a major peak during summer–early fall months (June–September) and a secondary peak during winter–early spring periods (January–April). The Florida winter peak is shorter in duration and magnitude than the summer increases (Livingston, 1984a) although this pattern varies from year to year. A drought period usually occurs during October and November with a lesser one in April and May. The mean annual rainfall in the region approximates 152 cm, which ranges from 163 cm

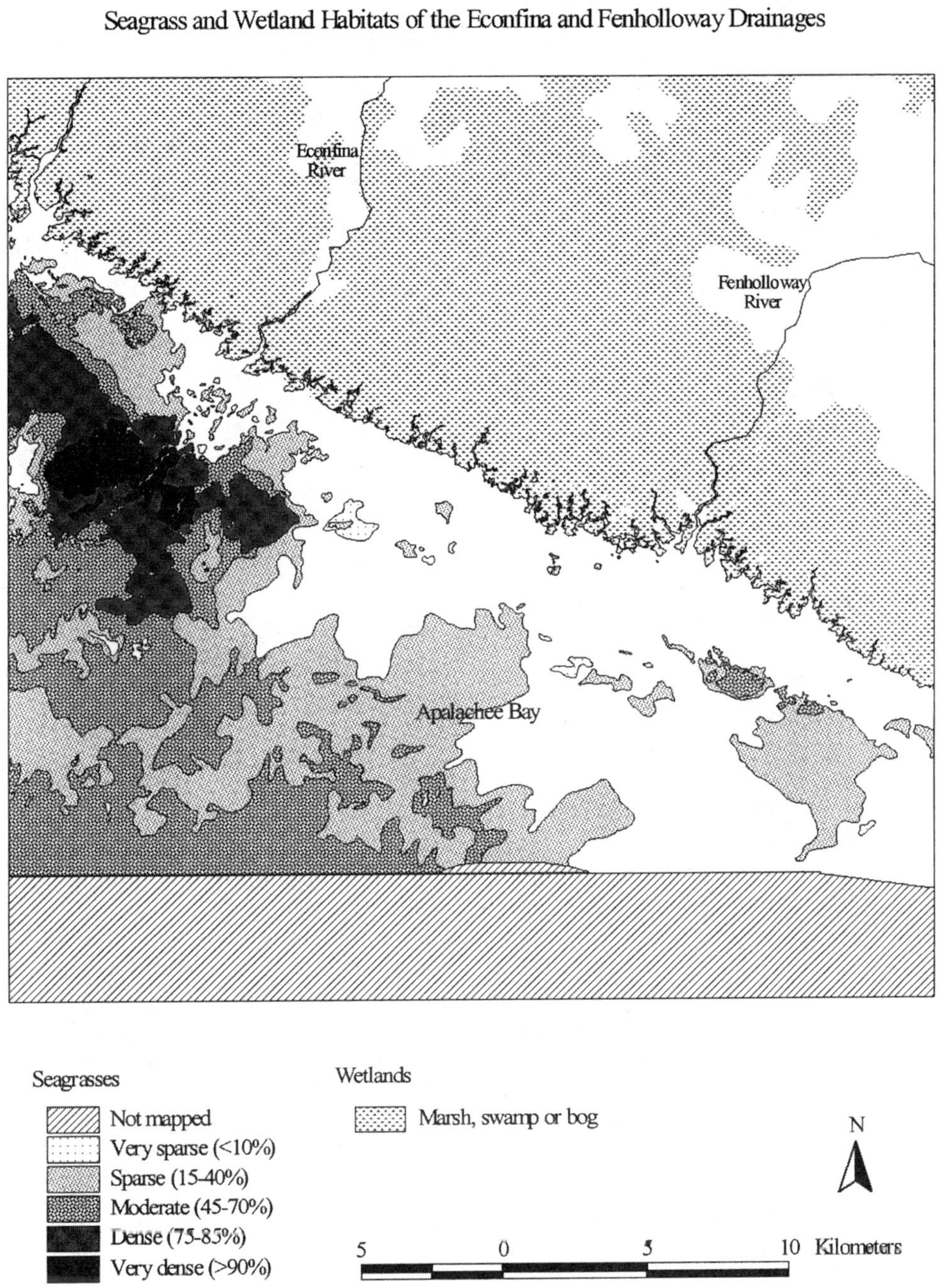

Figure 2.2(d) Apalachee Bay showing the Econfina and Fenholloway drainages with sampling stations for studies from June 1971 through the present.

in western areas to 142 cm in the east. There is wide variation in regional patterns of rainfall in time and space. Georgia rainfall has two peaks, summer and winter. The Georgia rainfall pattern is different from that of Florida in that both Georgia peaks are more or less equal. Thunderstorm

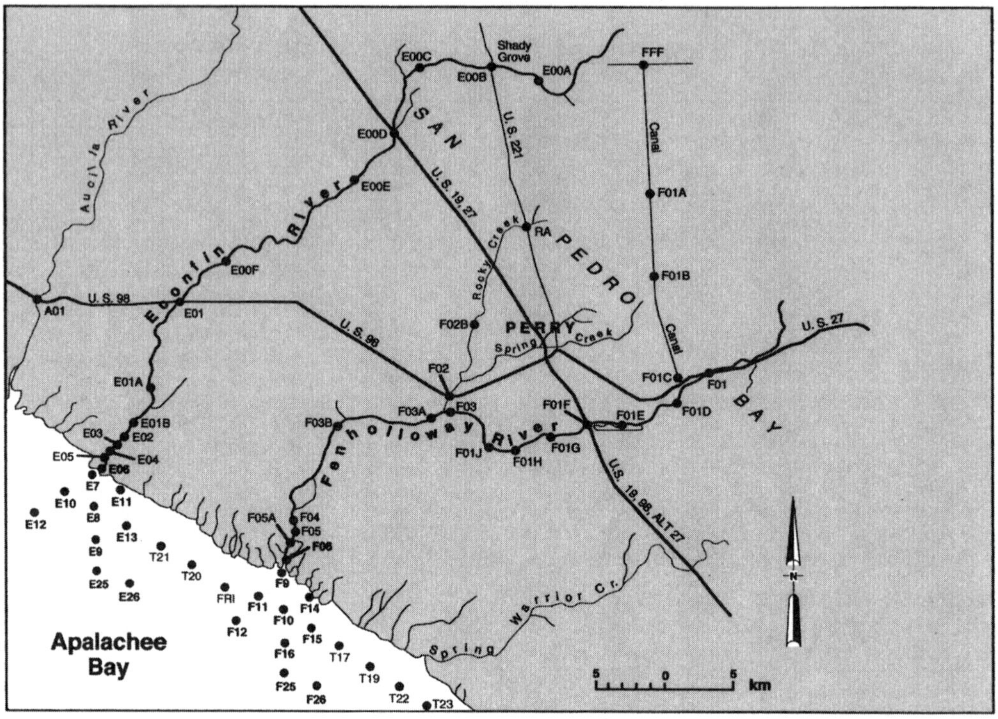

Figure 2.2(d) (continued)

activity is maximal during summer months along the Florida Gulf Coast. There is a trend toward increased precipitation in coastal areas of the eastern Gulf Coast of north Florida except during dry years, when the maximum rainfall occurs inland (Wolfe et al., 1988). There is, however, considerable spatial variability in rainfall patterns, with distinct differences along relatively short distances. Even within a given drainage basin, there can be considerable spatial/temporal differences in precipitation (Livingston, 1984a; 1986a). Periodic heavy summer rainfall is often associated with storm activity. Winter rains usually occur with frontal systems from the north, and are generally of longer duration than the summer storms. There is also the heavy rainfall that is often associated with hurricanes whereby large amounts of precipitation and associated flooding occur during relatively short periods of time.

C. Wind

Gulf air circulation is mainly anticyclonic around a high-pressure region during periods from March through September (Wolfe et al., 1988). During the winter, air masses of continental origin move through the north Florida area. An average of 30 to 40 such polar air masses penetrate the Gulf coast in a given year. Storms are usually formed along the winter cold fronts and tend to be more pervasive than the summer storms (Livingston, 1984a). Hurricane occurrence in the northeast Gulf is relatively frequent. During the period from 1885 to 1985, about 48 storms came ashore in the Panhandle region. Most such activity has occurred in the area from the Ocklochonee drainage to St. Andrews Bay (Panama City, Florida). Relatively little hurricane activity has occurred along the northwest peninsula in Apalachee Bay (Fernald and Patton, 1984). Size, duration, and direction of a given storm, together with the physiographic characteristics of the coastal basin, are determinants of the physical effects of such storms on the Gulf Coast.

From October through February, there is a western anticyclonic cell that is established along the Gulf of Mexico. The movement of this cell often results in low-level westerlies over the north Florida area. During this period, the prevalent wind direction is northerly, due partly to the high wind speeds that follow the winter fronts (Wolfe et al., 1988). However, examination of long-term wind records reveals considerable variation on a day-to-day basis. Winds tend to be highest during the months of January through March. In the spring (March–May), the winds come mainly from the south or southeast depending on the region. During the summer months (June–August), the prevailing winds are mixed, coming from the southeast (in eastern sections of the Florida Panhandle) or the southwest (in western sections). In shallow estuarine areas, winds can have a major influence on tidal current patterns. Such effects are due to the orientation, physiography, and spatial dimensions of a given river-estuarine or coastal system. Southerly winds tend to augment the astronomical tides and can cause sustained high water without the usual ebb during some months of the year, whereas northerly winds can increase substantially the effects of a major tidal efflux (Livingston, 1984a).

IV. RIVER FLOW

In the Panhandle region of Florida, the groundwater and surface water bodies are often directly connected by rock channels (Wolfe et al., 1988). The Karst topography that is so prevalent in this area is characterized by porous limestone that is penetrated by various solution channels and sandy substrates. The connections to the surface are typically in the form of sinkholes and springs, which can act as direct connections between the surface waters and groundwater aquifers. Rainwater drains rapidly into the Surficial Aquifer, which is composed mainly of coarse-to-fine sands intermixed with various forms of surface gravels. The Surficial Aquifer, which is recharged locally, is often separated from the underlying Floridan Aquifer by a clay-confining (impermeable) layer. The condition of this Surficial Aquifer, in terms of previous (recent) rainfall conditions, has an effect on the surface flows in the stream systems of the region. Some areas, such as the Perdido drainage basin, have relatively flashy flows in response to precipitation incidents. Periodic flash flooding occurs in the nonalluvial streams such as Elevenmile Creek.

There is considerable variability in the freshwater fluxes in the various subbasins of the north Florida region (Livingston, 1984a; 1989). Most of the major river systems are composed of a series of tributaries that drain the various subbasins into a main stem. The subbasins often follow differing hydrological cycles (Livingston, 1989; 1990) than the main stem, depending on regional rainfall patterns and the highly variable physiographic conditions in the different regions. Some of the drainage basins of the various north Florida areas extend into Alabama and Georgia, and the temporal patterns and volumes of river discharge to associated estuaries depend on the cumulative climatological conditions and seasonally varying evapotranspiration rates in the respective watersheds.

Flooding plays a major role during winter and early spring months in the north Florida drainages. Winter–spring flooding in the alluvial streams is a product of reduced evapotranspiration rates during winter rainfall peaks along the heavily vegetated drainage basins (Livingston, 1984a). The reduced surface flows during summer periods are due to increased evapo-transpiration in vegetated floodplains, even though the Florida rainy season usually peaks during this period. Surface flow to the Gulf is minimal during the fall drought period. Alluvial systems such as the Perdido, Apalachicola, and Choctawhatchee have flow patterns that are more correlated with seasonal rainfall patterns in Georgia and Alabama (where high percentages of the respective drainage basins are located) than in Florida. Nearshore coastal regions are affected by both surface water and groundwater runoff. Bays and estuaries with direct freshwater runoff have highly variable salinities that are dependent on the interaction of rainfall, river flow, tidal effects, and wind conditions with the local physiographic features of the respective coastal systems. The smaller streams along the Big Bend area are associated with relatively limited estuarine areas, whereas the alluvial river-dominated estuaries are proportionately larger due to the higher flow rates of the river. In the case of the Apalachicola system, salinities can be affected by river flow up to 240 km offshore (Livingston, 1984a).

V. TIDES

Astronomical tides in the western Panhandle from the Apalachicola to the Perdido systems are diurnal, having relatively small amplitudes that vary between 0.37 and 0.52 m (Livingston, 1984a). Tides from Apalachee Bay to the Apalachicola estuary are mixed and semidiurnal, having two (unequal) highs and lows each tidal day (24 hours and 50 min). Tides in this region range from 0.67 to 1.16 m. Local tidal conditions vary considerably, depending on various complex forcing features. Wave height and energy density are moderate (12.25 to 306.25 J/m^2) between the Ocklochonee and Perdido estuaries (Fernald and Patton, 1984). The entire coastal region in the NE Gulf of Mexico area is characterized by low energy conditions (1.96 to 12.25 J/m^2).

VI. AQUATIC HABITATS

Habitat conditions in the Gulf coastal areas of north Florida are determined by different factors. The barrier estuaries of the Florida Panhandle are associated with a series of rivers from the Perdido to the Apalachicola. Wind tends to be a major factor in the determination of water circulation and salinity in most of the shallow Gulf coastal estuaries (Livingston et al., 2000). Tidal effects vary from one area to another although the general effect is to enhance mixing in the nearshore waters. In areas where river flow is a factor, frictional effects on the saltwater regimes are combined with salt-wedge circulation and geostrophic forces to affect the estuarine salinity distribution. The physical dimensions of a given river-estuarine basin are a key element in the determination of the distribution and nature of aquatic habitats. The size and depth of a given bay along with the location of passes to the Gulf are important factors in stratification potential and salinity distribution. Thus, freshwater influxes and basin characteristics such as depth and areal dimensions determine the spatial and temporal habitat characteristics of the Gulf coastal estuaries of North Florida.

The alluvial bay systems of the Florida Panhandle are dominated by freshwater input from the primary rivers of the region. River flow patterns determine important features such as salinity, color and turbidity, primary productivity, and the rich blend of associated fisheries. Dominant species include oysters, penaeid shrimp, blue crabs, and finfishes such as anchovies and sciaenids (spot, croaker, spotted and sand seatrout). Submerged aquatic vegetation (SAV) is restricted to areas less than 1 m deep in most of the alluvial river-estuaries. Much of the bay bottom in such systems is composed of unvegetated mud and oyster bars. High phytoplankton productivity and imported organic matter from the river contribute to the sedimentary microbial activities that form the basis of the detritivorous food webs (Livingston, 1984a). Wind and tidal disturbance of the shallow system enhance secondary production, which is then utilized by a seasonal succession of euryhaline dominants. The result is some of Florida's most productive inshore systems, which are typified by the Apalachicola Bay system. Apalachee Bay, on the other hand, receives less surface freshwater runoff than the alluvial systems. Seagrass beds dominate the clear, nutrient-poor waters of Apalachee Bay with phytoplankton as a secondary source of organic carbon. Salinity is generally higher here; species richness and diversity are high, and relative population dominance relatively low. Species such as pink shrimp, spotted seatrout, and various offshore finfishes (many of which are important to the sport fisheries in the region) constitute food webs that are based largely on the seagrass habitat. In this case, the seagrasses provide both the major habitat and the most important source of organic matter for the system. Livingston (1990) gives a comprehensive review of the habitats of the coastal areas of the northwest coast of Florida.

An analysis was made of the dimensions of the various subsystem habitat types (Livingston, 1997a). Statistical analyses of the data were carried out according to transformation schedules that were developed from a series of databases taken since 1970 in the northeast Gulf of Mexico. The estuaries analyzed included the Econfina and Fenholloway estuaries (Apalachee Bay), the Apalachicola Bay system, the Choctawhatchee Bay system, the Pensacola Bay system, and the Perdido River–Bay system. Riverine wetlands are best developed in the alluvial systems such as the Perdido, Escambia, Choctawhatchee, and Apalachicola, which are the major sources of freshwater surface flow to the Gulf in the region. The coastal wetlands are most highly developed in the saltwater marshes of Apalachee Bay. The Apalachicola system has the most extensive forested/shrub wetland areas. In terms of open water/upland ratios, the Perdido has the lowest such ratio of the alluvial systems. This could mean that the freshwater runoff into Perdido Bay is proportionally greater than that of the Apalachicola and Choctawhatchee estuaries. The Apalachee Bay estuaries have very high wetlands/open water ratios compared with the alluvial estuaries. Choctawhatchee Bay has the least development of fringing wetlands and some of the highest erosion rates of the bays in the study area.

CHAPTER 3

Primary Study Sites

The inshore marine habitats of the northwest Florida coast (see Figure 2.1) represent a diverse, highly productive series of coastal systems. One of the important characteristics of these areas is high primary and secondary productivity. The multiple sources of nutrients, together with more or less continuous disturbance by wind and tidal currents, form the basis for an almost continuous supply of organic matter for the inshore food webs (Livingston, 1984a). Inputs from freshwater and saltwater wetlands (allochthonous and autochthonous), *in situ* phytoplankton productivity, and submerged aquatic vegetation all contribute to the rich coastal areas. In addition to organisms that spend their entire life cycles in such systems — such as most of the benthic infauna, shellfish, and spotted seatrout — there are many marine species that spawn offshore and migrate inshore during their larval and juvenile stages, utilizing the rich inshore waters as a nursery.

Many species, both freshwater and marine, spend at least part of their life histories in the nearshore marine environment. Penaeid shrimp (*Penaeus* spp.), blue crabs (*Callinectes sapidus*), oysters (*Crassostrea virginica*), and spotted seatrout (*Cynoscion nebulosus*) are just a few of the commercially important species that occupy this region. Most of the estuarine and coastal species are capable of living in rapidly changing conditions of salinity and other habitat variables. These numerically dominant populations utilize the abundant food resources of coastal systems while remaining relatively free from predation by the stenohaline (salinity-restricted) marine forms offshore. Thus, the inshore marine habitat of the northeast Gulf of Mexico serves as a physically stressed but highly productive nursery for developing stages of offshore forms, many of which form the basis of the sports and commercial species of the region. Gradients of physical variables, together with high primary production mediated largely by the phytoplankton, provide the basic components for the highly complex food webs of the inshore marine systems.

In a functional sense, inshore marine systems encompass a biological continuum from upland, freshwater drainage basins to the offshore Gulf. These marine systems are characterized by a combination of habitats with certain common physical, chemical, and biological attributes as indicated above. However, each such system represents a unique combination of these habitats and primary productivity differs widely both within a given system and among the various coastal areas. There is thus considerable ecological individuality of the different estuarine and coastal systems of the northeastern Gulf of Mexico. The biological attributes of each system vary from system to system, having been determined by adaptive responses of indigenous species, from phytoplankton to fishes, to unique combinations of local habitat and productivity. Seasonal and interannual cycles of controlling factors, together with intermittent storms and hurricanes, provide the highly variable background for the area-specific characteristics of these estuarine and coastal systems. It is therefore necessary to analyze each estuary separately with respect to these unique characteristics.

I. APALACHEE BAY (ECONFINA AND FENHOLLOWAY RIVER–ESTUARIES)

Apalachee Bay (see Figure 2.2d) remains in relatively natural condition due to the almost complete lack of development and human population in the primary drainage areas. A broad, shallow shelf and the absence of major rivers characterize this bay. The dominant habitat feature of Apalachee Bay is an extensive series of seagrass beds that extends to Florida Bay in the south (Iverson and Bittaker, 1986). The one area of significant anthropogenous effects in an otherwise pristine system is the Fenholloway River-estuary where pulp mill discharges have caused adverse effects due to high dissolved organic carbon (DOC) and water color, high biochemical oxygen demand (BOD), low dissolved oxygen (DO), and high nutrient loading (ammonia and orthophosphate) (Livingston, 1980; 1982; 1984b; 1985a,b; 1988; Livingston et al., 1998b). The Econfina River remains one of the most pristine blackwater streams along the coast, and is an ideal reference area for studies in the Fenholloway system. Both the Econfina and Fenholloway drainages have similar dimensions. They share a common origin (San Pedro swamp) and a common meteorological regime. River flow characteristics in the two drainages are comparable in rate and seasonal variation. The reference Econfina system thus has the same origins and basic features of the Fenholloway system with the exception that there is no pulp mill located within the Econfina drainage basin.

Our research group has studied the Econfina and Fenholloway drainages and associated coastal areas since 1970. Zimmerman and Livingston (1976a, b; 1979) found that seagrasses in the bay area off the Fenholloway River were characterized by reduced cover relative to similar areas in the Econfina reference system. In a combined field descriptive/experimental and laboratory experimental study, Livingston et al. (1998b) showed that pulp mill effluents in the Fenholloway River were associated with increased loading of DOC, water color, and nutrients to offshore areas relative to the unpolluted (Econfina River) reference system. This loading resulted in changes of water quality factors and light transmission characteristics in the polluted system. Sediments in offshore Fenholloway areas were characterized by increased silt/clay fractions and altered particle size relative to reference sites. Field analyses showed that the best predictors of submerged aquatic vegetation (SAV) distribution in the Fenholloway offshore areas were photic depths, qualitative aspects of wavelength distributions (light transmission data), and water quality factors such as color, DOC, and chlorophyll *a* which were inversely related to seagrass biomass. Experimental analyses using mesocosms of dominant seagrasses (*Thalassia testudinum, Syringodium filiforme,* and *Halodule wrightii*) indicated that direct contact with pulp mill effluents resulted in adverse growth responses at concentrations as low as 1 to 2%. Mesocosm experiments with light conditions that reproduced actual field conditions showed that light conditions in offshore areas affected by pulp mill effluents adversely affected growth indices of *S. filiforme* and *H. wrightii*. Field transfer experiments showed that growth indices of the three dominant seagrass species were adversely affected by water quality and sediment quality in the Fenholloway offshore receiving area (Livingston et al., 1998b). Data from the descriptive field analyses and the laboratory/field experiments were consistent with the conclusion that qualitative and quantitative light transmission characteristics represented the primary determinants of SAV distribution in offshore areas of the Gulf with sediment type, water quality, salinity, and depth as important modifying factors for such distribution.

Phytoplankton productivity in Apalachee Bay is generally low compared with the alluvial systems. Nutrient input is restricted to relatively small stream loading. Livingston (1993a) conducted a series of nutrient limitation experiments in the Econfina and Fenholloway systems (Livingston, unpublished data). Inshore areas of both systems were generally limited by nitrogen, whereas areas farther offshore were increasingly phosphorus limited. Relatively low TIN/TIP ratios in the inshore portions of the study areas during most of the experimental period also suggested nitrogen limitation. Farther offshore, higher TIN/TIP ratios indicated increased phosphorus limitation. The general increase of the TIN/TIP ratios from inshore to offshore Gulf areas off the respective river mouths provided further evidence of the increasing importance of phosphorus to offshore areas of the Gulf.

Estuarine phytoplankton populations in the Econfina system were more abundant than those in comparable areas of the Fenholloway system (Livingston, unpublished data). Estuarine and inshore portions of the Fenholloway system were low in phytoplankton species richness and diversity compared with the Econfina cognate stations. The highest numbers of phytoplankton were found in inshore portions of the Econfina and Fenholloway systems. Fenholloway offshore areas had higher numbers of phytoplankton than comparable Econfina areas. The suppression of phytoplankton species richness and diversity in the inshore Fenholloway system was most likely due to reduced light transmission, although other effects such as inhibition by high concentrations of ammonia could not be ruled out as an additional modifying factor.

Various studies (Hooks et al., 1975; Greening and Livingston, 1982; Dugan and Livingston, 1982) found that invertebrates taken in the Econfina offshore system were generally more abundant than those found in the Fenholloway system, and that macrophyte abundance was directly associated with invertebrate numbers. Livingston (1975) found that reduced fish species richness and diversity characterized the offshore Fenholloway system. These differences corresponded to previously described distributions of SAV and macroinvertebrates. The effects on fishes were also attributed to physiological impacts (respiration, growth, food conversion efficiency; Stoner, 1976) and/or behavioral responses such as avoidance of the mill effluents (Lewis, 1974; Livingston et al., 1976; Lewis and Livingston, 1977).

Changes in the Fenholloway seagrass beds were associated with altered food web patterns in terms of species shifts and feeding alterations relative to reference areas (Livingston, 1975; 1980). Feeding habits of the dominant fish species were different in the offshore Fenholloway system than those found in similar areas of the Econfina drainage. These differences were traced to basic changes in the habitat due to the reduced seagrass beds in the Fenholloway system (Clements and Livingston, 1983). Reduced SAV, together with increased phytoplankton productivity due to nutrients from the paper mill, appeared to alter the habitat conditions and fish trophic organization in the Fenholloway offshore area relative to the reference system. Subsequent improvement in water quality was associated with some recovery of the fish trophic organization in outer areas of the Fenholloway system (Livingston, 1982; 1984b). Livingston (1984b) showed that alterations of the seagrass beds were reflected in the trophic organization of the Fenholloway system. During the earlier period of the untreated mill effluents, plankton-feeding fishes in areas affected by discharges replaced grass-bed species. Subsequent water quality improvement due to the implementation of secondary treatment of the mill effluents was accompanied by trophic shifts that followed habitat changes in the outer portions of the Fenholloway estuary. Thus, areas affected by pulp mill effluents were characterized by altered food webs compared with the reference areas; such changes were directly influenced by shifts of the primary producers and altered benthic habitats as seagrasses were turned into unvegetated soft-sediment areas. In offshore Fenholloway areas, these food web shifts were also affected by increased phytoplankton production as dilution affected light transmission in areas of increased nutrients. The benthic food web species were thus replaced in part by planktivorous fish species.

II. APALACHICOLA RIVER AND BAY SYSTEM

The Apalachicola Bay system (see Figure 2.2c) remains in a relatively unaltered state with freshwater flow as the primary controlling variable of bay habitat conditions (Livingston, 1984a; Livingston et al., 1997). The Apalachicola River is one of the last major free-flowing, unpolluted alluvial systems remaining in the conterminous United States; the river drains a basin area of 45,405 km^2. The entire drainage area is ~53,095 km^2. The Apalachicola Bay system is a lagoon-and-barrier-island complex oriented on an east–west axis, parallel to the Gulf of Mexico. The bay system, comprising East Bay, Apalachicola Bay, St. Vincent Sound, and St. George Sound, is shallow (mean depth ≈ 2 m) with mean river flow rates of approximately 690 m^3/s (1958 to 1980; U.S. Army Corps of Engineers, Mobile, Alabama). Annual high flows average 3,000 m^3sec^{-1} (Leitman et al., 1983; 1991). The boundaries of the estuary are defined from the head of tide on the Apalachicola River located 40 km upstream of its terminus at East Bay (Gorsline, 1963) to its southern boundary at the Gulf of Mexico. Three barrier islands form the Gulfward extent of the bay. There are four natural openings to the Gulf: Indian Pass, West Pass, East Pass, and a pass to the east. An anthropogenic opening (Sikes Cut) was established in the western portion of St. George Island in 1954. The upper tidally influenced reaches of the system form an extensive delta region as the Apalachicola River enters the bay through a series of major tributaries.

Together with wind, tides, and local rainfall that provides runoff from Tate's Hell Swamp (Figure 2.2c) into the northeastern portions of East Bay, the Apalachicola River controls the salinity of the estuary (Livingston, 1984a). Studies conducted in the Apalachicola estuary (Livingston, 1984a) suggested an interaction among several dynamic physical mechanisms operating on different timescales. The functional attributes of the estuary include the seasonally fluctuating Apalachicola River, local rainfall, the physiographic structure of the receiving basin, wind speed (duration and direction), and exchanges between the estuary and the Gulf. Hydrodynamic processes are further complicated by bathymetric modifications due to dredging activities that include the opening and maintenance of Sikes Cut and the Intracoastal Waterway that extends from the mouth of the river to St. George Sound to the east. The degree of interaction among these variables produce salinity profiles in the primary estuary that range from near-freshwater conditions to near-dominance by saline Gulf waters. Meeter and Livingston (1978) and Meeter et al. (1979) showed a strong correlation of Apalachicola River flow with the spatial and temporal distribution of salinity throughout the bay.

The Apalachicola estuary accounts for ~90% of Florida's commercial oyster fishery and 10% nationally (Whitfield and Beaumariage, 1977). By coupling hydrodynamic modeling with descriptive and experimental biological data, Livingston et al. (2000) determined the effects of natural droughts and potential anthropogenous freshwater diversions on oyster production in Apalachicola Bay. Oyster mortality due to stenohaline predators was a major determinant of oyster production in the Apalachicola estuary. High salinity, relatively low velocity current patterns, and the proximity of a given oyster bar to entry points of saline Gulf water into the bay were important factors that contributed to increased oyster mortality. By influencing salinity levels and current patterns throughout the bay, the Apalachicola River was important in controlling such mortality. Other factors influenced oyster production in the Apalachicola system. The highest oyster densities and maximum overall bar growth were found in the vicinity of the confluence of high-salinity water moving westward from St. George Sound and river-dominated (low-salinity) water moving south and eastward from East Bay. Areas of confluence enhanced primary production rates that were associated with high rates of oyster bar growth. Oyster production rates in the Apalachicola system thus depend on a combination of variables that are directly and indirectly associated with freshwater input by the river as modified by wind, tidal factors, and the physiography of the bay. River flow reduction, whether through naturally occurring droughts, through increased upstream anthropogenous (consumptive) water use, or a combination of the two, could have serious adverse consequences for oyster populations.

The macroinvertebrate and fish fauna of the Apalachicola estuary are affected by complex, long-term (interannual) changes in river flow and the bay response to drought and flood conditions. Livingston et al. (1997) found that when a specific threshold of freshwater reduction was reached during a prolonged natural drought, clarification of the normally turbid and highly colored river–estuarine system led to rapid changes in the pattern of primary production which, in turn, were associated with major changes in the trophic structure of the system. Increased light penetration due to the cessation of river flow was an important factor in the temporal response of bay productivity and herbivore/omnivore abundance. There was a dichotomous response of the estuarine trophic organization with herbivores and omnivores responsive to river-dominated physical/chemical factors whereas the various levels of carnivory responded to biological factors such as prey availability. Trophic response time could be measured in months to years from the point of the initiation of low flow conditions. The reduction of nutrient loading during the drought period was postulated as a major cause of the loss of productivity of the river-dominated estuary during and after the drought period. Recovery of such productivity with resumption of increased river flows was likewise a long-term event. There was a range of cyclic periodicity among the various trophic levels in the bay. Responses to regular interventions such as droughts varied among the different food web components with lag periods measured from months to years.

III. THE CHOCTAWHATCHEE RIVER AND BAY SYSTEM

The Choctawhatchee River system is the third largest in Florida in terms of freshwater discharge. The river is formed by a series of tributaries in Alabama and Florida. The average flow rate ranges between 156 and 198 m^3/s with low flows approximating 96 m^3/s and high flows of 736 m^3/s (U.S. Geological Survey, Tallahassee, Florida, unpublished data). The Floridan Aquifer provides the major source of water to the river system, which includes numerous streams, lakes, and springs. The Choctawhatchee Bay system (see Figure 2.2b) is a drowned river plain surrounded by shallow shelf slopes and inshore bayous (Livingston, 1986a, b). The bay is aligned along an east–west axis with a major alluvial river basin to the east and a dredged Gulf pass (East Pass) to the west. Freshwater input is primarily associated with the Choctawhatchee River; secondary inflows from a series of bayous are located primarily in northern sections of the bay. Saltwater input comes through East Pass. During a particularly heavy freshwater flooding in 1929, historical evidence indicates that a channel was opened by a group of citizens through the barrier island to form East Pass. The "blowout" of East Pass during the heavy flooding at this time was accompanied by increased salinity throughout Choctawhatchee Bay. Increased salinity was associated by local observers with the loss of emergent (marsh, swamp) and submergent (seagrass) vegetation throughout the bay. Choctawhatchee Bay currently has one of the least well developed fringing (emergent) vegetation zones of the various Gulf estuaries in the region (Reyer et al., 1988; see Figure 2.2b).

Choctawhatchee Bay has shallow shelf areas with relatively steep slopes that eventually level out to form a level bay bottom. Average depths of the bay are ~3 m in eastern sections and ~8 m in western sections. Over the past 30 to 35 years, there has been a major increase of urbanization in western parts of the bay (Livingston, 1986a, b). Eastern sections remain relatively undeveloped. A complete survey of the Choctawhatchee River and Bay was conducted during the mid-1980s (Livingston, 1986a, b; 1987b; 1989; 1991d). Penaeid and periclimenid shrimp, blue crabs, and brief squid, all of which are important species in terms of the overall food web of the bay, dominated macroinvertebrate assemblages. Invertebrate distribution was associated with specific habitat characteristics (depth, water column stratification, hypoxia, sediment characteristics, seagrass cover, salinity gradients, nutrient dynamics, and storm water runoff). Macroinvertebrates and fishes, the backbone of the sports and commercial fisheries in the Choctawhatchee system, were concentrated at the river mouth, in the unpolluted bayous, and in the shallow seagrass habitats of the western sections of the bay. Major fishery species in the Choctawhatchee system included mullet, spotted seatrout, flounder, cigarfish, herrings, blue runner, ladyfish, pompano, king mackerel, red snapper, pinfish, Atlantic croaker, bluefish, and pigfish. Shellfish fisheries included penaeid shrimp, blue crabs, squid, and oysters. Shrimp were caught in the western bay nurseries, whereas the oysters were concentrated in shallow southeastern sections.

The Choctawhatchee Bay system is represented by a series of complex habitats. Habitat quality is strongly influenced by freshwater input (river, bayou drainages) and salinity stratification. According to historical remembrances of long-term residents, the blowout (i.e., anthropogenous opening) of East Pass during the heavy flooding in the late 1920s was accompanied by higher salinity in Choctawhatchee Bay, which in turn was associated with losses of emergent (marsh, swamp) and submergent (seagrass) vegetation. According to anecdotal information, these plants had, at the time, been well developed in this area. These changes were associated with losses of fish productivity and a reduction of the fishery potential of the bay. Storms, overfishing, and the use of toxicants such as DDT also were also purported to have had adverse impacts on the estuarine biota over the past 50 years. Although such recollections are often contradictory and without objective scientific verification, they are consistent with established accounts of existing problems in the Choctawhatchee system.

Analysis of aerial photographs from 1955 and 1985 indicated that there was an overall loss of almost 20% of the SAV, shoalgrass (*H. wrightii*), in western sections of the bay during this period. The spatial and temporal pattern of such loss would indicate that seagrass deterioration is possibly due to changes in the optical characteristics of the water column (i.e., possible changes in color, turbidity) although causative factors remain unknown. The opening of East Pass probably led to increased salinity in the bay, enhanced vertical thermohaline stratification, and associated water quality conditions in the system. These effects include periodic hypoxia and anoxia at depth in various parts of the system. Stabilization of the water column was associated with the development of liquid mud or nephlos (i.e., suspensions of fine particulate matter above the sediment/water interface) (Livingston, 1986a, b). Considerable parts of the deeper sections of the bay have been affected by the opening of the East Pass entrance to the Gulf. The combination of the resulting salinity stratification in central and western sections of the bay together with low turnover rates and associated habitat deterioration due to hypoxia and liquid mud development have led to relatively low secondary production in the main stem of the system (Livingston, 1986a, b). The introduction of saline water may also have added to the observed acceleration of erosion around the periphery of the bay by reducing emergent vegetation. Critical, highly productive habitats include the few remaining estuarine wetlands (emergent vegetation), the shallow submerged aquatic vegetation (*H. wrightii* beds), the river mouth and associated freshwater wetlands, and the various (unpolluted) bayous.

Other human activities associated with urbanization (e.g., storm water runoff and sewage spills) have contributed to the deterioration of natural aquatic resources in the Choctawhatchee Bay system. According to Livingston (1987b), discharges of sewage and storm water associated with unrestricted land development in western sections of Choctawhatchee Bay were responsible for water quality deterioration in a series of bayou areas in the northern sections of the bay and in Destin Harbor (Old Pass Lagoon) to the south. The situation in Old Pass Lagoon represents a classic case of cultural eutrophication (i.e., too many nutrients from human sources). The combination of the loading of nutrients and organic matter together with a lack of flushing periodically contributed to observed severe water quality problems (hypoxia, anoxia) and associated reductions of natural productivity in the lagoon. This situation was associated with high nutrients (N, P) in the water column, and silt-laden sediments (especially in eastern sections of the lagoon). Biological conditions included high phytoplankton productivity, low numbers of herbivores (zooplankton, meroplankton), a depauperate infaunal invertebrate community through central/deep portions of the lagoon, and relatively few epibenthic invertebrates and fishes (Livingston, 1986a, b; 1987a, b). The observed degradation of the biology of Old Pass Lagoon resulted from changes in the natural food webs of the lagoon caused by increased nutrients and organics and stagnant conditions (poor flushing, especially in eastern sections of the lagoon). It was suggested (Livingston, 1987a, b) that actions be taken to reduce input from sewage and urban storm water that accounted for the observed loading of nutrients, organic matter, and pollutants to the lagoon and some of the more polluted northern bayous.

The control and elimination of water pollution, due primarily to municipal sewage, storm water runoff, industrial wastes, and agricultural runoff should have been the central focus for future land planning in the region. Factors such as nutrients, organic matter, toxic agents, and disease-producing organisms (microbes such as bacteria and viruses) in both surface water and groundwater posed a threat to the Choctawhatchee River and Bay system in the mid-1980s. However, little was done to correct these observed problems. Recent anecdotal information (K. Spencer, personal communication, Destin Log, 1999–2000) indicates that continued urban development in western sections of the bay has been associated with increased levels of urban runoff and sewage spills. The anecdotal data also are consistent with the usual signs of advanced hypereutrophication. This includes hypoxia at depth, fish kills throughout the river–bay system, and reduced secondary production. These conditions have been accompanied by extensive red tide (*Gymnodinium breve*) blooms both in the bay and in offshore areas. Most of the existing information concerning the blooms was taken after the blooms had matured. Although virtually no systematic supporting ecological were taken, some public officials attributed the blooms to increased water temperature and salinity in the area. A prolonged drought during 1999–2000 was considered a causative factor. No mention was made of anthropogenous nutrient loading. In addition to the usual signs of cultural eutrophication, there was a more ominous problem in the form of the deaths of sea turtles and bottlenose dolphins in the Choctawhatchee system and along the coast between Choctawhatchee Bay and St. Andrews Bay (K. Spencer, personal communication, Destin Log, 1999–2000). About 121 dead dolphins were noted in this area between August 1999 and mid-January 2000. The dolphin deaths were such that they created a major disposal problem. The deaths of some fish-eating terrestrial mammals such as raccoons and foxes in the region were also observed in various parts of the western Choctawhatchee Bay area. Although there are no definitive answers as to causation of the mammal deaths, researchers found red tide toxins in the dolphins' systems (K. Spencer, personal communication, Destin Log, 1999–2000). Compared to an estimated annual death rate of three to five dolphins per year, the current die-offs were considered unusual.

Most of the experts attributed the algal blooms to *G. breve* populations moving inshore from Gulf areas. According to one source, the fish kills in the Bayou areas of the north bay were due to blooms that originated in the Gulf. "If it's back in the bay, it had to come from the pass" according to one such expert. There were massive kills of freshwater fishes (bass, bream, etc.) in the Choctawhatchee River where "tests failed to show the algae was present" (K. Spencer, Destin Log, 1999–2000). However, although tests for *G. breve* were conducted, virtually no systematic and comprehensive phytoplankton data were taken in the Choctawhatchee system. The general lack of reliable scientific information eliminated any basis for determination of causative factors relating to the blooms. In addition, there was little scientific evidence of what caused the *G. breve* blooms in the first place. As in the case of the Apalachicola system, no effort was made by public authorities to carry out ecosystem-level analyses to evaluate the current status of the Choctawhatchee River–Bay system. In addition, there is currently no management plan to correct known problems with nutrient loading to the bay. The twin aspects of urban storm water runoff and sewage spills in areas associated with the rapid proliferation of urban development are thus the primary suspects in the unusual series of mammalian deaths. Recent events justified the warnings made over 14 years ago (Livingston, 1986b; 1987b).

IV. THE PENSACOLA RIVER AND BAY SYSTEM

The Pensacola Bay system is located in the Coastal Plain province, and is associated with a series of river basins (the Escambia, Blackwater, and Yellow) (Figure 2.2a). Thorpe et al. (1997) have provided a comprehensive review of the scientific studies of the Pensacola Bay system. There are three primary rivers associated with the Pensacola Bay system: the Escambia, Blackwater, and Yellow (Figure 2.2a). The Pensacola Bay system includes five interconnected estuarine components: Escambia Bay, Pensacola Bay, Blackwater Bay, East Bay, and Santa Rosa Sound (Figure 2.2a). The Escambia River system extends northward about 386 km from the north end of Escambia Bay. It runs through Alabama as the Conecuh River. The drainage area of the Escambia River basin includes 10,880 km^2, with ~90% of the basin in Alabama. The Blackwater River basin includes

2230 km^2 of which ~81% is in Florida. The Yellow River basin extends 177 km from Blackwater Bay to a point northeast of Andalusia, Alabama. This basin includes an area of about 3535 km^2 and ~64% of that is located in northwest Florida.

Escambia Bay is located east of the City of Pensacola, with the Garcon Point peninsula to the east and the Escambia River delta to the northwest (Figure 2.2a). The primary source of fresh water to the bay is the Escambia River; other sources include the Pace Mill Creek and Mulatto Bayou drainage basins in the upper bay, and the Bayou Texar, Bayou Chico, and Bayou Grande basins in the lower bay. Blackwater Bay receives freshwater inflow from the Blackwater River, with East Bay immediately south of the Blackwater Bay system. East Bay receives freshwater flows from the Blackwater River–Bay system, the Yellow River, and the East Bay River. East Bay is bordered to the south by the Gulf Breeze peninsula. Pensacola Bay receives flows from Escambia Bay, East Bay, and Bayou Grande, and is bordered to the north by the City of Pensacola and to the south by Santa Rosa Island. Pensacola Bay empties into the Gulf of Mexico through a pass at its southwestern terminus. Santa Rosa Sound is a lagoon between the mainland and Santa Rosa Island. This sound connects Pensacola Bay with Choctawhatchee Bay to the east. The various components of the Pensacola Bay system are characterized by a relatively shallow shelf, a sloped area, and deeper plain. Upper Escambia Bay and Blackwater Bay are relatively shallow. Depth in the central parts of both systems tends to increase southward, with the greatest depths in Pensacola Bay.

Thorpe et al. (1997) noted the relative paucity of published literature concerning the rivers of the Pensacola Bay system. Generalized information has been provided by the U.S. Army Corps of Engineers (1980), Fernald and Patton (1984), Wolfe et al. (1988), Nordlie (1990), and Hand et al. (1996). Thorpe et al. (1997) stated that industrial and domestic discharges as well as non-point-source pollution affect the Escambia River. The lower river is dredged, and there are two dams upstream in Alabama. Bass (1990; 1991) indicated that the Escambia River had had an improving fish biota from 1983 through 1988, and that the electrofishing catch per unit effort (numbers) was comparable to that of other alluvial rivers such as the Choctawhatchee and Apalachicola. Available data for the lower Escambia River indicated that this part of the system was not affected by anthropogenous activities. Hand et al. (1996) characterized water quality in the Escambia River as generally good. However, Thorpe et al. (1997) indicated that the river tributaries appear to be threatened and/or impaired by encroaching urban and agricultural activities and that agricultural development and road runoff has affected various creeks in the Escambia drainage basin.

The history of Escambia Bay reflects long-term adverse impacts due to urban runoff, nutrient loading from point and non-point sources, and toxic waste loading from urban activities and industrial sources. The bay response has included hypereutrophication, extensive fish kills, and massive habitat loss (Livingston, 1999). Collard (1991a, b) published extensive reviews of the scientific literature concerning the Pensacola Bay system noting a history of decreased biological integrity since the mid-1950s. Upper Escambia Bay, with reduced circulation due to both natural and anthropogenous conditions, and anthropogenous nutrient loading was the subject of extremely debilitating habitat destruction and massive fish kills (Livingston, 1997a; 1999) that occurred during the late 1960s and early 1970s. An active oyster fishery was largely destroyed. Most of the seagrasses of the Pensacola Bay system disappeared. Local sports fisheries were no longer viable. The fish kills led to restrictions on nutrient loading that were accompanied by limited recovery of water quality and secondary production and reductions in the overall fish mortality rates. The oyster industry did not fully recover and the only seagrass beds left are in the upper Blackwater system.

Relative to the upper bay, lower Escambia Bay and Pensacola Bay have improved circulation and turnover rates due to the physiographic characteristics of this part of the system and proximity to open Gulf waters. However, the non-point-source pollution from the City of Pensacola and a Naval Air Station, together with major pollution of the bayous (Texar, Chico, Grande) in this region, continues to present a serious problem in terms of multisource contamination that is difficult to evaluate and even more difficult to mitigate. Non-point-source pollution appears to be a major part of the current problem with the Escambia Bay system although sewage discharge continues to cause serious water quality problems in local areas. Storm water outfalls are particularly numerous in western sections of Escambia Bay. In addition, septic tank numbers are extremely high in the vicinity of lower Escambia Bay.

The bayous on western parts of Escambia Bay collect urban runoff and have a long history of poor water and sediment quality with accompanying fish kills. Bayou Texar has been well studied (Moshiri, 1976; 1978; 1981; Moshiri and Crumpton, 1978; Moshiri et al., 1978; 1979; 1980; 1987). The Pensacola Bay system is one of the most heavily polluted bays in Florida in terms of toxic substances (Seal et al., 1994). High levels of metal contamination were found in Bayou Grande (Cd, Pb, Zn) and Bayou Chico (Cr, Zn). Bayou Chico was also contaminated with polynucleated aromatic hydrocarbons (PAHs) and polychlorinated biphenyls (PCBs). In Pensacola Bay, PAHs and PCBs were found in sediments close to shore and in central parts of the bay. Phenolic compounds were found at one site near Pensacola Harbor. Long et al. (1997), in a study of sediment toxicity in four bays of the Florida Panhandle, found that the greatest toxicity of all the bays analyzed occurred in Bayou Chico, with the other developed bayous in the Pensacola area showing "relatively severe toxicity." Toxicity of the Pensacola Bay sediments was attributed to high-molecular-weight PAHs, zinc, DDD/DDT isomers, total DDT, and the highly toxic organochlorine pesticide known as dieldrin. Similar toxicity of sediments in upper Escambia Bay were not associated with known toxic agents.

Livingston (1997a; 1999) found that whole-water phytoplankton data supported the view that Escambia Bay was in a moderately eutrophic state. However, subdominant phytoplankton populations known to be noxious bloom species in Perdido Bay were found in mid and lower Escambia Bay. Urban storm water runoff was considered a factor in the maintenance of these populations. Compared with other alluvial river-dominated Gulf estuaries such as the Apalachicola, Choctawhatchee, and Perdido systems, the Pensacola Bay system had relatively low secondary productivity. There was relatively low biomass of infauna, epibenthic invertebrates, and fishes along eastern sections of upper Escambia Bay, in lower Escambia Bay, and in Pensacola Bay. The main concentrations of infauna (polychaete worms), epibenthic macroinvertebrates (brown shrimp, blue crabs), and fishes (sciaenids) were found in the relatively shallow parts of upper Escambia Bay and Blackwater Bay. Upper Escambia Bay was, in fact, the center of primary and secondary production in the entire Pensacola Bay system, a probable result of nutrient loading from the Escambia River. The trophic organization emanating from the higher primary production of the upper bay was the main determinant of the distribution of dominant, commercially important populations in the Pensacola Bay system. Habitat factors, such as sediment type, dissolved oxygen, and salinity distribution, while important in determining the general distribution of organisms in the bay system as a whole, had effects that could only be understood within the context of the distribution of primary (phytoplankton) production. However, when compared with unpolluted systems such as Apalachicola Bay, secondary production in upper Escambia Bay was relatively low.

V. THE PERDIDO RIVER AND BAY SYSTEM

The Perdido Bay system (Figure 2.2a) is a shallow to moderately deep inshore body of water that is oriented along a northeast–southwest axis, approximately perpendicular to the Gulf of Mexico. Perdido Bay has a length of 53.4 km, an average width of 4.2 km, and an average depth of 2.2 m. The primary source of freshwater input to the estuary is the Perdido River system that flows southward about 96.5 km, draining an area of about 2937 km^2 (Livingston, 1998). Freshwater inflow peaks during March and April and is lowest in October and November. The tidal range averages 0.2 to 0.3 m. The Elevenmile Creek system (including Eightmile Creek) is a small drainage basin (about 70 km^2) that receives input from municipal and industrial point sources (historically, the St. Regis Paper Company; now the Champion International Corporation Pensacola Mill) and non-point runoff. Elevenmile Creek drains into the estuary about 1.6 km east of the Perdido River entry point in the northernmost portion of the system. The relatively small Bayou Marcus Creek drains a residential area of western Pensacola with input from urban storm water runoff and a sewage treatment plant.

The bay can be divided into four distinct geographic regions: the lower Perdido River, upper Perdido Bay (north of the route 98 bridge), lower Perdido Bay (bounded to the south by a line between Ross Point and Inerarity Point), and the Perdido Pass complex (Figure 2.2a). The Gulf Intracoastal Waterway (GIWW) runs through the lower end of Perdido Bay about 5.6 km northeast of Perdido Pass. The U.S. Army Corps of Engineers (USACOE) maintains the Perdido Pass channel at about 4 m as part of the GIWW (U.S. Army Corps of Engineers, 1976). As in the Choctawhatchee Bay system, there is a shelf around the periphery of the Perdido estuary that usually does not exceed 1 m in depth. This shelf can extend up to 400 m in width. The upper bay is relatively shallow, and depth tends to increase southward. The deepest parts of the estuary are located at the mouth of the Perdido River and in the lower bay off Ross Point.

The high ratio of drainage area to estuarine area (518 km^2), together with relatively steep gradients and high incidence of flash flooding contribute to rapid responses of the estuarine salinity regime to regional rainfall incidents (Livingston, 1998). Prior to the opening of Perdido Pass in the early 1900s, Perdido Bay was a largely freshwater system, covered with lily pads and used for various purposes as a body of fresh water (Brush, 1991). Access to the Gulf was restricted by the shallow, shifting body of fresh water. At the time of an early survey (1767), the pass had a depth of ~2 m on its bar. During an outbreak of malaria in the early 1900s, it is said that a group of men dug out the mouth of the bay with shovels opening the bay to the Gulf and making it the saltwater system it is today. This action resulted in the creation of the Ono Island/Old River complex. The history of Perdido Bay is thus comparable to that of the Choctawhatchee Bay system (Livingston, 1986a, b), which was also opened at the mouth by another group of citizens with shovels in 1929.

Paper mill construction was begun at the head of Elevenmile Creek in 1939 as the Florida Pulp and Paper Company in Cantonment. Construction was completed in 1941 and a second mill was constructed toward the end of World War II. In April 1970, the pulp mill (St. Regis Company) was discharging ~120 million l d^{-1}. Problems concerning high color and foam were reported. In 1984, Champion International Corporation acquired the St. Regis Company including what was then called the Pensacola Mill. The Champion Mill underwent a major conversion that increased production levels leading to continued problems associated with dissolved oxygen, un-ionized ammonia, transparency (color), specific conductance, iron, zinc, and biological integrity of the receiving aquatic systems including Elevenmile Creek and Perdido Bay.

Since the turn of the century, there has been a steady increase in the human population in the Perdido basin. By the 1980s, Perdido Key was undergoing rapid residential and commercial development. A series of condominiums was built on the Key. The problems of sewage treatment and storm water runoff remain unresolved in various residential areas around the bay to the present day. Agricultural runoff, based largely in Alabama, contributes to nutrient loading in western parts of the lower bay.

Schropp et al. (1991) found that concentrations of dissolved organic carbon, Kjeldahl nitrogen, nitrate-nitrite, orthophosphate, particulate carbon, total nitrogen, total phosphorus, and total suspended solids were found to be uniformly higher in Elevenmile Creek than in the other tributary streams. Dissolved nutrients were removed from the water column in the upper bay during most of the year whereas particulate carbon, nitrogen, and phosphorus were produced throughout the year. Sediment nutrient concentrations in Perdido Bay showed little or no deviation from nutrient concentrations found in what are considered to be "natural" sediments. Isphording and Livingston (1989) found that wind events, salinity stratification, and temperature conditions were associated with observed changes in water quality and hydrological conditions. Low bottom dissolved oxygen was associated with high temperature and salinity stratification. The Perdido River was a major source of sediments to the bay. Concentrations of carbonate carbon and organic carbon in the sediments of Perdido Bay were comparable to those in other estuaries of the northeast Gulf of Mexico. No anomalies were noted in the sediment pH and Eh (redox) values of Perdido Bay relative to other gulf estuarine systems. Metal concentrations (Cd, Co, Cu, Cr, Ba, Zn, Fe, Va, Al) in sediments of Perdido Bay were comparable to those found in other Gulf estuaries. Relatively low organic carbon and high percentages of coarser-grained sediments indicated only limited sites for the accumulation and attachment of contaminants that enter the bay.

According to a U.S. Fish and Wildlife Service Report (1990), a series of aerial photographs of Perdido Bay was analyzed from the periods 1940–1941 (prior to the establishment of the Pensacola pulp mill), 1979, and 1987. The data indicated that, in the upper bay, four small areas of submerged aquatic vegetation (SAV) were noted. These beds had disappeared by 1979. However, a few years later, four new and somewhat larger beds had developed, with the largest located between Grassy Point and Lillian, Alabama. By 1987, this band had become three separate and smaller areas, with another somewhat expanded area near Lillian. South of the Highway 98 Bridge, two new (small) areas of SAV had developed to replace a slightly larger area that had occurred in 1979. No SAV was noted in eastern portions of upper Perdido Bay during any of the study periods. The U.S. Fish and Wildlife Service (1990) concluded that the preponderance of SAV in Perdido Bay since 1940–1941 occurred in the lower bay. The SAV in the lower bay decreased by more than half from 1940–1941 to 1979. After a subsequent increase in 1987, there was a net loss of SAV (48%) over the period of study. Land-use changes, navigation projects, and tropical storms were cited as possible causative factors. Logging (turn of the century), silvaculture operations, farming, and commercial and residential development also might have been implicated in the reductions of SAV in lower Perdido Bay. Increased nutrient levels and accompanying phytoplankton and epiphytic algae were also considered as possible causes due to discharges of nutrients and suspended materials. Dredging of the GIWW in the 1930s and later was also noted with pass enlargement and open water spoil disposal as postulated factors in the decline of SAV in the lower bay.

The U.S. Fish and Wildlife Service report (1990) was ground-truthed by Bortone (1991). He concluded that the previous aerial surveys in 1987 were not accurate. Some beds, such as the one off Grassy Point on the west side of the upper bay, were extensive in the ground-truthing but were not shown in the aerials. Other beds, in lower bay areas, were shown in the aerials but could not be substantiated by the diving survey. According to the Bortone survey, no beds were noted on the east side of upper Perdido Bay. A series of seagrass beds was noted on the west side of upper Perdido Bay from Grassy Point to the vicinity of the 98 Bridge. *Vallisneria americana* dominated these beds. Scattered beds were noted in midportions of the bay below the bridge past Dupont Point. A bed was also noted in Tarkiln Bay southeast of Dupont Point. Most of the SAV distributions that were ground-truthed were in the lower bay (Bortone, 1991).

A series of laboratory and field seagrass experiments (Livingston, 1992) indicated that light transmission characteristics, comparable to the immediate receiving area in Perdido Bay (station P23), were not inhibitory to the growth of *V. americana*. Direct exposure of *Vallisneria* to pulp mill effluents at concentrations comparable to the end of Elevenmile Creek were inhibitory to the growth of this species, whereas concentrations of mill effluents at levels comparable to the immediate bay receiving runoff from Elevenmile Creek were stimulatory to *V. americana* growth. Transplant experiments demonstrated that *Vallisneria* from other estuaries can grow in upper Perdido Bay. The limiting depth for such growth was 0.8 to 1.0 m. Depth limitations of *Vallisneria* were due to light limitation, a finding that was consistent with the experimental results and the general distribution of this species in upper Perdido Bay. The relatively stable distribution of the *V. americana* beds in western parts of the upper bay over the recent past is consistent with the experimental results. Davis et al. (1999) found that a limited *V. americana* transplant by R. J. Livingston in 1991 had developed into a relatively large bed by 1997–1998. Davis et al. (1999) concluded that *V. americana* beds in upper Perdido Bay were recruitment limited rather than constrained by water quality, toxic substances, light inhibition, or unsuitable substrate.

Detailed, long-term analyses of the Perdido drainage system have been carried out for the past 11 years (Livingston et al., 1998a). This has included baywide analyses at fixed stations with long-term monitoring areas shown by circles (Figure 2.2a). The results of an initial 3-year study (Livingston, 1992) showed that the shallow periphery of the bay was characterized by sandy substrates. The shelf areas graded down a slope to the deeper parts of the bay. Sediments here were characterized as liquid mud, a stable mixture of water and suspended particles of a flocculent nature that appeared to be a slurry of particulate matter and bay water. The sediment orthophosphate levels in the lower end of Elevenmile Creek and the nearby bay areas tended to be the highest such concentrations found in the Perdido Bay system. Stable isotope relationships ($\Delta^{15}N$, $\Delta^{13}C$) for the sediments (Coffin and DeFuentes, 1992; 1999) indicated that the source of solubilized particulate organic matter in the bay had unusually light isotopic ratios and therefore was autochthonous in origin. In lower parts of the bay, sediment $\Delta^{13}C$ signatures were similar both to the coastal end member and Elevenmile Creek, indicating that the source of organic matter to sediments in lower Perdido Bay was phytoplankton. During most times and in most portions of the bay, chlorophyll *a* was relatively low in the Perdido Bay system compared with other reference bay systems such as Apalachicola Bay and Choctawhatchee Bay.

This database will serve as a subject of detailed review with regard to long-term changes in phytoplankton assemblages relative to nutrient loading trends. These data will be related to observed changes in food web phenomena and associated losses of secondary production over the 11-year period of observation.

CHAPTER 4

The Perdido Bay System: Eutrophication Processes and Plankton Blooms

I. INTRODUCTION

Perdido Bay (Figure 2.2a) in the northeastern Gulf is somewhat unique in that it is a relatively small system with only one primary source of pollution in the upper bay, a pulp mill. Pulp mill effluents tend to have relatively high concentrations of nutrients. Although there is some agricultural runoff into the Perdido River, such activity is concentrated principally to the west and south of the upper bay. Most of the land in the upper Perdido drainage basin is lightly populated. The highest agricultural and urban runoff is thus located in the lower bay. This combination of factors established upper Perdido Bay as an excellent place to carry out research concerning nutrient loading and eutrophication since the upper bay is small and relatively free of pollutants other than a single point source of nutrients (Livingston, 1998). Accordingly, there was an emphasis on the determination of the effects of nutrient loading and seasonal/interannual responses of water and sediment quality and associated biological interactions with a series of mill-related changes in nutrient loading over an 11-year period of study. Field descriptive data were used in conjunction with field and laboratory experiments and modeling efforts to determine the response of the Perdido system to this single source of anthropogenous nutrient loading.

Perdido Bay is not completely free of human impacts. The opening of Perdido Pass and the resulting salinity stratification from the entry of high-salinity water into the bay have caused a situation similar to that in Choctawhatchee Bay. Deeper parts of both bays are strongly salinity stratified. Stratification results in subhaloclinal hypoxia during warm periods and the creation of liquid mud at depth (Livingston, 1997a). These effects, together with seasonal and interannual changes in Perdido River flow, provided the background for the eutrophication analyses. By closely following nutrient loading from all sources (including the pulp mill) over a prolonged period, information was gathered concerning the origin, incidence, and impacts of a series of plankton blooms that occurred from 1993 through 1999. At the same time, the interactions of the phytoplankton assemblages with other levels of biological organization in the bay were determined. The results of the study of upper Perdido Bay were compared with analyses of the effects of non-point nutrient sources from agricultural and urban areas on the lower bay (Wolf Bay and Ono Island–Perdido Key; Figure 2.2a). These analyses were then compared to studies in other river-dominated estuaries in the northeastern Gulf of Mexico with respect to the effects of various forms of anthropogenous nutrient loading. The information from the northern Gulf Coast was then reviewed within the context of studies of plankton blooms and their effects on biological production in diverse coastal systems around the world. By using the relatively simple Perdido Bay system as a model for long-term trends of eutrophication, it was possible to compare this model with similar data taken in more complex coastal systems to test specific hypotheses of causation.

A. The Perdido Bay Study

The Perdido Bay analyses were carried out continuously from October 1988 through September 1999. Detailed outlines of protocols for field and laboratory operations are given in Appendix I. Information has been collected using the same quantitative field methodology that has been used in the analyses of eight river–estuarine areas along the northeastern Gulf of Mexico (Figure 2.1). These studies span a 31-year study period. Field methods were quantified for various forms of data collection (Livingston, 1976; 1979; 1984a; 1987c; 1992; 1998; 1999; Livingston and Joyce, 1997; Livingston and Ray, 1989; Livingston et al., 1974; 1976; 1997; 1998a; Stoner et al., 1982). Taxonomic consistency was maintained by continuous verification by appropriate experts.

The Perdido Study (Livingston, 1992) was composed of the following elements:

1. *Hydrographic analyses*: (a) Definition of the physical boundaries of Perdido Bay; (b) Establishment of the physical classification of the estuary; (c) Evaluation of the flushing times of nutrients; (d) Evaluation of mixing/stratification processes; (e) Analysis of event-driven processes that had impacts on the hydrography and water quality of the bay.
2. *Basinwide nutrient loading studies*: (a) Evaluation of river and non-point freshwater runoff that included pollutant loads; (b) Establishment of the composition and relative magnitude of nutrient loads from contributing subbasins; (c) Determination of the average and extreme nutrient loads over monthly, quarterly, and annual periods; (d) determination of seasonal and interannual variability of nutrient loading to the bay.
3. *Nutrient limitation experiments*: Mesocosm experiments were carried out to determine which nutrients were limiting to phytoplankton in Perdido Bay. These studies were related to descriptive nutrient studies, spectroradiometric determinations of light transmission, and productivity studies.
4. *Sediment analyses*: Characterizations were made concerning particle size distribution, percent organics, nutrients, and toxic agents (metals, chlorinated organics, polynucleated aromatic hydrocarbons) of sediments throughout the Perdido estuary. Nutrient transfer experiments were run to determine the storage/release capacity of the bay sediments. Sediment oxygen demand analyses were made relative to seasonal changes in temperature, salinity, dissolved oxygen, and the stratification of the water column. These data were used in subsequent modeling efforts.
5. *Carbon, nitrogen, and phosphorus stable isotope analyses*: Experiments were carried out to determine the origins of the dissolved organic carbon, nitrogen, and phosphorus pools in Perdido Bay. Carbon, nitrogen, and phosphorus stable isotope analyses of bacteria were coupled with short-term oxygen consumption experiments and ultimate biological oxygen demand (BOD) determinations to examine the origin of these compounds found in the bay and to determine the potential effects of mill discharges on dissolved oxygen concentrations in the bay.
6. *Descriptive water quality studies*: Seasonal and interannual changes of the vertical and horizontal distribution of physical/chemical components of the basic water quality factors in the Perdido Bay system were analyzed at monthly to quarterly intervals over the 11-year period of study. These data were used in the modeling efforts.
7. *Biological studies*: Determinations were made concerning seasonal and interannual changes of phytoplankton, zooplankton, infaunal macroinvertebrates, epibenthic macroinvertebrates, and fishes in the Perdido River–Bay system. These data were used to transform all biological collections into trophic equivalents (biomass m^{-2}).
8. *Ecological models*: Models were developed to relate the field and laboratory components to the research questions. Water circulation in the model was determined from the Pritchard technique that is based on a balance of ocean-derived salinity. This analysis determined vertical flows and dispersion as well as longitudinal flows. A WASP3 model was calibrated with water quality data to identify critical conditions with respect to dissolved oxygen distribution in Perdido Bay. Other models were developed to quantify the relationships of the influence of the Gulf on the bay in terms of water mass movements and nutrient loading.
9. *Statistics*: Statistical methods were developed to analyze interannual trends over the 11-year period of observation.

Field samples were taken at fixed stations (Figure 2.2a) at monthly to quarterly intervals in all the rivers of the Perdido drainage basin and throughout Perdido Bay from October 1988 through September 1999. The following factors were included in this sampling program:

1. Physical–chemical factors were taken at the surface and at 0.5 m intervals through the water column. This analysis included temperature, conductivity, salinity, dissolved oxygen, and pH. Water depths and Secchi depths were routinely analyzed. Field notes of habitat conditions were also taken. Baywide water samples were taken synoptically with two to three boats over a single tidal period. Basic habitat conditions throughout the bay were monitored periodically by divers.
2. Surface and bottom water samples were examined in the laboratory for the following: NCASI color, ratio and regular turbidity, chlorophyll *a*, chlorophyll *b*, chlorophyll *c* (with phaeophytin corrections), sulfide, particulate organic matter (POM), particulate inorganic matter (PIM), orthophosphate (PO_4), particulate organic phosphorus (POP), dissolved organic phosphorus (DOP), total dissolved phosphorus (TDP), total phosphorus (TP), unionized ammonia (NH_3), nitrite (NO_2), nitrate (NO_3), particulate organic nitrogen (PON), dissolved organic nitrogen (DON), total organic nitrogen (TON), total nitrogen (TN), biochemical oxygen demand (BOD), dissolved organic carbon (DOC), particulate organic carbon (POC), inorganic carbon (IC), total carbon (TC), silica (SiO_2), total suspended solids (TSS), total dissolved solids (fresh water only) (TDS), dissolved inorganic matter (DIM), dissolved organic matter (DOM).
3. Whole-water and net phytoplankton were taken along with net zooplankton. Infaunal macroinvertebrates were taken with cores and/or ponars, and epibenthic macroinvertebrates and fishes were taken with seines in freshwater areas and around the edges of the bay and with otter trawls in the bay. All biological studies were carried out using quantitative sampling techniques; all samples were characterized to species wherever possible. The field biological data were then transformed into trophic units based on long-term stomach content data in various estuarine areas along the northeastern Gulf coast.

B. Phytoplankton Bloom Status

The designation of bloom status in phytoplankton assemblages is somewhat subjective and is dependent to a considerable degree on the natural history characteristics of an individual species. Some of the older literature used water discoloration as a definition of an algal bloom. Another definition included the concept of high numbers of cells and low species diversity (Fryxell and Villac, 1999). Anderson (1996) included in the concept of blooms the concentration(s) of one or more species that cause harm to other species or that cause accumulations of toxins in a way that causes harm to those who may eat the toxic species. Many algal specialists consider a bloom to be defined as a population density in excess of 1×10^6 cells l^{-1}. This definition leads to other complications regarding which measure to use in the bloom definition; numbers, biomass, cell volume, or indices calculated on the basis of chlorophyll *a* concentrations (Fryxell and Villac, 1999).

In our analyses, there were subdominant plankton populations that occurred over a range from 1×10^4 to 1×10^6 cells l^{-1}. These populations were found to be significant in terms of the overall composition of the phytoplankton community. Determination of relative dominance requires a complete community analysis to the species level. The methods and timing of the plankton collections (whole-water and net samples taken at monthly intervals) precluded a fully representative determination of bloom dynamics because of the scale of bloom variation relative to the intervals of sampling frequency. Organisms that have major changes of numerical abundance in relatively short periods (minutes to hours) are more difficult to sample quantitatively than those with longer reproductive intervals.

Another research question concerned how the community characteristics of the phytoplankton assemblages as a whole were affected by periodic outbreaks of bloom species. This aspect was related to changes in associated food webs. The size, shape, and biomass of any given microalgal species may not be related to the relative impact on water quality and/or toxicity. However, the relative dominance (numbers or biomass) of a set of bloom species can be an important indicator of the effects of such blooms on the overall phytoplankton community structure. This measure, in turn, can be related to important changes in associated food webs and secondary production. Phytoplankton community indices such as relative dominance and species richness do not change rapidly; therefore, they are more representative of community organization than blooms, which often occur over relatively short intervals and are less well sampled in terms of quantitative field collections.

This book uses the conventional definition of a bloom as a population density that exceeds 1×10^6 cells l^{-1}. However, as will become evident, this measure of bloom status is not necessarily the most important part of the phytoplankton analysis. There are associated indices of the phytoplankton assemblages such as species richness and species diversity that may be important as determinants of the influence of plankton blooms on the trophic response of the system at various levels of biological organization. In addition, relative dominance, based on plankton volume and biomass, were determined as indices associated with the species richness and diversity of the phytoplankton community. The emphasis of the research was therefore on (1) the qualitative and quantitative aspects of the species-specific composition of the phytoplankton community relative to long-term changes of nutrient loading and (2) associated effects on the trophic organization of the system.

II. RIVER FLOW TRENDS

The role of freshwater input in determining the productivity of river-dominated estuaries has been the subject of considerable review (Snedaker et al., 1977; Cross and Williams, 1981). The different effects of alluvial river flow on physical–chemical variables in receiving estuaries have been described (Schroeder, 1978). Under natural river flow conditions, the combination of generally high levels of estuarine primary production combined with reduced predation by stenohaline marine organisms have created conditions that allow rapid growth and enhanced productivity of eurytopic populations which are adapted to rapidly changing environmental conditions (Livingston, 1984a; 1991b). Direct nutrient enrichment by river loading, advected nutrient-rich offshore water into the estuary, nutrient out-welling from estuarine wetlands, and rapid recycling of nutrients *in situ* contribute to the generally high primary production of river-dominated estuaries (Nixon, 1981; 1995). Freshwater input controls long-term nutrient dynamics in river-dominated coastal systems (Livingston, 1981b; 1984a; Peterson and Howarth, 1987; 1988; Baird and Ulanowicz, 1989). River-derived nutrients and particulate organic matter, together with wind, tidal subsidies, and associated phytoplankton productivity, are considered essential to the naturally high production of organic carbon in temperate estuarine systems (Nixon, 1980; Livingston, 1984a).

The significance of organic input, direct or indirect, to a given coastal system also depends on the geomorphology of the drainage basin, the magnitudes of tidal input, and freshwater input from upland areas. Such input is then altered by complex feedback phenomena by the biological components of the receiving area (Haines, 1979; Odum et al., 1979; Nixon, 1980; 1981; Welch et al., 1982; Kemp and Boynton, 1984; Livingston, 1993). There is a direct connection between the primary production of the estuary and the secondary food webs. The importance of phytoplankton as a major source of organic carbon in coastal systems is now being recognized (Coffin and Cifuentes, 1999). The dominant animal associations of shallow Gulf estuaries include infaunal macroinvertebrates, epibenthic macroinvertebrates, and fishes (Sheridan, 1978; 1979; Laughlin, 1979; Sheridan and Livingston, 1979; 1983; Livingston, 1984a; 1991b). Interactions of these groups are mediated by competition for habitat and by predator–prey relationships (Virnstein, 1977; Peterson, 1982; Livingston, 1982, 1984b; Whitlatch and Zajac, 1985; Livingston et al., 1997). These biological processes undergo complex changes in space and time. Temporal cycling varies within the different levels of biological organization with plankton cycles measured in terms of minutes to years while the upper levels of the food web follow primarily seasonal/annual and interannual

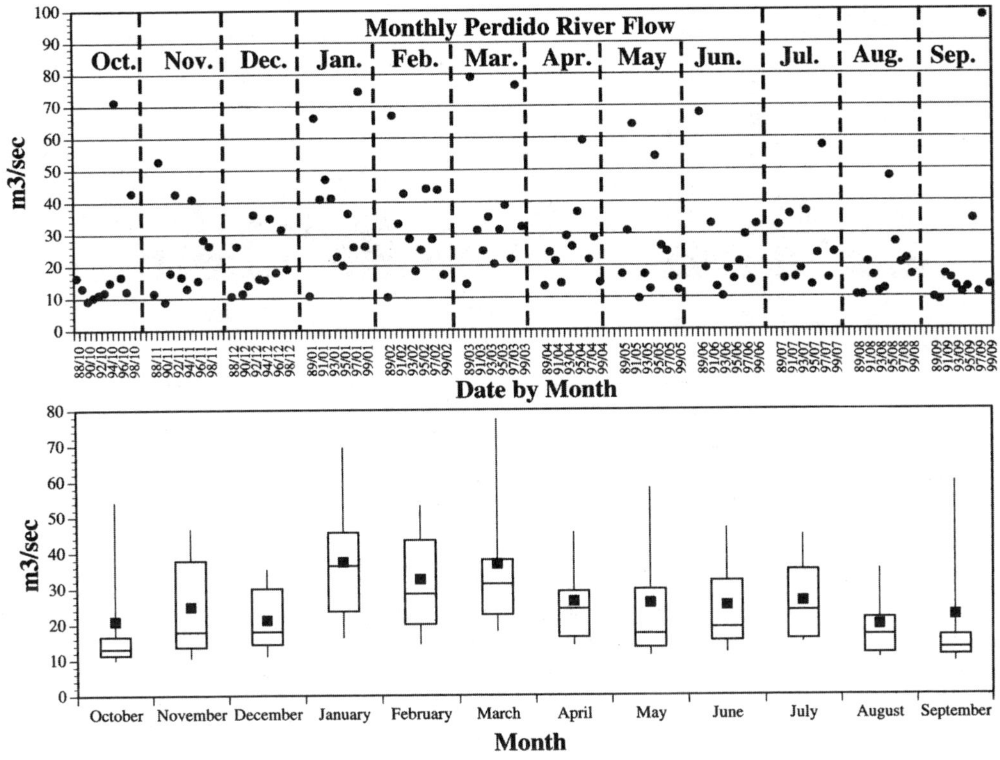

Figure 4.1 Monthly Perdido River flows (m³ s⁻¹) shown by month over the 11-year sampling period. Data are also given as means with the 10th, 25th, 75th, and 90th percentiles.

cycles. Seasonal and interannual cycles of alluvial river flows thus play a major role in determination of the trends of primary and secondary productivity of Gulf estuarine and coastal systems (Livingston et al., 1997).

A. Seasonal and Interannual River Flow Patterns

There were seasonal patterns of Perdido River flow (Figure 4.1), with low flows from August through December and high flows from January through March. There were two basic components that contributed to the long-term alluvial river flow patterns. One was general background flow based on regular seasonal progressions of meteorological conditions with winter–early spring highs and late summer–fall lows. The second was based on episodic influxes of upland runoff due to storms and hurricanes. This accounted for the observation that there were occasional peak flows that occurred during most months of the year. Over the 11-year study period, there were five major storms (Figure 4.2): Tropical Storm Alberto, 3 July 1994; Hurricane Erin, 3 August 1995; Hurricane Opal, 5 October 1995; Hurricane Danny, 19 July 1997; and Hurricane Georges, 28 September 1998. Two-way ANOVAs, run by year and season using monthly averages within each season as replicates, indicated significant ($P = 0.05$) differences between winter–early spring and late summer–fall flows. March and September flows had the highest interannual variation. Prolonged periods of low flow occurred during 1988–1989, 1993–1994, and 1998–1999. The 1993–1994 drought was the most prolonged period of low flows during the study period. Relatively heavy river flow events occurred during the winter–spring of 1990, over a prolonged period from spring 1995 through winter 1996, during winter 1998, and during the five storms. There were significant ($P = 0.05$) differences among

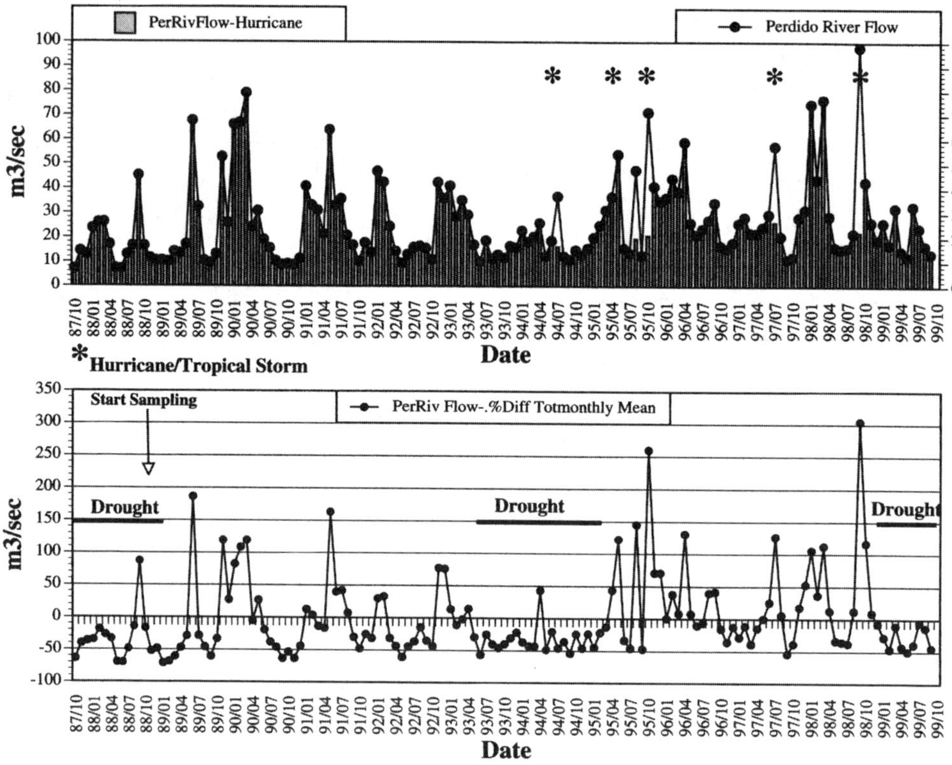

Figure 4.2 Monthly Perdido River flows (m^3 s^{-1}) plotted by month over the 11-year sampling period and showing major hurricanes. Monthly Perdido River flows also plotted as percent differences from the grand mean allowing clearer delineation of the drought periods.

the annual flows over the study period. River flows during drought periods were significantly ($P = 0.05$) different from those during the peak years of 1990 and 1995–1996.

A trend analysis of monthly river flows (Figure 4.2) indicated that storm-related increases in river flows were most noticeable during 1995 and 1997–1998. With the exception of the 1994 tropical storm, most storm events occurred during periods of summer low flows. Droughts were defined by relatively flat river curves during winter–early spring flood periods. Meeter et al. (1979) reviewed the long-term climatological changes in the Apalachicola system. A spectral analysis of long-term trends of Apalachicola River flow and Georgia rainfall indicated drought–flood cycles of 6 to 7 years. Cross-spectral analyses of these two variables showed that Apalachicola River flow and precipitation in Georgia were in phase. This is similar to the pattern noted in the Perdido River system whose primary basin extends into Alabama. The interannual trends of the drought–flood sequences represent an important variable that is related to water quality in Perdido Bay.

B. Riverine Influence on Bay Stratification and Flushing

The upper Perdido Bay is defined as the area above the Highway 98 Bridge, whereas mid-bay areas are from the bridge to Dupont Point and the lower bay is from Dupont Point to Inerarity Point (see Figure 2.2a). Niedoroda (1992) analyzed the relationship of Perdido River flow with physical processes in Perdido Bay. Salinity was inversely correlated with river discharge. During relatively wet years, surface and bottom salinity of Perdido Bay decreased, but the lower layer decreased the most. Flushing times of middle and lower bay areas were similar and were somewhat less than those of the upper bay. Most of the differences in flushing times were attributed to differences in the width of the bay. The upper layer of the middle bay had flushing times between 0.3 and 4.5 days. Corresponding values for the whole lower layer in the middle bay ranged between 0.3 and 6 days. The computed flushing times for a representative cell in the lower bay ranged between 0.1 and 3 days for the upper layer with much longer values (up to 15 days) for the lower layer. Thus, turnover rates were lowest in subhaloclinal areas of the lower bay.

The explanation for systematic patterns of flushing behavior in the upper and lower layers of the upper bay was in the interaction between the magnitude of Perdido River input and the distribution of salinities (Niedoroda, 1992). Moderate discharge events tended to flush out the upper layer of the upper bay while not destroying the lower layer. This led to increased salinity stratification and the sharply reduced entrainment of the lower layer into the upper layer, thus reducing flow in the bottom layer. Major storms flushed the entire upper bay with fresh water. During such discharge events, the lower layer disappeared and the upper layer had a very short flushing time. Afterward, the return of the bottom layer to the upper bay would take no more than a month. However, because the surface layer extended over the whole depth, flushing times during these periods were similar to those of low-flow conditions. There was thus a shifting balance between river input and the intrusion of saline conditions up the bay. Under prolonged conditions of very low flow, high salinities were carried up the bay by the bottom flow and highly stratified conditions developed throughout the bay. Although there were several competing processes that controlled the cell volumes and rates of inflow, the upper layer flushing times remained relatively similar under a wide range of river discharges.

Relative trends in flushing times were more significant than absolute values (Niedoroda, 1992). There was a balance between river discharge and the intensity of stratification that varied over roughly annual periods. During the dry season, salinity stratification was maximal in the upper bay and minimal in the lower bay. Conversely, as river input increased, stratification in the upper bay decreased to zero as it was flushed. High river inputs led to marked decreases in the layer-averaged salinities in the lower bay, resulting in increased salinity contrasts between upper and lower layers. This, in turn, reduced upward entrainment of bottom water and probably contributed to decreased rates of bottom water circulation in the lower bay. Perdido Pass represents the primary source of salt water to Perdido Bay. It is possible that Gulf water enters only briefly during each tidal cycle when the river flow is low. During periods of elevated river flow, the outward bias of currents in the pass may restrict the amount of Gulf water, which can again enter the bay during a rising tide. Thus, changes in the configuration of the inlet, such as those that occur naturally as a result of sand transport along the Gulf beaches, can exert major influences on the distribution and circulation of the water masses, and, accordingly, the flushing times in all regions of the bay. Overall, Perdido Pass controls salinity relationships in Perdido Bay.

The strong stratification of the bay together with the complexity of turnover rates represent important variables in the evaluation of water quality, phytoplankton production, and subhaloclinal hypoxia during warm seasons of the year. The relatively narrow bay, with a trending NE–SW axis, tended to minimize vertical mixing thus contributing to the relatively strong halocline in the system. Hagy (1996) used models to determine residence times of nitrogen, carbon, and silica for the Patuxent River estuary. He estimated that seasonal differences of freshwater residence ranged from 44 to 76 days. Hagy (1996) found riverine nitrogen sources during winter–spring and recycled/seaward nitrogen sources during summer. There was evidence that this situation exists in Perdido Bay (Niedoroda, 1992). High Patuxent River flows usually were associated with low phytoplankton biomass. Low flows were associated with high algal biomass in the upper estuary. Hypoxia was "hyperbolically related to river flow and bottom layer oxygen consumption." As in the long-term analyses of the Perdido system (Livingston, 1999), phytoplankton production and phytoplankton biomass were not always associated in time and space. Correlations were found between water–year river flow and annual hypoxia; these relationships were considered to be driven by oxygen demand and not the physical effects of river flow on water column stratification. The complexity of the exchanges with Chesapeake Bay added to the variability noted in the response of the Patuxent estuary to river discharge and nutrient loading. The Perdido Bay system is far simpler in structure, and nutrient loading and river flow characteristics are closely associated.

III. NUTRIENT DYNAMICS

A. Nutrient Loading

Nutrient loading is an important process in the determination of the ecological response of a coastal system to freshwater inflow (Flemer et al., 1997). Because of the complex ecological processes of estuarine/coastal systems, a simple resolution of the relationship between nutrient loading and hypereutrophication is not likely. Rabalais (1992) estimated that 23 of 58 estuaries in the Gulf of Mexico, including Perdido Bay, have problems or potential problems of overenrichment. Based on projections of human population growth for the northern Gulf, nutrient overenrichment problems are likely to increase in the Gulf region in the future. Cultural eutrophication due to overenrichment of estuaries and coastal areas by human activities remains the most important single factor in the deterioration of these important habitats.

A detailed analysis was carried out concerning nutrient loading to the Perdido system over the 11-year study period. Nutrient loading models, based on a ratio estimator developed by Dolan et al. (1981), were run to determine nutrient loading into upper Perdido Bay (Appendix I). The U.S. Geological Survey (Tallahassee, Florida, M. Franklin, personal communication) provided river flow data. Nutrient sampling was carried out in the various rivers leading into the bay (Figure 4.3). Monthly nutrient concentrations from these rivers were used to estimate loading. Overland runoff

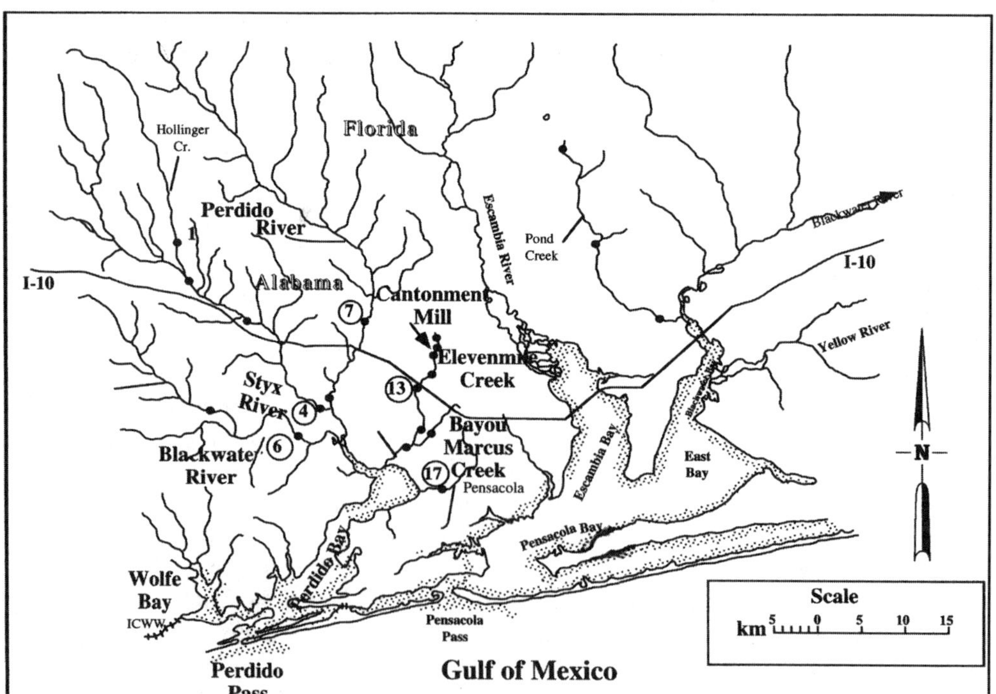

Figure 4.3 Map of the freshwater drainages in the Perdido system showing sampling stations for nutrient loading calculations.

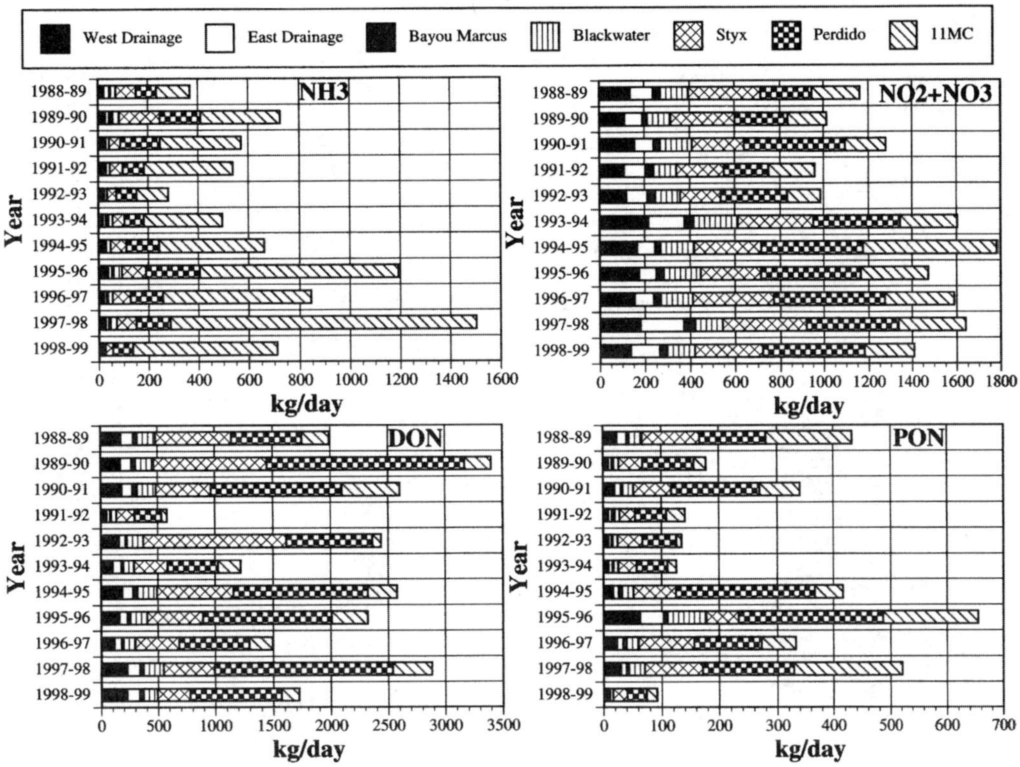

Figure 4.4(a) Annual loading (kg d^{-1}) of nitrogen, phosphorus, and carbon compounds and silica over the 11-year sampling period in Perdido Bay.

(non-point) from unmonitored areas was estimated using models devised by A. Niedoroda (personal communication). Loading models for the 11-year database were run on monthly, quarterly, and annual intervals.

Analysis of the annual and quarterly loading for the Perdido system (Figure 4.4a, b) indicated that the Perdido River system (Perdido–Blackwater–Styx Rivers) was the primary source of nitrate/nitrite, dissolved organic nitrogen, and particulate organic nitrogen to the upper bay. The Perdido River system was the main source of dissolved organic carbon, particulate organic carbon, and, with the exception of the second half of the sampling period, inorganic carbon. The Perdido River was also the main source of silica, organic phosphorus, and particulate organic phosphorus to the bay. Pulp mill effluents tend to be enriched by orthophosphate and ammonia due to a combination of factors associated with water treatment processes. Due to nutrient loading from the pulp mill, Elevenmile Creek was the primary source of ammonia during most of the study period. Ammonia loading from the creek was particularly high from 1995 through 1998. Elevenmile Creek was also the primary source of orthophosphate from 1993 through 1999 with the lowest such loading from 1989 through 1991. Nitrite/nitrate loading to the bay was generally higher in the Perdido River system than Elevenmile Creek. Peak loading of this nutrient occurred during 1992, 1995, and 1998; there was no overt interannual trend of the nitrite/nitrate loading to Perdido Bay from these sources.

Ammonia, nitrite/nitrate, and orthophosphate loading data were averaged quarterly and plotted as a function of time (Figure 4.5). By removing some of the variability, the loading patterns for these nutrients were more clearly outlined. The interannual loading trends of orthophosphate to Perdido Bay from the principal sources (Elevenmile Creek and the Perdido River system) followed

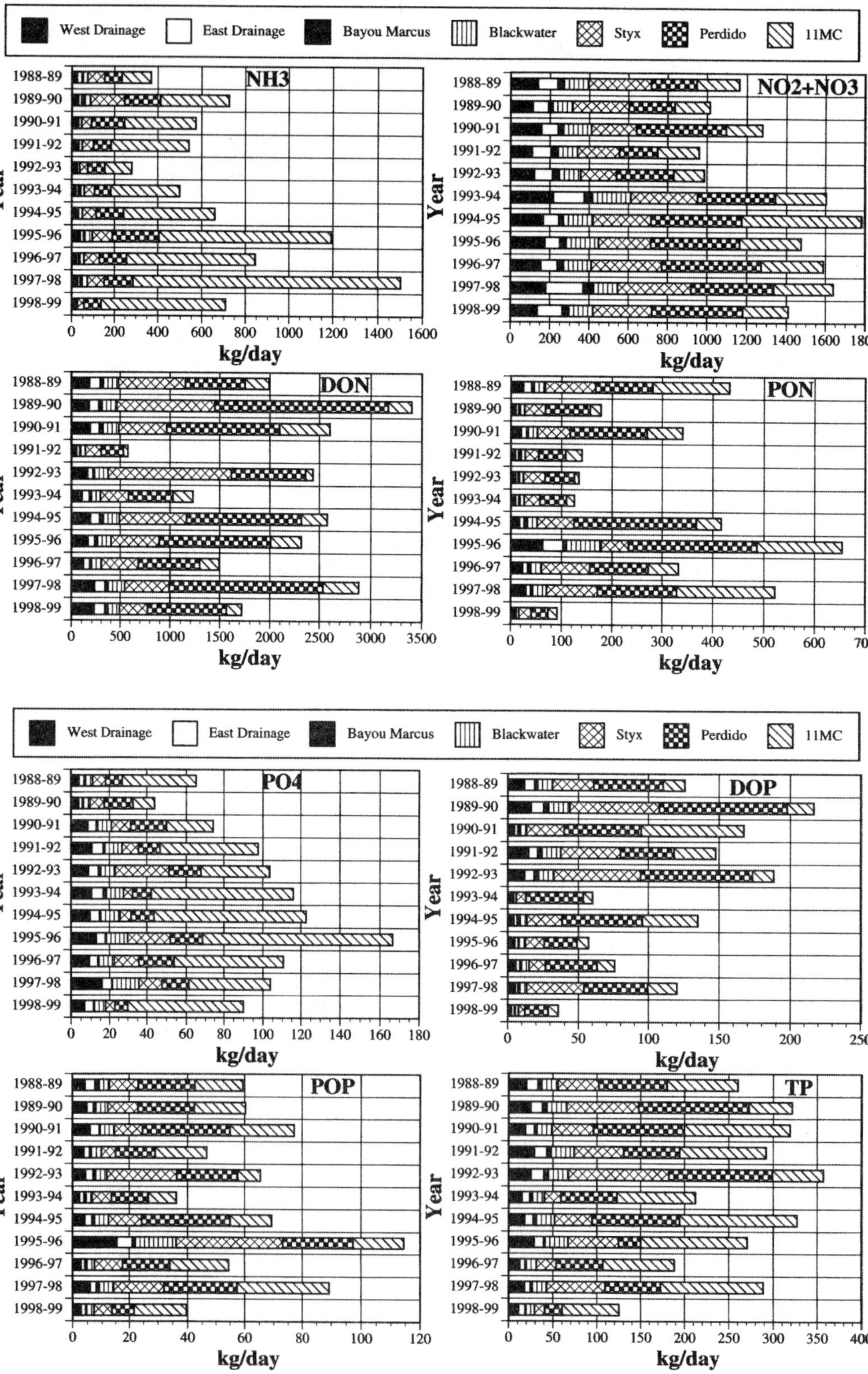

Figure 4.4(a) (continued)

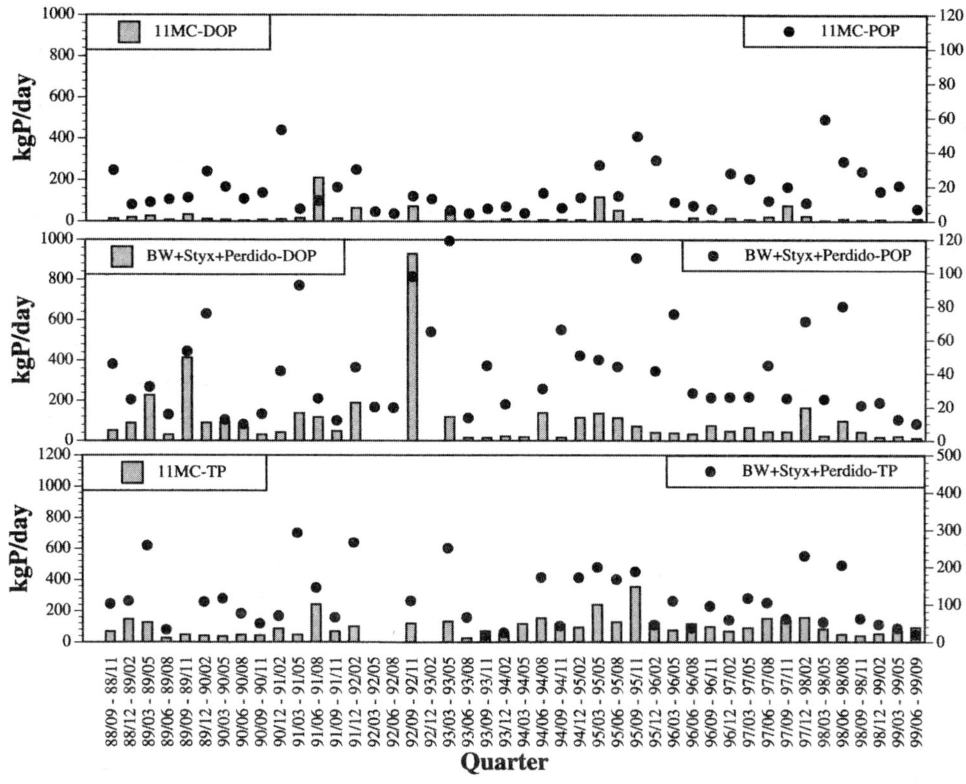

Figure 4.4(b) Quarterly loading (kg d^{-1}) of nitrogen, phosphorus, carbon compounds, and silica over the 11-year sampling period in Perdido Bay. This includes dissolved organic phosphorus (DOP), particulate organic phosphorus (POP), total phosphorus (TP), dissolved organic nitrogen (DON), particulate organic nitrogen (PON), total nitrogen (TN), silica (SiO_2), dissolved organic carbon (DOC), particulate organic carbon (POC), and inorganic carbon (IC).

a complex temporal pattern (Figure 4.5). After higher orthophosphate loading from Elevenmile Creek during fall 1988 to winter–spring 1989, such loading remained comparable to that of the Perdido River system from the second quarter 1989 through the fourth quarter 1991. There were periodic increases of orthophosphate loading from Elevenmile Creek during 1992–1993 that were accompanied by increased orthophosphate loading from the Perdido River. From the second quarter 1994 through the third quarter 1997, there were sustained increases of orthophosphate loading from Elevenmile Creek. High orthophosphate loading started during the drought of 1993–1994. During late summer 1997, the pulp mill reduced orthophosphate discharges to Elevenmile Creek. Quarterly orthophosphate loads reflected this effort from the final quarter 1997 to the second quarter 1999. There was a brief period of increased orthophosphate loading from the mill from late December 1998 through mid-January 1999. Following a few months of low orthophosphate loading from the mill, such discharge again increased in April 1999 and remained periodically high through August 1999. During September 1999, the mill resumed orthophosphate treatment and loading of this nutrient was again reduced. Mill orthophosphate loading was significantly ($P = 0.0212$, $r^2 = 0.505$) and positively associated with orthophosphate loading from Elevenmile Creek. Total organic phosphorus loading from the mill was not significantly associated with loading from Elevenmile Creek, whereas total phosphorus (TP) loading from the mill was marginally ($P = 0.018$, $r^2 = 0.173$) associated with TP loading from Elevenmile Creek.

Quarterly orthophosphate loading data for Elevenmile Creek were analyzed through a series of two-sample statistical analyses (Appendix II) to determine the significance of temporal differences

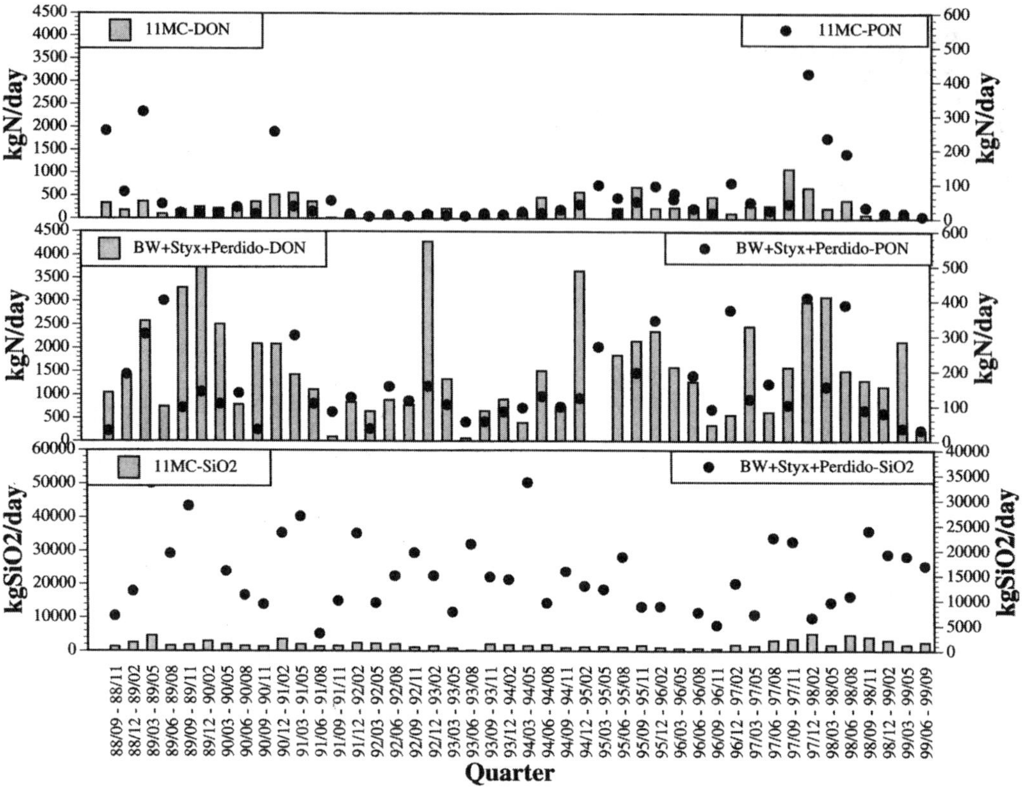

Figure 4.4(b) (continued)

of such loading. The comparisons were run with the Elevenmile Creek quarterly loading data for five separate 2-year periods (1989–1990; 1991–1992; 1993–1994; 1995–1996; 1997–1998). There were no significant differences in the means and variances between the following periods: 1995/1996 to 1993/1994; 1989/1990 to 1991/1992; 1997/1998 to 1989/1990; 1997/1998 to 1991/1992; 1997/1998 to 1993/1994; 1998/1998 to 1995/1996. There were significant differences between the means of the following periods (the highest orthophosphate means are given in the first 2-year period of each pair): 1993/1994 to 1989/1990, $p = 0.0191$; 1993/1994 to 1991/1992, $p = 0.0010$; 1995/1996 to 1989/1990, $p = 0.00014$; 1995/1996 to 1991/1992, $p = 0.02$; 1995/1996 to 1997/1998, $p = 0.05$. These analyses, together with regressions showing the relationship between mill orthophosphate loadings and Elevenmile Creek orthophosphate loadings, indicate that there was a significant increase of orthophosphate loading from Elevenmile Creek to upper Perdido Bay from late 1993 through the third quarter 1997. This loading was associated with discharges from the paper mill. The return of orthophosphate loading from Elevenmile Creek during fall 1997 to spring 1999 to levels comparable to the early years (1989–1990) allowed an evaluation of the effects of high orthophosphate loading on the bay since the pulp mill represented the chief source of this nutrient to the upper bay. The quarterly orthophosphate loading data indicated a separation of Elevenmile Creek values from those of the Perdido River system at about 50 kg Pd^{-1}.

Elevenmile Creek was the primary source of ammonia to upper Perdido Bay (Figure 4.5). Mill ammonia loading was only marginally ($p = 0.0115$, $r^2 = 0.189$) associated with ammonia loading from Elevenmile Creek. There were sustained increases of ammonia loading from Elevenmile Creek during the 1995 to 1998 period. Quarterly ammonia loading data for Elevenmile Creek were run

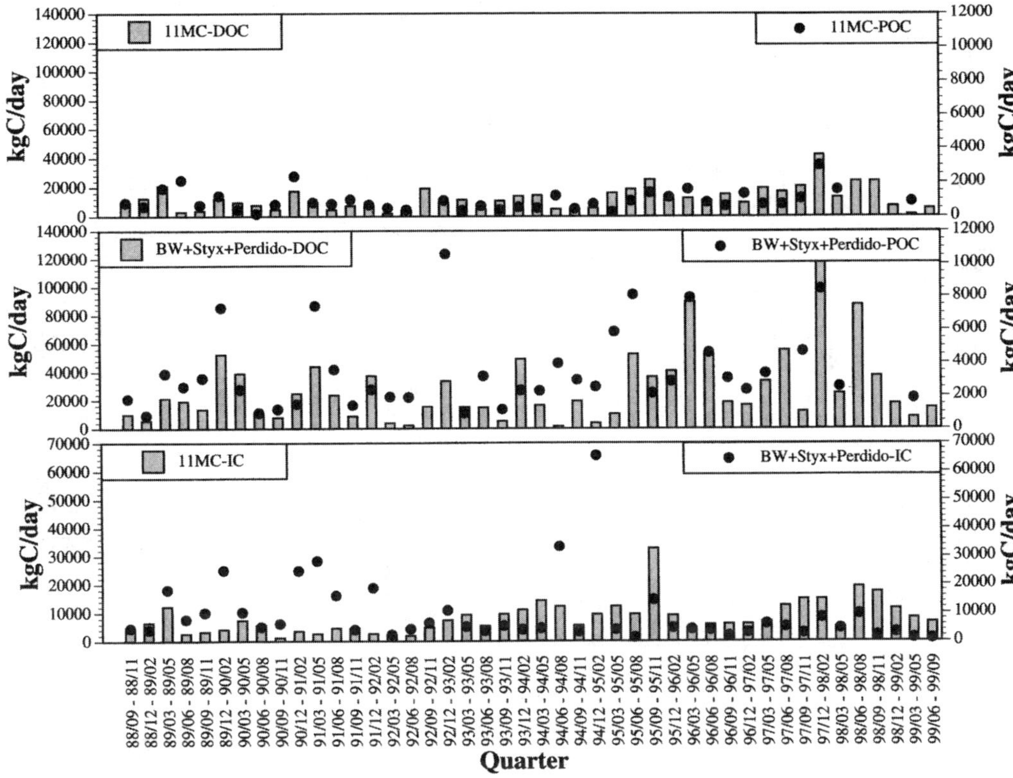

Figure 4.4(b) (continued)

through a series of two-sample statistical analyses similar to that for orthophosphate. There were no significant differences in the means and variances between the following periods: 1989/1990 to 1991/1992; 1989/1990 to 1993/1994; 1993/1994 to 1991/1992; 1995/1996 to 1991/1992; 1995/1996 to 1997/1998. There were significant differences between the means of the following periods (the highest ammonia means are given in the first 2-year period of each pair): 1995/1996 to 1989/1990, $p = 0.01$; 1995/1996 to 1993/1994, $p = 0.05$; 1991/1992 to 1989/1990, $p = 0.05$; 1997/1998 to 1989/1990, $p = 0.05$; 1997/1998 to 1991/1992, $p = 0.02$; 1997/1998 to 1993/1994, $p = 0.02$. This analysis indicated that ammonia loading from Elevenmile Creek to Perdido Bay was significantly higher during the period 1995 to 1998 than that of the preceding periods. In this case, the Elevenmile Creek ammonia data were distinguished from the ammonia loading from the Perdido River system at quarterly average loading rates exceeding 300 to 400 kg d^{-1}. Ammonia loading rates beneath these levels were comparable to loading rates of Elevenmile Creek and the Perdido River system during the period of relatively low ammonia and orthophosphate loading from 1989 to 1991.

B. Nutrient Concentration Gradients

Based on the nutrient loading information, mill-associated discharges of orthophosphate and ammonia were the chief anthropogenous sources of nutrients to upper Perdido Bay. Elevenmile Creek was characterized by generally high concentrations of orthophosphate (Figure 4.6a). There were periodic increases of orthophosphate (>0.1 mg l^{-1}) at station P22 from 1988 through 1991; by 1992, these concentrations were somewhat higher. From 1993 through the third quarter of 1997,

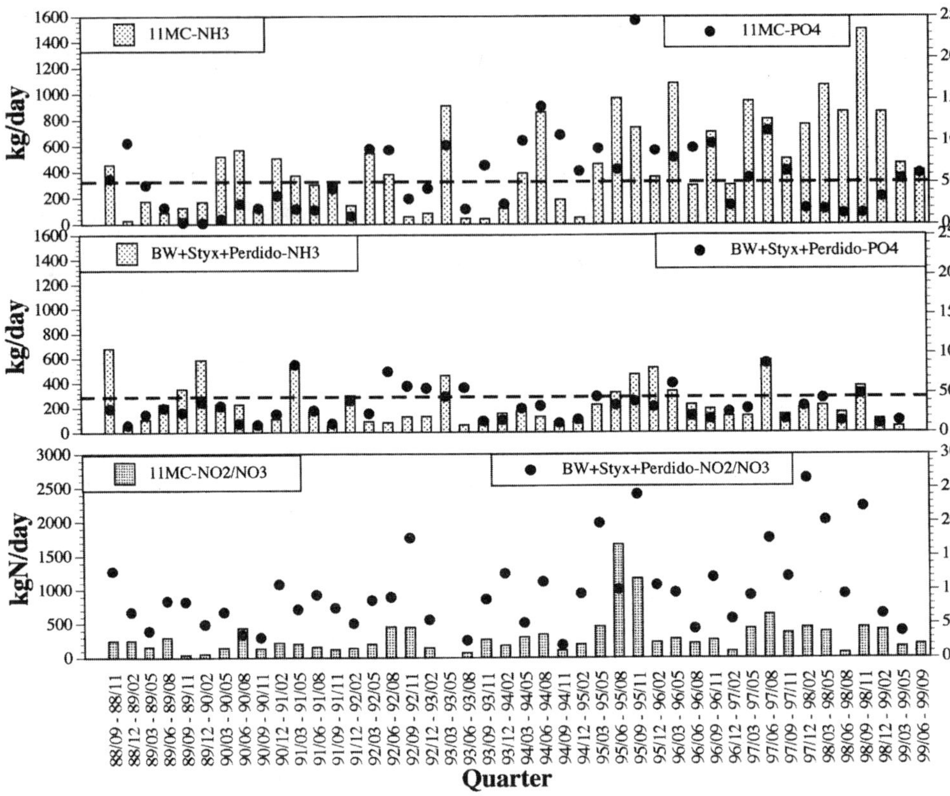

Figure 4.5 Quarterly loading of orthophosphate, ammonia, and nitrite-nitrate from the Perdido River system and Elevenmile Creek from October 1988 through September 1999. Dashed lines represent levels where loading from Elevenmile Creek exceeded those from the Perdido River system. Levels of exceedance for ammonia approximated 300 kg d^{-1} and levels of exceedance for orthophosphate approximated 50 kg d^{-1}.

there was a more consistent pattern of high orthophosphate concentrations at this station. Relatively high average orthophosphate concentrations in the bay occurred near the mouth of Elevenmile Creek (station P23) during the period of high orthophosphate loading to the bay. There was a precipitous decrease of this nutrient throughout the bay during late 1997–1998; this decrease was associated with reductions of mill orthophosphate discharges from August 1997 through March 1999. Orthophosphate levels during late 1997–late 1998 resembled those during the period 1989–1990. This trend was followed by increased orthophosphate concentrations during the spring and summer of 1999 during which time the mill resumed relatively high loading of orthophosphate to Elevenmile Creek. By September 1999, the mill again reduced orthophosphate loading to the bay, and upper bay concentrations of this nutrient reflected this change.

The high concentrations of ammonia and orthophosphate in Elevenmile Creek (station P22) were the result of loading from the mill and low creek flows relative to the Perdido River (Figure 4.6a). Relatively rapid dilution of water from Elevenmile Creek as it entered upper Perdido Bay, along with possible uptake of these nutrients by phytoplankton, established steep gradients of ammonia and orthophosphate concentrations between stations P22 and P23. By dividing the ammonia and orthophosphate concentrations at station P22 by those at station P23, a nutrient gradient index for each compound was established. Ratios of ammonia >8 were considered above the ammonia ratios found during the first 2 to 3 years of sampling. The same is true for ratios of orthophosphate that exceeded 14.

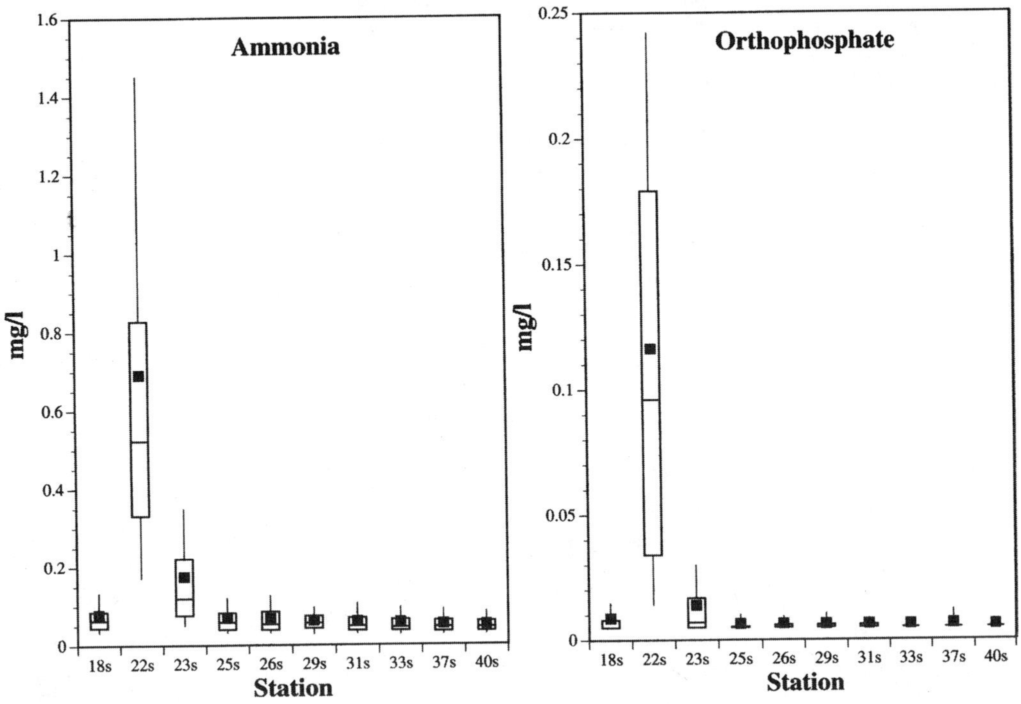

Figure 4.6(a) Ammonia (mg l^{-1}) and orthophosphate (mg P l^{-1}) taken at stations in Perdido Bay monthly from October 1988 through September 1999. Data are presented as means at all stations with the 10th, 25th, 75th, and 90th percentiles.

An effort was made to simplify the nutrient loading data. The monthly ammonia loading rates from the pulp mill were reviewed; monthly loading rates >600 kg d^{-1} were considered above the ambient loading during the first few years of sampling (Figure 4.6b). Such rates could be averaged over quarterly intervals so that average daily rates of ammonia loading should not exceed 300 to 400 kg d^{-1} (Figure 4.5), a more conservative figure that was consistent with possible effects of high concentrations of ammonia on phytoplankton assemblages (Livingston, unpublished data). A comparison was made of the high gradients of ammonia and orthophosphate at the entry point to upper Perdido Bay with the high loading rates of these nutrients from Elevenmile Creek (Figure 4.6c). There were some differences between nutrient loading and nutrient concentration ratios during various periods that could be explained by inadequate sampling frequency and/or high local variation of these highly variable factors. However, there was a general pattern of concurrence that was consistent with the use of these indices to represent general changes of nutrient loading in the Perdido system over the 11-year period of study.

C. Nutrient Limitation

Phytoplankton bioassay for nutrient addition is commonly used to determine limiting nutrients (Maestrini et al., 1984a, b). Understanding the kinetics of nutrient limitation of phytoplankton growth is important in assessing the effects of hypereutrophication (Sakshaug and Olsen, 1986). D'Elia et al. (1986) found that experiments concerning nutrient limitation of phytoplankton production in growth bioassays are useful in understanding how phytoplankton depend on different combinations of nutrients. Analysis of the stoichiometry of C:N:P fluxes following the approach of Smith (1991) can provide useful insights into potential nutrient limitation of estuarine primary

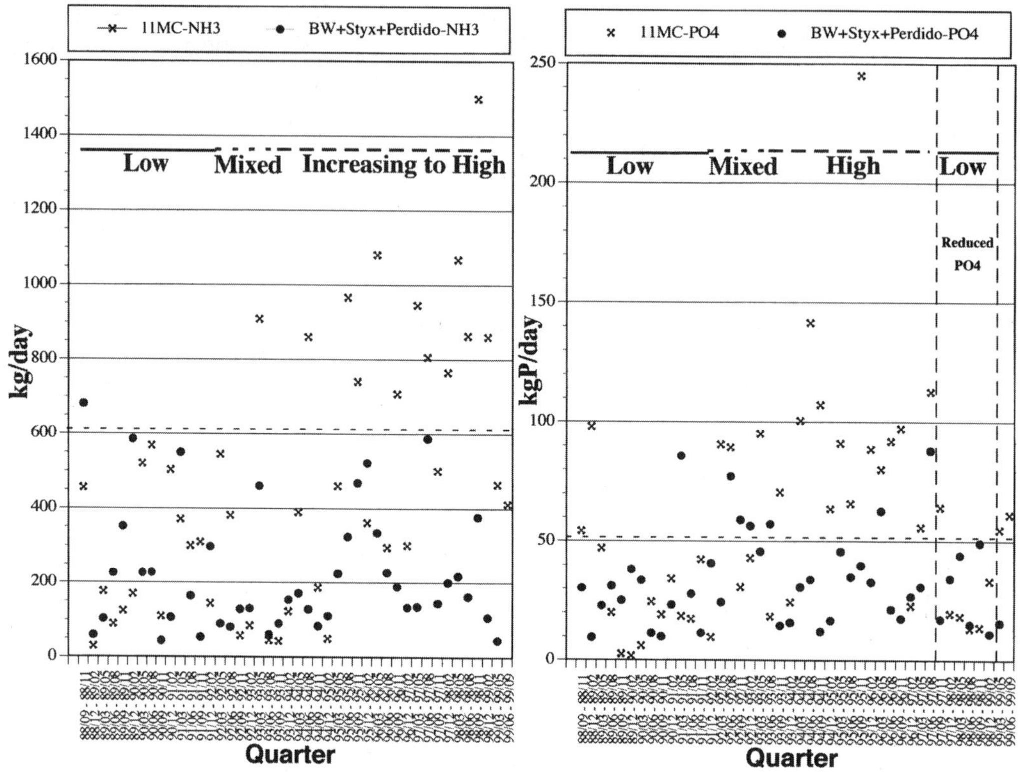

Figure 4.6(b) Simplified analysis of monthly ammonia and orthophosphate loading to Elevenmile Creek from the mill where only values above 600 kg d^{-1}ammonia and 50 kg d^{-1}orthophosphate are shown. Nutrient gradients of these compounds were calculated as ratios of ammonia and orthophosphate between stations P22 (mouth of Elevenmile Creek) and P23 (bay stations immediately adjacent to the mouth of Elevenmile Creek). Ammonia ratios >8 and orthophosphate ratios >14 were used to represent the periods of high nutrient gradients at the mouth of Elevenmile Creek.

production. Sediment stratigraphic analysis can provide evidence of eutrophication (Brush, 1984; 1991), but such information does not address the mechanisms of nutrient-limited phytoplankton productivity (Malone et al., 1996).

The question of phosphorus or nitrogen limitation has been addressed in many studies. However, nutrient bioassay experiments with estuarine phytoplankton communities in the Gulf of Mexico have been relatively uncommon compared with other coastal systems — e.g., Chesapeake Bay and tributary estuaries (D'Elia et al., 1986; Fisher et al., 1992); Narragansett Bay (Nixon et al., 1984); Baltic Sea region (Graneli, 1987; Kivi et al., 1993). Hannah et al. (1973), Estabrook (1973), Myers and Iverson (1981), and Montgomery et al. (1991) assessed effects of nitrogen and phosphorus additions on phytoplankton productivity in several Gulf coastal waters through the use of *in situ* bottle arrays and ^{14}C uptake. In these studies, phosphorus-limited phytoplankton productivity more than nitrogen. Smith and Hitchcock (1994), using bottle tests, determined that surface waters on the Louisiana shelf associated with the Mississippi River plume responded to phosphorus, nitrogen, and silica, but responses were a function of season and salinity. Phytoplankton biomass in Florida Bay was shown to be deficient in phosphorus relative to nitrogen based on Redfield N:P ratios of particulate organic matter and water column ratios of dissolved inorganic nitrogen and soluble reactive phosphorus (Fourqurean et al., 1993). Phosphorus was limiting in North Sea estuaries (Postma, 1985). Oligotrophic tropical systems may be phosphorus limited (Howarth et al., 1995)

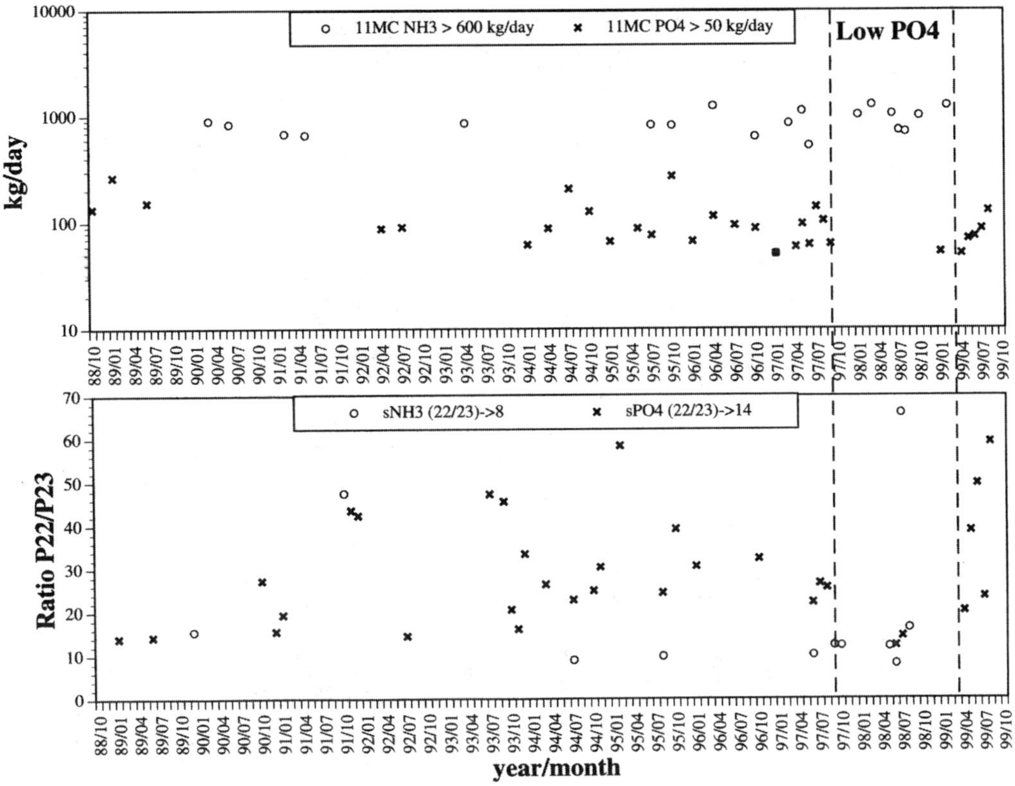

Figure 4.6(c) Summary of ammonia and orthophosphate loading with generalization of loading periods.

but such areas may be nitrogen limited farther from shore (Corredor et al., 1999). Nutrient limitation can also switch seasonally in areas such as Chesapeake Bay (Malone et al., 1996) and parts of the Gulf of Mexico (Rabelais et al., 1999). However, it is thought that relatively little of the phosphorus-limited production sinks to the bottom (Gilbert et al., 1995, Rabalais et al., 1999). Questions of nutrient limitation in the Baltic Sea are still under discussion (Hellstrom, 1996; Hecky, 1998).

Perdido Bay has been considered to be low to moderately enriched in nutrients compared with other Gulf estuaries (Macauley et al., 1995; Rabalais, 1992). A hydrographic and biological survey of Perdido Bay in 1974–1976 found the following mean concentrations: chlorophyll *a* (15 to 24 mg m^{-3}), ammonia-N, (3.2 to 4.4 μg atom l^{-1}), nitrate/nitrite-N (2.2 to 5.2 μg atom l^{-1}), and phosphate-P (0.7 to 1.5 μg atom l^{-1}) (Livingston, 1992). Further examination of the data showed that, during the summer, most stations were characterized by surface dissolved inorganic nitrogen:dissolved inorganic phosphorus (DIN:DIP) values that were less than 10. Thus, previous field research tended to indicate that phosphorus was a major limiting factor for phytoplankton production in upper Perdido Bay. Schropp et al. (1991) found that Elevenmile Creek, the Styx and Blackwater Rivers, and Bayou Marcus Creek were all affected by anthropogenous input of nutrients. Although sediment nutrient concentrations (TOC, TN, TP) in Perdido Bay were high, the authors found that they were within the natural concentration range of other Florida estuaries. According to Macauley et al. (1995), upper Perdido Bay had slightly increased dissolved nutrient concentrations that appeared to be related to significantly higher rates of carbon fixation and phytoplankton biomass. The authors found that the Perdido River diluted nutrients and color from Elevenmile Creek; the effects of these inputs appeared to be restricted to the area at the mouth of Elevenmile Creek.

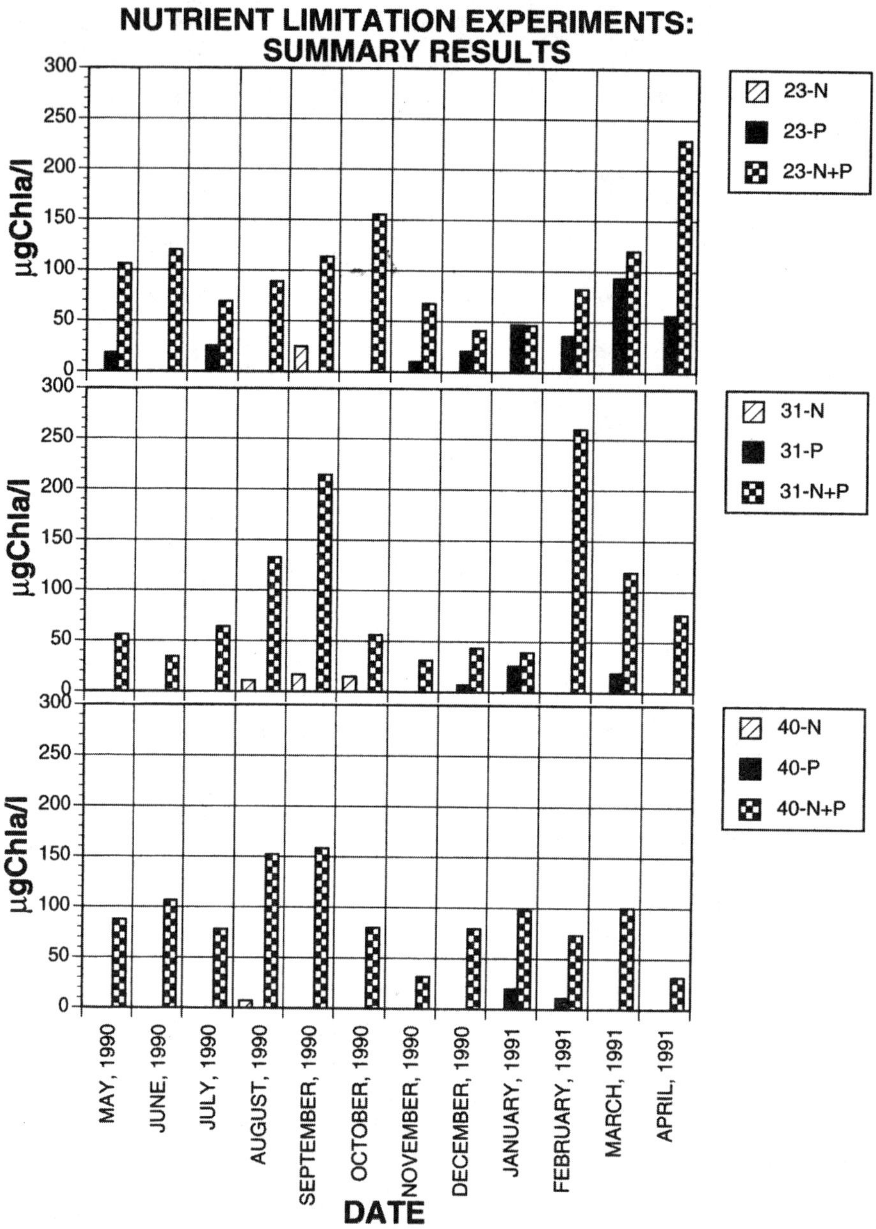

Figure 4.7 Averaged results of nutrient limitation experiments carried out with Perdido Bay water monthly from May 1990 through April 1991.

A series of nutrient limitation experiments was conducted with water taken from Perdido Bay during 1991 (Flemer et al., 1997). Six treatments in triplicate for each of three bay stations (P23, P31, P40) were established as follows: three control tanks, three P-enriched tanks at 10 μM PO_4-P above ambient, three N-enriched tanks at 50 μM NH_3-N above ambient, and three combined NH_3 + PO_4 (referred to as N+P) above ambient as described for single additions. A summary of the results is shown in Figure 4.7. Flemer et al. (1997) reported that the microcosm-based phytoplankton growth responses in Perdido Bay fell into three principal categories: (1) Primary P limitation, occurring mostly during the cooler months at upper (tidal brackish) and mid-bay (lower mesohaline) stations; (2) Primary N limitation, occurring mostly during warmer months in mid-bay areas and infrequently at upper and lower bay stations (upper mesohaline); and (3) Apparent N+P co-limitation, occurring mainly during warmer months in upper and mid-bay areas and during both warmer and cooler months in the lower bay. Primary orthophosphate limitation was associated with high dissolved inorganic nitrogen (DIN); DIN:DIP ratios ranged from 20 to 200. Conversely, primary N and N+P co-limitation were associated with decreasing DIN:DIP ratios. Phytoplankton assemblages did not appear to be strongly nutrient limited but, given a nutrient increase, these groups responded differentially to nitrogen and phosphorus, both seasonally and along the longitudinal salinity gradient. Grazing pressure, in concert with nutrient limitation, was advanced as an hypothesis to explain N+P co-limitation.

Phosphorus limitation in estuaries along the northeastern Gulf could be related to relatively low concentrations of phosphorus in the soils of this region. Although phosphorus appears to be more influential than nitrogen in controlling phytoplankton primary productivity in estuaries along this coast, the relative importance of other nutrients such as nitrogen cannot be discounted. The combination of phosphorus and nitrogen was usually more stimulatory to phytoplankton growth in Perdido Bay than either of these nutrients alone (Flemer et al., 1997). The application of a mathematical mass balance model of estuarine nutrient transport for the Ochlockonee River–estuary, based on Redfield ratios, suggested that nitrogen is, at times, more likely to be limiting than phosphorus (Kaul and Froelich, 1984). Similar data (Livingston, unpublished data) in the Choctawhatchee Bay system indicated mixed results that depended on the time of year and area of the bay. In Bayou Texar (Pensacola Bay system), an area noted for fish kills related to nutrient overenrichment and hypoxia, nitrate was reported to control primary productivity to a greater degree than phosphorus (Moshiri et al., 1987). Studies in Weeks Bay, off the lower Mobile Bay system (Dauphin Island Sea Laboratory, unpublished data), indicated that no single nutrient form was operable in controlling estuarine primary productivity in isolation of other physical, chemical, and biological variables and without regard to the role of other nutrients and limiting factors. Overall, nutrient limitation in Perdido Bay was seasonal, with phosphorus limitation during cold months and nitrogen and/or nitrogen + phosphorus limitation during warm months.

Despite the evidence concerning phosphorus limitation in Gulf estuaries, Howarth et al. (2000) stated that nitrogen usually controls eutrophication of coastal systems, adding that phosphorus can be important in certain situations and can become the limiting nutrient if nitrogen is managed. This conclusion differs from other views that phosphorus is the primary cause of coastal eutrophication (Tyrrell, 1999). The current emphasis on nitrogen as the primary limiting nutrient to coastal systems (Howarth et al., 2000) was based in part on mesocosm experiments in Narragansett Bay (Oviatt et al., 1995) that showed that only nitrogen additions led to phytoplankton increases of abundance and increased primary production. In addition, long-term "natural" experiments showed the importance of nitrogen as a limiting factor. The long-term response of a Swedish estuary (Himmerfjarden) to changes in nutrient loading from the sewage treatment plant (Elmgren and Larsson, 1997) indicated that, over a 17-year period, phytoplankton abundance and water clarity were related to total nitrogen concentrations. Another case of long-term changes of Laholm Bay in Sweden (Rosenberg et al., 1990) indicated nitrogen control.

The ratios of inorganic nitrogen:phosphorus reflect the relative importance of these nutrients (Graneli et al., 1990; Oviat et al., 1995) which, according to Howarth (1988) and Howarth et al. (2000), substantiates the conclusion that nitrogen limitation is the rule in most estuarine and coastal systems and that "nitrogen availability is the primary regulator of eutrophication in most coastal systems." However, Flemer et al. (1997) found that the situation regarding nutrient limitation may be more complex than some of the models offered by the above-mentioned studies. Seasonal changes of phosphorus and nitrogen limitation exist in the Perdido Bay system and such changes may be complicated by the interaction of these nutrients. This indicates that both nutrients should be considered important if the impacts of anthropogenous nutrification are to be controlled in this estuary with the timing of nutrient interactions playing a crucial role in the response of the phytoplankton. There should be an emphasis on timing of nutrient loading both in terms of seasonal changes and interannual trends. The complexity of the response of estuaries and coastal systems to nutrient loading underscores the necessity to remain circumspect with regard to models that emphasize the importance of nitrogen to the exclusion of other forms of nutrients.

IV. SEDIMENT QUALITY

A. Introduction

Sediments represent an important component in any evaluation of eutrophication in shallow coastal systems. Sediment chemistry is determined in part by water column production in the form of phytoplankton. Organic deposition, decomposition, nutrient regeneration, and nutrient transfer into and out of the sediments all play a role in the food web dynamics of a given estuarine system. The sediment/water interface is an important area of mineralization through microbial activity. The microbes are stimulated into production by various forms of energy subsidies such as river flooding, wind- and tide-induced currents, and other disturbances that include bioturbation (Federle et al., 1983a, b; 1986; Livingston, 1984a). The deposition of allochthonous (terrestrial detritus) and autochthonous (phytoplankton) organic matter into the sediments provides a rich matrix for microbial activity (Federle et al., 1983a). The bacteria, fungi, and microalgae transform dissolved nutrients and organic matter into high-quality (i.e., P- and N-enriched) food for the estuarine detrital food webs (Federle et al., 1983b). The microbiota represent the link between the organic carbon that accumulates in the sediments and the meiofauna and infaunal macroinvertebrates. Protozoans, oligochaete and polychaete worms, nematodes, small crustaceans, and bivalve and gastropod mollusks depend on microorganisms for food. In turn, the infaunal macroinvertebrates that feed on the microbiota and the enriched organic carbon produced by the microbes represent a major part of the biomass of the largely benthic food webs in shallow coastal areas (Livingston et al., 1997). The infauna, dominated by various forms of worms, crustaceans, and mollusks, are thus an important link between the microbial associations in the sediments and the upper components of the food webs that form the basis of the sports and commercial fisheries of these systems (Livingston, 1984a). In this way, sediments are an extremely important part of the overall high natural productivity of shallow coastal systems of the northeastern Gulf.

Water depth is an important habitat factor that affects sediment quality. The depth of an estuary has important effects on vertical exchanges of various factors such as nutrients and dissolved oxygen. Turbulent action and water column stability are strongly affected by depth characteristics of a given coastal system. Wind action in the shallow Apalachicola Bay system is associated with periodic peaks of phytoplankton production as inorganic nutrients, regenerated in the sediments, are mixed through turbulence into the euphotic zone (Livingston et al., 1974; Iverson et al., 1997). Perdido Bay is shallow with an average depth of 2.2 m. Except for the dredged GIWW and Perdido Pass in the lower bay, there is no well-defined channel in Perdido Bay. The U.S. Army Corps of Engineers maintains Perdido Pass at about 4 m as part of the GIWW. There is a dredged extension of this channel just north of Robinson Island (USACOE, 1976). However, there is considerable disparity in the distribution of depth throughout the estuary. There is a deep hole at the lower end of the Perdido River as it enters the bay (station P18). There is a shelf around the periphery of the bay that usually does not exceed 1 m in depth and can be up to 400 m in width. The upper bay (above the Highway 98 Bridge) is relatively shallow. The deepest parts of the estuary are located in the lower bay off Ross Point (Figure 2.2a).

Niedoroda (1992), in a comparison of depths in Perdido Bay between 1935 and 1991, found several changes over the 56-year interval. Deposition increased southward from about 0.5 to 1 m in central portions of the lower bay. No significant changes were evident in shallow (shelf) portions of the lower bay. Deposition was most evident in central parts of the lower bay with up to 1 m increases in the deepest areas. According to Niedoroda (1992), between the 1937 and 1991 surveys, depth changes of up to 1 m were common indicating that sediment provided by the drainages into upper Perdido Bay accumulated in the lower bay. A sediment reentrainment study was carried out in the upper and lower parts of Perdido Bay (Niedoroda, 1992) to characterize conditions necessary to resuspend deposited sediments due to wind events. North winds dominate during winter. During warm periods, southerly winds are prevalent, with more frequent occurrences of winds from the southwest. The axis of minimum strong wind activity is well aligned with the northeast/southwest axis of the bay. Because of the relationship of the physiography of the relatively narrow bay relative to wind effects, it was found that fine sand (5 phi) was very rarely resuspended and even very fine sand (8 phi) was reentrained only about 1% of the time. Higher rates of reentrainment occurred in areas in peripheral (shallow) parts of the bay (<2 m depth). Overall, the geographic position and the physiography of the bay, along with the general distribution of depth, added to the relative stability of the water column in Perdido Bay, with attending effects on the distribution of sediments in the estuary as a product of depositional processes, largely in the lower bay.

B. Physical Characteristics and Liquid Mud

The distribution of sediment quality of Perdido Bay is shown in Figure 4.8. Sandy conditions predominate in the river and creek approaches and in western sections of the upper bay. The shallow (<1 m) periphery of the bay (not shown) is also sandy. The upper bay has relatively low silt concentrations, although western areas appear to have some silt deposition. There are increased concentrations of silts and clays down the bay with peaks of fine elements occurring around station P40 off Inerarity Point. These areas coincide with the deepest parts of the bay. High silt-clay fractions are also evident in Wolf Bay. Deposition of high concentrations of organic matter tend to follow levels of phytoplankton production in the upper bay (see below) and silt-clay distributions in deeper parts of the bay.

Diving surveys indicate that sandy shelf areas grade down a slope to the deeper parts of the bay which are characterized by liquid mud or nephlos (i.e., suspended particles of a flocculent nature at varying depths above the more solid sediment/water interface). The floc appears to be composed of black particles about 2 to 3 mm in diameter. These particles are neutrally buoyant and the least disturbance sets them in motion. In upper Perdido Bay, in about 1.5 m of water, the bottom is composed of a brown flocculent material. At station P25 in the seagrass beds, the area is sandy with interspersed flocculent materials. At station P26, the substrate is sandy and is subject to wind effects. In deeper parts of the lower bay (stations P31, P37, and P40), the sediments are very soft and dark with heavy infiltration of liquid mud. At the surface, there are rust-colored particles that tend to settle into depressions giving the interface between the liquid mud and the water column a patchy appearance. Depths of the liquid mud in the lower bay vary but are often greater than 1 m. Liquid mud characterizes most of Wolf Bay. Sediments in lower Perdido Bay are sandy.

As a basis for comparison, a diving survey was also carried out at the head of Choctawhatchee Bay. The delta area of the eastern part of the bay sediments is composed of fine-grained brown sand. At stations farther down the bay, the bottom sediments are highly uniform with a fine flocculent layer above a sandy substrate. Liquid mud, gray-black in color, predominates in the deeper areas of the bay. In general appearance and consistency, the nephlos in Choctawhatchee Bay is almost identical to that in Perdido Bay. In both systems, a persistent salt wedge is usually associated with the presence of the liquid mud. No currents are evident in the liquid mud layer. Thus, in salinity-stratified areas in both bays, there is extensive development of a layer of fine particles that remain suspended above the sediments as long as water column turbulence remained low. The halocline

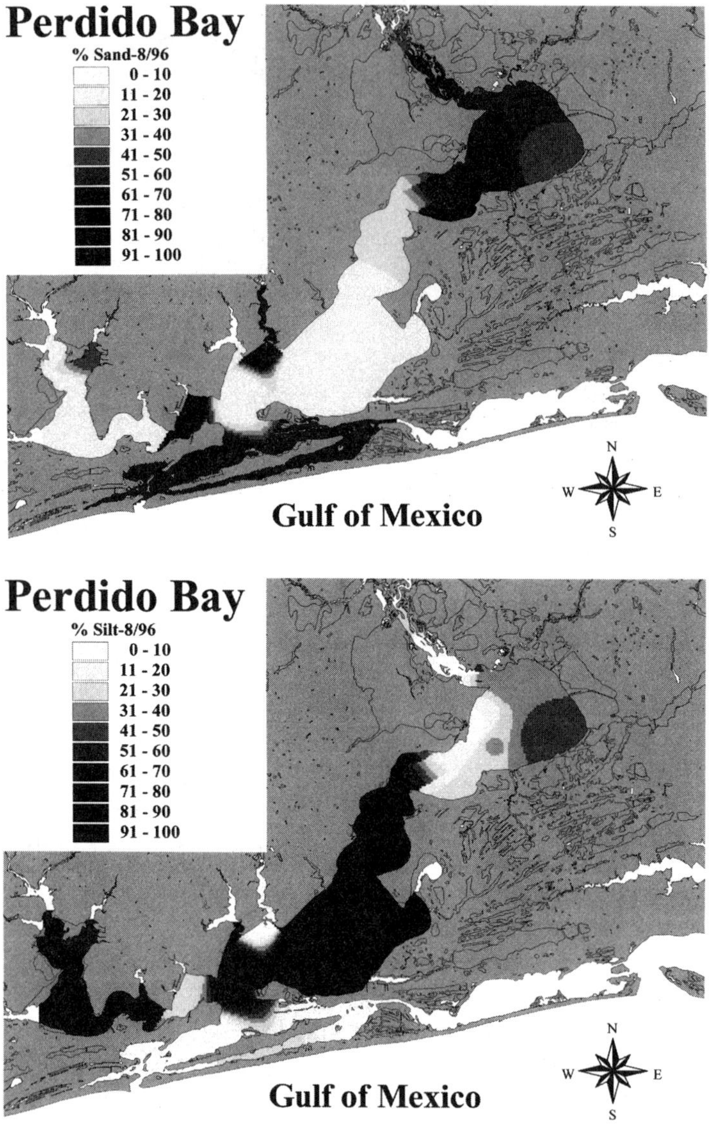

Figure 4.8 Sediment characteristics in Perdido Bay in August 1996. % sand (upper); % silt (lower).

in both Choctawhatchee Bay and Perdido Bay is relatively stable in areas deeper than 2 m. In such areas, the liquid mud is not affected by vertical turbulence even during storms with wind velocities exceeding 25 kt. Displacement of the fine-grained liquid mud interfaces in both estuaries occurs only during periods of high river flow at which time the halocline is removed and vertical mixing is established with the subsequent breaking up of the nephlos interface.

C. Sediment Nutrients

Sediment nutrients were analyzed from samples taken in Perdido Bay in 1989–1991, and these data were compared to samples taken in Choctawhatchee Bay during 1987 (Livingston, 1992). The inorganic nitrogen component in both systems is dominated by ammonia nitrogen. In the upper

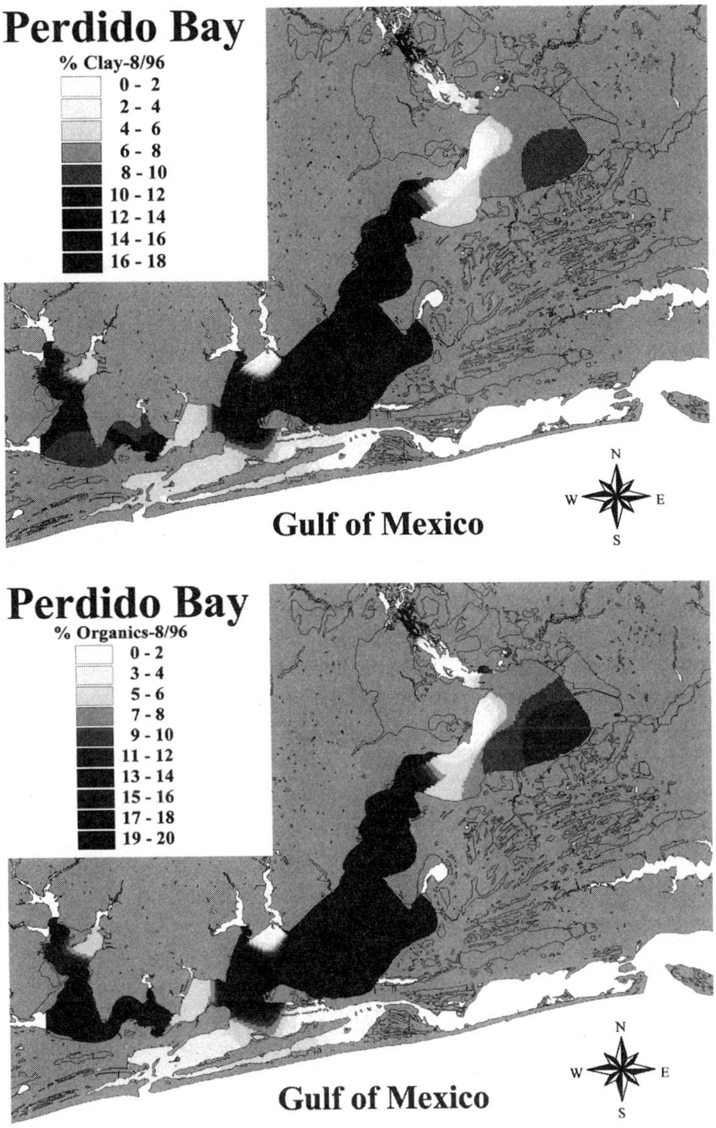

Figure 4.8 (continued) Sediment characteristics in Perdido Bay in August 1996. % clay (upper); % organics (lower).

and middle portions of Perdido Bay, nitrogen levels are comparable to those in upper Choctawhatchee Bay. However, in lower Perdido Bay, the sediment ammonia levels are somewhat higher than those in upper and middle Choctawhatchee Bay locations. Sediment orthophosphate levels in lower Elevenmile Creek (stations P21, P22) and in nearby bay areas (stations 23, 24) tend to be the highest in Perdido Bay. These levels are also higher than those in the Choctawhatchee reference sediments. The data indicate that runoff from Elevenmile Creek increased phosphorus concentrations in sediments of the immediate bay receiving area. The distribution of sediment carbon concentrations in upper Perdido Bay is similar to that encountered in upper Choctawhatchee Bay. However, stations in middle (P31) and lower (P37, P40) Perdido Bay are characterized by higher concentrations of carbon than the other portions of the system. This carbon is in the particulate organic form.

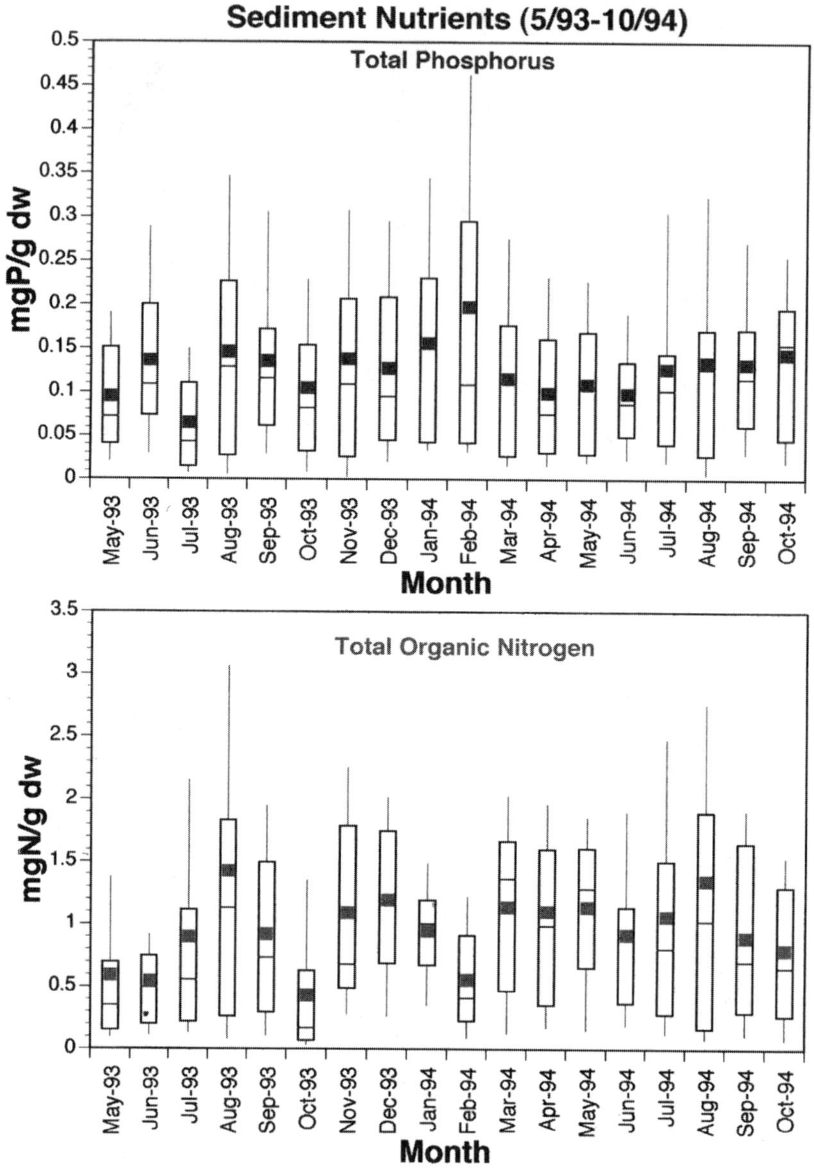

Figure 4.9 Sediment nutrients in Perdido Bay shown as means with the 10th, 25th, 75th, and 90th percentiles of all stations (Figure 2.2a) monthly from May 1993 through October 1994.

Detailed analyses of sediment nutrients were carried out monthly from May 1993 through October 1994. The results (Figure 4.9) indicate that there was phosphorus deposition in the sediments during winter high river flows. However, there were relatively rapid reductions of sediment phosphorus during succeeding months indicating that the sediments may act as a temporary reservoir for phosphorus. The nitrogen data indicated increased TON during summer months (August); again, there was no indication that the sediments acted as a long-term reservoir for high concentrations of nitrogen. The highest sediment phosphorus concentrations in Perdido Bay were found at stations P26, P29, P31, P33, and P37 and in parts of Wolfe Bay (Figure 4.10). These areas coincided generally with depositional areas in deeper sections of the bay. Sediments in the upper and lower parts of Perdido Bay had relatively lower phosphorus concentrations. Sediment nitrogen

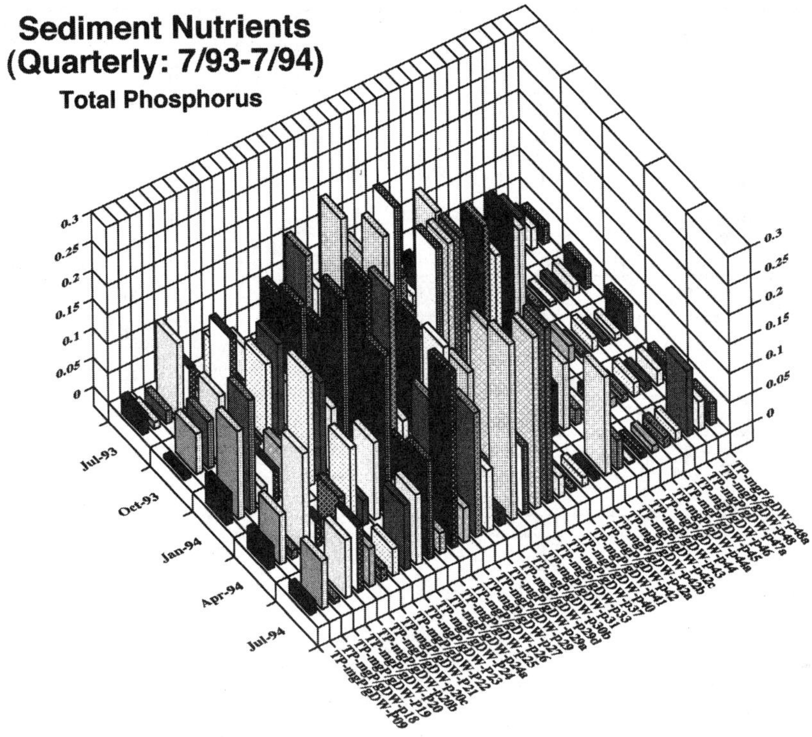

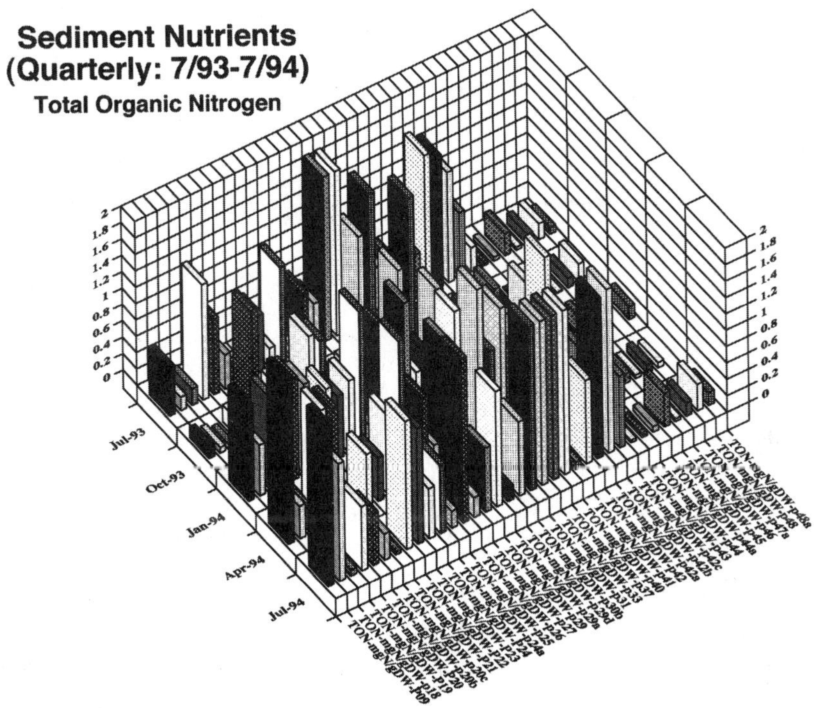

Figure 4.10 Sediment total phosphorus (upper) and sediment total organic nitrogen (lower). Data are given by station as quarterly averages from July 1993 through July 1994.

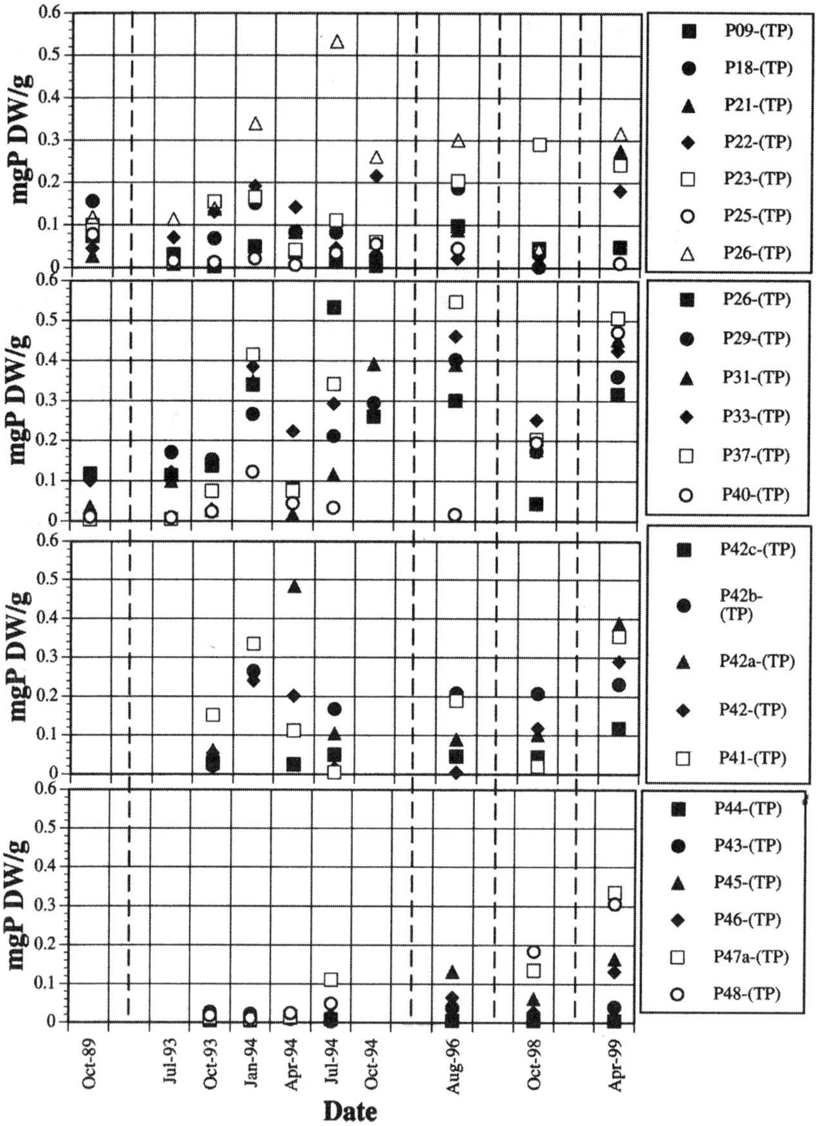

Figure 4.11 Sediment total phosphorus in Perdido Bay from October 1989 through April 1999. Dashed lines represent periods between samplings.

(Figure 4.10) did not follow this pattern. There were relatively even distributions of sediment nitrogen in most parts of the bay with the lowest such concentrations occurring in the lower bay.

An analysis of the long-term changes of sediment nutrients (Figure 4.11) indicate that sediment phosphorus was relatively low at most stations during 1989 and 1993. However, there was an increase in sediment phosphorus in 1994 at stations P26, P29, P31, P33, P37, and in Wolf Bay. Stations in the Perdido River and the western upper bay (grassbeds) did not show these increases with time. These trends continued in 1996. However, after the reductions of P-loading by the mill during late 1997–1998, there were noticeable declines in sediment phosphorus at all stations except P23 (mouth of Elevenmile Creek). By April 1999, after a resumption of mill phosphorus loading, there was again another increase in sediment phosphorus at various stations including some in the lower bay. The pattern of sediment phosphorus concentrations in upper Perdido Bay thus tended to follow that of the mill phosphorus loading over the 11-year period of analysis. This retention of phosphorus in sediments could be related to the patterns of nutrient limitation found in Perdido Bay, a situation that also could be affected by the degree of eutrophication. Other studies (Howarth et al., 1995) have indicated that there is a decrease in the ability of the sediments to adsorb and store phosphorus as the system becomes more eutrophic. This did not seem to be the case in Perdido Bay, however.

D. Stable Isotope Analyses of Carbon Cycling

Coffin and Cifuentes (1992; 1999) conducted a stable isotope analysis of the carbon cycling ($\Delta^{13}C$) and nitrogen cycling ($\Delta^{15}N_{NO_3}$) in Perdido Bay. Surface and bottom water samples were taken at stations throughout Perdido Bay for analysis of the stable carbon and stable nitrogen isotope relationships. Suspended sediments were collected by filtering water samples. During the 1991 study, the bay was highly stratified, with low river flows. Potential sources of organic matter and nutrients included terrestrial runoff (forests, agricultural lands), urban runoff, municipal and industrial wastes, and water from coastal regions through advection from the mouth of the estuary.

A comparison of the mean and standard deviation of $\Delta^{15}N_{NO_3}$ in Elevenmile Creek, the Perdido River, and the upper and lower estuary showed similar values in all regions except Elevenmile Creek (Coffin and Cifuentes, 1999). Here, the isotopic values were substantially lighter. In spite of high nitrogen loading in Elevenmile Creek, the isotopic data indicate that the mill was not the major source of nitrogen to the Perdido estuary. The stable isotope signatures from Elevenmile Creek were not observed in Perdido Bay, suggesting that the creek did not contribute significant amounts of organic matter to the bay. This was consistent with estimates that organic matter from the creek was a relatively small percentage of the total input to the bay (Hydroqual, 1992). Dilution effects from the Perdido River and the lack of any differential degradation rates between mill-derived particulates and river-derived particulates support the view that it is highly unlikely that carbon and nitrogen in the sediments of Perdido Bay are derived directly from the Pensacola mill. In addition, inorganic nitrogen concentrations in the bay are likely to originate from the coastal region. Water volume from Perdido Pass into the bay is substantially greater than that from the Perdido River and Elevenmile Creek. Therefore, there is evidence that the coastal region is a significant source of nitrogen to the bay, a conclusion that is consistent with results from the nutrient limitation experiments carried out in Perdido Bay (Flemer et al., 1997).

A large number of the samples taken from Perdido Bay were ^{15}N-depleted and ^{13}C-depleted compared with values from other riverine and coastal areas (Coffin and Cifuentes, 1999). There appeared to be another source of organic matter to the bay in addition to above-named sources. Having characterized the obvious sources of allochthonous (e.g., Perdido River, pulp mill, sewage treatment plant) suspended particulate matter (SPM), it appeared that one source of SPM with unusually light isotopic ratios had to be autochthonous. One likely source of nitrogen contained in SPM was likely to be inorganic nitrogen that was assimilated into the SPM by microorganisms. There was evidence that primary production was linked to an important fraction of the ^{13}C-depleted SPM, and that phytoplankton activity was responsible for such production. Organic production was high during spring months based on phytoplankton assimilation rates of ^{13}C-depleted dissolved inorganic carbon (DIC). A seasonal analysis indicted that there were seasonal trends of organic carbon generation with a rapid transition from primary to heterotrophic activity from April to May. Although simple analyses of $\Delta^{13}C_{DOC}$ and $\Delta^{13}C_{POC}$ indicated that a primary source of organic carbon was terrestrial, phytoplankton appeared to be an important source of the organic carbon pool in Perdido Bay.

The paradigm for stable carbon isotope ratios of organic matter in river-dominated estuaries is a transition from ^{13}C-depleted river input to ^{13}C-enriched material. The shift in $\Delta^{13}C$ is postulated to involve dilution of the terrestrial organic matter and increased primary production from the coastal area. Measurements of the delta^{13}C in SPM, DOC, and bacteria through the Perdido estuary (Coffin and Cifuentes, 1999) indicates that terrestrial input is a major (allochthonous) carbon source. However, a more thorough analysis of the stable carbon and nitrogen isotope ratios in particulate and dissolved matter combined with analysis of the physical transport system indicates a different identification of the primary carbon source (i.e., phytoplankton). A similar study (Chanton and Lewis, 1999) in the Apalachicola Bay system found that primary production and heterotrophic activity were balanced. A comparison of estimates of allochthonous carbon transport and production in Perdido Bay with autochthonous production (Macauley et al., 1995) is an additional indication that phytoplankton production in Perdido Bay is an important source of organic carbon to the carbon pool.

The quality of organic matter that accumulates in sediments and is available to bottom-dwelling organisms has been shown to vary along the salinity gradient of a given estuary. High C:N terrestrial detritus dominates in freshwater parts of the estuary whereas phytoplankton detritus dominates at the marine end (Hopkinson et al., 1998; 1999). Canuel and Zimmerman (1999) found that suspended and sedimentary organic matter in southern Chesapeake Bay was derived from autochthonous sources that included fresh and detrital phytoplankton, zooplankton, and bacteria. The dominant factor that determined temporal variation of the sedimentary pool was phytoplankton productivity. Enrichments of POC, chlorophyll *a*, total fatty acids, total sterols, and various biomarkers specific to phytoplankton were found at the surface and the bottom of the study area. Thus, particulate organic matter produced in a given estuary is the result of various processes that impinge on sediment quality and food supplies for benthic organisms. The studies in Perdido Bay, when viewed within the context of recent analyses of the importance of phytoplankton in the determination of organic matter and the sedimentary environment, indicate that there is a direct connection between phytoplankton in the water column and benthic processes that are important to the overall food web of the bay.

E. Toxic Agents

Sediments represent the focal point of the concentration of various forms of long-lasting toxic substances that accumulate in coastal systems. A series of analyses was carried out to determine the distribution of anthropogenous toxic substances in the sediments of Perdido Bay. Livingston (1992) and Isphording and Livingston (1989) found generally low sediment metal concentrations in Perdido Bay. Shropp et al. (1989) carried out sediment analyses for 83 priority organic pollutants. The only quantified contaminants were PAHs at stations in some of the tributaries and in drainage ditches around an oil products facility. The PAHs represent a family of planar organic compounds composed of benzene units. A preliminary survey for PAHs was carried out in October 1996 (Livingston, unpublished data). Results indicate that the highest PAH concentrations in Perdido Bay occurred near a marina in lower Perdido Bay. Brim (1993) did a study of the toxic agents in Perdido Bay sediments. He found that Perdido Bay was generally free of toxic compounds (pesticides, metals, PAHs, dioxin). However, field toxicity tests with bay water and sediments indicated that upper and lower bay sediments reduced feeding activity in some estuarine organisms. This was taken as a measure of reduced habitat quality. Pore water from these sediments also caused considerable toxicity in test organisms even though no high concentrations of any toxic agents were noted. This toxicity decreased considerably overnight. Since no contaminants were present in the sediments or the pore water, the cause of the toxicity was not determined. The lower bay appeared to be more adversely affected than the upper bay.

A screening survey was carried out concerning the concentrations of organochlorine compounds associated with mill effluents in sediments and animals taken from Elevenmile Creek and Perdido Bay (Livingston, 1992). The EOCl (extractable, organically bound chlorine or halogen) represents the highly lipophilic, potentially bioaccumulable compounds that are characterized by having more than a 1000-fold greater concentration in 1-octanol than in water at equilibrium. Low concentrations of EOCl (1.1 $\mu g\ g^{-1}$) were noted in sediments in the Perdido River–estuary and at the mouth of Elevenmile Creek (1.2 $\mu g\ g^{-1}$). The simplest explanation was that Elevenmile Creek (and the Pensacola mill) constituted the source of the EOCl found in the sediments in the upper portions of the bay. However, no EOCl concentrations were detected in the sediments taken in Elevenmile Creek or other upper bay stations. Low levels of EOCl were noted in the sediments of the mid (0.74 $\mu g\ g^{-1}$) and lower (3.0, 1.8 $\mu g\ g^{-1}$) bay. It is possible that EOCl compounds, attached to small particles, settled out in this part of the bay. In this case, the mill might constitute a source for such concentrations.

Bioavailability of toxic agents represents an important qualifying factor when assessing the ecological significance of toxic agents in sediments. Analyses for EOCl in invertebrate and fish tissues taken from Perdido Bay indicated no evidence of biological concentration or food web magnification of EOCl in the bay (Livingston, 1992). The concentrations of EOCl in fishes and invertebrates taken from Perdido Bay were similar to those in animals taken from Apalachicola Bay. When compared with concentrations of EOCl near pulp mills in other areas, the concentrations in sediments and animals of Perdido Bay and Apalachicola Bay were relatively low (Livingston, 1992).

Overall, various studies have indicated that sediments in Perdido Bay are generally free of high concentrations of toxic agents such as heavy metals, PAHs, pesticides, and chlorinated compounds (including dioxin). Alexander et al. (1999), using radiochemical chronologies of sediment contaminants in the Savannah River Estuary, found that toxicants such as organochlorine pesticides, metals, PCBs, and PAHs tended to peak during the 1960s and 1970s. This condition has been found in other coastal areas as well, a situation the authors attributed to increasingly effective control of toxic agents. With the possible exception of PAHs in urban areas, mercury distributions in various aquatic systems, and certain forms of pesticides (i.e., herbicides, fungicides), this may be true. However, Alexander et al. (1999) also found that nutrients have been increasing in sediments in various coastal areas since 1965. This finding coincides with increased incidence of harmful algal blooms (HABs) in coastal systems around the world.

V. WATER QUALITY: SPATIAL/TEMPORAL TRENDS

A. Introduction

There is considerable scientific information concerning the relationship of water quality to eutrophication processes. Estuarine and coastal environments are highly variable because of the continuous interaction of freshwater runoff from land and climatological conditions. Shallow inshore marine habitats from Apalachee Bay to Perdido Bay are physically stressed systems due to natural water quality cycles (Livingston, 1990). Water temperature is an important limiting factor during both summer and winter periods when wind mixing of the water column leads to rapid air–water equilibration (Livingston, 1984a). In north Florida, seasonal variation is considerable, with temperatures ranging from 4 to 33°C over any given 12-month period. Water temperatures can go as low as 2 to 3°C during particularly cold winters. There is some evidence that peak summer temperatures (32 to 33°C) may also stress inshore organisms in north Florida estuaries (Livingston, unpublished data). Salinity can also be a significant limiting feature with extreme spatio-temporal variation of this factor in alluvial river-dominated systems (Livingston, 1991c). In addition to absolute salinity values, the rate of change of both salinity and temperature is often an important determinant of the distribution of organisms from both direct and indirect (i.e., predator–prey interactions) causes (Livingston et al., 2000). However, in many inshore areas, estuarine organisms are euryhaline and eurythermal, and can thus withstand rapid changes of salinity and temperature over varying time periods. Thus, although salinity and temperature can be limiting to stenohaline and stenothermal marine organisms as predators and potential competitors for food and habitat, these factors may not be directly limiting for many forms of estuarine organisms.

Dissolved oxygen (DO) has been considered an important limiting factor in inshore marine systems (Santos and Bloom, 1980; Santos and Simon, 1980a, b). Cultural eutrophication (nutrient excess leading to overproduction of microalgae and associated trophic imbalances) is common in estuaries near human population centers (Livingston, 1987a; 1997a). This condition is characterized by exaggerated fluctuations of dissolved oxygen from supersaturation during the day to hypoxia and anoxia at night (Livingston, 1997a). The eutrophication process involves complex trophic interactions that often are system-specific (Livingston, 1997a). Hypereutrophication can lead to periodic hypoxia and anoxia (Livingston, 1997a). According to Rabalais et al. (1991), there is a "dead zone" of over 9500 km^2 in the Gulf of Mexico due to hypoxia. However, DO levels in coastal areas are often subject to considerable natural variability, largely because of fluctuations of temperature, salinity, basin stratigraphy, productivity, and associated biological conditions. Thus, periodic hypoxia in Gulf estuaries often reflects natural conditions (Turner et al., 1987; Seliger and Boggs, 1988). Episodes of naturally low DO ("jubilees") occur in Gulf coastal areas (Loesch, 1960; May, 1973).

Considerable effort has been made to establish appropriate DO criteria or standards for the protection of natural systems (Doudoroff and Shumway, 1967). However, it is possible that there is no single appropriate water quality standard in estuaries along the northeastern Gulf. The development of water quality factors such as DO should be carried out within established limits of the natural cycling of water quality factors such as temperature and salinity. High nutrient/organic-C loading and low water exchange rates (stagnation), combined with anthropogenous salinity stratification and natural cycles of seasonal temperature changes, can produce extreme changes in available DO. Restricted water circulation, altered phytoplankton production, and trophic anomalies can influence the frequency and severity of such episodes (Seliger and Boggs, 1988; Breitburg, 1990; 1992; Livingston, 1997a). Lenihan and Peterson (1998) showed that salinity stratification, depth considerations, and periodic hypoxia/anoxia combined to affect oyster mortality and associated fishes and invertebrates. Interactive factors were operational in the specific patterns of the response to different forms of disturbance. Based on the complexity of the interactive controlling factors, the authors indicated the need for more integrative approaches to ecosystem management.

Salinity stratification and the restriction of vertical mixing in estuaries have been associated with hypoxic conditions (Schroeder, 1978; Schroeder and Wiseman, 1986; Livingston, 1997a). River flow is a dominant controlling mechanism of salinity stratification in Perdido Bay (Niedoroda, 1992), and wind stress has relatively little effect on such stratification due to the geographic placement and physiography of the system. Smith et al. (1992) found that deep-water hypoxic conditions in Chesapeake Bay resulted from the cumulative effects of biochemical mechanisms that were modified by physical stratification. Point and non-point nutrient inputs are known to support phytoplankton biomass in excess of the oxygen assimilation capacity of various coastal systems. However, the exact relationship of nutrient loading to hypoxic/anoxic conditions is often not well understood due to factors such as biochemical feedback loops and interannual changes of climatic factors (i.e., drought–flood cycles) and productivity (Livingston, 1993a).

Various field studies have been carried out to evaluate the impact of hypoxia and anoxia on estuarine populations and communities. Pearson (1980) and Pearson and Rosenberg (1978) published detailed studies on the effects of organic enrichment and low DO on marine systems. Van Es et al. (1980) demonstrated a direct response of meio- and microfauna in tidal flats along distinct gradients of organic enrichment and oxygen saturation. However, the authors could not isolate the impact of gradients of DO from the effects of other natural forcing factors such as salinity (absolute value and fluctuations), sediment composition, and relative emersion time. The interaction and covariance of important ecological features of an estuary often confound direct correlations between the oxygen regime and the biological organization of the system. Santos and Bloom (1980) and Santos and Simon (1980a, b) found that annual defaunation in Hillsborough Bay (Tampa Bay system) was due to hypoxia. A stochastic recolonization response of the soft-bottom macroinvertebrate community was demonstrated following the recurrent defaunation event. Accounts of the association of extensive kills with low DO are not uncommon in the literature (Dauer et al., 1992). However, distinct causal relationships of the interacting physical, chemical, and biological factors that lead to such events have not been well documented, and little attention has been paid to the interaction of sediment quality and low DO as interacting controls of the biological response to changing habitat conditions. In addition, relatively little attention has been paid to the adaptive response of estuarine organisms to hypoxia, both natural and anthropogenous.

The search for water quality indicators of phytoplankton activity has been both extensive and inconclusive. Nutrients are necessary for useful productivity. Nutrient concentrations are often mediated by phytoplankton activity, especially in river-dominated estuaries. Mutual feedback mechanisms make it difficult to separate the causes of hypereutrophication from the effects when using nutrient concentrations as indicators. In terms of nutrient fluxes into the estuary, there is a transition from nutrient-controlled natural growth rates (i.e., maximum response mode; Caperon et al., 1971) to nutrient-induced increases in phytoplankton biomass production that result in water quality problems (i. e., noxious algal blooms, hypoxia, anoxia; Malone et al., 1996). Justic (1987), comparing nutrient data for the Mississippi River plume and northern Adriatic Sea, suggested both DIN and DIP concentrations of $<0.1\ \mu M$ as potentially limiting for phytoplankton growth rates. They suggested that DIN:DIP ratios <10 indicate potential N limitation and that ratios >30 indicate potential P limitation, a ratio somewhat on the high side for P limitation (Boynton et al., 1982). The DIP minima may apply to the Perdido Bay situation (Flemer et al., 1997) as low DIP concentrations suggest incipient P-limitation. The DIN concentrations are usually greater than the threshold value of 1.0 μm throughout Perdido Bay, especially in the upper bay near riverine influence (Flemer et al., 1997). During the nutrient limitation experiments, the lowest DIN:DIP ratios (e.g., 15 to 30) occurred in mid and lower bay areas from July through November, but these values were still within a N and P level of sufficiency. Ratios decreased below the threshold value of 10 only in October and November in the mid and lower bay. From these observations, DIN limited growth rates on a relatively infrequent basis (Flemer et al., 1997).

Ambient nutrient concentrations in a system such as Perdido Bay depend on complex interactions with the phytoplankton community. The questions of nutrient co-limitation (Fuhs et al., 1972; Gerhart and Likens, 1975; Norin, 1977; Caraco et al., 1987; Suttle and Harrison, 1988; Paasche and Erga, 1988; Pennock and Sharp, 1994) and incipient co-limitation (Powers et al., 1972; Graneli, 1987; Kivi et al., 1993) are directly related to nutrient concentrations in the bay. Nutrient co-limitation of phytoplankton growth rates has been reported in various estuarine environments. Flemer et al. (1997) hypothesized that N+P co-limitation in Perdido Bay may at times involve escape of the phytoplankton assemblages from grazing pressure. However, since the nutrient limitation experiments in Perdido Bay were carried out during a period that was relatively free of bloom production, it is possible that such generalizations may not apply in case of excess nutrient loading and resulting hypereutrophication. The complications associated with time-dependent differences in nutrient limitation may thus complicate the use of nutrient concentrations as indicators of phytoplankton activity or bloom distributions in space and time. Based on the proven complexity of the interaction of phytoplankton assemblages and nutrient concentrations, the use of such concentrations as indicators of phytoplankton activity and bloom generation should be analyzed with a full complement of species-specific plankton data to avoid spurious correlations and false assumptions. Such assessments are not common in the scientific literature. Field descriptive nutrient concentrations as indicators of bloom generation in estuarine and coastal systems should thus be used with caution.

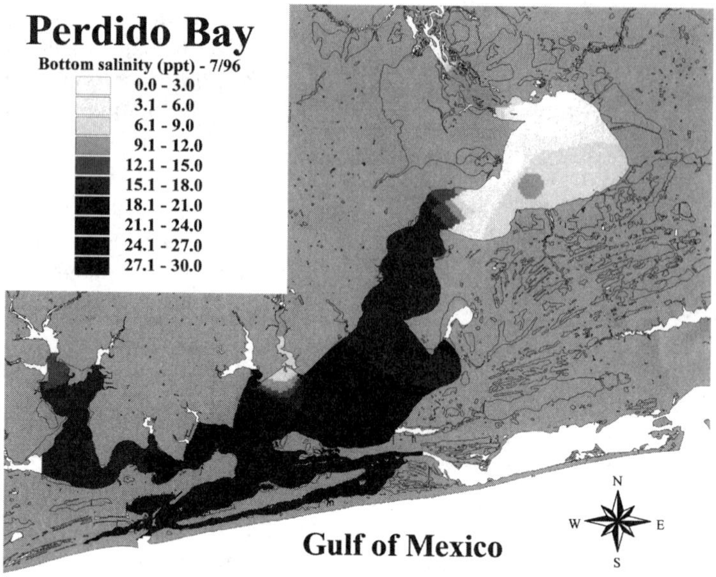

Figure 4.12 Spatial dimensions of bottom salinity (ppt) in Perdido Bay in July 1996.

B. Temperature and Salinity

There is relatively little published information concerning interannual variation of water quality factors over periods exceeding 10 or more years in coastal systems. Water temperature in Gulf coastal systems undergoes systematic, long-term (>10 years) variation (Livingston et al., 1997). Water temperature in Perdido Bay (Livingston, 1998) followed both seasonal and interannual cycles. Within-bay temperature variance tended to be highest during winter–spring months. Seasonal temperature ranges were highest during periods of high river flow. Winter low and summer high temperatures tended to be higher during (drought) periods of reduced river flow and lower turnover rates in the bay. The water temperatures of Perdido Bay were thus affected by the drought–flood cycles of Perdido River flow. This pattern of increased winter and summer temperature during droughts could be an important factor in phytoplankton production cycles.

The spatial dimensions of bottom salinity in Perdido Bay (Figure 4.12) illustrate the influence of the artificial Perdido Pass as an entry point for high-salinity Gulf water into the bay. Salinity was highest in the lower bay adjacent to the pass. Bottom salinity gradients into Wolf Bay and Perdido Bay followed the physiographic contours with the deeper parts of lower Perdido Bay having relatively high bottom salinity. The upper bay, highly influenced by the Perdido River, had low salinity relative to the rest of the system. The most stratified parts of the system (Figure 4.13) included deep areas at the mouth of the Perdido River in the upper bay and areas from Inerarity Point to the lower bay. The relatively shallow upper bay is the least stratified part of the system. There were distinct seasonal changes of surface salinity, with high salinity usually occurring during summer–fall months of low river flow. Low salinity was associated with winter–spring peak river flows. Interannual salinity trends followed the long-term drought sequence with the highest seasonally adjusted salinities occurring during periods of low river flow. The influence of hurricanes was evident with periodic low salinities that corresponded to storm frequency during the latter part of the study. Spatial variance of surface salinity was generally high during most periods, especially during droughts. Thus, interannual changes in the drought–flood sequence affected long-term salinity characteristics of Perdido Bay.

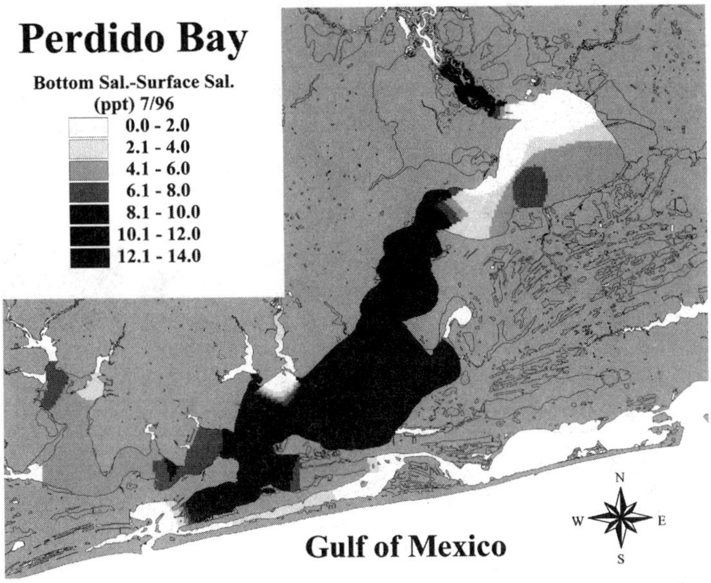

Figure 4.13 Spatial dimensions of salinity stratification in Perdido Bay in July 1996.

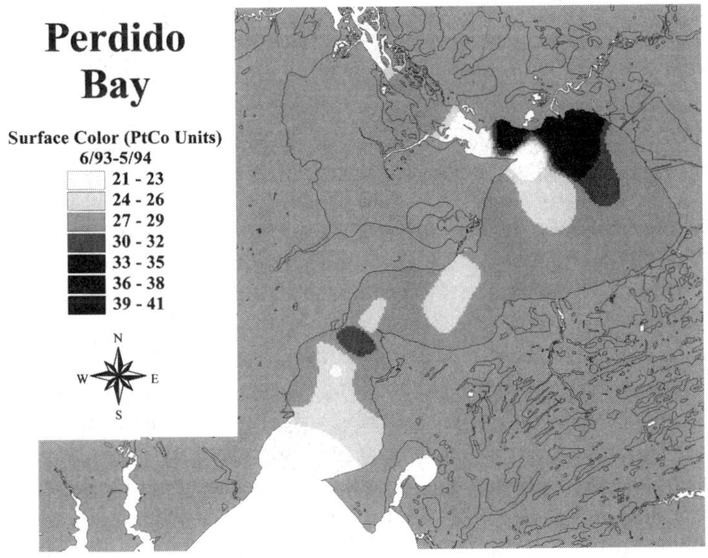

Figure 4.14 Spatial dimensions of surface color (Pt-Co units) in Perdido Bay averaged from monthly data taken from June 1993 through May 1994.

C. Color, Turbidity, and Secchi Depths

Spatial distribution of water color in Perdido Bay (Figure 4.14) reflected the combination of Elevenmile Creek discharge, wind effects, and eastward drift due to Perdido River flows. The upper bay tended to be more highly colored than the rest of the system. Discharge from Elevenmile Creek had some influence on water color in the northernmost parts of Perdido Bay. The Perdido River was a primary source of water color in the upper bay. The local influence of Elevenmile Creek on

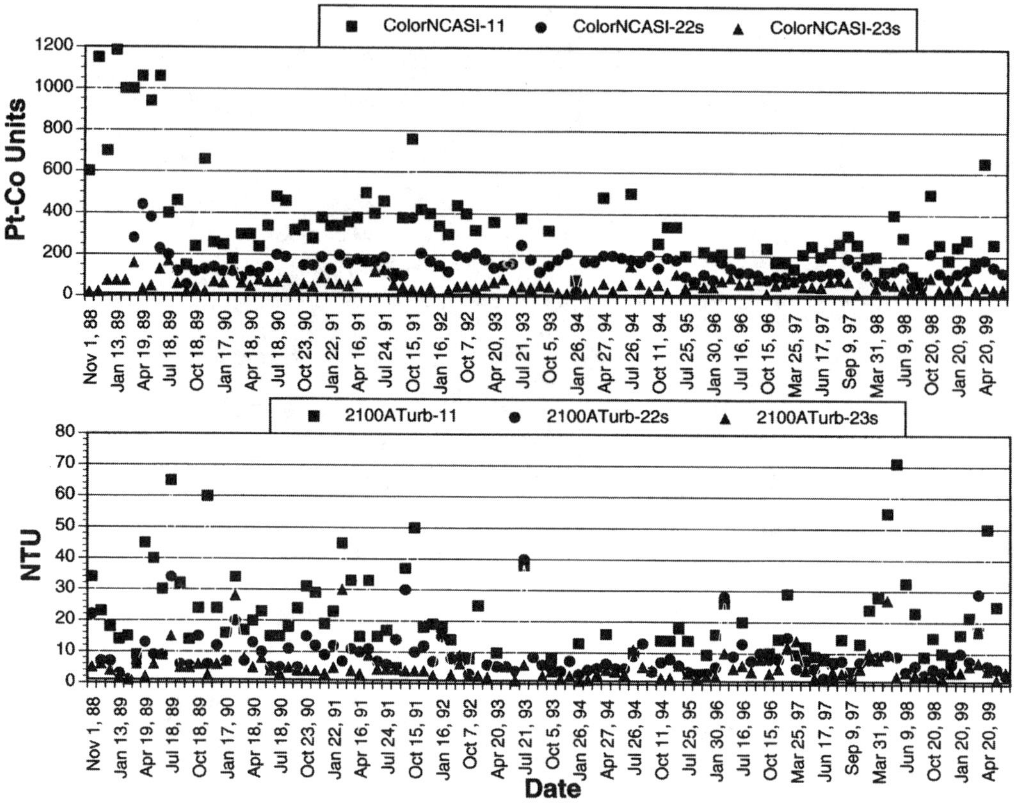

Figure 4.15 Surface water color (Pt-Co units) (upper) and turbidity (NTU) (lower) at stations 11, 22, and 23 monthly from October 1988 through September 1999.

water color was illustrated in a review of color concentrations at the mill outfall (station P11) relative to the lower reaches of Elevenmile Creek (Figure 4.15). However, the color reduction effort by the pulp mill during 1989 was reflected in reduced water color in receiving areas of Elevenmile Creek.

Turbidity was higher in Elevenmile Creek than in the bay (Figure 4.15). Turbidity tended to be lowest during the drought of 1993–1994. Light penetration (Secchi depths) increased with distance from the upper bay (Figure 4.16), and was generally lower in areas immediately affected by Elevenmile Creek. This indicated that Secchi depths, mediated by color and turbidity discharges from Elevenmile Creek, were influenced by mill discharges into upper Perdido Bay. Average Secchi depth differences were on the order of 0.3 to 0.4 m, which indicates relatively minor changes relative to natural background variability of this variable. Secchi depths usually peaked during the fall-early winter seasonal drought periods, and were greatest during interannual drought periods.

D. Light Distribution and Primary Productivity

Spectroradiometric data taken over the 1990–1991 period (Figure 4.17) indicated that seasonal changes in light penetration were most evident in Elevenmile Creek and in areas of upper Perdido Bay that directly receive freshwater flows (stations P18 and P23). The highest extinction coefficients (K_d [photosynthetically active radiation or PAR]) were found in Elevenmile Creek during winter–spring months. The seasonal pattern of extinction coefficients in the bay tended to be somewhat different than that in the creek. Seasonal changes of light extinction coefficients in upper Perdido

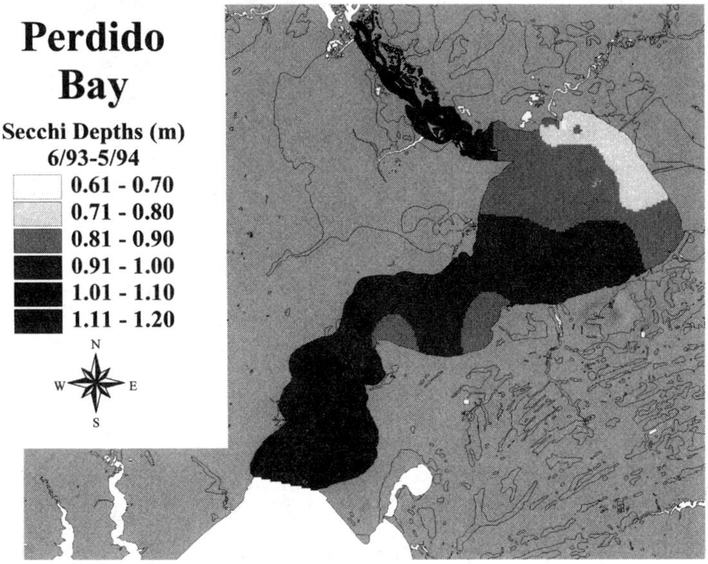

Figure 4.16 Spatial dimensions of Secchi depths (meters) in Perdido Bay averaged from monthly data taken from June 1993 through May 1994.

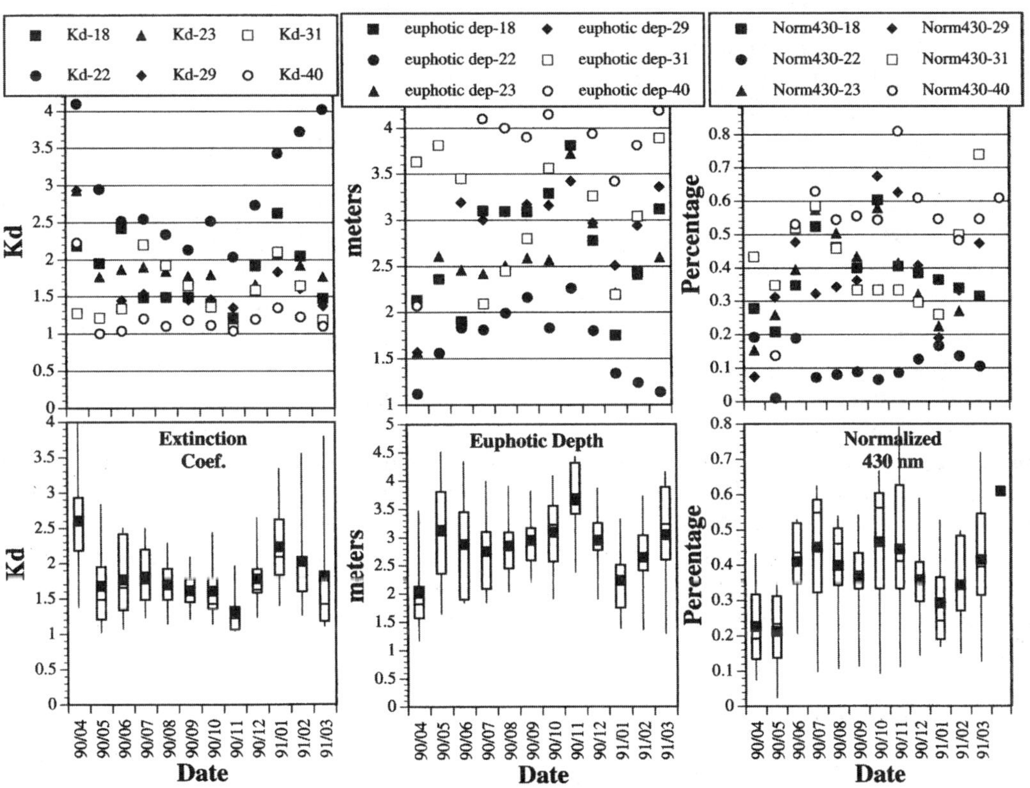

Figure 4.17 Extinction coefficients, euphotic depths (m), and normalized light (430 nm) at stations in Perdido Bay. The data are presented as baywide means (with the 10th, 25th, 75th and 90th percentiles) monthly from April 1990 through March 1991.

Bay followed the pattern at the mouth of the Perdido River rather than that at Elevenmile Creek. This trend was less apparent at stations more distant from the river. These data indicated that the light transmission effects of Elevenmile Creek discharges were restricted to immediate receiving areas in upper Perdido Bay, and that input from the Perdido River affected light transmission in upper and midle parts of the bay. These effects were inversely related to distance from the river mouth. There was minimal light transmission at 430 nm in Elevenmile Creek. However, this trend was not evident in the bay, again indicating that pulp mill influence on light transmission in Perdido Bay was most pronounced in Elevenmile Creek and in areas immediately adjacent to the creek mouth. Other parts of the bay were not significantly influenced by discharge from the creek. Lower Perdido Bay was clearly out of the direct influence of freshwater runoff from both the river and the creek during periods of low flow. During periods of high river flow, however, the K_d levels in the lower bay were comparable to those in the Perdido River, which indicated that, during such periods, light attenuation features in the lower bay were affected by seasonal changes of freshwater runoff.

Gross and net productivity levels (Figure 4.18) were highest in the upper bay during July at depths of 0.1 to 0.5 m. Respiration rates tended to have no consistent relationship to depth. As a result of the relatively high respiration rates, there was no net productivity in the upper bay during flooding in May and July 1991. Net productivity became positive in the upper bay during August and peaked in September 1991. As the bay cleared and temperatures remained high, net productivity became positive in the upper bay. Despite the relatively clear water in December 1991, net productivity was lower in the upper bay at this time. However, net productivity was positive at greater depths in December than in previous months, which reinforces the possibility that increased light transmission through the water had a positive influence on net production.

The data indicate that light transmission affected productivity in upper Perdido Bay. In the mid-bay areas, there was a different pattern with peak gross productivity during July. Gross productivity in this area of the bay was considerably higher than that in the upper bay, whereas the respiration rates were generally similar. This is further evidence of at least some light limitation of primary productivity in the upper bay. Net productivity was positive in mid-bay areas down to 1.5 m during every month of analysis. However, the highest levels of net productivity occurred from July to September 1991. Apparently, mid-bay areas had high enough nutrient concentrations to allow high productivity while there was enough light to sustain a relatively high rate of net productivity. Gross and net productivity in the lower bay was considerably lower than that in mid-bay areas, and such rates were similar to those in the upper bay. Significant levels of productivity occurred at depths of 4.5 m, which was a direct reflection of the lower light attenuation coefficients in the lower bay. There was relatively little response of gross productivity in the lower bay to the temporal changes of freshwater runoff. By December 1991, there was a marked decrease of overall productivity. Respiration was highest during August and September, with no consistent depth distribution. Respiration tended to be inversely related to the normalized (ratio of measured light to incident light) PAR extinction coefficients. Overall, increased light availability in the lower bay was accompanied by lower nutrient levels, which resulted in low levels of primary productivity.

The data indicate that productivity in Perdido Bay was dependent on seasonal changes in water temperature and spatial distributions of light and nutrients with peak net production occurring in mid-bay areas where light availability and nutrient concentrations were optimal for production down to depths of 1.0 to 1.5 m.

E. Dissolved Oxygen, Biochemical Oxygen Demand, and pH

The relationship of dissolved oxygen and phytoplankton production has been widely discussed in various studies (Livingston, 1997a; 1998; 1999). In the Perdido system during warm periods, there was a basic dichotomy between the DO regimes above and below the halocline (Livingston, 1998). Surface DO in the Perdido River–estuary was usually high compared with bottom DO.

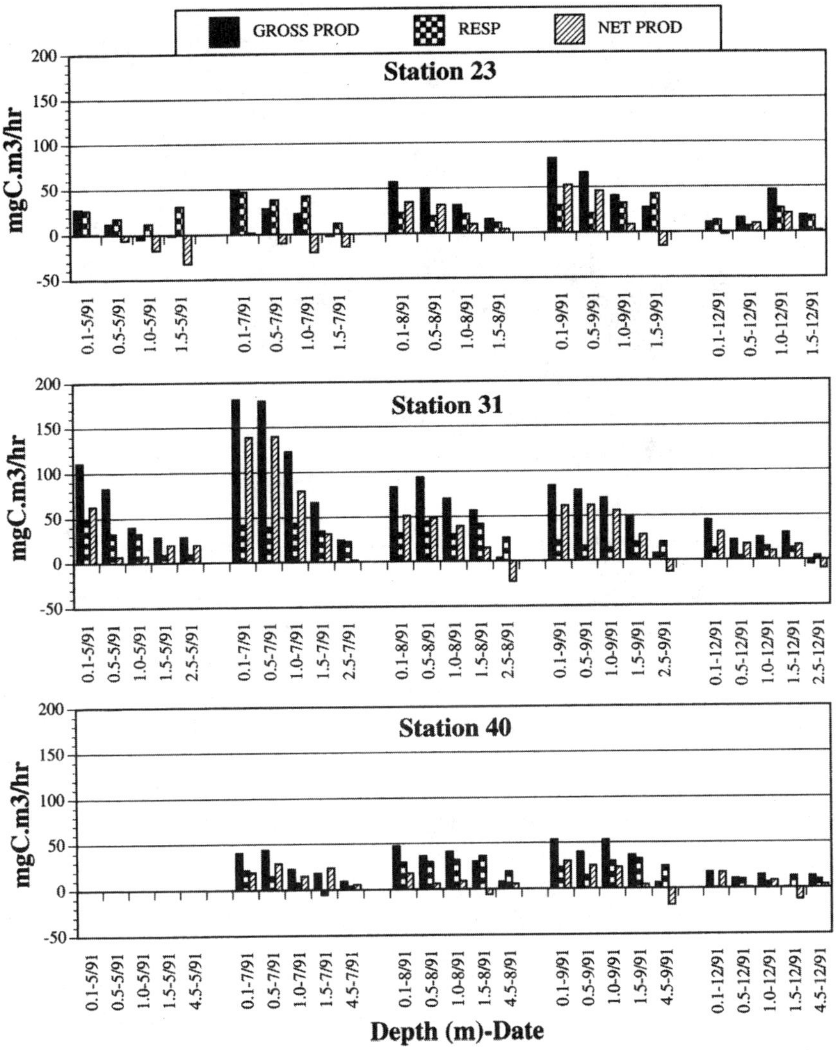

Figure 4.18 Gross and net productivity (mg C $m^{-3}h^{-1}$) and respiration (mg C $m^{-3}h^{-1}$) at depths of 0.1, 0.5. 1.0, and 1.5 m in Perdido Bay (station P23, P31, P40) during May, July, August, September, and December 1991.

Differences between surface and bottom DO were less pronounced in Elevenmile Creek because of the relatively lower levels of surface DO. In shallow estuarine areas off Elevenmile Creek, there was relatively little difference between the surface and bottom DO levels. Dissolved oxygen was usually >4 mg l^{-1} in water shallower than 2 m. Bottom hypoxia was most evident during warm periods in eastern sections of upper Perdido Bay and in the deeper sections of the bay from the Highway 98 Bridge to Inerarity Point (Figure 4.19). Hypoxic conditions were also evident in Wolf Bay at depth and in areas of the lower bay that were adjacent to intensive urban development.

The association of salinity stratification with low DO has been well established (Livingston, 1997a). Hypoxia at depth in Perdido Bay was most pronounced in deeper parts of the bay (Figure 4.20). The most pronounced hypoxic conditions were found in the relatively deep holes of the lower Perdido River where there was maximal salinity stratification. In areas where there was less difference between the surface and bottom salinities (shallow shelf areas of upper Perdido Bay), there was less pronounced hypoxia. Relatively low DO prevailed in the deeper parts of the

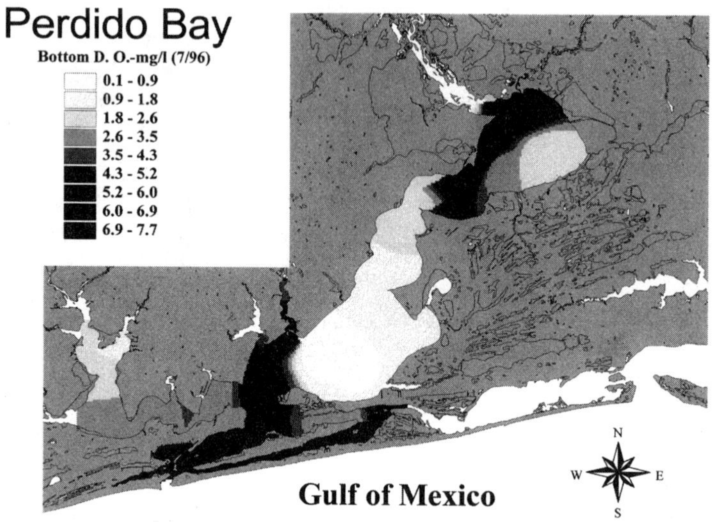

Figure 4.19 Spatial dimensions of bottom dissolved oxygen (mg l^{-1}) in Perdido Bay in July 1996.

lower bay where salinity stratification was maximal. Depth and salinity stratification were thus closely associated with hypoxia at depth in Perdido Bay. The least incidence of subhaloclinal hypoxia occurred in shallow (upper bay) stations (P23) and in the grassbeds in western sections of upper Perdido Bay (P25). In the lower bay (station P40), bottom DO was usually higher due to the direct effects of more oxygenated (high-salinity) water moving up the bay from the Gulf. There was, however, considerable interannual variation in the incidence of hypoxia with percent stratification (Figure 4.20). The most pronounced hypoxia at the deeper stations occurred during 1995–1996; this was not associated with increased stratification which, in fact, occurred during periods of high river flow from 1996 through 1998. Relatively low incidence of hypoxic conditions at depth occurred during 1990–1991 and 1992–1993. Although salinity stratification was an important factor in the reduction of sub-haloclinal DO, the long-term trends of DO indicated that other factors were also important determinants of benthic hypoxia in Perdido Bay.

The regular seasonal changes in surface DO during the first 3 years of analysis (1988 to 1991) differed considerably from the seasonal patterns of DO during the succeeding 6 years (1992 to 1997) (Figure 4.21). Peaks usually occurred during winter–spring periods. By 1998, there was a general reduction in these peak levels. The surface DO data thus coincided to some degree with the nutrient loading trends from Elevenmile Creek. The relatively high DO concentrations during winter–spring periods from 1993 through 1996 ended during a period when orthophosphate loading from the creek was reduced. With the exception of Elevenmile Creek, surface DO in the bay was generally above 5 mg l^{-1}. There was also a general trend of higher surface DO at stations P26, P29, P31, and P33 over the observation period. The general trends of surface and bottom DO were similar over the 11-year study period. Spatial variance of bottom DO at any given time was usually greater than that at the surface. The long-term bottom DO pattern resembled that of the surface DO with regular seasonal changes during the first 3 years followed by periods of increased bottom DO during winter–spring periods from 1992 to 1997. However, during the period from 1994 to 1997, bottom DO concentrations tended to be lower during warm periods of the year. This trend returned to former cycles during late 1997–1998, with lower DO conditions during the warm months of 1999. These data indicate that the long-term trends of both surface and bottom DO coincided with nutrient loading trends. Long-term trends of bottom oxygen anomalies (Figure 4.22) tended to follow the surface DO trends which gave further evidence that bottom DO in Perdido Bay was controlled to some degree by factors other than physical changes (i.e., phytoplankton, see below).

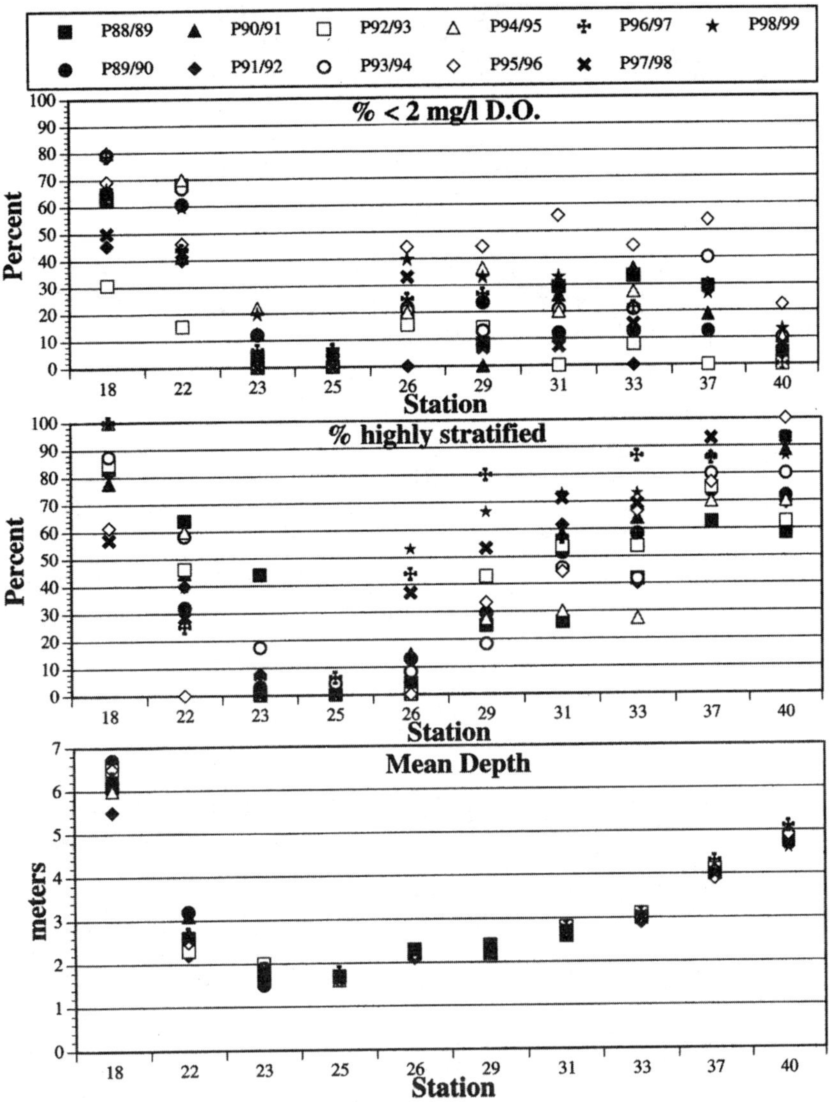

Figure 4.20 DO data averaged over 1-year periods and expressed as percentages shown against percentage of highly stratified waters and mean station depth in Perdido Bay from 1988 through 1999.

The long-term DO regime in Perdido Bay can be explained in part by reviewing changes in biochemical oxygen demand (BOD) over the study period (Figure 4.23). During the period of reduced bottom dissolved oxygen (1994 to 1997), there were general increases in BOD at depth. This trend was followed by reductions of BOD from late 1997 to 1999 with subsequent increases during the late summer of 1999. The pH trends (Figure 4.24) followed cyclic seasonal changes during the first several years, with relatively high pH during the initial drought. After resumption of increased river flows during 1990–1991, there was a reduction of surface pH. By 1992–1993, there was an increase in pH that remained relatively stable until a trend of reduced pH in late 1997 to early 1998. Following another period of high pH during summer–fall 1998, there was a lessor reduction of pH followed by resumption of higher pH during 1999. The pH of the water appeared to follow the long-term trends of nutrient loading (and phytoplankton response to nutrient loading; see below).

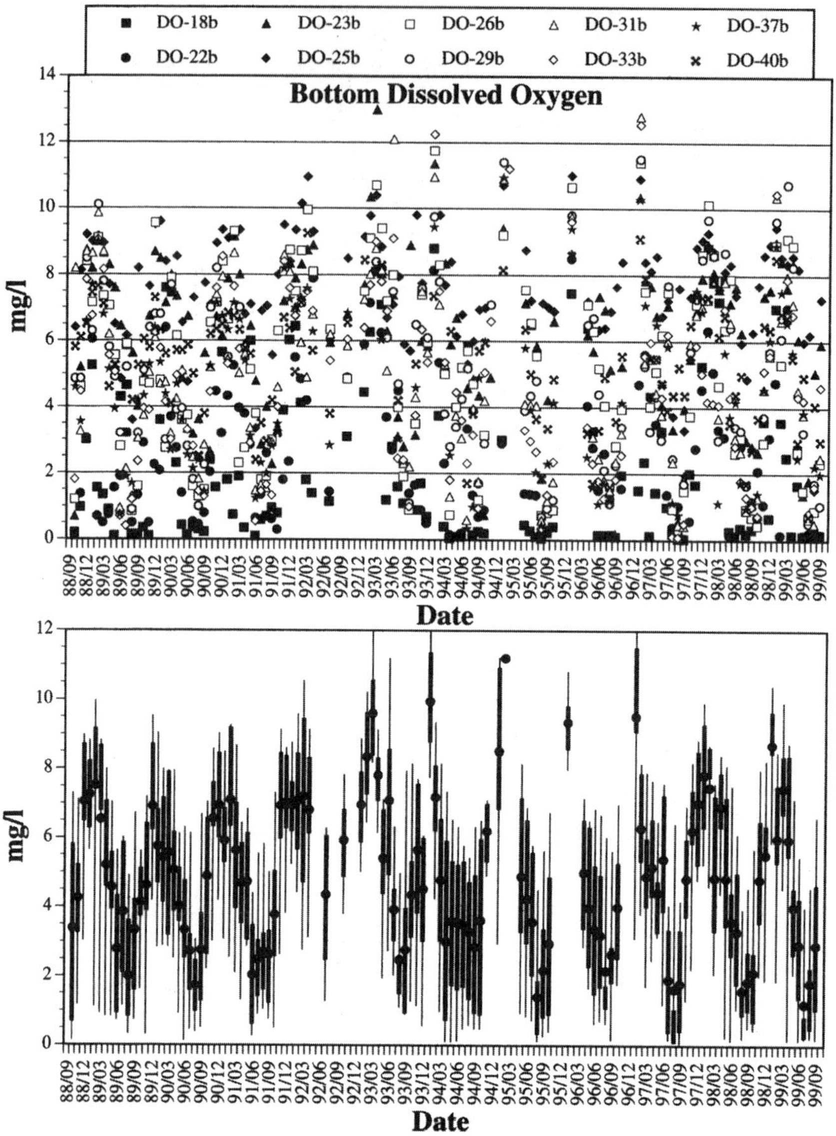

Figure 4.21 Bottom DO (mg l^{-1}) given as station-specific values and means with the 10th, 25th, 75th, and 90th percentiles of all stations monthly from October 1988 through September 1999.

F. Nutrients and Chlorophyll *a*

Based on the nutrient loading information, mill-associated discharges of orthophosphate and ammonia were the chief anthropogenous sources of nutrients to upper Perdido Bay. Relatively high average orthophosphate concentrations in the bay occurred near the mouth of Elevenmile Creek during the period of initial high loading to the bay in 1993–1994 (Figure 4.6a). Orthophosphate gradients were highest at the mouth of the creek. There were periodic increases of orthophosphate (>0.1 mg l^{-1}) at station 22 from 1988 through 1991; by 1992, these concentrations were somewhat higher (Figure 4.25). From 1993 through the third quarter of 1997, there was a more consistent pattern of high orthophosphate concentrations at the mouth of Elevenmile Creek. There was a

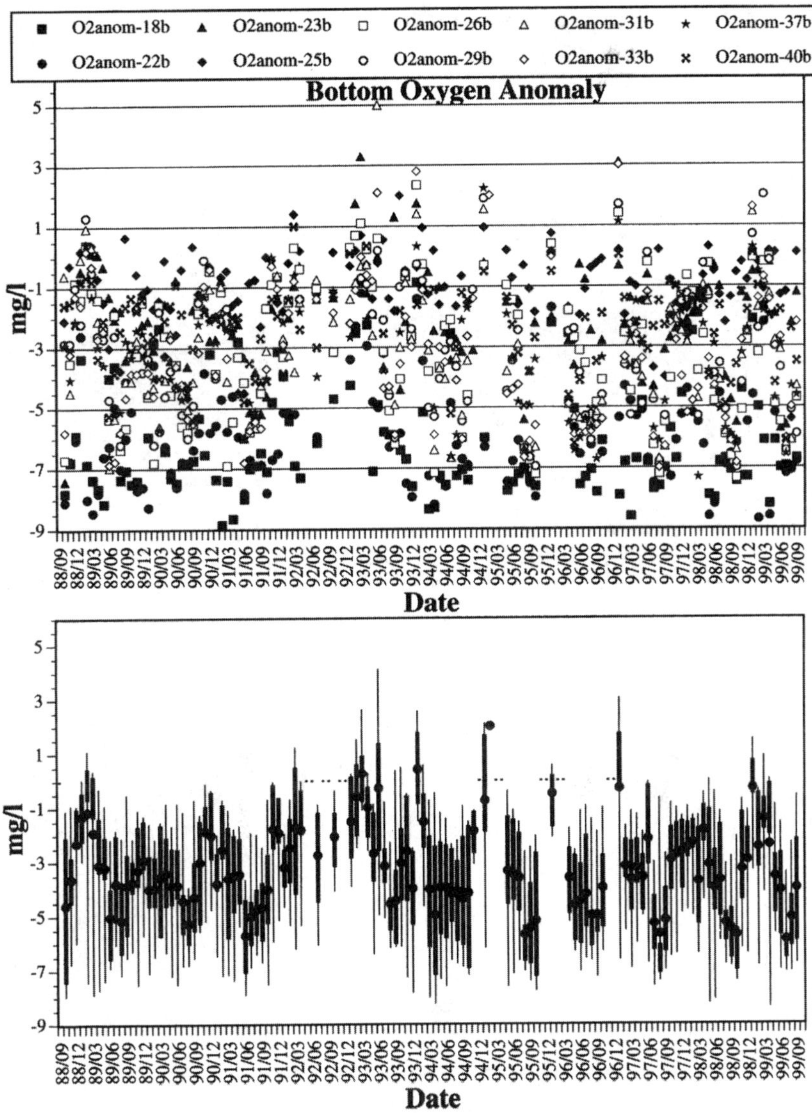

Figure 4.22 Bottom DO anomaly (mg l^{-1}) given as station-specific values and as baywide means (with the 10th, 25th, 75th, and 90th percentiles) monthly from October 1988 through September 1999.

precipitous decrease of this nutrient throughout the bay during late 1997–1998; this decrease was associated with reductions of mill orthophosphate discharges from August 1997 through March 1999. Orthophosphate levels from late 1997 to late 1998 resembled those during the period 1989–1990. This trend was followed by increased orthophosphate concentrations during the spring and summer of 1999 at which time the mill resumed relatively high loading of orthophosphate to Elevenmile Creek. By September 1999, the mill again reduced orthophosphate loading to the bay, and upper bay concentrations of this nutrient reflected this change.

Ammonia gradients at the mouth of Elevenmile Creek were high (Figure 4.6a). Long-term ammonia trends of ammonia in the bay (Figure 4.26) were characterized by high concentrations in Elevenmile Creek (station P22) and the immediate receiving area of the bay (station P23). There were periodic increases of ammonia during 1990 and 1991. By 1993, mean ammonia concentrations in the bay were somewhat lower; this pattern of reduced ammonia continued with further decreases

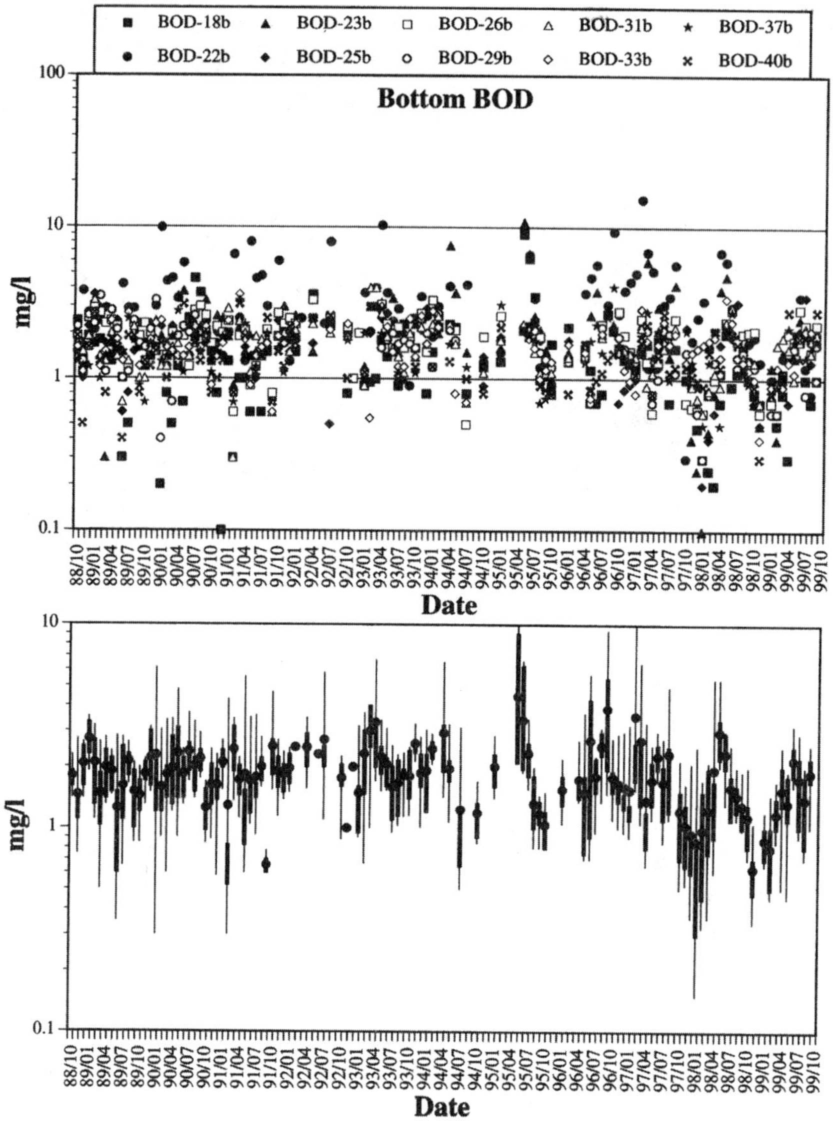

Figure 4.23 Bottom biochemical oxygen demand (mg l^{-1}) given as station-specific values and as baywide means (with the 10th, 25th, 75th, and 90th percentiles) monthly from October 1988 through September 1999.

from late 1997 through early 1998. Increased ammonia concentrations were noted during summer–fall months of 1998 with reduced ammonia during 1999. Thus, higher ammonia loading from Elevenmile Creek from 1995 to 1998 was accompanied by periodic reductions of ammonia concentrations in the upper bay. This situation indicated that variables other than loading were involved in the trends of ammonia concentrations in the bay.

Silica concentrations in Perdido Bay (Figure 4.27) were relatively high during the period 1988 to 1991. The highest silica concentrations occurred in Elevenmile Creek during this period. However, during 1992–1993, there was a general decline of silica throughout the bay. There was a gradient of high to low silica from the upper to the lower bay. There was a brief resumption of high silica during late 1993 to early 1994 that was followed by prolonged reductions of this compound from 1994 through early 1997. There was a general increase of silica during 1997. The reduced orthophosphate loading in late summer 1997 was accompanied by a brief reduction of

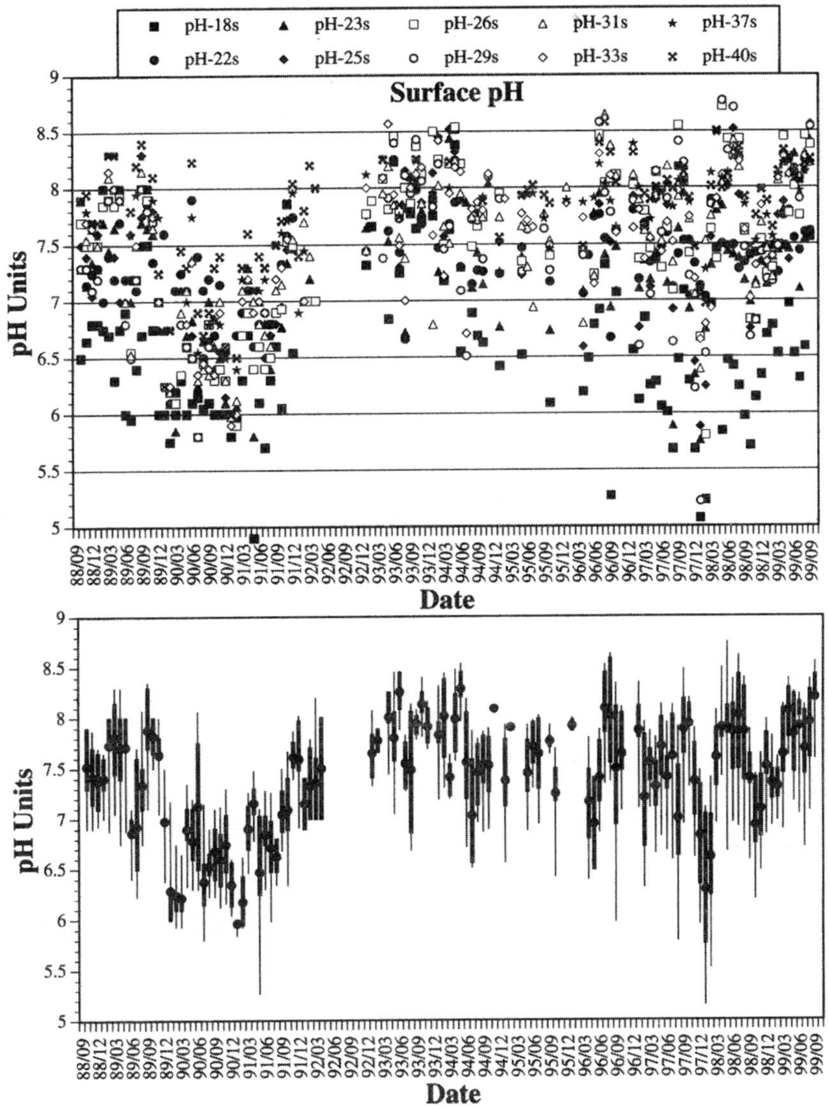

Figure 4.24 Surface pH given as station-specific values and means with the 10th, 25th, 75th, and 90th percentiles of all stations monthly from October 1988 through September 1999.

silica in the bay that was followed by general increases in 1998. Silica concentrations at this time exceeded those during the first 3 years of sampling. By 1999, silica concentrations decreased to a spring low that was followed by modest increases during summer–fall months. A comparison of silica concentration trends with silica loading from the Perdido River and Elevenmile Creek (Figure 4.4b) indicated that bay silica concentrations over time were not a function of silica loading. The 1992–1993 decline of silica concentrations in the bay coincided with increased orthophosphate loading from Elevenmile Creek (Figure 4.5). During the sustained orthophosphate loading from 1993 to early 1997, silica concentrations in the bay were low. There was a general increase of silica concentrations during the period of reduced orthophosphate loading from late summer 1997 through spring 1999. Drought incidence, river flooding, or silica loading from the Perdido River or Elevenmile Creek could not explain the long-term trends of silica concentrations in Perdido Bay. Thus, silica concentrations in Perdido Bay reflected interactions with other variables that changed through time. The connection of baywide silica concentrations with temporal changes of orthophosphate

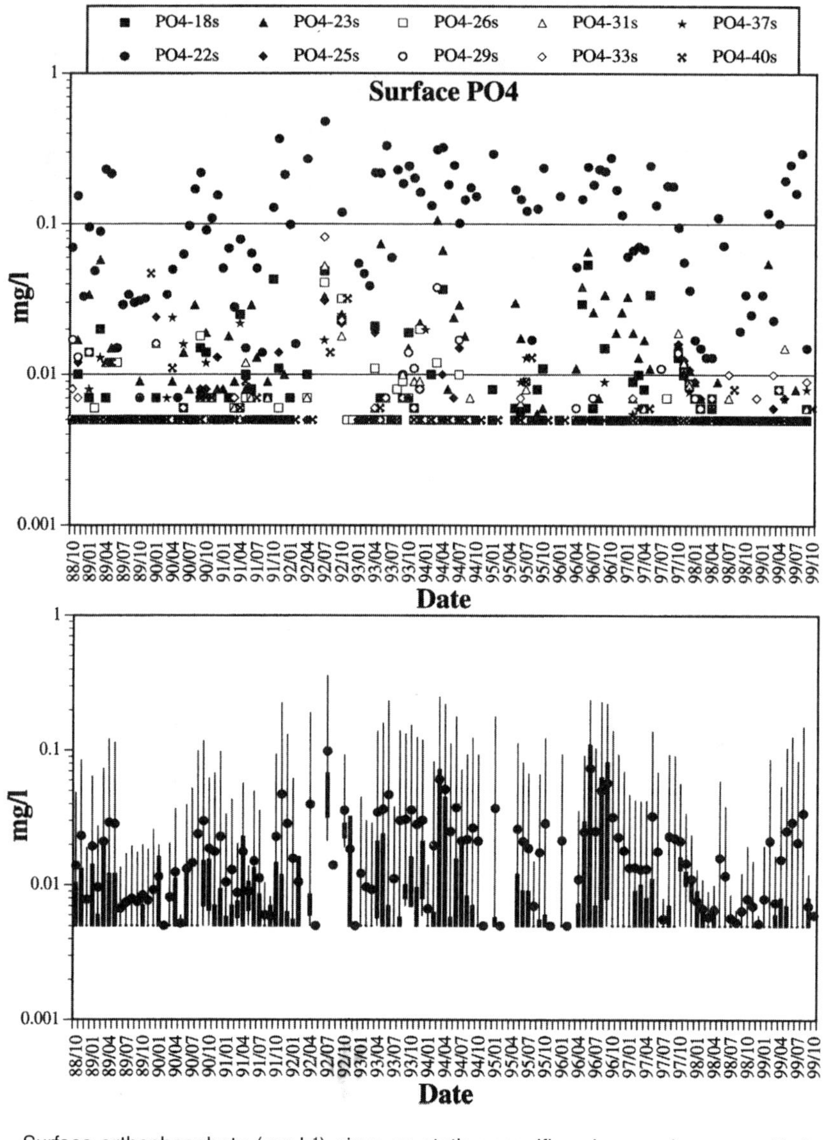

Figure 4.25 Surface orthophosphate (mg l^{-1}) given as station-specific values and means with the 10th, 25th, 75th, and 90th percentiles of all stations monthly from October 1988 through September 1999.

loading indicated possible control by interannual trends of the phytoplankton community (see below).

Annual averages of chlorophyll *a* in Perdido Bay during 1993–1994 (Figure 4.28) were highest at the mouth of Elevenmile Creek and in eastern sections of the upper bay. Relatively low concentrations of chlorophyll *a* occurred in western sections of the bay (i.e., the *Vallisneria americana* seagrass beds). The long-term chlorophyll *a* trends in Perdido Bay (Figure 4.29) indicated seasonal trends with peaks during summer–fall periods. The highest concentrations of chlorophyll *a* were noted at station 23 (at the mouth of Elevenmile Creek) during the droughts of 1989, 1994, and 1999. There was no association of long-term changes of chlorophyll *a* with interannual trends of ammonia, orthophosphate, or silica. Seasonal trends of chlorophyll *a* were more pronounced than interannual cycles. Thus, the long-term trends of chlorophyll *a* concentrations in Perdido Bay did not follow the nutrient loading changes over the 11-year period of study. Chlorophyll *a* trends appeared to be more associated with the drought–flood cycles.

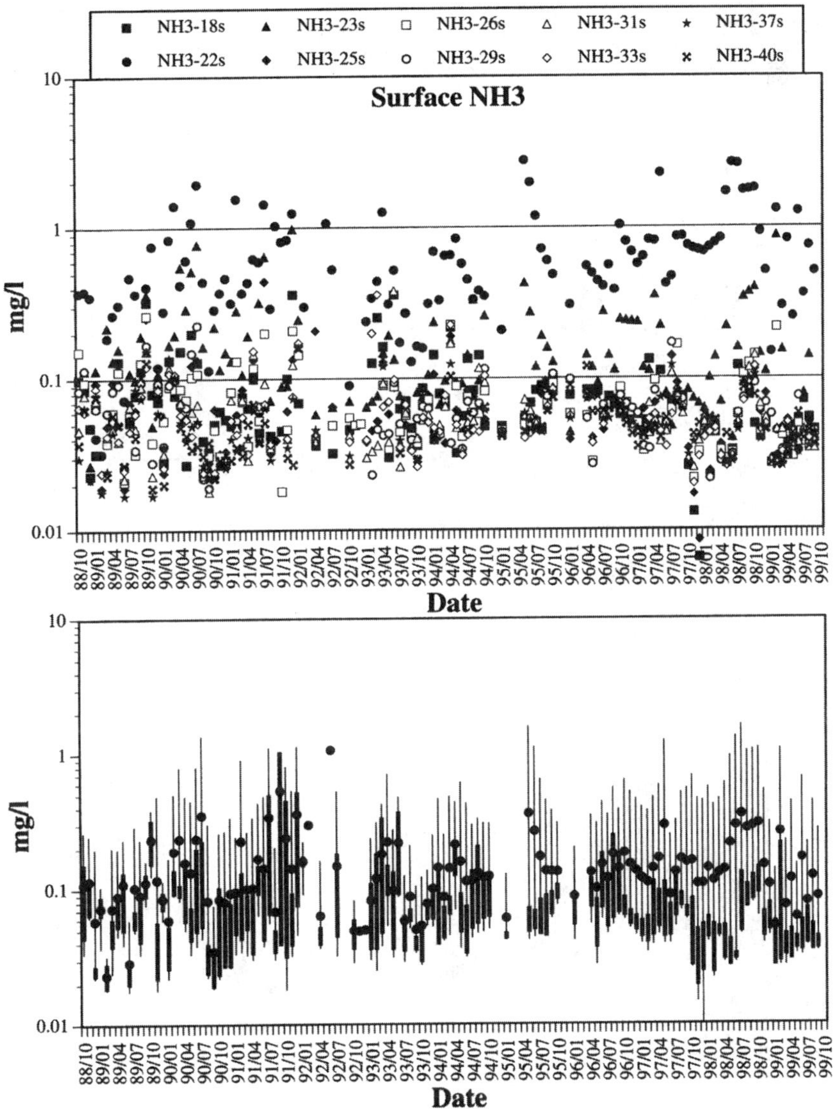

Figure 4.26 Surface ammonia (mg l^{-1}) given as station-specific values and means with the 10th, 25th, 75th, and 90th percentiles of all stations monthly from October 1988 through September 1999.

VI. LONG-TERM PHYTOPLANKTON TRENDS IN PERDIDO BAY

A. Introduction

The increasing global significance of harmful algal blooms (HABs; Hallegraeff et al., 1995) has led to corresponding increases in autecological analyses of individual bloom species. Although HABs represent natural phenomena, there is a general perception that, over the past two decades, there has been an increase in the frequency, intensity, and geographic distribution of HAB incidents. Increased nutrient loading due to anthropogenous sources such as urban and agricultural runoff (non-point sources) and industrial and municipal waste disposal (point sources) has been associated with this trend. It is thought that altered nutrient ratios due to human activities in coastal waters

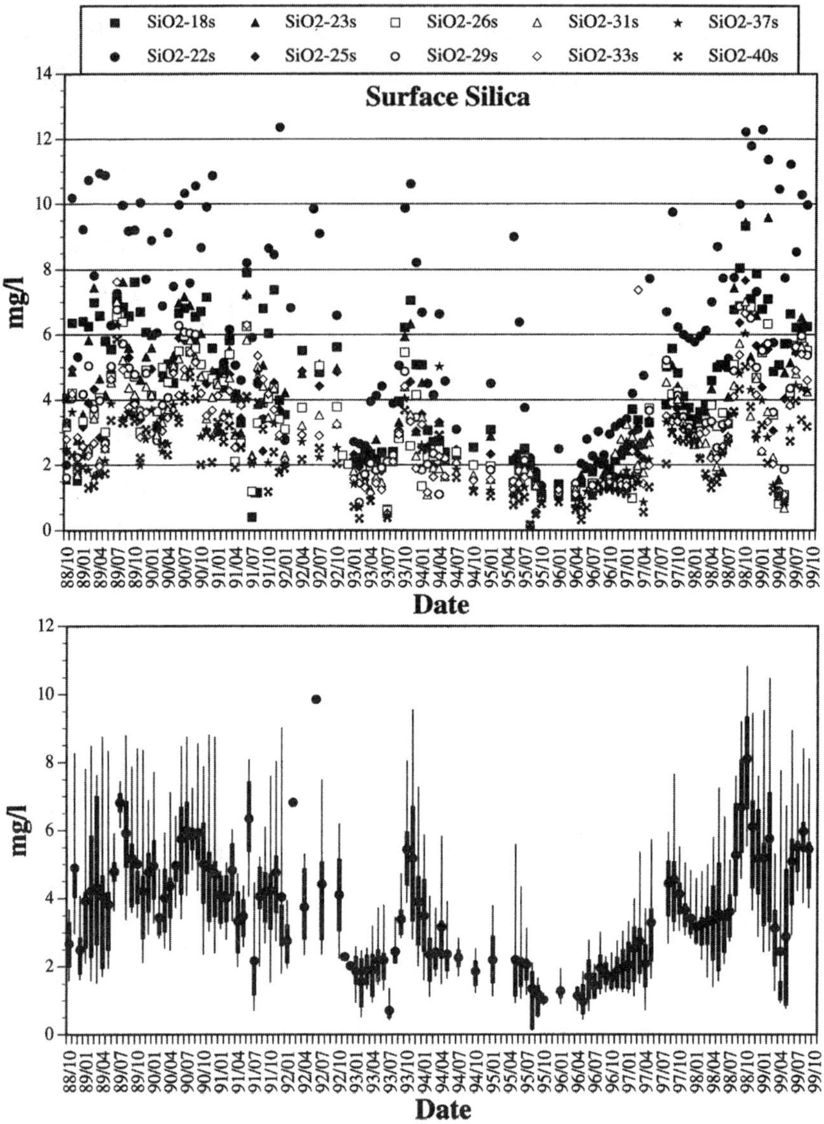

Figure 4.27 Surface silica (mg l^{-1}) given as station-specific values and means with the 10th, 25th, 75th, and 90th percentiles of all stations monthly from October 1988 through September 1999.

have led to blooms of nuisance species that replace the naturally occurring phytoplankton species (Ryther, 1954; Ryther and Dunstan, 1971; Smayda, 1990). Although there are numerous studies concerning the occurrence of bloom species, the origin of such blooms, along with the effects of the blooms on phytoplankton community structure and associated food webs, have been virtually ignored (Mallin, 1994).

The ecological and oceanographic aspects of the propagation and natural history of individual bloom species have been studied in depth (Anderson, 1997). However, there are virtually no long-term, quantitative ecological studies on HAB interactions with phytoplankton assemblages. Such studies will be needed to establish the actual impacts of HABs on secondary production (Smayda, 1997b, c). Horner et al. (1997) emphasized that questions concerning the sources of bloom populations, the environmental conditions that encourage such blooms, and how inshore blooms may be transported offshore remain largely unanswered. The use of biomass indicators such as

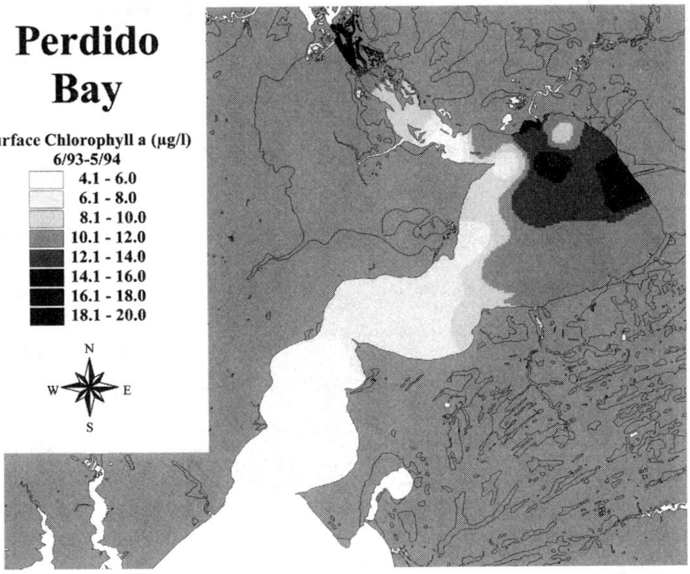

Figure 4.28 Spatial dimensions of surface chlorophyll *a* (µg l^{-1}) in Perdido Bay averaged from monthly data taken from June 1993 through May 1994.

chlorophyll *a* as substitutes for, or indicators of, phytoplankton community dynamics has been generally accepted even though such indices are considered in some quarters as highly problematic (Smayda, 1997a, b). Traditional models such as the spring diatom bloom phenomenon in temperate coastal waters remain largely inferential and have little to do with bloom sequences. Long-term monitoring of phytoplankton and/or zooplankton abundance are considered important (Cullen et al., 1997), but are rarely carried out with enough supporting background data to determine the contributing causes to the onset of a bloom or blooms. The little information that is available usually is dependent on single-station collections rather than systemwide collections. Thus, the influence of key habitat variables and biological interactions in the origin and sustenance of plankton blooms remains largely unknown.

Comprehensive analyses of phytoplankton processes have been restricted to general questions concerning water column associations of nutrient loading, water quality, and chlorophyll *a* concentrations. There is little in the way of analyses of the quantitative, long-term aspects of the effects of nutrification (natural and anthropogenous) on phytoplankton communities. The connection between such phytoplankton assemblages and coastal secondary production and food webs has not been made. Questions concerning bloom impacts have been largely restricted to individual species (usually populations of economic interest) and the interactions of phytoplankton (usually represented by chlorophyll *a*) with zooplankton (Buskey and Stockwell, 1993). Most studies of toxic or nuisance plankton usually begin after the initiation of a given bloom so that little attention is given to the environmental or biological factors that actually caused the bloom in the first place (Buskey et al., 1997). The difficulties involved with quantification and species identification are usually used to explain the lack of comprehensive analyses of phytoplankton even though this group of organisms is central to the study of eutrophication and hypereutrophication.

The long-term (11-year) studies in Perdido Bay were carried out in a uniform and consistent manner with the collection of water and sediment quality data taken simultaneously with samples of phytoplankton, zooplankton, benthic microphytes, infaunal macroinvertebrates, epibenthic macroinvertebrates, and fishes. The descriptive field data were supplemented with laboratory and field experiments so that processes such as nutrient limitation, the origin and distribution of nitrogen, phosphorus, and carbon particulate matter, and the response of the food web to phytoplankton trends could be evaluated before and after the initiation of the phytoplankton blooms. The phytoplankton

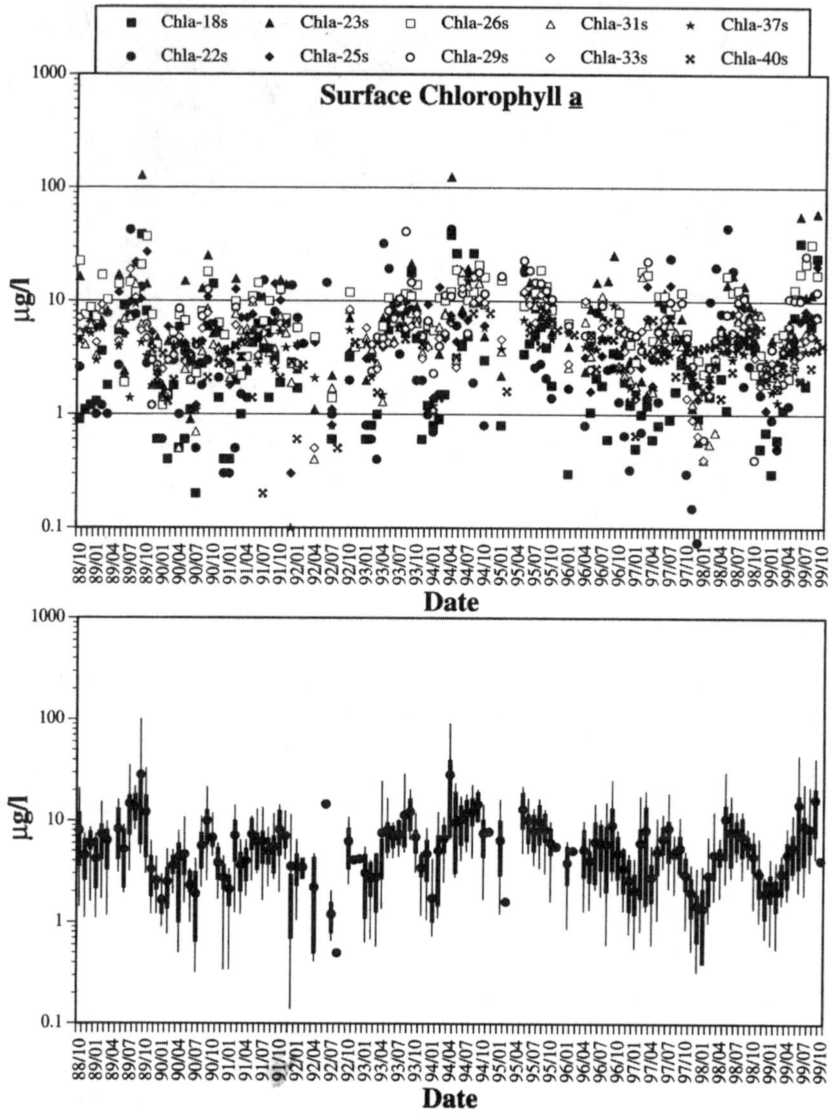

Figure 4.29 Surface chlorophyll *a* (µg l^{-1}) given as station-specific values and means with the 10th, 25th, 75th, and 90th percentiles of all stations monthly from October 1988 through September 1999.

were collected using both net and whole-water methods and were numerically quantified to species levels wherever possible. The combination of drought and flood sequences, single-factor changes in nutrient loading phenomena, and relatively simple habitat responses to natural and anthropogenous events allowed a determination of the key factors that were associated with the initiation and succession of plankton blooms in Perdido Bay.

Some caution should be used in the interpretation of the results of the phytoplankton studies. The primarily monthly sampling pattern was inadequate as a quantitation of short-term bloom sequences, given the fact that determinations of phytoplankton reproduction rates should be measured in periods of hours to days. This consideration does not appear to be of general concern as semiannual or even annual sampling is considered adequate by many investigators (Dauer et al., 2000; Howarth et al., 2000). However, using a monthly sampling interval is not adequate as a truly quantitative measure of plankton blooms, given the relative short time intervals of such events. On the other hand, when blooms are taken consistently using monthly sampling, it can be assumed that such phenomena are relatively common and widespread.

Based on the recognition of probable sampling errors, various indices were tested as indicators of the phytoplankton community structure. These indices were viewed as more stable in time than the relatively evanescent plankton blooms. Most authors use numerical counts as the chief method of analysis (Anderson and Garrison, 1997). However, there are problems with the use of blooms being defined solely by cell numbers (D. M. Anderson, personal communication). For this reason, numerical counts were transformed into biomass figures. Phytoplankton species richness (as a monthly or cumulative [annual] index) was tested as an indicator of long-term trends. Relative dominance (numerical and biomass) was also used as an index of phytoplankton community structure in addition to the usual count identification of bloom incidence (i.e., $>10^6$ cells l^{-1}). So-called bloom species were determined without reference to any form of toxic effects since such impacts were not analyzed in this study. Overall, long-term changes of the community structure of plankton assemblages were determined and placed within the context of trends of nutrient loading, nutrient gradients, water and sediment quality, and changes in the primary food webs of the Perdido Bay system. Plankton community trends were compared with long-term changes of food web factors and secondary production in the Perdido system. These findings were then compared with the potential influence of altered phytoplankton community structure on food webs and secondary production in other estuarine and coastal systems in the northeastern Gulf of Mexico. The overall results of our 31 years of research in Gulf coastal areas were then compared with worldwide eutrophication studies.

B. Chronology of Phytoplankton Trends in Perdido Bay

Phytoplankton trends in Perdido Bay should be viewed within the context of several factors. Seasonal and interannual cycles of temperature and river flow are important in terms of the estuarine habitat changes. Related changes in bay turnover rates are directly related to phytoplankton production. Salinity gradients and stratification of the water column should also be taken into account. The drought–flood cycles over the 11-year period of observation were accompanied by changes in all these variables resulting in long-term water quality trends that could also affect phytoplankton development in the bay. Interannual changes of nutrient loading from Elevenmile Creek and corresponding concentration gradients in entry areas (stations P22, P23) were also considered in the phytoplankton analyses.

The changes in the phytoplankton community that took place in Perdido Bay over the 11-year observation period form the basis of the eutrophication potential of this system. The general trends of phytoplankton bloom occurrence in Perdido Bay can be summarized as follows:

1. 1988 to 1991

The period 1988 to 1990 was characterized by a drought that was broken by high river flows during winter 1990. Orthophosphate concentrations in the upper bay indicated dilution effects due to Perdido River flows into the bay and rapid nutrient utilization at the mouth of Elevenmile Creek where chlorophyll *a* concentrations were higher than anywhere else in the bay (Livingston, 1998). Chlorophyll *a* concentrations throughout the bay were relatively low from 1988 through 1991. Sediment orthophosphate levels in the lower end of Elevenmile Creek (stations P21, P22) and the nearby bay areas (stations P23, P24) tended to be the highest in the bay. Light transmission characteristics of the water were directly related to the spatial and temporal pattern of phytoplankton activity in Perdido Bay. Net and whole-water phytoplankton collections in Perdido Bay gave no indication of hypereutrophication during the first 3 years of the study (Livingston, 1998). Periodic increases of nutrient loading from Elevenmile Creek were associated with increased phytoplankton activity in the immediate receiving areas of the bay. However, no blooms were noted during the first 3 years of study.

2. 1992 to 1994

Peak Perdido River flows usually occurred during winter–spring periods. Drought conditions prevailed with no major winter–spring peak flows during 1993–1994 (Figure 4.2). At this time, water temperature and salinity were higher than usual during winter–spring months. Water clarity and light penetration in the bay increased due to reduced color. By 1994, there was prolonged, high orthophosphate loading by the pulp mill; such sustained loading was correlated with the first phytoplankton blooms (Table 4.1, Figure 4.30). During 1994, a succession of bloom species was noted that included the diatoms *Leptocylindrus danicus* Cleve, *Miraltia throndsenii* Marino et al., and *Cyclotella choctawhatcheeana* Prasad. During July 1994, Perdido River flow increases during a summer storm led to reduced salinities in the bay. This was followed by a 2-year hiatus of the phytoplankton blooms.

3. 1995 to 1997

A 5 to 6 month analysis of the water/sediment quality, net and whole-water phytoplankton, zooplankton, and infaunal macroinvertebrates was carried out in upper Perdido Bay and Soldier Creek during the summer–fall of 1996 (Livingston, 1997b, c; 1998). The whole-water phytoplankton community was characterized by significantly ($p > 0.05$) higher numbers at stations P23 and P37. A series of blooms was noted (stations P23, P37, SC1; Table 4.1, Figure 4.2a) during June, July, and August 1996. There were two novel phytoplankton blooms with massive cell concentrations encountered in the upper bay stations during summer 1996: (1) a diatom bloom (*Synedropsis* sp.) in June and July 1996 and (2) a raphidophyte flagellate (*Heterosigma akashiwo* (Hurlburt) bloom in July 1996. The *Synedropsis* bloom continued through August 1996. This diatom species was never observed during previous investigations of Perdido Bay. Preliminary studies concerning the bloom populations revealed that *Synedropsis* represents a genus of araphid diatoms known to be associated with sea-ice in polar waters (Hasle et al., 1995). Analysis of the longitudinal gradients of cell numbers indicated that these blooms originated in upper bay areas, and moved from the bay into Soldier Creek. Over 20 million cells l^{-1} of *Synedropsis* sp. were found at station P37 in July 1996; the bloom extended into Soldier Creek with its origin in the bay. During February 1997, Perdido Bay was characterized by the dominance of the dinoflagellate *Prorocentrum minimum* (Pavillard) Schiller. The distribution of this species included near-bloom concentrations in mid-areas of the upper bay. During spring–summer 1997, massive blooms of *Miraltia throndsenii*, *Heterosigma akashiwo*, and *Synedropsis* sp. occurred throughout the bay. By fall 1997, *H. akashiwo* dominated the bay blooms, primarily in the upper bay. During this period, there was continued high orthophosphate loading from the mill with increasing ammonia loading.

4. 1998 to 1999

With the exception of a *Merismopedia tenuissima* Lemmermann bloom in Elevenmile Creek during November 1997, there were no blooms in Perdido Bay during late 1997 and winter–early spring 1998 (Table 4.1). This period coincided with the phosphorus control program initiated by the paper mill. Due to our methods of preservation, it was not possible to determine if cells of the freshwater blue-green alga *Merismopedia* were alive in the bay. Consequently, when this species was found in the bay, the numbers found were reported as tentative and were not considered true blooms (Table 4.1). The same is true for the freshwater green alga *Pedinophora* sp. and the freshwater diatom *Urosolenia eriensis* (Smith) Round. During spring and summer 1998, there was periodic high loading of ammonia into the bay by the mill although orthophosphate loading

Table 4.1 List of Noted Phytoplankton Blooms (>1,000,000 cells l^{-1}) Taken in the Whole-Water Samples from Perdido Bay over the 11-year Sampling Period (10/88 to 9/99)

Station	Date	Depth	Species	No. Cells	l^{-1}	Chl a	Chl b	Chl c	Chl date
26	22-Feb-94	S	*Leptocylindrus danicus* Cleve	1,950,000	1.95 × 106	10.7	<0.2	2.7	2/22/94
29	22-Feb-94	S	*Leptocylindrus danicus* Cleve	1,640,000	1.64 × 106	4.0	<0.2	0.5	2/22/94
23	22-Mar-94	S	*Miraltia throndsenii* Marino et al.	1,250,000	1.25 × 106	7.9	<0.2	1.5	3/22/94
26	22-Mar-94	S	*Miraltia throndsenii* Marino et al.	4,780,000	4.78 × 106	10.1	<0.2	2.1	3/22/94
29	22-Mar-94	S	*Miraltia throndsenii* Marino et al.	5,620,000	5.62 × 106	11.3	<0.2	2.4	3/22/94
23	12-Apr-94	S	*Cyclotella choctawhatcheeana* Prasad	2,430,000	2.43 × 106	124.4	<0.2	17.9	4/27/94
26	12-Apr-94	S	*Cyclotella choctawhatcheeana* Prasad	8,620,000	8.62 × 106	12.1	0.6	2.2	4/27/94
29	12-Apr-94	S	*Cyclotella choctawhatcheeana* Prasad	8,530,000	8.53 × 106	11.1	<0.2	2.1	4/12/94
33	12-Apr-94	S	*Cyclotella choctawhatcheeana* Prasad	7,650,000	7.65 × 106	6.5	<0.2	<0.2	4/27/94
23	24-Jun-96	S	*Synedropsis* sp.	5,750,000	5.75 × 106	14.8	0.2	1.3	6/24/96
SC1	17-Jul-96	S	*Synedropsis* sp.	2,280,000	2.28 × 106	6.9	<0.2	0.6	7/16/96
23	18-Jul-96	S	*Synedropsis* sp.	2,660,000	2.66 × 106	4.8	2.2	1.2	7/16/96
23	18-Jul-96	S	*Heterosigma akashiwo* (Hada) Hada	23,930,000	23.93 × 106	4.8	2.2	1.2	7/16/96
37	18-Jul-96	S	*Synedropsis* sp.	20,680,000	20.68 × 106	7.3	<0.2	0.3	7/16/96
23	12-Aug-96	S	*Synedropsis* sp.	3,340,000	3.34 × 106	15.3	<0.2	1.5	8/12/96
37	13-Aug-96	S	*Synedropsis* sp.	2,970,000	2.97 × 106	4.7	2.5	<0.2	8/13/96
31	24-Mar-97	S	*Miraltia throndsenii* Marino et al.	11,190,000	11.19 × 106	6.2	0.3	1.7	3/24/97
31	24-Mar-97	B	*Miraltia throndsenii* Marino et al.	26,180,000	26.18 × 106	4.4	<0.2	0.8	3/24/97
37	26-Mar-97	S	*Miraltia throndsenii* Marino et al.	67,810,000	67.81 × 106	7.0	<0.2	1.6	3/25/97
40	26-Mar-97	S	*Miraltia throndsenii* Marino et al.	18,590,000	18.59 × 106	6.9	<0.2	0.8	3/25/97
25	27-Mar-97	S	*Miraltia throndsenii* Marino et al.	12,740,000	12.74 × 106	13.7	<0.2	0.6	3/25/97
25	27-Mar-97	S	*Heterosigma akashiwo* (Hada) Hada	11,800,000	11.80 × 106	13.7	<0.2	0.6	3/25/97
25	27-Mar-97	B	*Miraltia throndsenii* Marino et al.	13,950,000	13.95 × 106	8.8	<0.2	0.5	3/25/97
25	27-Mar-97	B	*Heterosigma akashiwo* (Hada) Hada	9,230,000	9.23 × 106	8.8	<0.2	0.5	3/25/97
26	27-Mar-97	S	*Miraltia throndsenii* Marino et al.	10,830,000	10.83 × 106	17.2	<0.2	2.5	3/25/97
26	27-Mar-97	S	*Heterosigma akashiwo* (Hada) Hada	1,370,000	1.37 × 106	17.2	<0.2	2.5	3/25/97
26	27-Mar-97	B	*Miraltia throndsenii* Marino et al.	12,330,000	12.33 × 106	2.8	0.2	<0.2	3/25/97
29	28-Mar-97	S	*Synedropsis* sp.	1,050,000	1.05 × 106	22.6	<0.2	4.0	3/25/97
29	28-Mar-97	S	*Miraltia throndsenii* Marino et al.	21,920,000	21.92 × 106	22.6	<0.2	4.0	3/25/97
29	28-Mar-97	B	*Miraltia throndsenii* Marino et al.	36,140,000	36.14 × 106	17.9	<0.2	2.8	3/25/97
29	28-Mar-97	B	*Heterosigma akashiwo* (Hada) Hada	8,760,000	8.76 × 106	17.9	<0.2	2.8	3/25/97
23	13-May-97	S	*Heterosigma akashiwo* (Hada) Hada	7,670,000	7.67 × 106	10.1	<0.2	0.6	5/13/97
23A	13-May-97	S	*Heterosigma akashiwo* (Hada) Hada	4,690,000	4.69 × 106	6.9	<0.2	1.0	5/13/97
25	15-Jul-97	S	*Heterosigma akashiwo* (Hada) Hada	1,390,000	1.39 × 106	13.9	<0.2	1.3	7/15/97
26	15-Jul-97	S	*Synedropsis* sp.	1,320,000	1.32 × 106	8.4	<0.2	0.9	7/15/97

Table 4.1 (continued) List of Noted Phytoplankton Blooms (>1,000,000 cells l^{-1}) Taken in the Whole-Water Samples from Perdido Bay over the 11-year Sampling Period (10/88 to 9/99)

Station	Date	Depth	Species	No. Cells	l^{-1}	Chl a	Chl b	Chl c	Chl date
26	9-Sep-97	S	*Heterosigma akashiwo* (Hada) Hada	1,540,000	1.54 × 106	12.0	<0.2	1.4	9/9/97
24A	9-Sep-97	S	*Heterosigma akashiwo* (Hada) Hada	45,280,000	45.28 × 106	132.9	<0.2	16.1	9/9/97
22	19-Nov-97	S	*Merismopedia tenuissima* Lemmermann	1,070,000	1.07 × 106	—	—	—	—
31	14-Apr-98	S	*Cyclotella choctawhatcheeana* Prasad	1,650,000	1.65 × 106	5.1	<0.2	0.9	4/14/98
37	14-Apr-98	S	*Cyclotella choctawhatcheeana* Prasad	2,270,000	2.27 × 106	3.7	<0.2	0.7	4/14/98
40	14-Apr-98	S	*Cyclotella choctawhatcheeana* Prasad	1,300,000	1.30 × 106	5.1	<0.2	0.8	4/14/98
22	12-May-98	S	*Pedinophora* sp. (4 fl. in mucus)[a]	1,020,000	1.02 × 106	44.2	5.3	4.8	5/12/98
23	12-May-98	S	*Cryptomonas* sp.	1,976,355	1.98 × 106	14.1	0.6	1.5	5/12/98
25	12-May-98	S	*Cyclotella choctawhatcheeana* Prasad	4,590,000	4.59 × 106	3.7	<0.2	0.4	5/12/98
26	12-May-98	S	*Heterosigma akashiwo* (Hada) Hada	2,000,000	2.00 × 106	16.8	<0.2	1.5	5/12/98
26	12-May-98	S	*Cyclotella choctawhatcheeana* Prasad	2,140,000	2.14 × 106	16.8	<0.2	1.5	5/12/98
29	12-May-98	S	*Cyclotella choctawhatcheeana* Prasad	3,800,000	3.80 × 106	7.7	<0.2	0.5	5/12/98
31	12-May-98	S	*Cyclotella choctawhatcheeana* Prasad	6,210,000	6.21 × 106	4.6	<0.2	0.3	5/12/98
37	12-May-98	S	*Cyclotella choctawhatcheeana* Prasad	5,140,000	5.14 × 106	4.5	<0.2	0.4	5/12/98
40	12-May-98	S	*Cyclotella choctawhatcheeana* Prasad	1,670,000	1.67 × 106	2.4	<0.2	0.3	5/12/98
23	9-Jun-98	S	*Heterosigma akashiwo* (Hada) Hada	7,470,000	7.47 × 106	17.1	<0.2	1.3	6/9/98
23	9-Jun-98	S	*Cryptomonas* sp. (Teleaulax-like)	1,683,579	1.68 × 106	17.1	<0.2	1.3	6/9/98
25	9-Jun-98	S	*Synedropsis* sp.	2,680,000	2.68 × 106	7.8	<0.2	0.6	6/9/98
26	9-Jun-98	S	*Synedropsis* sp.	1,660,000	1.66 × 106	12.1	<0.2	1.2	6/9/98
29	9-Jun-98	S	*Synedropsis* sp.	1,510,000	1.51 × 106	5.3	<0.2	0.6	6/9/98
31	9-Jun-98	S	*Synedropsis* sp.	1,470,000	1.47 × 106	3.6	<0.2	0.5	6/9/98
25	21-Jul-98	S	*Synedropsis* sp.	1,240,000	1.24 × 106	9.4	<0.2	0.3	7/21/98
25	21-Jul-98	S	*Cyclotella choctawhatcheeana* Prasad	1,070,000	1.07 × 106	9.4	<0.2	0.3	7/21/98
29	21-Jul-98	S	*Synedropsis* sp.	1,370,000	1.37 × 106	10.0	<0.2	0.5	7/21/98
26	25-Aug-98	S	*Heterosigma akashiwo* (Hada) Hada	1,260,000	1.26 × 106	10.9	<0.2	0.7	8/25/98
25	20-Oct-98	S	*Urosolenia eriensis* (Smith) Round[b]	2,974,851	2.97 × 106	6.1	0.5	1.1	10/20/98
26	20-Oct-98	S	*Urosolenia eriensis* (Smith) Round[b]	1,223,109	1.22 × 106	2.3	<0.2	<0.2	10/20/98
29	20-Oct-98	S	*Urosolenia eriensis* (Smith) Round[b]	2,438,724	2.44 × 106	0.4	<0.2	<0.2	10/20/98
31	20-Oct-98	S	*Urosolenia eriensis* (Smith) Round[b]	1,550,946	1.55 × 106	7.8	<0.2	1.1	10/20/98
22	15-Dec-98	S	*Merismopedia tenuissima* Lemmermann[c]	5,100,000	5.10 × 106	0.7	0.2	<0.2	12/15/98
23	15-Dec-98	S	*Merismopedia tenuissima* Lemmermann[c]	2,990,000	2.99 × 106	2.9	<0.2	0.7	12/15/98

22	20-Jan-99	S	*Merismopedia tenuissima* Lemmermann[c]	1,220,000	1.22 × 106	0.9	<0.2	0.3	1/20/99
23	20-Jan-99	S	*Prorocentrum minimum* (Pavillard) Schiller	1,060,000	1.06 × 106	4.6	<0.2	1.5	1/20/99
18	20-Apr-99	S	*Cyclotella choctawhatcheeana* Prasad	3,050,000	3.05 × 106	5.8	<0.2	0.6	4/20/99
23	20-Apr-99	S	*Cyclotella choctawhatcheeana* Prasad	20,670,000	20.67 × 106	3.7	<0.2	0.5	4/20/99
25	20-Apr-99	S	*Cyclotella choctawhatcheeana* Prasad	8,350,000	8.35 × 106	2.2	<0.2	0.4	4/20/99
26	20-Apr-99	S	*Cyclotella choctawhatcheeana* Prasad	36,380,000	36.38 × 106	2.1	<0.2	0.4	4/20/99
29	20-Apr-99	S	*Cyclotella choctawhatcheeana* Prasad	22,630,000	22.63 × 106	10.6	<0.2	1.3	4/20/99
31	20-Apr-99	S	*Cyclotella choctawhatcheeana* Prasad	12,470,000	12.47 × 106	6.4	<0.2	0.9	4/20/99
37	20-Apr-99	S	*Cyclotella choctawhatcheeana* Prasad	4,110,000	4.11 × 106	4.7	<0.2	0.7	4/20/99
40	20-Apr-99	S	*Cyclotella choctawhatcheeana* Prasad	3,550,000	3.55 × 106	3.4	<0.2	0.4	4/20/99
23	20-Apr-99	S	*Miraltia throndsenii* Marino et al.	2,330,000	2.33 × 106	3.7	<0.2	0.5	4/20/99
26	20-Apr-99	S	*Miraltia throndsenii* Marino et al.	6,400,000	6.40 × 106	2.1	<0.2	0.4	4/20/99
29	20-Apr-99	S	*Miraltia throndsenii* Marino et al.	2,270,000	2.27 × 106	10.6	<0.2	1.3	4/20/99
26	20-Apr-99	S	*Nitzschia reversa* Wm. Sm.	2,100,000	2.10 × 106	2.1	<0.2	0.4	4/20/99
23	20-Apr-99	S	*Miraltia* sp.	1,690,000	1.69 × 106	3.7	<0.2	0.5	4/20/99
26	20-Apr-99	S	*Miraltia* sp.	5,400,000	5.40 × 106	2.1	<0.2	0.4	4/20/99
29	20-Apr-99	S	*Miraltia* sp.	2,270,000	2.27 × 106	10.6	<0.2	1.3	4/20/99
25	12-May-99	S	*Cyclotella choctawhatcheeana* Prasad	2,188,120	2.19 × 106	4.5	<0.2	0.3	5/12/99
26	12-May-99	S	*Cyclotella choctawhatcheeana* Prasad	7,699,629	7.70 × 106	12.7	<0.2	1.3	5/12/99
29	12-May-99	S	*Cyclotella choctawhatcheeana* Prasad	5,095,304	5.10 × 106	10.3	<0.2	1.0	5/12/99
31	12-May-99	S	*Cyclotella choctawhatcheeana* Prasad	6,095,429	6.10 × 106	2.9	<0.2	0.3	5/12/99
37	12-May-99	S	*Cyclotella choctawhatcheeana* Prasad	1,015,000	1.02 × 106	2.5	<0.2	0.4	5/12/99
18	23-Jun-99	S	*Heterosigma akashiwo* (Hada) Hada	1,258,824	1.26 × 106	32.7	<0.2	4.1	6/23/99
23	23-Jun-99	S	*Chattonella cf. subsalsa* Biecheler	2,390,000	2.39 × 106	56.3	<0.2	6.8	6/23/99
42b	14-Jul-99	S	*Gymnodinium* sp.	1,657,586	1.66 × 106	30.5	<0.2	3.4	7/14/99
22	24-Aug-99	S	*Merismopedia tenuissima* Lemmermann[c]	2,900,000	2.90 × 106	5.9	0.2	<0.2	8/24/99
23	24-Aug-99	S	*Merismopedia tenuissima* Lemmermann[c]	3,180,000	3.18 × 106	11.6	<0.2	0.7	8/24/99
26	24-Aug-99	S	*Heterosigma akashiwo* (Hada) Hada	2,689,994	2.70 × 106	32	0.2	0.6	8/24/99
22	8-Sep-99	S	*Merismopedia tenuissima* Lemmermann[c]	10,010,000	10.01 × 106	7.4	0.7	0.3	9/8/99
22	8-Sep-99	B	*Merismopedia tenuissima* Lemmermann[c]	2,710,000	2.71 × 106	87.0	254.1	<0.2	9/8/99

Note: Also shown are the chlorophyll *a*, *b*, and *c* concentrations during these bloom periods. Data are listed by station and date.

[a] Freshwater green alga: not able to tell if alive.
[b] Freshwater diatom: not able to tell if alive.
[c] Freshwater blue-green alga: not able to tell if alive.

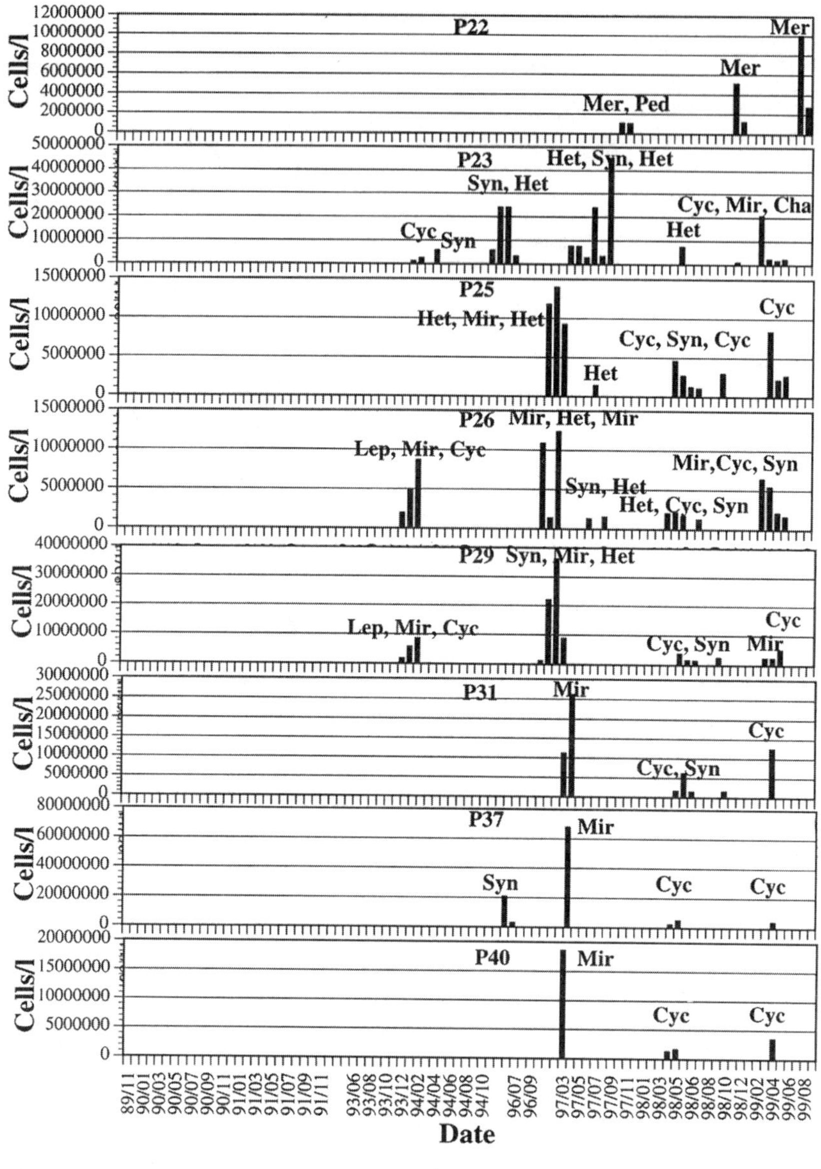

Figure 4.30 Review of plankton blooms taken as whole-water samples from Perdido Bay during the period October 1988 to September 1999. The following abbreviations were used: *Leptocylindrus danicus* Cleve = Lep; *Prorocentrum minimum* = Pro; *Miraltia throndsenii* = Mir; *Cyclotella choctawhatcheeana* Prasad = Cyc; *Synedropsis* sp. nov. = Syn; *Heterosigma akashivo* = Het; *Merismopedia* = Mer; *Chattonella* = Cha; *Urosolenia* = Uro; *Pedinophora* = Ped.

remained low. At this time, blooms of *Cyclotella choctawhatcheeana* (April, May), *H. akashiwo*, and *Synedropsis* sp. (June through August 1998) occurred at various stations in Perdido Bay. During late December 1998 and early January 1999 (3 weeks), the mill resumed high orthophosphate loading to Elevenmile Creek. This was accompanied by *M. tenuissima* blooms in Elevenmile Creek during December 1998 and January 1999, and *Prorocentrum minimum* blooms in immediate receiving areas of the bay (station P23) in January 1999. During winter–early spring 1999, there was a brief period of low orthophosphorus loading that was followed by high loading of this nutrient to Elevenmile Creek from April through August 1999. Massive blooms of Cyclotella choctawhatcheeana throughout the bay accompanied this loading in April 1999 along with blooms of

Miraltia throndsenii and *Nitzschia reversa* Wm. Sm. During June, blooms of the raphidophyte *Chattonella cf. subsalsa* Biecheler were noted in the upper bay. In July 1999, there were near blooms of *H. akashiwo* in the upper bay and *Gymnodinium* spp. blooms in Wolf Bay. During August, blooms of *H. akashiwo* were noted in the upper bay. By September 1999, the mill again reduced its orthophosphate loading to the bay. At this time, there were *Merismopedia tenuissima* blooms throughout the water column in lower Elevenmile Creek. The extremely high concentrations of chlorophyll *b* at depth indicated that this bloom could have originated at the bottom during this period of generally low orthophosphate and ammonia concentrations in Elevenmile Creek.

Several conclusions can be drawn from these data. The primary plankton blooms in Perdido Bay started during the 1993–1994 drought at which time there was prolonged high orthophosphate loading by the mill to Elevenmile Creek (Table 4.1, Figure 4.30). No blooms were noted during the early years of net and whole-water sampling. Bloom initiation occurred during a major drought at which time turnover rates in the bay were low; water temperature was higher than usual; water clarity was high; and salinity in the upper bay was relatively high. During this period, orthophosphate concentration gradients at the mouth of Elevenmile Creek were consistently high. These orthophosphate gradients generally coincided with the increased mill loading and were located in areas where most of the blooms originated. In addition, there was evidence that some of the blooms (i.e., *Miraltia throndsenii* at various stations on 24 March 1997, 27 March 1997, 28 March 1997; *H. akashiwo* at station 25 on 27 March 1997; *Merismopedia tenuissima* at station 22 on 8 September 1999) started at or near the sediment/water interface. Some species (*Miraltia throndsenii*, *Synedropsis* sp., and *Cyclotella choctawhatchee*) showed baywide distributions from blooms that originated in the upper bay (Figure 4.30). Thus, although the most intensive bloom activity occurred in the upper bay (between stations P23 and P26), some of these blooms extended throughout the bay. There was evidence that some of the *Gymnodinium* blooms could have originated in Wolf Bay (i.e., parts of lower Perdido Bay affected by nutrient loading from postulated agricultural activities). There was no significant relationship among chlorophyll *a*, *b*, and *c* concentrations and phytoplankton bloom numbers (Table 4.1; $r^2 = -0.006$, $p = 0.4845$; $r^2 = -0.004$, $p = 0.5773$; $r^2 = -0.0006$, $p = 0.854$, respectively).

C. Bloom Species

High nutrient loading from various sources makes the estuarine habitat more favorable for algal growth and toxic effects (Burkholder et al., 1992). The various bloom species documented in the Perdido Bay system since 1994 (Table 4.1) represent a taxonomically diverse group dominated by diatoms, raphidophytes, dinoflagellates, and blue-green algae. The numerically dominant Perdido Bay phytoplankton bloom species have highly variable structural components (Figure 4.31). The chain-forming diatom *Leptocylindrus danicus* has relatively large cells (29.0 to 47.3 µm in length; biomass of 404 picog ash-free dry weight [AFDW] cell^{-1}). This species has numerous well-developed chloroplasts. The diatom *C. choctawhatcheeana* has relatively small cells (3.0 to 5.0 µm in length; 40 picog AFDW cell^{-1}) with few chloroplasts and little chlorophyll. The raphidophyte *H. akashiwo* has numerous chloroplasts. This species is relatively large (length, 8.8 to 12.5 µm, 207 picog AFDW cell^{-1}). The diatom *Miraltia throndsenii* has very small chloroplasts and not much chlorophyll. This species is also very small (length, 5.4 µm; 19 picog AFDW cell^{-1}). The diatom *Synedropsis* sp. has only two chloroplasts, whereas *C. choctawhatcheeana* more but not as many as *H. akashiwo* or *L. danicus*. *Synedropsis* sp. is somewhat larger (length, 20.0 to 23.1 µm; 26 picog AFDW cell^{-1}) than *C. choctawhatcheeana.* The dinoflagellate *P. minimum* is photosynthetic with relatively small chloroplasts. This species is relatively large (length, 17.5 to 22.0 µm; 518 picog AFDW cell^{-1}). The raphidophyte *Chattonella subsalsa* is also relatively large (length, 23 to 43 µm; 1933 picog AFDW cell^{-1}). The colony-forming blue-green alga *Merismopedia tenuissima* is common in stagnant eutrophic fresh waters, mainly fertilized fishponds. This species also occurs in brackish waters with some capacity to tolerate low salinities, particularly during the warm season of the year (Komarek

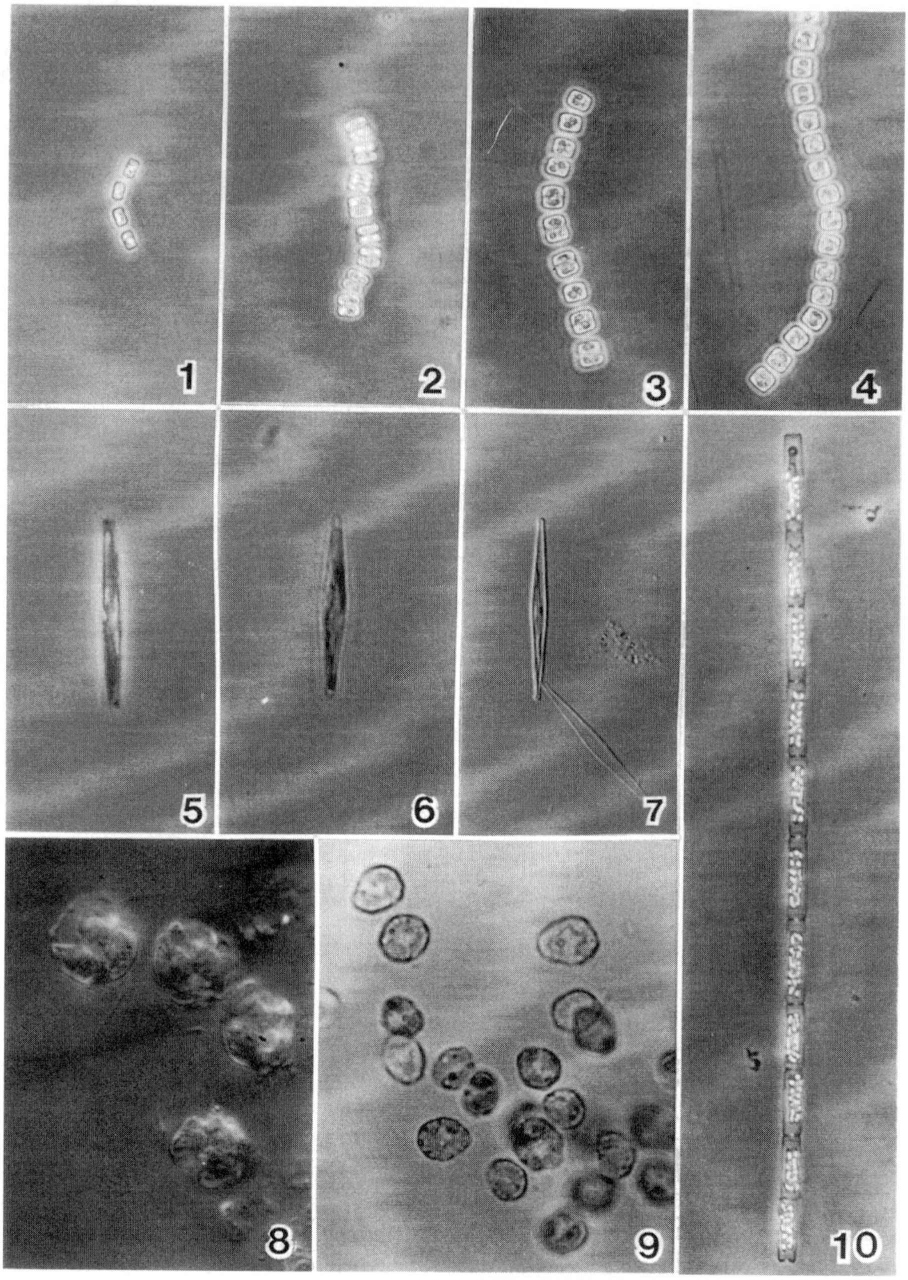

Figure 4.31 Pictures of species associated with blooms noted in the Perdido Bay system after January 1994. Species include the following: *Leptocylindrus danicus* Cleve, *Prorocentrum minimum* (Pavillard), *Miraltia throndsenii* Marino et al., *Cyclotella choctawhatcheeana* Prasad, *Synedropsis* sp. nov., *Chattonella* cf. *subsalsa*, and *Heterosigma akashivo*. Hurlburt. (Plates I to VI figures were developed by Dr. A. K. S. K. Prasad in coordination with the Scanning Electron Microscope Unit at Florida State University, Tallahassee.) Plate I (1–10) Bloom-forming phytoplankton species of Perdido Bay, Florida. All photomicrographs, except 7, taken from Lugol's-preserved materials. (1–4) Chain-forming centric diatom, *Cyclotella choctawhatcheeana* Prasad. Original magnification ×650. (5–7) The pennate diatom, *Synedropsis* sp. (new species). (7) Acid-cleaned material. (8, 9) The raphidophyte flagellate, *Heterosigma carterae* (Hurlburt) Taylor. Note the flagellated cells in 8 and the non-motile cells in 9 and 10. Chain-forming centric diatom, *Leptocylindrus danicus* Cleve. Original magnification ×650.

Figure 4.31 Plate II (1–7) Scanning electronmicrographs of *Cyclotella choctawhatcheeana* Prasad. 1. Lugol's preserved. Chain formation. Newly formed sibling cells embedded in mucus. (2–7) Acid-cleaned. (2–4) Exterior of the valve showing marginal zone of radial striae and the tangentially undulated central area. (5–7) Interior of the valve showing a ring of marginal strutted processes (fultoportulae), a single labiate process (rimoportula) and a subcentral strutted process. Note the presence of two satellite pores around marginal processes and three pores on the subcentral ones.

and Anagnostidis, 1998). However, the small size of the cells (1 to 2 μm) makes it difficult to determine viability; therefore, *M. tenuissima* was reported primarily in the freshwater entry points to upper Perdido Bay. The fact that there is considerable variability among the various bloom species with respect to size, biomass, and chloroplast development could be related to the lack of a statistical relationship between the chlorophyll observations and the incidence and intensity of the blooms in Perdido Bay (Table 4.1).

The diatom *L. danicus* has a worldwide distribution in a variety of tropical coastal environments (Hargraves, 1990). It is often a major constituent of spring phytoplankton outbursts as well as fall growing periods (Marshall, 1988). This species has been associated with red tide events in Japan (Fukuyo et al., 1990). According to Fryxell and Villac (1999), *L. danicus* is believed to be harmful to caged sea trout (*Cynoscion regalis*), Atlantic salmon (*Salmo salar*), and smolt of coho salmon (*Oncothuynchus kisutch*) (Clement and Lembeye, 1993). Another diatom, *Miraltia throndsenii*

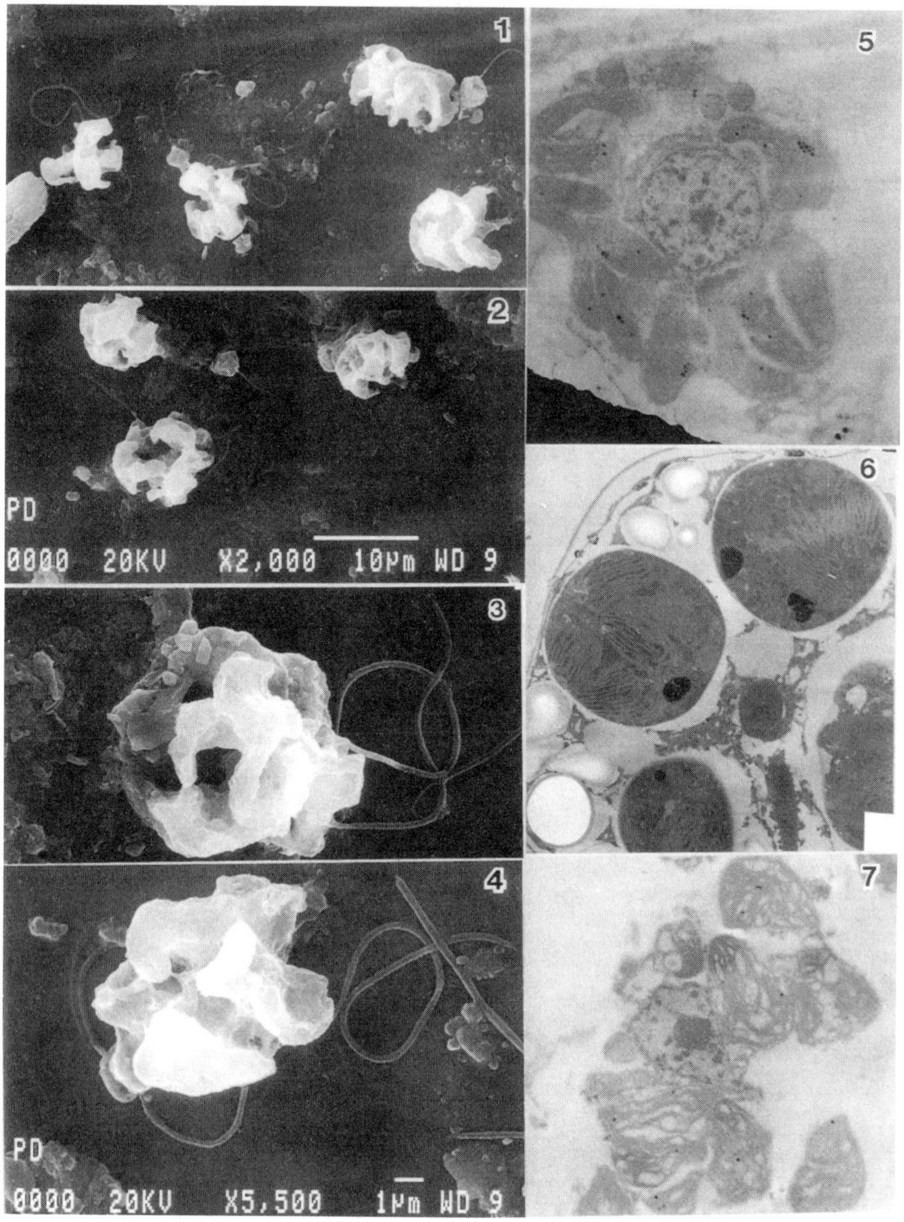

Figure 4.31 Plate III (1–7) The raphidophyte flagellate, *Heterosigma akashiwo* (Hurlburt) Taylor. (1–4) Lugol's-preserved material. Scanning electronmicrographs of air-dried material. Note the bi-flagellate condition in 3 and 4. (5–7) Transmission electronmicrographs of ultra thin sections showing the cellular organization. Note the presence of many chloroplasts surrounding the central nucleus in 4. Material fixed in glutaraldehyde/Osmium tetraoxide and embedded in Spurr's resin. (5 and 7) Original magnification ×6500. (6) Original magnification ×10000.

Marino, Montresor & Zingone (Marino et al., 1987) has been associated with blooms stimulated by effluents from sewage and industrial outfalls (Marino et al., 1987). *Miraltia throndsenii* was shown to be congeneric with *Chaetoceros ehrenberg* on the basis of fine structure of vegetative cell and resting spore morphology and the resulting combination was *C. throndsenii* (Marino, Montresor & Zingone) Marino Montresor & Zingone (Marino et al., 1992). According to Livingston and Prasad (in manuscript), the diatom *Cyclotella choctawhatcheeana* responds to a combination

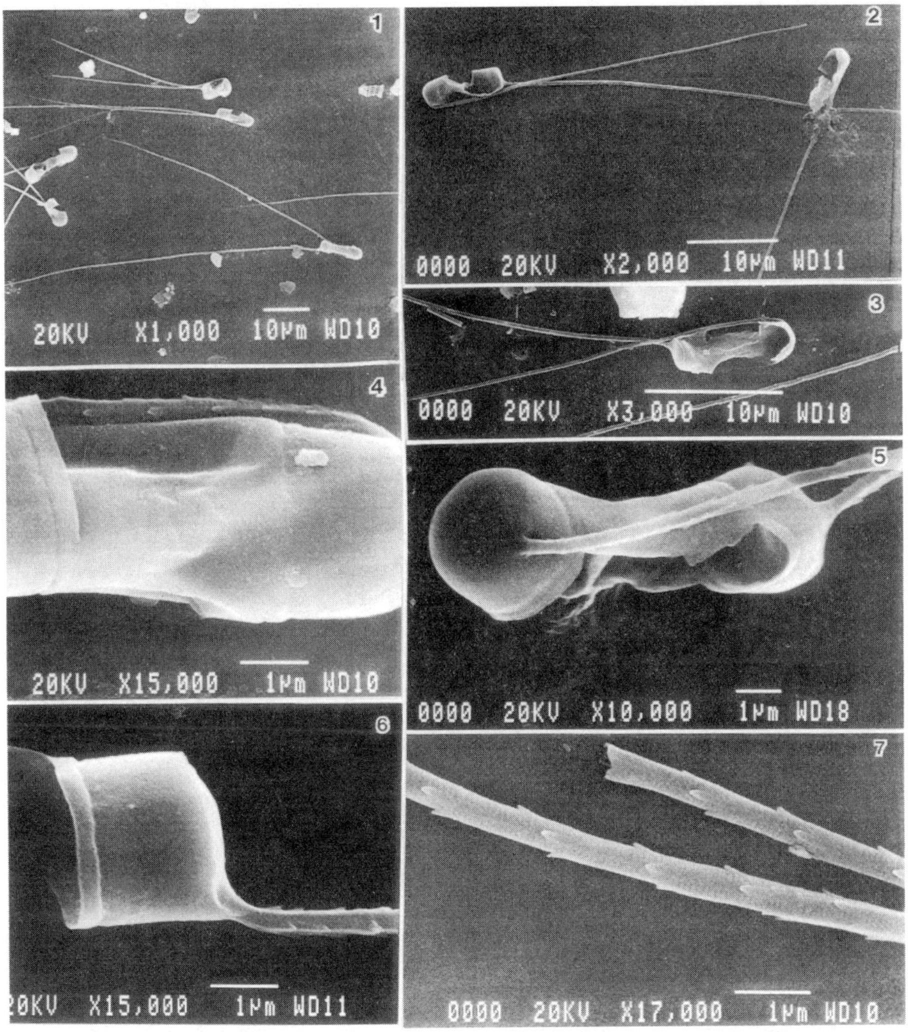

Figure 4.31 Plate IV (1–7) Scanning electronmicrographs of acid-cleaned material of centric diatom, *Miraltia throndsenii* Marino et al. from Perdido Bay, Florida. Note the single seta from one valve curved and running parallel to that of the other valve.

of environmental factors that range along inshore–offshore gradients of light and nutrients with salinity as a possible modifying factor. This species was first described from samples taken in the Choctawhatchee Bay system (Prasad et al., 1990) where short chains of *C. choctawhatcheeana* were found during each month of the year. Small blooms were observed at some stations during November and December with one such bloom appearing in April and persisting until June. Salinity ranges for *C. choctawhatcheeana* in Choctawhatchee Bay were >20‰ at temperatures >25°C. Numbers decreased sharply as temperatures climbed to 30°C. Optimum salinities ranged from 15 to 20‰. This distribution indicated that *C. choctawhatcheeana* can survive and grow under varying salinity conditions. Carvalho et al. (1995) explained the noted differences in salinity preferences of this diatom as a product of possible competition that may have forced *C. choctawhatcheeana* into lower salinities. Cooper (1995a, b), working in Chesapeake Bay, found that *C. choctawhatcheeana* had increased abundances that coincided with increased sedimentation, turbidity, and eutrophication of the estuary. Overall, this species has a high tolerance for wide ranges of salinity and is stimulated by increased nutrient loading. It has been suggested that *C. choctawhatcheeana* may be

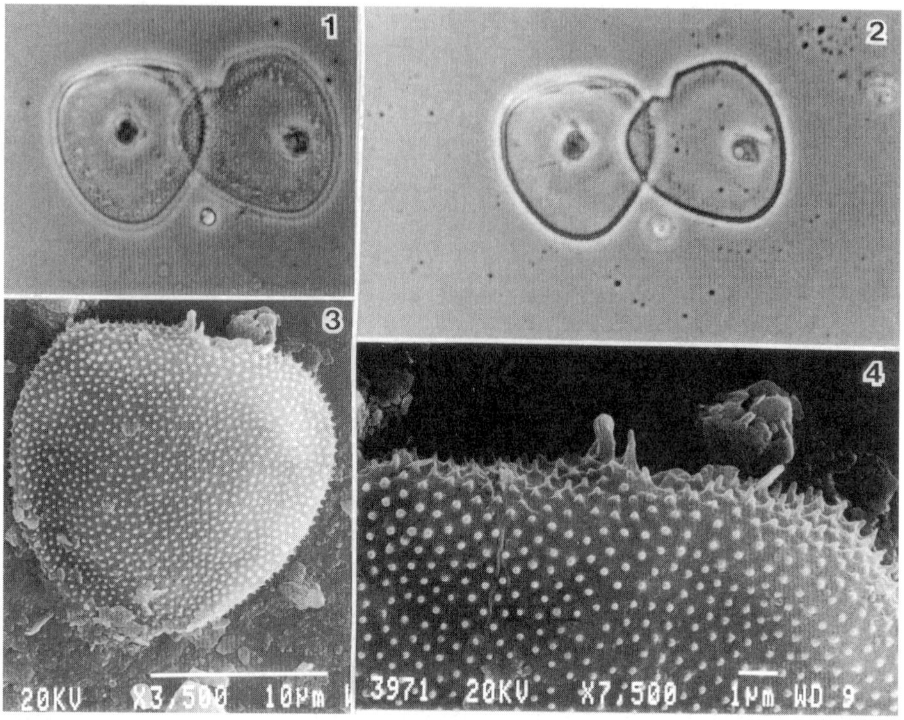

Figure 4.31 Plate V (1–4) The dinoflagellate, *Prorocentrum minimum* (Pavillard) Schiller from Perdido Bay, Florida. Lugol's preserved material (1 and 2). Photomicrographs focused on the terminal spine and the spinules on the surface of the valve. (3 and 4). Scanning electronmicrographs.

a good indicator of estuarine habitats that are characterized by various anthropogenous effects (Cooper, 1995a).

Between 1965 and 1976, the number of confirmed worldwide red tide outbreaks (raphidophytes, dinoflagellates) increased sevenfold concurrent with a twofold increase in nutrient loading mainly from untreated sewage and industrial waste (Hallegreaff, 1995). During the last two decades, there have been many red tide blooms along various coasts, especially in gulfs and inland seas. Among the most notable of the many kinds of flagellates that have been recognized as components of these blooms is the species *H. akashiwo* which frequently causes heavy and extensive red tide events (Hara and Chihara, 1987). During recent years, considerable attention has been focused on species of the genus *Heterosigma* as the cause of fish kills in New Zealand, Chile, and British Columbia (Chang et al., 1990). Smayda (1997c) detailed the ecophysiology and bloom dynamics of *H. akashiwo* which is well known for ichthyotoxic blooms (Honjo, 1993). Honjo (1993) found that *H. akashiwo* blooms were associated with river runoff, low bottom oxygen, and wind-induced turbulence of bottom sediments. Growth occurs from overwintering motile forms and/or germinating resting spores. This species has a high growth potential and can form a red tide in a short time. It excretes allelopathic substances that can suppress diatom growth and may dramatically reduce cell numbers in other phytoplankton species. In addition to making vertical migrations, *H. akashiwo* can explore diverse spatial habitats in search of favorable growth conditions. Nutrient limitation stimulates migration tactics. There are relatively high nutrient requirements for *H. akashiwo*. Bactivorous feeding is stimulated by P-limitation with Fe and Mn needed for bloom continuance. Nutrient enrichment supports blooms of this species with dependence on river flows in various regions. Temperature, nutrients, and competition appear to be limiting for bloom generation in *H. akashiwo*. Smayda (1997a, b) indicated multifactorial control of blooms which, once released,

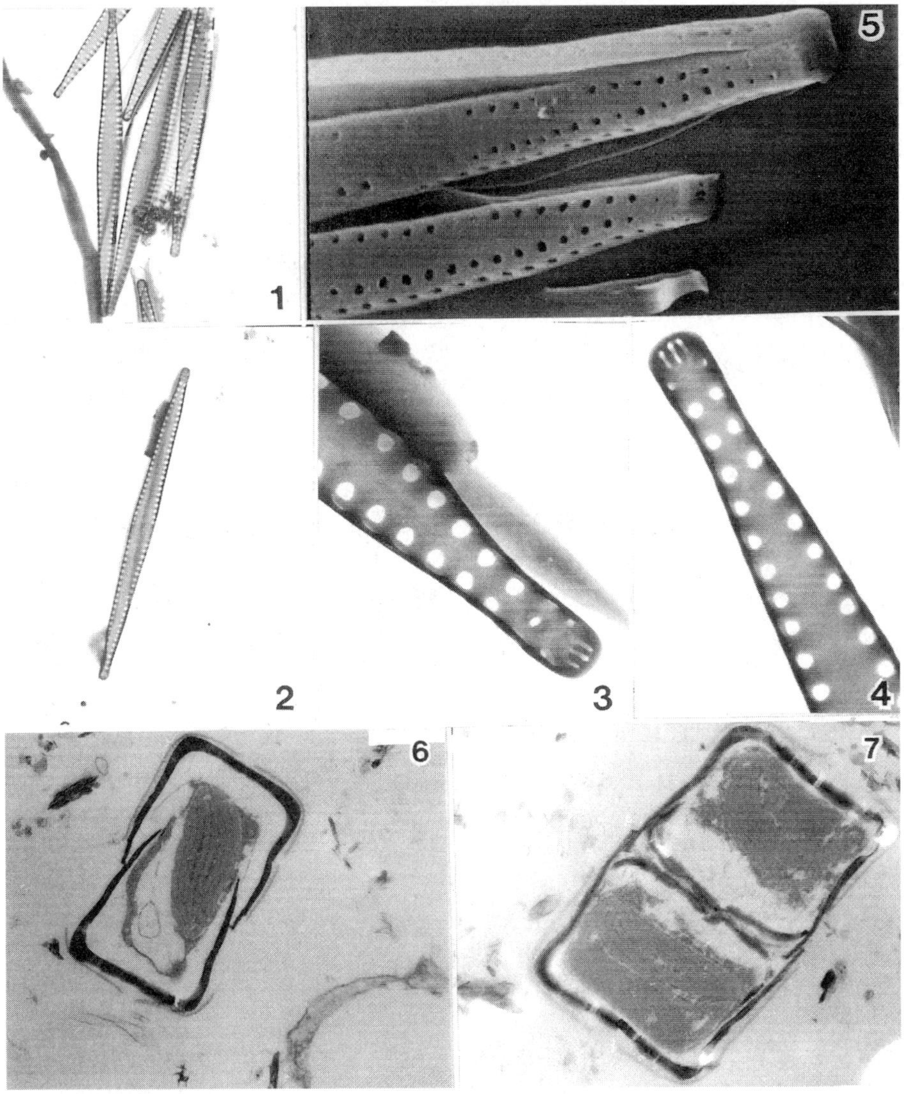

Figure 4.31 Plate VI (1–7) The pennate diatom, *Synedropsis* sp. (1–4) Acid-cleaned material. (1–4) Transmission electronmicrographs of valves showing marginal rows of areolae, slit fields on the polar mantles and labiate process. (5) Scanning electronmicrograph of the valve apices. Note single rows of areolae on the valve mantle. (1 and 2) Original magnification ×3000. (3 and 4) Original magnification ×20000. (5) Original magnification ×15000. (6 and 7) Original magnification ×25000. (6 and 7) Transmission electronmicrographs of ultrathin sections of single cell (5) and the recently formed sibling cells (7), showing the cellular organization.

have a "remarkably high degree of broad spectrum allelochemical allelopathic antagonisms" that are used against competing species and potential grazers.

The fish-killing mechanism of raphidophyte blooms is still poorly understood. Both physical clogging of gills by mucus excretion as well as gill damage by hemolytic substances such as polyunsaturated fatty acids may be involved (Shimada et al., 1983; Chang et al., 1990). There is accumulating evidence that the production of superoxide radicals represents the primary mechanism of fish mortality. Imai et al. (1997) reported details concerning life cycle and bloom dynamics of *Chattonella*, a genus with two known fish-killing species (*C. antqua* and *C. marina*). Nutrients and competitors (mainly diatoms) appear to affect the development of *Chattonella* populations. Anything

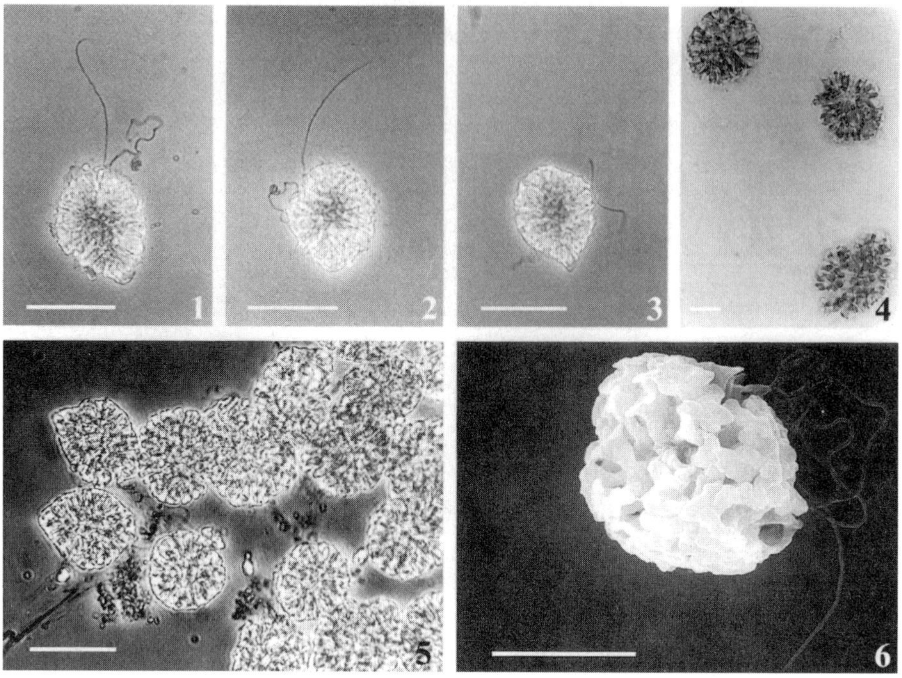

Figure 4.31 Plate VII (1–6) Autotrophic, bloom-forming Raphidophyte, *Chattonella* cf. *subsalsa* from Perdido Bay, Florida. Lugols preserved. (1–3) Light microscope (LM) images of cells with two flagella. Differential interference contrast (DIC). (4) LM. Brightfield illumination. Three cells showing numerous plastids. (5) LM. Phase contrast optics. (6) Scanning electron microscope (SEM) image of air-dried flagellate cell.

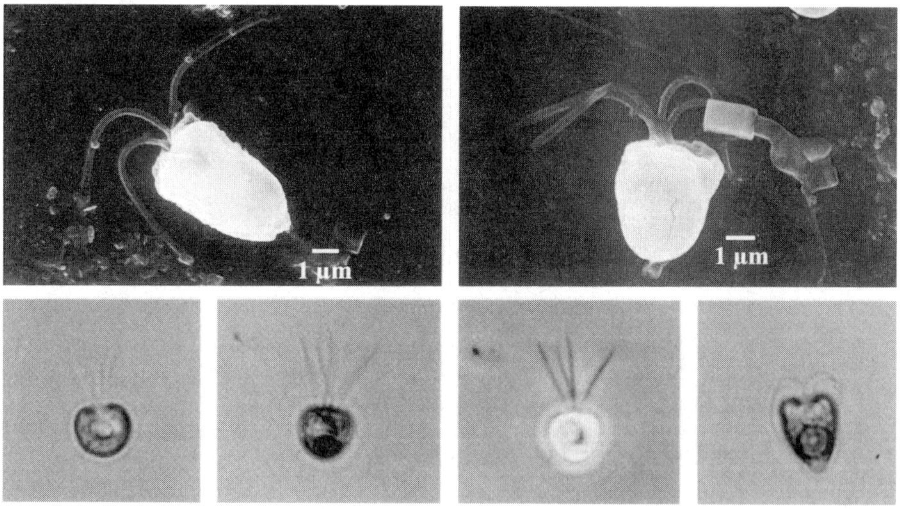

Figure 4.31 Plate VIII. 1-6. Autotrophic, undetermined prasinophytes (*Pyraminonas* and *Pyramimonas*-like forms from Perdido Bay, Florida. Lugol's preserved. (1,2) SEM. Air-dried material. (3–6) Brightfield illumination. (5) Phase contrast optics.

interfering with diatom proliferation could give *Chattonella* an advantage. During a severe phytoplankton outbreak in the Seta Sea in 1972, a raphidophyte red tide killed 14 million cultured yellow tail fish. Effluent controls were then initiated to reduce the organic carbon loading and the discharge of phosphates from household detergents. Following a time lag of 4 years, the frequency

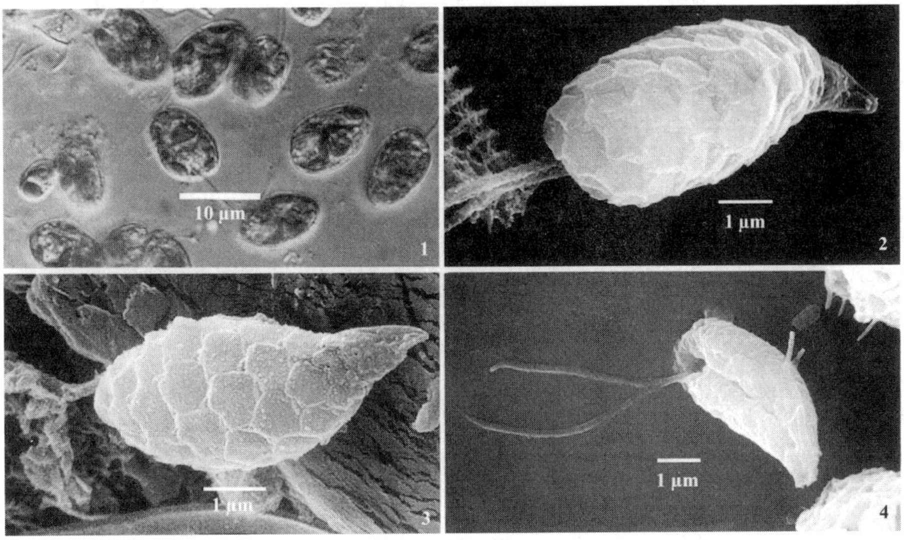

Figure 4.31 Plate IX (1–4) Autotrophic, undetermined cryptophytes from Perdido Bay, Florida. Lugol's preserved. (1) *Cryptomonas* sp. DIC. (2–4) Undetermined cryptophytes with affinities to members of *Plagioselmis*. SEM. (2,3) Dorsal view of the flagellate with hexagonal plates of the cell covering (pellicle). (4) Ventral view showing the insertion of subequal flagella.

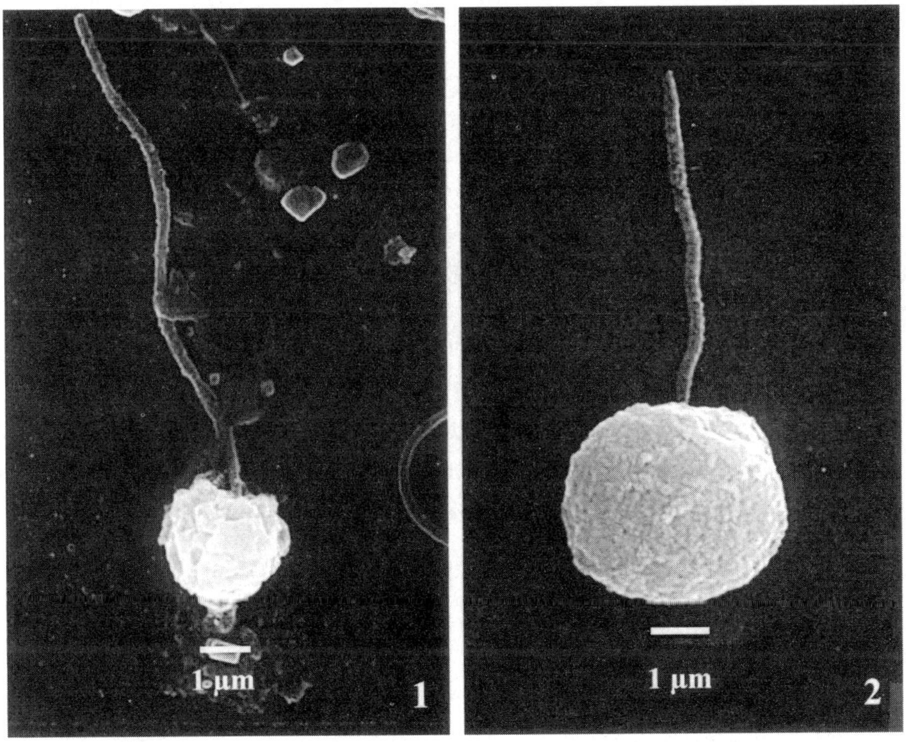

Figure 4.31 Plate X (1–2) Undetermined nanoflagellates from Perdido Bay, Florida. Lugol's preserved and air-dried material. SEM.

of red tide events in the Seta Sea then decreased by about twofold to a more stable level (Hallegreaff, 1995). A similar pattern of long-term loading of coastal waters was evident for the North Sea in Europe (Smayda, 1990). Since 1955, the phosphorus loading of the River Rhine has increased

7.5-fold, and nitrogen levels have also increased. This has resulted in a significant sixfold decline in Si:P ratios because long-term reactive silicate concentrations (a nutrient derived from natural land weathering) have remained constant. There is considerable concern that altered nutrient ratios in coastal waters may favor blooms of nuisance flagellate species which replace the normal spring and autumn blooms of siliceous diatoms (Smayda, 1990).

The dinoflagellate *P. minimum* is a bloom species that is mostly estuarine and cosmopolitan in cold temperate to tropical waters. It is a toxic species associated with postulated shellfish poisoning and fish kills. Nakazima (1965) indicated poisonous effects on shellfish feeding on *Prorocentrum* sp. The toxic substance venerupin has been associated with *P. minimum* although there is some question concerning this association. Hansen (1997) reported that *P. minimum* can ingest certain cryptophytes (*Cryptomonas* sp.) and ciliates. Mixotrophy (being photosynthetic and phagotrophic) is widespread among dinoflagellates with a variety of feeding mechanisms found in this group. The mechanisms of prey selectivity are not well understood but such selectivity is not by size alone. The control of plankton organisms by phagotrophy is considered to be highly variable (Granelli and Carlsson, 1997). However, losses of plankton to pigmented flagellates must be considered according to these authors. Sellner et al. (1995) indicated that *P. minimum* did not adversely affect oysters (*Crassostrea virginica*) in Chesapeake Bay, which effectively reduced these bloom-forming dinoflagellates. However, Lassus and Berthome (1988) reported that *P. minimum* caused mortalities in old oysters. Woelke (1961) found that this species caused oyster (*Ostrea iurida*) mortalities and cessation of feeding at high densities. Wikfors and Smolowitz (1993) found that increased abundance of *P. minimum* could cause shell losses. This observation was based on feeding experiments where toxins from this species caused slow growth and mortality in shellfish. Okaichi and Imatomi (1979) reported that in 1942 shellfish poisoning by *P. minimum* killed 114 out of 324 human victims of poisoned shellfish. The authors isolated the toxic substances believed operational in those deaths. Cardwell et al. (1979) found that *Gymnodinium splendens*, another dinoflagellate, was acutely toxic to oyster larval stages.

D. Long-Term Changes at the Division/Class Level

The long-term, species-specific phytoplankton data (whole-water) were grouped by division and class. Abundances of the major divisions or classes of these microalgae (following van den Hoek et al., 1995) were expressed as percentages of the total number taken by station (Figure 4.32). A comparison was made between the period of relatively low nutrient loading from Elevenmile Creek (November 1988 to January 1991) and the period of high phosphorus and/or ammonia loading (July 1996 to September 1999).

The class Bacillariophyceae (diatoms) are usually more diverse in areas characterized by relatively high nutrient levels. Their silica-based cell wall is unique. In the Perdido system, the diatoms were dominant at most stations during the first 2 years of study. During certain periods, relatively low percentages of diatoms were noted primarily in the lower Perdido River (station P18) and lower Elevenmile Creek (station P22). From 1996 to 1999, diatom dominance dropped precipitously, especially in the upper bay. There was some dominance by the diatoms in the lower bay during this period, although such dominance was considerably lower than that during the first 2 years of sampling. There was a period of increased diatom dominance during winter–spring and fall 1998, the period of low orthophosphate loading by the pulp mill. A second brief increase of diatom dominance was noted during early 1999, another period of low orthophosphate and ammonia loading by the mill. However, by spring 1999, diatom percentage abundance was again extremely low with the resumption of high orthophosphate loading by the mill. In general, the relative abundance of diatoms was inversely related to nutrient loading from Elevenmile Creek.

The division Dinophyta (dinoflagellates) is primarily marine. Dinoflagellates are flagellated, with extensive horns, ridges, and wings. About half of the dinoflagellate species lack chlorophyll, and are obligate phagotrophs (van den Hoek et al., 1995). Some dinoflagellates are facultatively

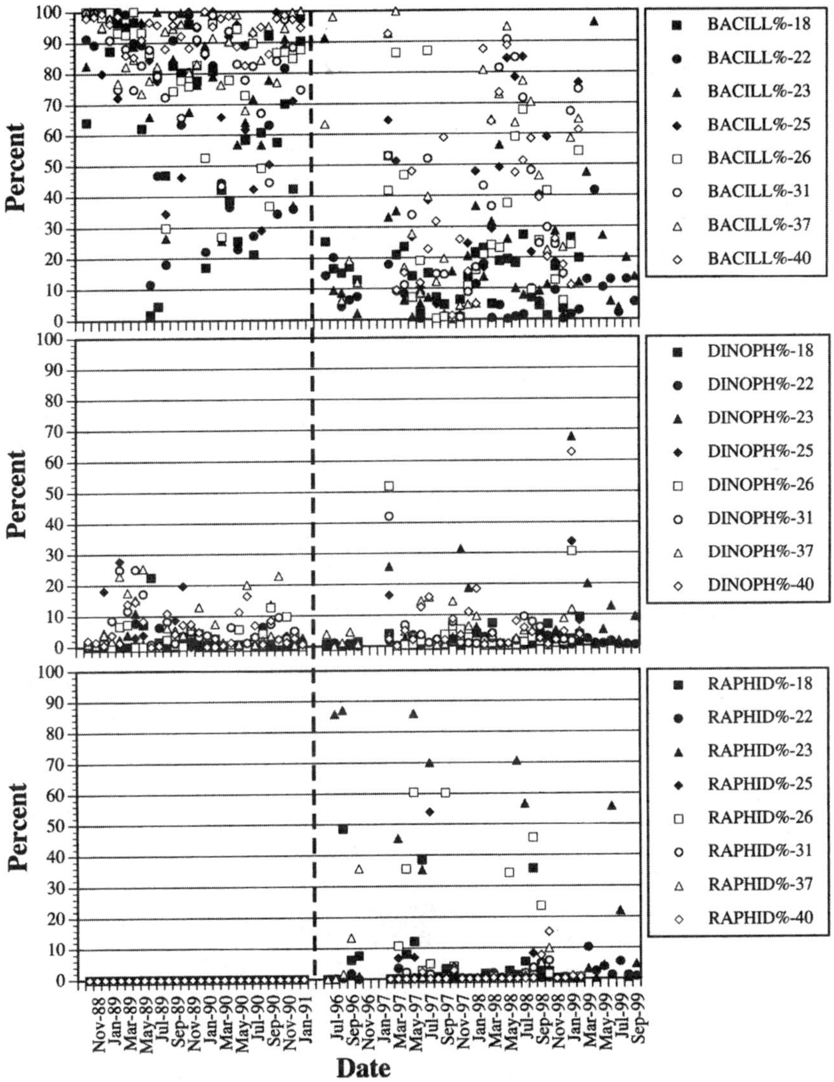

Figure 4.32 Percentage counts of phytoplankton (grouped by division or class) taken as whole-water samples from eight stations in Perdido Bay. Comparisons were made between collections taken from 11/88 to 1/91 and collections taken from 7/96 to 9/99. The following abbreviations were used: Class Bacillariophyceae = Bacill; division Dinophyta = Dinoph; class Raphidophyceae = Raphid; class Chrysophyceae = Chryso; division Cryptophyta = Chrypto; division Chlorophyta = Chloro; division Cyanophyta = Cyanop; division Euglenophyta = Euglen.

heterotrophic, and are able to switch from autotrophy to heterotrophy in response to ambient environmental conditions. Some dinoflagellates are considered bloom species with various forms being toxic. Dinoflagellate percentages were generally low during the first 2 years of sampling (Figure 4.32). During the last 3 years of sampling, there were increases in the percentage composition of the dinoflagellates in Perdido Bay, usually during winter periods due to increased numbers of *P. minimum*. Dinoflagellates were most dominant at stations P23, P26, and P25 in the upper bay. Species of the class Raphidophyceae (the raphidophytes) are often quite large in size (50 to 100 μm), and are represented by a variety of bloom species that occur in coastal waters. The raphidophytes were generally absent during the first 2 years of sampling (Figure 4.32). However, from 1996 to

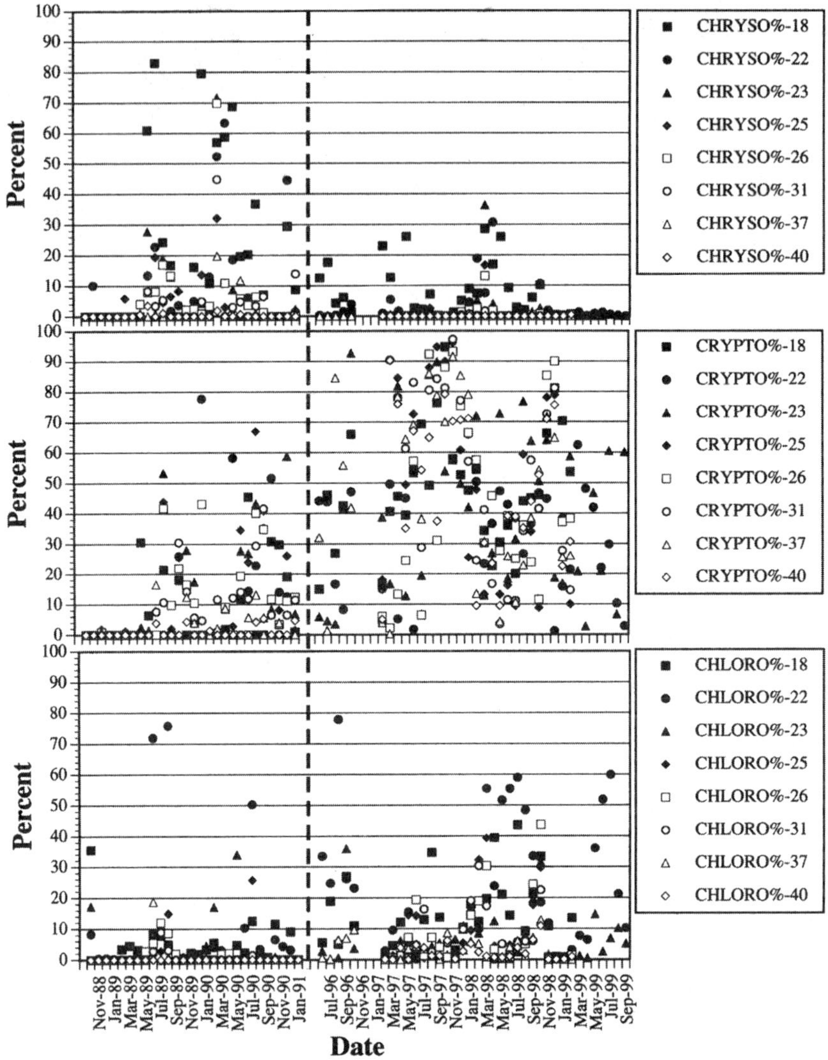

Figure 4.32 (continued)

1999, this group was dominant during warm months with major concentrations occurring at stations 18, 23, and 26 in the upper bay. Raphidophyte blooms were generally associated with *H. akashiwo* and *Chattonella subsalsa* (June 1999).

The class Chrysophyceae (the golden brown algae) has relatively few marine species, and individuals appear to prefer oligotrophic waters where they are good competitors for phosphorus (van den Hoek et al., 1995). In Perdido Bay, the chrysophytes were dominant largely in the upper bay during the period from 1989 through 1990 (Figure 4.32). Chrysophyte dominance decreased to relatively low levels during the last 3 years of sampling with some return to dominance in the upper bay during late 1997 and early 1998 (i.e., the period of reduced orthophosphate loading by the mill). Extremely low percentages of chrysophytes were present during the period of renewed nutrient loading in 1999. The cryptophytes (division Cryptophyta) are flagellated microalgae that are known to be bloom-forming. During the first 2 years of sampling, the cryptophytes had minimal dominance with some increase during 1990, particularly in Elevenmile Creek and the upper bay (Figure 4.32). However, during the last 3 years, the cryptophytes became highly dominant throughout the entire bay. Peak abundance occurred during 1997. There was reduced percent dominance

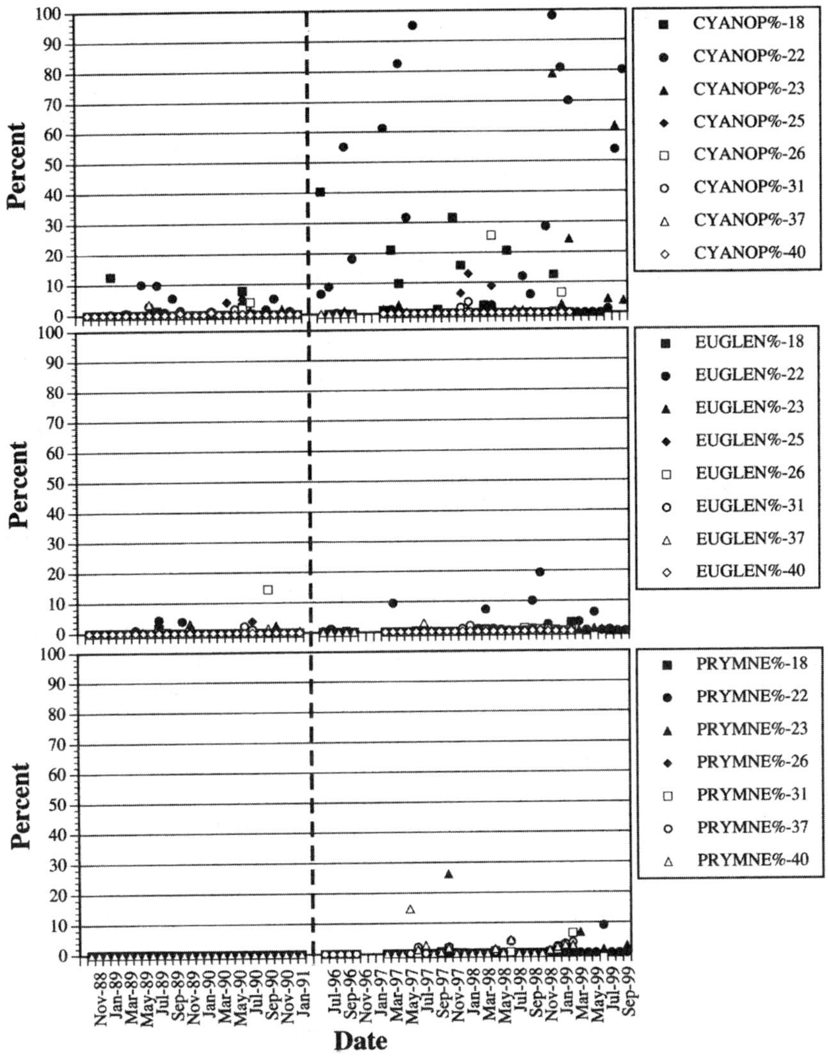

Figure 4.32 (continued)

of this group during the period of low phosphorus loading in 1998. Cryptophyte relative abundance tended to be the mirror image of the chrysophytes in Perdido Bay over the period of study, with both groups appearing to be good indicators (positive and negative) of nutrient loading rates in upper Perdido Bay.

The division Chlorophyta (green algae) (Figure 4.32) is characterized by carotenoids as well as chlorophyll *a* and *b*. This group was occasionally dominant in Elevenmile Creek during periods of low orthophosphate loading (1989–1990). There was a general increase of chlorophyte dominance in Elevenmile Creek during spring–summer months of 1998 (i.e., periods of high ammonia and low orthophosphate loading). The division Cyanophyta (cyanobacteria or blue-green algae) is represented by some marine forms that are often members of the picoplankton (0.2 to 2.0 μm). Many of these forms can grow well at low light levels. In Perdido Bay, the blue-green algae were not well sampled and identified because of their small size; this group was represented mainly in Elevenmile Creek and occasionally in the bay during 1996, late 1998, and summer 1999 (Figure 4.32). The euglenophytes and pyaminesiphytes were not well represented in Perdido Bay and euglenoid species were present only occasionally in the bay.

Overall, high nutrient loading tended to favor the cryptophytes, cyanophytes, dinoflagellates, and raphidophytes. These groups are well represented by prominent bloom species in coastal and marine waters. During periods of high winter orthophosphate loading, the dinoflagellates were numerically dominant, whereas summer ammonia loading was associated with increased dominance by the raphidophytes. The chrysophytes, and to a lesser degree, the chlorophytes, tended to be more abundant during periods of low nutrient loading. The diatoms were good indicators of the long-term nutrient loading patterns in Elevenmile Creek with reduced percent abundance during periods of high nutrient loading and associated dinoflagellate and raphidophyte blooms.

E. Silica and Diatom Abundance

Silica input to coastal systems comes primarily from weathering of upland soils. Eutrophication of an estuary can lead to reduced silica due to incorporation by diatoms and eventual storage in sediments (Howarth et al., 2000). Reduced silica, if accompanied by increased nitrogen, can also lead to formation of harmful algae if diatoms are reduced by silica limitation and associated competition is reduced (Rabalais et al., 1996). Smayda (1989) indicated that decreased silica availability relative to nitrogen or phosphorus has been correlated with increased incidence of harmful algal blooms. Reduced diatom abundance could also lower sedimentation and therefore counteract the effects of cultural eutrophication.

The spatial distribution of silica in Perdido Bay showed distinct longitudinal (i.e., north–south) gradients (Figure 4.33) with the highest concentrations in Elevenmile Creek and the lowest concentrations in the lower bay. There were also temporal changes of silica concentrations in the bay with reductions from 1993 through 1997 that paralleled the orthophosphate loading trends from Elevenmile Creek. The relatively stable seasonal trends during the period 1988 to 1991 gave way to reductions of silica during 1992 with precipitous decreases occurring during 1993. After a brief period of higher silica during the late fall of 1993, there was a sustained series of low silica concentrations throughout the bay that were not related to silica loading in the upper bay (Figure 4.4). During the period of reduced orthophosphate loading from late 1997 to spring 1999, there were increasing summer–fall silica peaks with relatively high levels of silica in the bay during fall 1998. By fall 1999, after resumption of spring and summer orthophosphate loading by the mill, the silica levels again dropped. This was accompanied by associated diatom (*Cyclotella choctawhatcheeana*) blooms throughout the bay.

Diatom numbers tended to increase during late winter-spring periods; these numbers varied inversely ($p = 0.05$) with the silica concentrations during the first 3 years of study (Figure 4.34). There were considerable increases of the diatom bloom species *C. choctawhatcheeana* during the period 1993 to 1996 that coincided generally with the relatively low silica concentrations in the bay. This pattern was repeated during 1999. It is likely that this nonconservative nutrient was taken up by phytoplankton with increasing reductions as surface water moved south. During the period of reduced orthophosphate loading from late summer 1997 through spring 1999, there were complex interactions of diatom numbers with silica concentrations (Figure 4.34). Silica remained low during the period of increasing diatom numbers during fall–winter of 1997 and spring 1998. The spring diatom peak occurred during reduced silica levels and the reductions of diatom numbers during succeeding summer–fall months were associated with increasing silica concentrations that peaked during fall 1998. As diatom numbers increased during the winter–spring 1999 period, with a major peak during April, silica concentrations declined. There was some subsequent recovery of the silica concentrations, although silica values remained relatively low during the summer–fall of 1999 as diatom numbers decreased.

The long-term changes of silica concentrations in Perdido Bay appeared to have an inverse relationship with diatom distribution in the bay. Diatom blooms were generally associated with

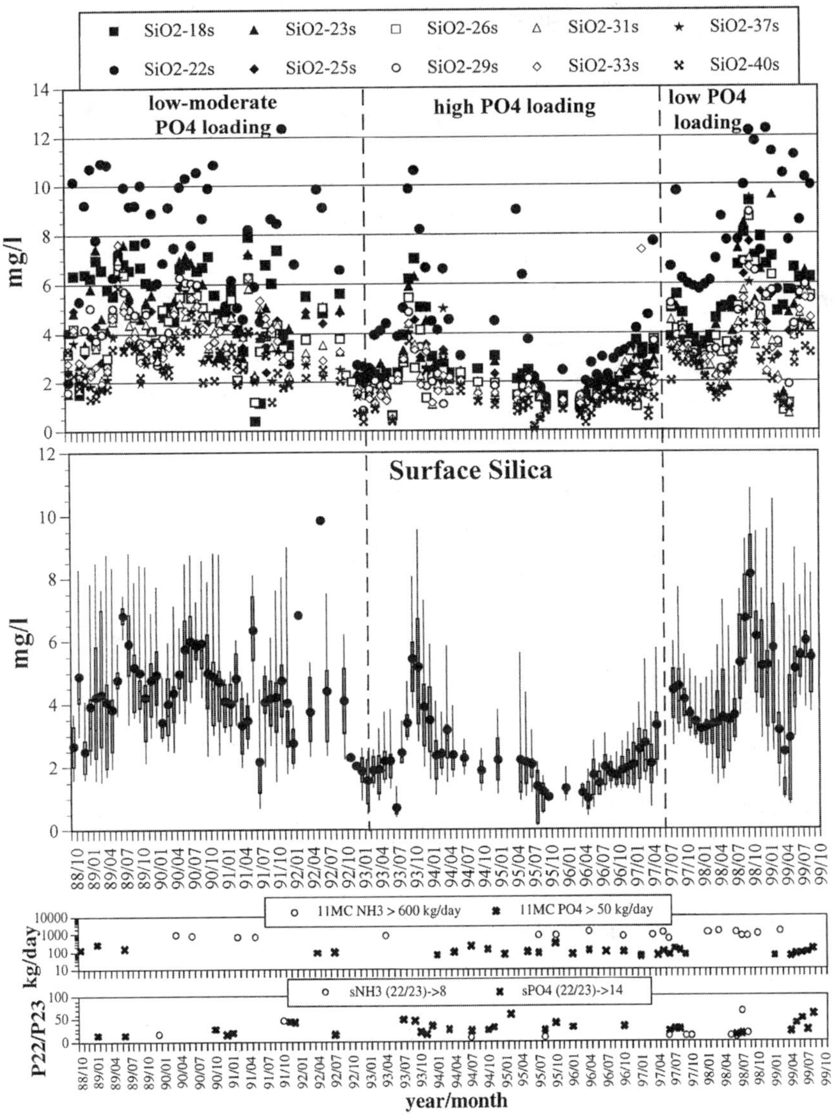

Figure 4.33 Silica concentrations (mg l^{-1}) in Perdido Bay taken monthly from October 1988 through September 1999. Station-specific and averaged data are shown and compared to mill loads of ammonia (>600 kg d^{-1}) and orthophosphate (>50 kg d^{-1}). Comparisons were also made of silica concentrations and ratios that exceeded 8 for ammonia and 14 for orthophosphate. Data are presented as means of all stations with the 10th, 25th, 75th, and 90th percentiles.

lower concentrations of silica throughout the bay. The long-term silica trends were superimposed over the longitudinal gradients of decreased silica with distance from the river sources. Interannual changes of silica concentrations in space and time thus reflected diatom activity that, in turn, tended to follow nutrient loading trends and possible interactions with other bloom species. It is possible that reduction of silica could have eventually become limiting to blooms of some of the diatom species. It is also possible that silica reduction could have contributed to the long-term succession from early diatom dominance in 1993–1994 to increasing dominance of the raphidophytes and dinoflagellates in succeeding years.

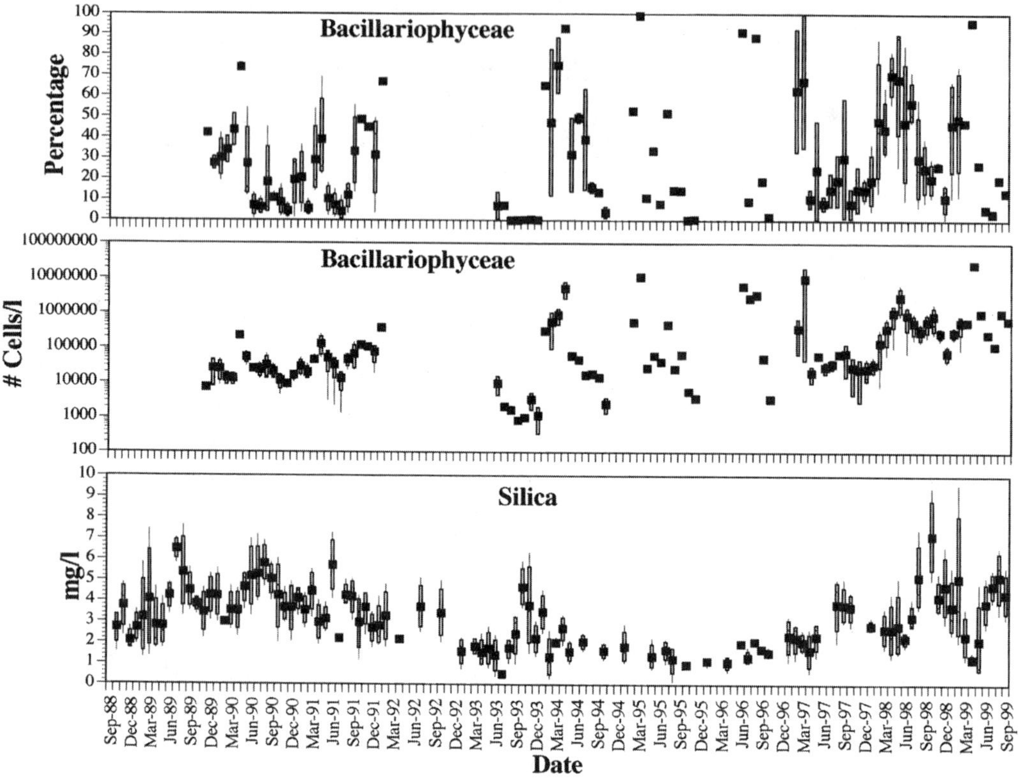

Figure 4.34(a) Total numbers of cells l^{-1} of diatoms (Bacillariophyceae) and percentages of total diatom numbers relative to mean silica concentrations taken over the 11-year study period. Data are presented as means of all stations with the 10th, 25th, 75th, and 90th percentiles.

F. Net Phytoplankton and Zooplankton

A comparison was made of the net phytoplankton taken during the years of relatively low nutrient loading (1988 to 1991) and the period characterized by the initiation of consistently high orthophosphate loading (1993 to 1994) (Figure 4.35; Table 4.2). During the drought of 1993–1994, there was a series of phytoplankton blooms dominated by diatom species. When compared with the phytoplankton taken during the drought of 1988–1989, there were increased phytoplankton numbers and corresponding reductions of phytoplankton species richness at all stations except in seagrass areas of upper Perdido Bay (Figure 4.35). Numerical abundance of the net phytoplankton was significantly higher in 1993–1994 during all months. Species richness was significantly lower in 1994 in 7 out of the 9 months analyzed. There was an orderly progression of the relative abundance of various bloom species. During fall 1993, *Falcula hyalina* was dominant. This species was followed by *L. danicus* blooms in January–February 1994. During March 1994, *M. throndsenii* was dominant. There were extensive blooms of *C. choctawhatcheeana* during April 1994. With the exception of April 1994, orthophosphate concentration increases at station P23 (receiving runoff from Elevenmile Creek) and associated blooms were not followed by substantial increases in chlorophyll *a* concentrations. The only significant difference of chlorophyll *a* occurred in January and was actually higher during the nonbloom year (1989). The relatively high chlorophyll *a* concentration at station P23 in April 1994 (Figure 4.35) was the only sign of bloom activity throughout the study period. The pH of the water was higher during 1993–1994 than 1988–1989. These data suggest that there was an orderly succession of phytoplankton blooms that co-occurred

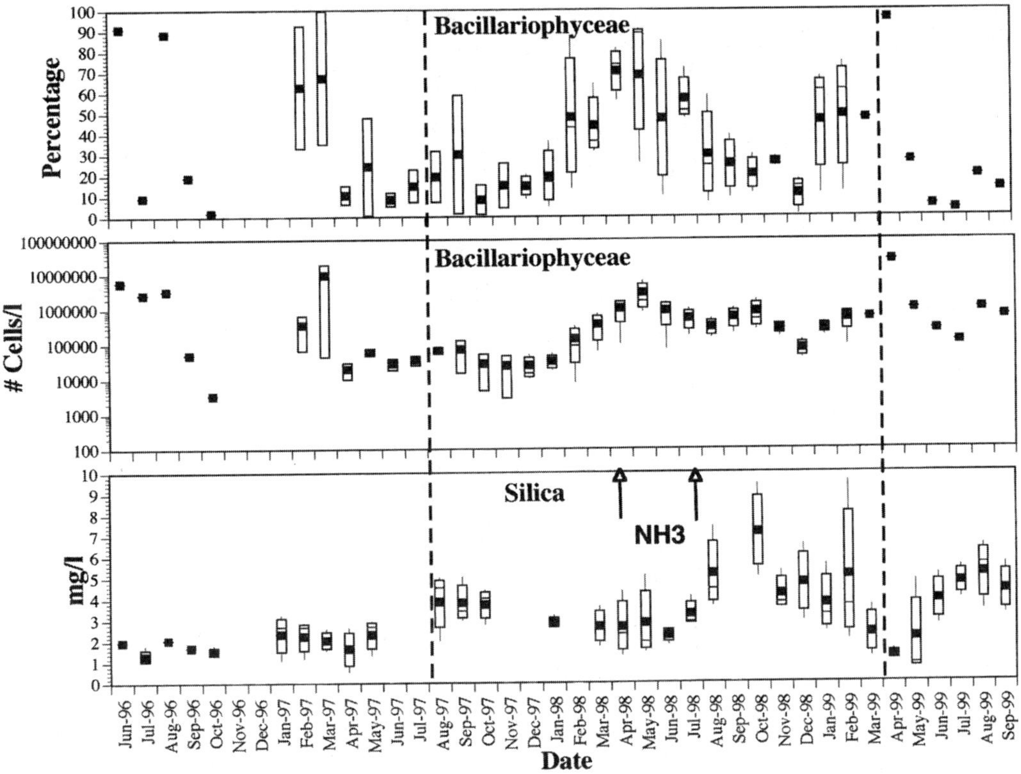

Figure 4.34(b) Total numbers of cells l^{-1} of diatoms (Bacillariophyceae) and percentages of total diatom numbers relative to mean silica concentrations taken over the period from June 1996 through September 1999. Data are presented as means of all stations with the 10th, 25th, 75th, and 90th percentiles.

with the 1993–1994 drought conditions and was loosely associated with sustained increases of orthophosphate loading by the pulp mill.

Analysis of the zooplankton (Figure 4.36) indicated that, with the exception of February 1994, increased numbers and reduced species richness of the phytoplankton during the sustained increases of orthophosphate loading were accompanied by increased zooplankton abundance. Peak zooplankton numbers at stations P23, P26, and P31 were associated with increased numbers of *C. choctawhatcheeana* and *M. throndsenii* during spring 1994. The zooplankton dominant, *Acartia tonsa*, tended to have comparable numbers between 1988–1989 and 1993–1994. Zooplankton species richness was higher during fall 1993 than at any other time (Figure 4.36). Although zooplankton species richness tended to go down during February and March 1994, there was no general reduction of this index during the 1993–1994 bloom period.

Increased zooplankton numbers were associated with high numbers of phytoplankton from January through March 1994. Significant increases of zooplankton numbers occurred during April and May 1994 in Perdido Bay. Zooplankton species richness was significantly higher from September 1993 through January 1994 than that observed in the nonbloom period (1988–1989); however, relatively lower numbers of zooplankton species were noted during the succeeding bloom months which could have been associated with the high dominance of the leading species *A. tonsa*. (Figure 4.36). The inverse relationship of zooplankton numbers and zooplankton species richness in terms of significant differences between the two study periods could be related to possible interspecific competition. Thus, the diatom blooms of 1993–1994 appeared to have little adverse effect on the zooplankton which, if anything, increased in numbers during the bloom period.

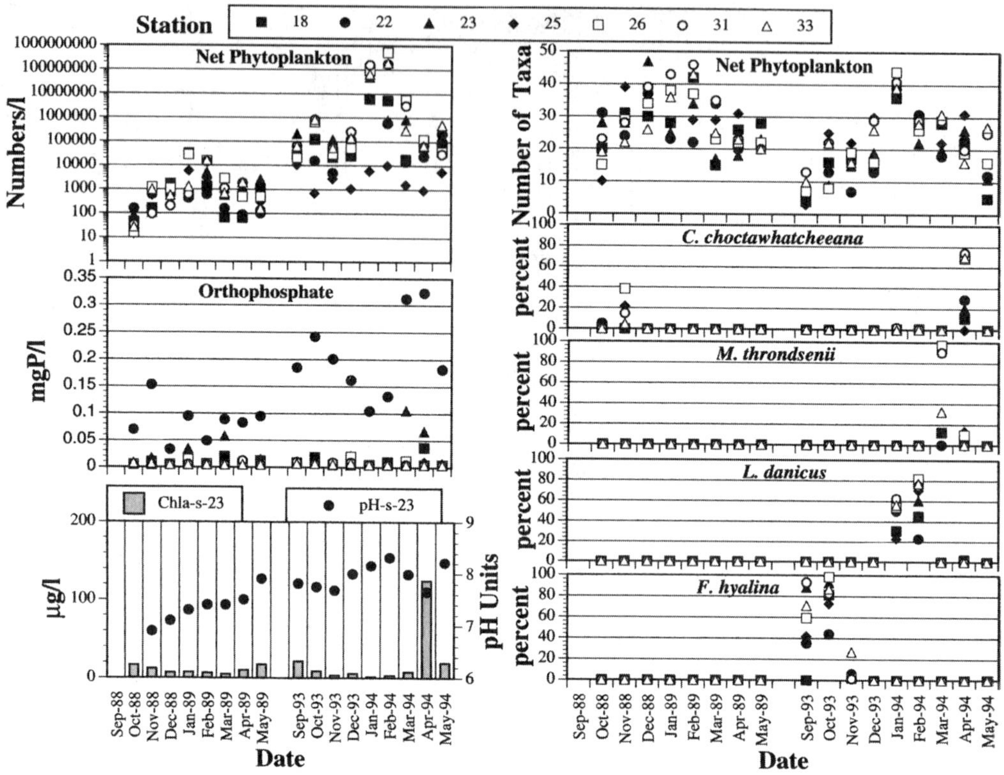

Figure 4.35 Net phytoplankton numbers l^{-1} and species richness, orthophosphate concentrations, surface chlorophyll *a* and pH, and percent numerical dominance of net phytoplankton species at stations in Perdido Bay during the relatively low nutrient loading years (1988 to 1991) and the beginning of the consistently high orthophosphate loading period (1993 to 1994).

Infaunal abundance was significantly lower in the bay during the phytoplankton blooms from February through May 1994. Infaunal species richness was significantly lower during bloom months of January, April, and May 1994 than that during the same period in 1989. The main exception to this baywide trend was in the seagrass beds in the upper bay (station P25) where there were no apparent differences of infaunal numbers and species richness between 1988–1989 and 1993–1994 (Figure 4.36). The bivalve *Rangia cuneata*, normally a drought dominant form at upper bay stations P23, P25, and P26, had virtually disappeared at stations P23 and P26 during 1993–1994; relatively high numbers of this species were found in the seagrass bed area (station P25). This species feeds on phytoplankton and particulate organic matter (Olsen, 1973), and reduction of this species in those areas of Perdido Bay most affected by the phytoplankton blooms could indicate an alteration of the trophic organization of the bay due to changes in the phytoplankton community.

Overall, the plankton blooms of 1994 were significantly and negatively associated with phytoplankton species richness, infaunal numbers and species richness, and herbivorous bivalves. Zooplankton numbers, on the other hand, appeared to be positively associated with the blooms that were dominated by various diatom species. The data also suggest that during the initial phytoplankton blooms in Perdido Bay, the seagrass habitat in western sections of the upper bay were relatively unaffected relative to the unvegetated sedimentary environment. The resumption of increased Perdido River flow during early summer 1994 was associated with a cessation of phytoplankton bloom activity.

Table 4.2 Statistical Comparison of Monthly Data Taken from September 1988 to May 1989 and from September 1993 to May 1994

All Stations by Month Comparisons	Chlor a (μg/l)	Netphyt #Cells/l	Netphyt #taxa	Zoopl. #/l	Zoopl. #taxa	Infauna #/m2	Infauna #taxa
(9-88)/(9/93)	6.8/12.0	**293/62,907**	**24.8/7.4**	1510/1070	**1.2/7**	406/1023	**1.35/2.71**
	WN	**WS**	**WS**	WN	**WS**	WN	**WS**
(10-88)/(10/93)	8.6/7.3	**534/3,702**	20.8/16.4	2492/1761	**1.6/5.6**	**298/801**	**4.42/2.29**
	WN	**WS**	WN	WN	**WS**	**WS**	**WS**
(11-88)/(11/93)	5.0/3.2	**533/42,165**	**28.7/16.1**	528/2160	**1/4.3**	514/751	3.57/2.42
	WN	**WS**	**WS**	WN	**WS**	WN	WN
(12-88)/(12/93)	5.3/4.5	**902/97,584**	**43.7/21.2**	1205/1837	**1.3/3.4**	511/613	3.86/2.85
	WN	**WS**	**WS**	WN	**WS**	WN	WN
(1-89)/(1/94)	**4.4/1.7**	**9892/54,470,000**	**32.8/39**	2037/3988	**2/4.6**	789/740	**5.57/3.85**
	WS	**WS**	**WS**	WN	**WS**	WN	**WS**
(2-89)/(2/94)	7.6/5.8	**8028/690,102**	**36.1/27.7**	1626/511	2.4/2.3	**1272/470**	5.29/4.29
	WN	**WS**	**WS**	WN	WN	**WS**	WN
(3-89)/(3/94)	6.0/6.0	**884/1,424,150**	26.9/25.3	**2500/8679**	3.1/2.6	**740/365**	5.57/3.57
	WN	**WS**	WN	**WS**	WN	**WS**	WN
(4-89)/(4/94)	8.2/34.2	**876/64,900**	**24.7/21**	**1296/2719**	3.4/4	**1047/534**	**5.28/3.71**
	WN	**WS**	**WS**	**WS**	WN	**WS**	**WS**
(5-89)/(5/94)	8.9/11.6	**702/152,034**	**23.3/15**	2111/3347	3.8/5.3	**2398/1587**	**5.43/2.71**
	WN	**WS**	**WS**	WN	WN	**WS**	**WS**

Note: Stations (P18, P22, P23, P25, P26, P31, P33; Figure 2.2a) were used as replicates for determinations of the statistical relationships of chlorophyll *a* (Chlor a), net phytoplankton numbers of cells l^{-1} (Netphyt #cells/l), net phytoplankton species richness (Netphyt #taxa), zooplankton numbers l^{-1} (Zoopl. #/l), zooplankton species richness (Zoopl. #taxa), infaunal numbers m^{-2} (Infauna #/m2), and infaunal species richness (Infauna #taxa) taken between 1988–1989 and 1993–1994. Data were analyzed using means testing methods listed in Appendix II.

WN = not significant (p = 0.05), WS = significant (p = 0.05), CV = RHO > critical value.

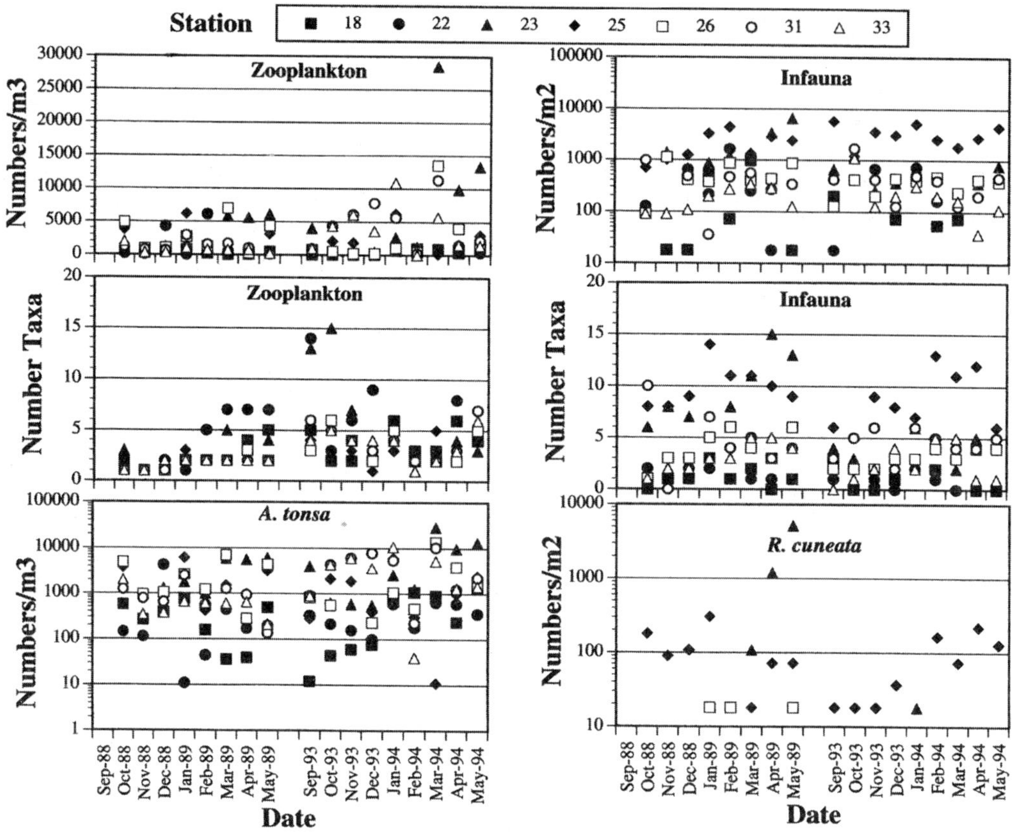

Figure 4.36 Zooplankton numbers m^{-3} and zooplankton species richness, infaunal numbers and infaunal species richness, numbers m^{-3} of dominant zooplankton species *Acartia tonsa* and numbers m^{-2} of dominant infaunal species *Rangia cuneata* at stations in Perdido Bay during the relatively low nutrient loading years (1988 to 1991) and the beginning of the consistently high orthophosphate loading period (1993 to 1994).

G. Whole-Water Phytoplankton

An analysis was made of the long-term trends of whole-water phytoplankton in Perdido Bay. The period from third quarter 1997 through first quarter 1999 was characterized by low orthophosphate loading from the mill and was considered an ecosystem experiment whereby the effects of reduced loading could be compared against previous periods of high orthophosphate loading. Because of the relative similarity of results between the net phytoplankton and whole-water phytoplankton collections, the remainder of the study was carried out with collections of whole-water plankton which were considered more representative of overall community structure.

1. Long-Term Changes in Phytoplankton Distribution

There were seasonal trends of whole-water phytoplankton species richness (Figure 4.37), with peaks usually occurring during winter–spring months and reduced species richness during summer–fall periods. From 1989 to 1991, phytoplankton species richness peaked during winter–spring 1990. There was a general trend of higher species richness in the lower bay relative to upper and mid-bay areas. There was a considerable seasonal range of species richness indices during this period that could have been affected to some degree by periodic increases of ammonia and orthophosphate loading from the pulp mill. However, following increased orthophosphate loading

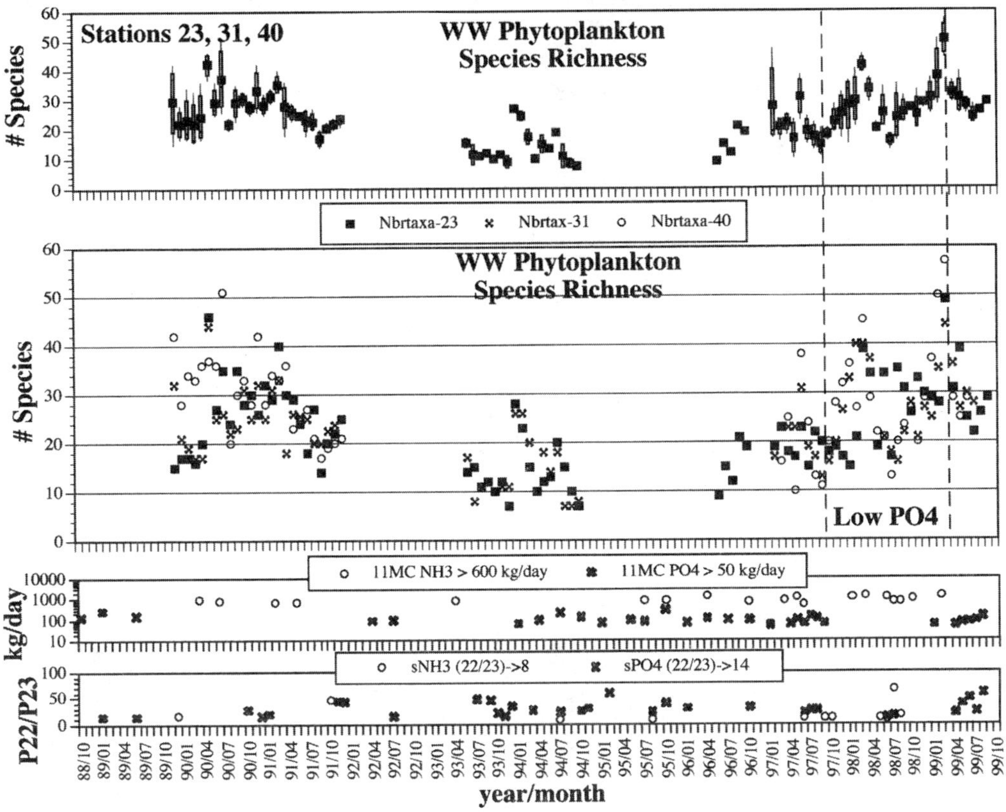

Figure 4.37 Species richness of whole-water phytoplankton taken at stations P23, P31 and P40 in Perdido Bay over the 11-year study period. Data are presented as station-specific means with the 10th, 25th, 75th, and 90th percentiles. Data are compared to ammonia (>600 kg d^{-1}) and orthophosphate (>50 kg d^{-1}) loading rates by the Cantonment paper mill along with nutrient (ammonia and orthophosphate) gradients expressed as concentration ratios (P22/P23).

during 1993–1994, there were sharp reductions of phytoplankton species richness during all seasons in upper and mid-bay areas. The seasonal species succession and annual patterns of species richness of the whole-water plankton were similar to such trends of the net phytoplankton.

From 1996 through summer 1997, there was continued low phytoplankton species richness in Perdido Bay. Orthophosphate and ammonia loading from Elevenmile Creek were high during this period. Following the initiation of reduced orthophosphate loading by the pulp mill in late summer–fall 1997, there was a general increase of phytoplankton species richness that peaked during the winter of 1998. Species richness levels during this period were almost comparable to those during 1990–1991. The general reduction of species richness during summer–fall 1998 occurred during a period of high ammonia loading from the mill. By the following winter, there was a short period of high orthophosphate loading during December 1998 to January 1999 that was accompanied by *P. minimum* blooms in the upper bay. Areas affected by the blooms were characterized by low plankton species richness. After continued low orthophosphate and ammonia loading during the winter–early spring of 1999, there was another general increase of phytoplankton species richness to levels comparable to, or even slightly higher than, those noted during 1990–1991. There was a rapid decline of species richness after renewed orthophosphate loading (and baywide blooms) during spring–summer of 1999. There was thus a stepwise increase in the phytoplankton species richness during succeeding winter–spring periods of low orthophosphate loading during the 1998–1999 period.

Table 4.3 Statistical Comparison of Whole-Water Phytoplankton Numerical Abundance, Biomass, and Species Richness

Station/Year Comparisons	wwphyt # Cells/l	wwphyt µg/l	WW#taxa
23 (90-91)/23 (93/94)	454,020/906,350 WN	51.6/112.8 WN	**30.1/14.1** **WS**
23 (90-91)/23 (96/97)	454,020/3,725,537 WN	51.6/599.2 WN	**30.1/18.2** **WS**
23 (90-91)/23 (98/99)	**454,020/4,432,000** **WS**	**51.6/721.9** **WS**	30.1/31.0 WN
31 (90-91)/31 (93/94)	906,350/1,138,750 WN	93.0/179.4 WN	**27.3/15.0** **WS**
31 (90-91)/31 (96/97)	—		—
31 (90-91)/31 (98/99)	**857,787/4,128,704** **WS**	**93.0/260.2** **WS**	27.3/27.4 WN
40 (90-91)/40 (93/94)	—		—
40 (90-91)/40 (96/97)	—		—
40 (90-91)/40 (98/99)	**374,762/1,967,499** **WS**	**32.1/150.6** **WS**	31.9/30.8 WN

Note: From samples taken at Stations P23, P31, and P40 over 12-month periods from July 1990 to June 1991 (prebloom period), from July 1993 to June 1994 (initial sustained blooms), from July 1996 to June 1997 (continued blooms), and from July 1998 to June 1999 (period of low orthophosphate loading). Data were analyzed using means testing methods listed in Appendix II.

WN = not significant (p = 0.05), WS = significant (p = 0.05), CV = RHO > critical value.

The seasonal variation tended to obfuscate the interannual changes of whole-water phytoplankton species richness. A statistical analysis was run (Appendix II) to determine significant difference among the various whole-water phytoplankton indices (numbers of cells l^{-1}, biomass, and species richness; Table 4.3). There were significant reductions of species richness during the 1993–1994 period and the 1996–1997 period compared with the nonbloom year of 1990–1991. By 1998–1999, no significant differences were noted. This was due in part to increased species richness during November and December 1998 and March and May 1999 relative to these months during the nonbloom period. A similar pattern was noted at stations P31 and P40 due, in large part, to increased species richness during January, February, and March 1999. These increases were observed largely during cold months, periods of high incidence of orthophosphate as a limiting nutrient to upper Perdido Bay. Thus, the response of the system to the period of low orthophosphate loading was consistent with the experimental determination of limiting nutrients. These results tend to support the significance of species richness as a phytoplankton community index denoting assemblage response to blooms stimulated by anthropogenous nutrient loading.

The long-term trends of phytoplankton cell numbers during 1990–1991 (Figure 4.38) tended toward spring peaks. These peaks were accentuated during the 1994 blooms in upper and mid-bay areas. Increased phytoplankton numbers during April 1994 reflected a major bloom of *C. choctawhatcheeana*. The bloom period of 1996–1997 was characterized by extremely high cell counts. The lack of blooms during winter 1998 (i.e., low orthophosphate loading, high ammonia loading) could indicate that ammonia was not a limiting nutrient during such periods, a result that would be consistent with the nutrient limitation experiments. Increased cell numbers during spring–summer 1998 (Figure 4.38) followed increased ammonia loading from Elevenmile Creek. The resumption of the orthophosphate loading in spring 1999 was followed by some of the highest

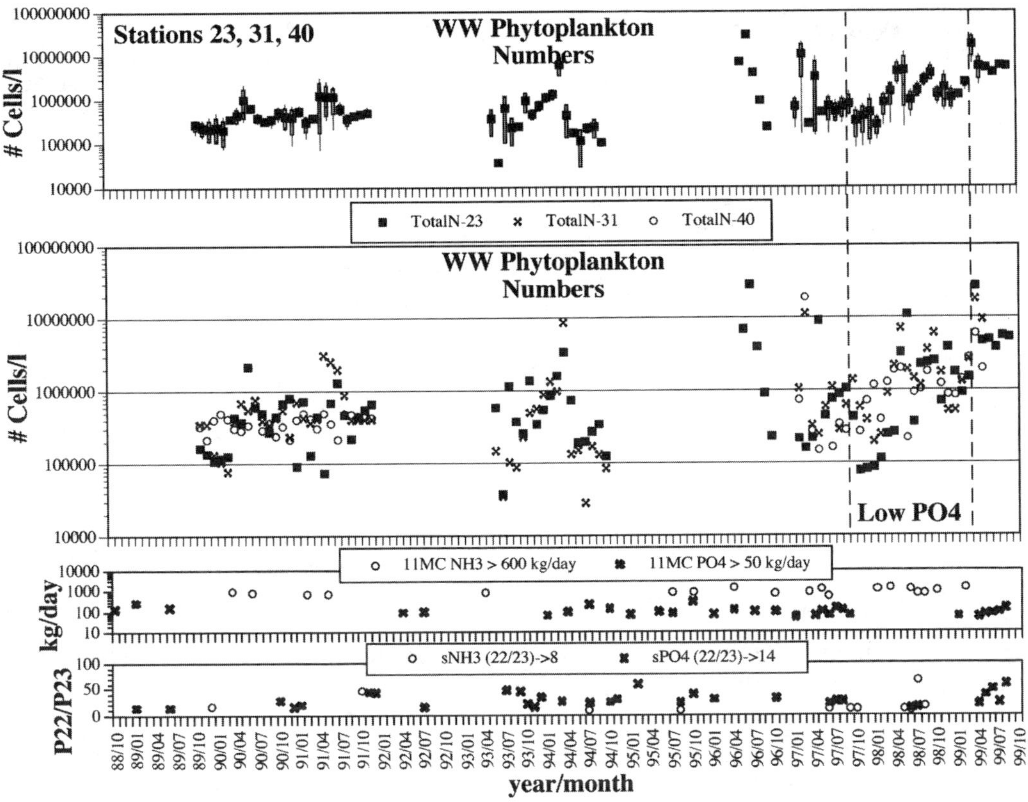

Figure 4.38 Total numbers (cells l^{-1}) of whole-water phytoplankton taken at stations P23, P31, and P40 in Perdido Bay over the 11-year study period. Data are presented as station-specific means with the 10th, 25th, 75th, and 90th percentiles. Data are compared to ammonia (>600 kg d^{-1}) and orthophosphate (>50 kg d^{-1}) loading rates by the Cantonment paper mill along with nutrient (ammonia and orthophosphate) gradients expressed as concentration ratios (P22/P23).

cell counts taken during the study. The immediate bay receiving area of nutrient loading from the mill (station P23) was characterized by increased phytoplankton numbers from the initial bloom period to the spring of 1999. These observations are corroborated by the results of the statistical analyses (Table 4.3). Cell numbers were generally higher during the 1993–1994 and 1996–1997 periods but not significantly so. By 1998–1999, there were significant increases of phytoplankton cell numbers compared with the nonbloom years. These increases were most apparent at all three stations from April through July; this coincides with increased ammonia loading during summer 1998. This result is consistent with the hypothesis that ammonia is most effective as a limiting nutrient during warm months of the year.

Phytoplankton biomass (Figure 4.39) showed slight increases during winter months in 1990–1991. There was a major increase of phytoplankton biomass during early winter (1994) blooms of *P. minimum*. However, most of the phytoplankton blooms during this period were dominated by low-weight diatom species. The lack of correspondence of peak numbers and biomass over the entire study period could reflect differences in species-specific biomass as the smaller-celled diatom bloom species were replaced by the larger-celled raphidophytes during summer months and dinoflagellates in winter months. The seasonal pattern of high winter phytoplankton biomass was repeated during 1996–1997, especially in the upper bay (Figure 4.39). By winter–spring 1998, biomass was generally low. By summer 1998, biomass peaked, again reflecting the summer ammonia loading. During winter-early spring 1999, following low orthophosphate

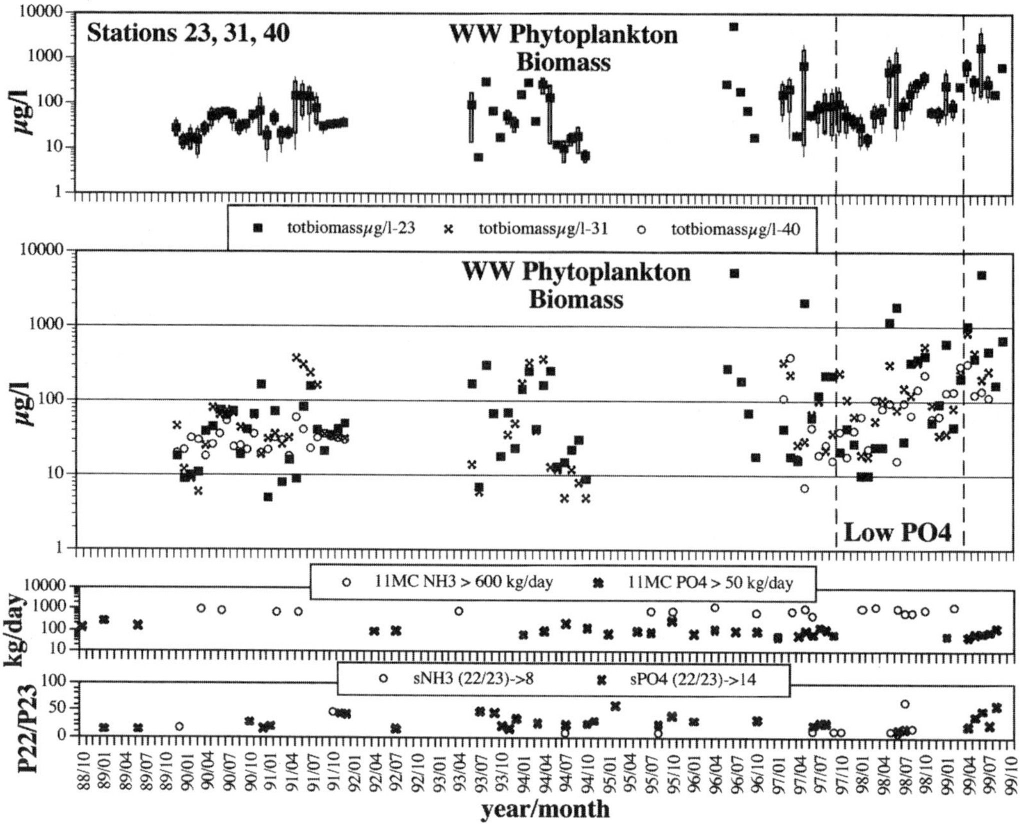

Figure 4.39 Total biomass (μg l^{-1}) of whole-water phytoplankton taken at stations P23, P31, and P40 in Perdido Bay over the 11-year study period. Data are presented as station-specific means with the 10th, 25th, 75th, and 90th percentiles. Data are compared to ammonia (>600 kg d^{-1}) and orthophosphate (>50 kg d^{-1}) loading rates by the Cantonment paper mill along with nutrient (ammonia and orthophosphate) gradients expressed as concentration ratios (P22/P23).

loading, there was relatively low phytoplankton biomass. Phytoplankton biomass was again relatively high throughout the spring–summer 1999 months of high orthophosphate loading. Statistical analyses of the biomass data (Table 4.3) gave similar results to those concerning phytoplankton numbers. There was a sequential biomass increase with time becoming statistically significant as a difference with the nonbloom period by 1998–1999. The primary differences of biomass occurred from April through August at station P23 and during spring and fall months at stations P31 and P40. There was thus a general trend whereby winter periods of high phosphorus loading and summer periods of high ammonia loading were accompanied by increased phytoplankton biomass. High ammonia loading during winter periods was not usually followed by such increases.

The trends of numbers and biomass over the study period were strongly affected by individual bloom species. Dominant phytoplankton species distributions over the 11-year study period are shown in Figure 4.40. Some species such as *P. minimum*, *C. choctawhatcheeana*, *L. danicus*, and *M. throndsenii* were present in the bay before and after nutrient loading was increased in 1993–1994. Each of these species had specific seasons, or even months, when they were most likely to form blooms. The diatom *C. choctawhatcheeana* was a spring species with most blooms occurring in April (Figure 4.40a). This species peaked during April 1994, April–May 1998, and April–May 1999. The spring 1998 blooms occurred during a major reduction in the numbers of other bloom species that could reflect release following lowered competitive pressure. The dinoflagellate

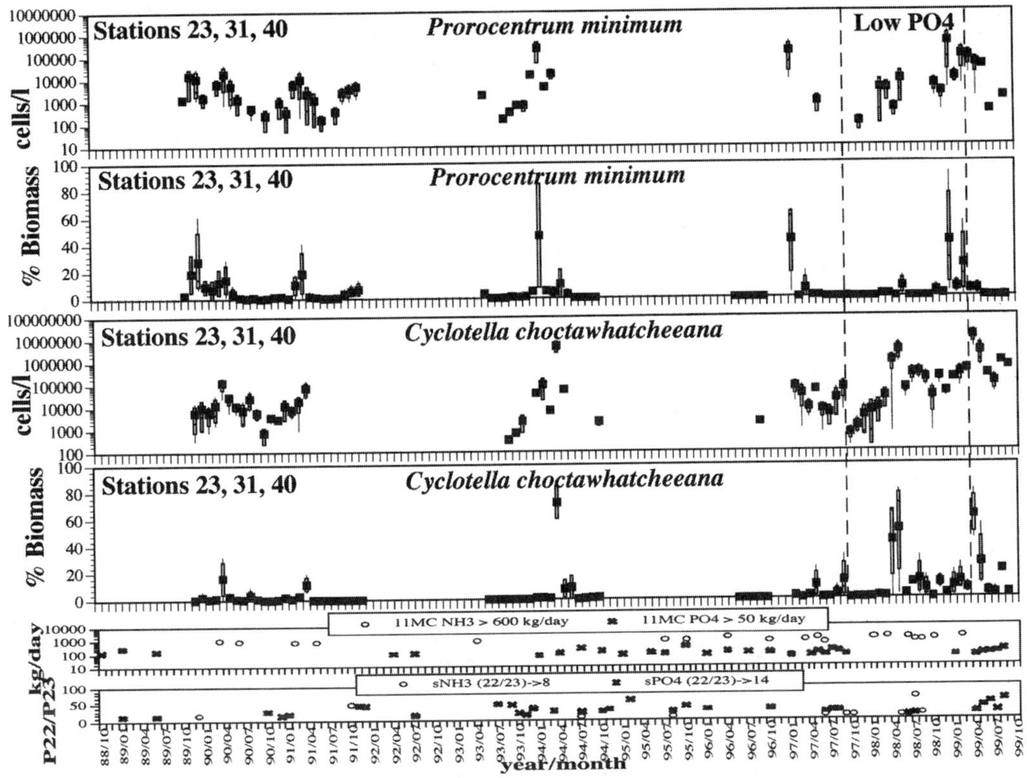

Figure 4.40(a) Numbers of cells l^{-1} and percent dominance by weight of *Prorocentrum minimum* and *Cyclotella choctawhatcheeana*.

P. minimum usually increased during high phosphorus loading periods of the winters of 1993–1994, 1996–1997, and 1998–1999. The *P. minimum* bloom of January 1999 at station P23 followed a relatively brief period of increased orthophosphate loading. It is possible that this dinoflagellate had been sensitized to orthophosphate loading since similar loading levels during the early years of sampling did not stimulate blooms. The lack of blooms by this species during the low orthophosphate loading period (1997–1998) further supports the hypothesis that high orthophosphate loading during the winter stimulates *P. minimum* blooms. Unlike the smaller celled diatoms, these blooms accounted for relatively high percentages of the total phytoplankton biomass. Relatively high numbers of *P. minimum* remained in the upper bay during the high orthophosphate loading of spring 1999.

The diatoms *M. throndsenii* and *Miraltia* sp. were present in the bay during 1990–1991 (Figure 4.40b). *Miraltia throndsenii* formed blooms during spring periods of high orthophosphate loading in 1994, 1997, and 1999. *Miraltia* did not form blooms prior to the increased orthophosphate loading in 1993–1994 or during the 1998 period of low orthophosphate loading. There was a spring 1999 bloom of this species during periods of high orthophosphate loading. *Leptocylindrus danicus* bloomed only during the February 1994 orthophosphate loading period (Figure 4.40b; it was almost completely absent during 1998 (low orthophosphate loading), although it was present during the preceding and succeeding years. Some bloom species were not present at all in our collections until after nutrient loading to upper Perdido Bay was increased; these included the diatom *Synedropsis* sp. which formed blooms during spring–summer months, and the raphidophytes *Heterosigma akashiwo* and *Chattonella subsalsa* which were bloom dominants during spring–summer

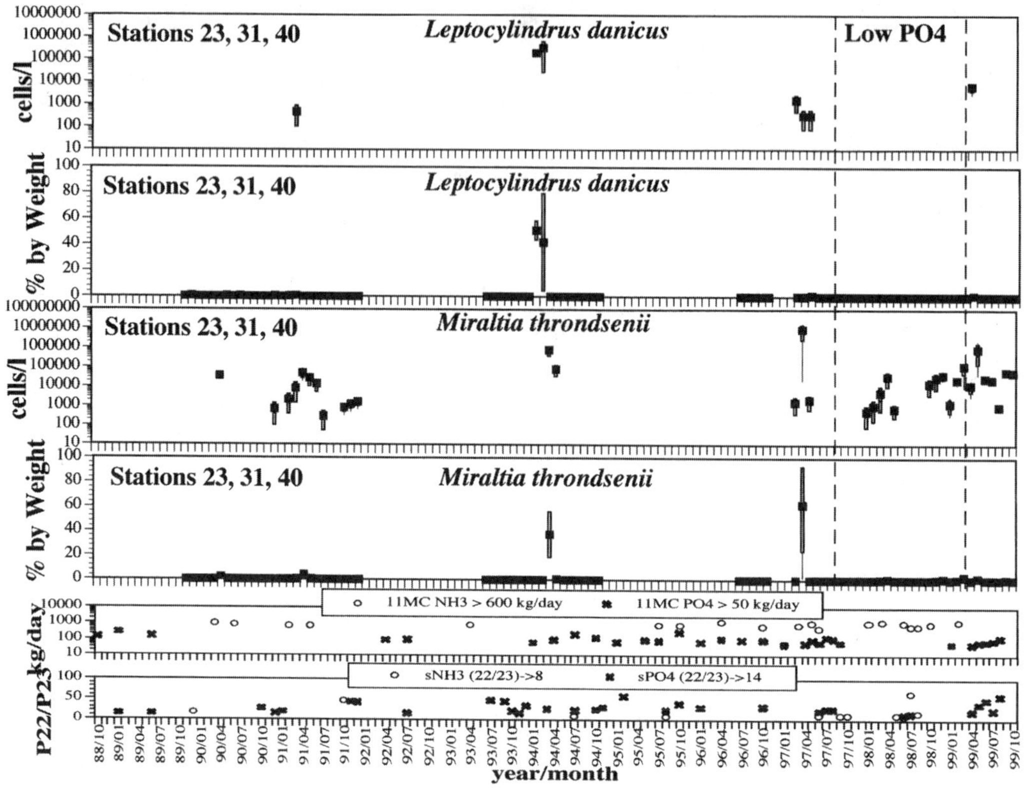

Figure 4.40(b) Numbers of cells l^{-1} and percent dominance by weight of *Leptocylindrus danicus* and *Miraltia throndsenii.*

periods (Figure 4.40c). *Synedropsis* sp. and *H. akashiwo* were dominant and formed blooms or near blooms during the summer of 1998 when there was low orthophosphate loading but high ammonia loading. Both species were dominant during the summers of 1996, 1997, and 1998 when ammonia loading was highest. They were not present during the low ammonia loading of summer 1994 and only minimally present during summer 1999 (low ammonia loading). The data thus indicate that ammonia loading during spring–summer periods stimulated *Synedropsis* sp. and *H. akashiwo.*

The freshwater species *Merismopedia tenuissima* first appeared in the Perdido system during 1996 and was were usually found in high numbers during late summer–fall–winter periods (1998–1999) (Figure 4.40d). *Urosolenia eriensis*, another freshwater diatom, tended to form blooms during fall periods (1998) of low orthophosphate and high ammonia loading (Figure 4.40d). The freshwater species *Pedinophora* sp. was found only once in lower Elevenmile Creek during May 1998 (Table 4.1). The raphidophyte *C. subsalsa* only appeared as a bloom in June 1999 at station P23 (Table 4.1). The dinoflagellate *Gymnodinium splendens* was present in the bay during summer–fall periods (Figure 4.40e). The highest numbers of this species occurred during the high orthophosphate loading of summer 1999. Other species of *Gymnodinium* had a similar pattern of occurrence (Figure 4.40e).

These data are generally consistent with results of the nutrient limitation experiments that showed that orthophosphate was often limiting during cold periods, whereas ammonia was often limiting during warmer months of the year. This remains a loose generalization since both nutrients interact in complex ways (Flemer et al., 1997) that are not completely understood at this time. However, the order and species specificity of the bloom periods, both seasonal and interannual,

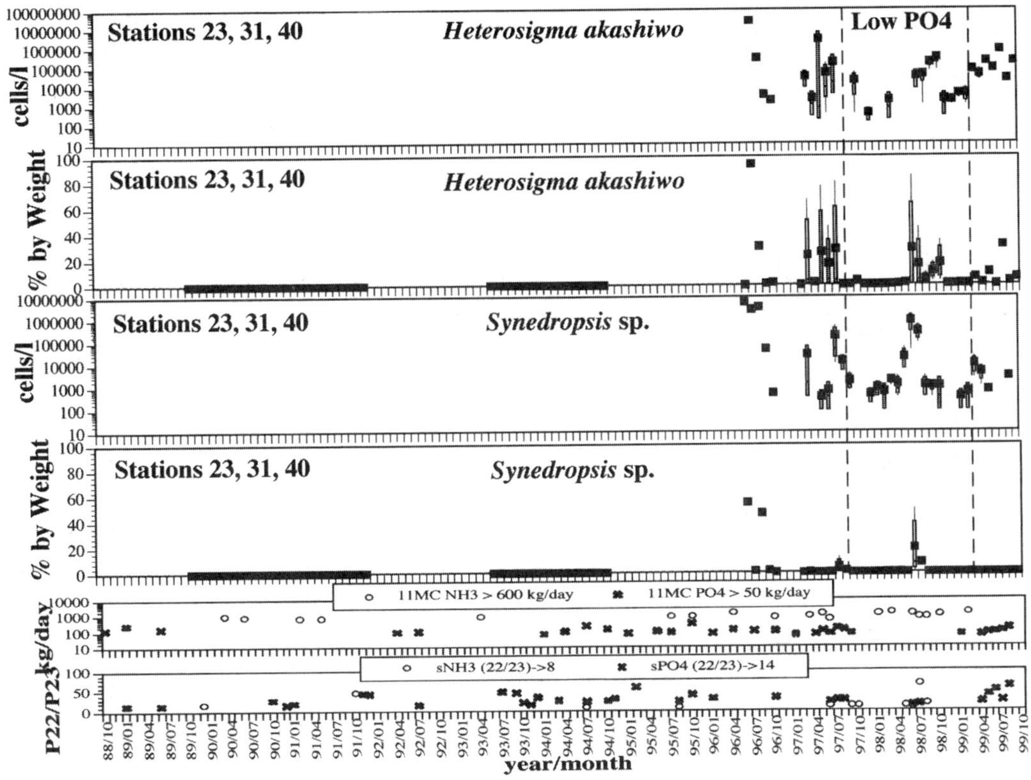

Figure 4.40(c) Numbers of cells l^{-1} and percent dominance by weight of *Heterosigma akashiwo* and *Synedropsis* sp.

suggested that there was a seasonal response of the phytoplankton assemblages during increased nutrient loading to upper Perdido Bay. Thus, phytoplankton blooms appeared to be affected by both the type and amount of the nutrients loaded to the bay as well as the seasonal period of such loading. There also appeared to be a sensitization of the bloom response (i.e., *Prorocentrum minimum*, *Cyclotella choctawhatcheeana*) over the period of exposure whereby these species required less nutrient stimulus to form blooms with time. The general increase of both phytoplankton numbers and biomass with time could reflect a generalized sensitization of certain bloom species after prolonged periods of high nutrient loading to the bay. Although the blooms in any given year were seasonally sequential, new bloom species were added through time as increased anthropogenous nutrient loading continued. Shifts of plankton bloom species representation in Perdido Bay over time indicated that any given bloom probably depended on which species were present when the nutrient loading occurred. This would add an element of chance to what in other ways resembled a well-ordered phytoplankton response to the trends of nutrient loading. Because of the inherent variability of the nutrient loading sequences and the long-term patterns of drought–flood conditions in the bay, the bloom sequences should be viewed as general trends rather than specific events with the overall changes in species richness and species-specific bloom occurrences changing in time as a function of complex interspecific interactions that were not accounted for in this study.

Cumulative phytoplankton species richness (12-month totals) was plotted against averaged ammonia and orthophosphate gradient indices (Figure 4.41). Cumulative species richness in upper, mid and lower bay areas was reduced during periods of increased orthophosphorus gradients (June 1993 to May 1994; June 1996 to May 1997) relative to the 1990–1991 period. Cumulative species richness throughout the bay during the period of reduced orthophosphate loading and increased

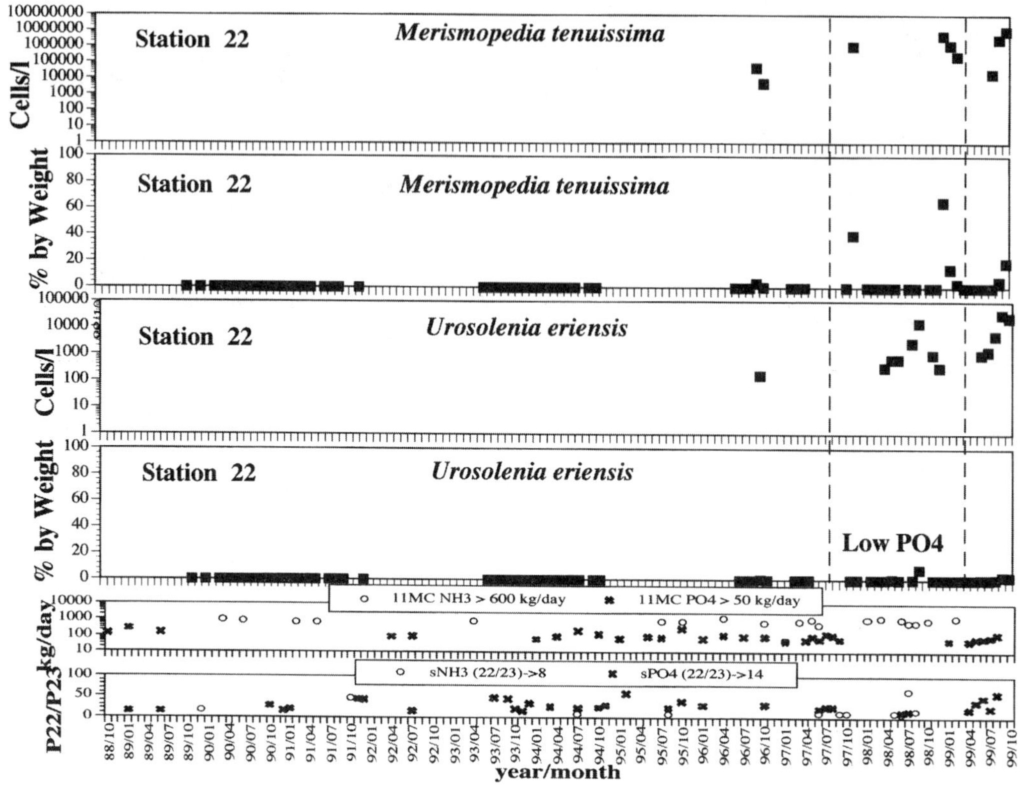

Figure 4.40(d) Numbers of cells l^{-1} and percent dominance by weight of *Merismopedia tenuissima* and *Urosolenia eriensi.*

ammonia loading (October to May 1998) increased to levels that were intermediate between the prebloom period and the bloom years. These data suggest a partial recovery of cumulative species richness during the first 12 months of reduced orthophosphate loading from the mill. This result is consistent with the pattern of monthly species richness from these periods (see Figure 4.37).

2. Plankton Response to Nutrient Loading

The variable nature of the nutrient loading combined with the evanescent nature of bloom formation of the plankton species account for much of the variability in the data. Many factors contribute to the bloom sequences. Nutrient loading is an important component of the initiation and maintenance of the changes in the order of dominance of the plankton assemblages. However, loading rates of ammonia and orthophosphate from the Perdido River system that were periodically comparable to those of Elevenmile Creek did not result in blooms. The primary difference between nutrient loading from the Perdido River and Elevenmile Creek was the high concentrations of orthophosphate and ammonia in Elevenmile Creek and the gradients of these nutrients as water from the creek entered the bay. Many of the blooms originated between stations P22 and P23 where the nutrient gradients were highest. It is therefore likely that changes in the concentration ratios as nutrients entered the bay had some effect on the initiation of the blooms and were therefore associated with changes in the dominance characteristics of the plankton assemblages in the bay. This does not preclude the importance of overall nutrient loading as a stimulus for bloom formation, but it provides a possible explanation for initiation of the blooms in upper Perdido Bay.

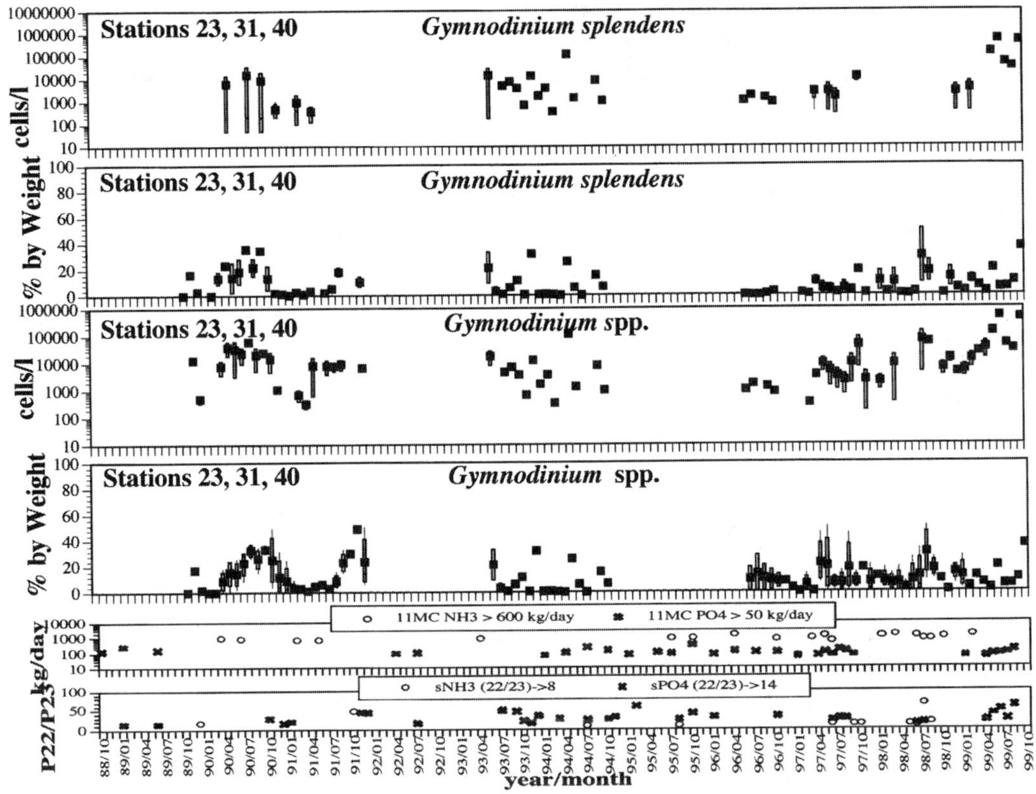

Figure 4.40(e) Numbers of cells l^{-1} and percent dominance by weight of *Gymnodinium splendens* and *Gymnodinium* spp. Data were taken as whole-water phytoplankton samples from Perdido Bay over the 11-year study period, and represent means of all stations with the 10th, 25th, 75th, and 90th percentiles. Data are compared to ammonia (>600 kg d^{-1}) and orthophosphate (>50 kg d^{-1}) loading rates by the Cantonment paper mill along with nutrient (ammonia and orthophosphate) gradients expressed as concentration ratios (P22/P23).

A number of variables could contribute to changes in the phytoplankton community over time. These include (1) the exact timing of nutrient delivery; (2) which nutrient (or nutrients) was (were) being loaded at a given time; (3) interactions among the various nutrients; (4) bay habitat conditions relative to the interannual drought–flood sequence; and (5) the nutrient requirements of the species present at the time of the nutrient loading. The natural history of the bloom species was a factor in the often species-specific seasonal and interannual responses to increased nutrient loading and concentrations. Adding to the complexity of bloom initiation were the possible effects of other compounds that were not monitored (e.g., urea) that can be utilized by species such as *P. minimum* (D. Anderson, personal communication; Howarth et al., 2000). It is also likely that some species that formed bottom blooms derived their nutrients from the sediments. The fact that there was a seasonal component to phytoplankton community indices, such as species richness, indicated seasonal changes in the phytoplankton dominance hierarchies. In addition, there appears to have been a long-term component whereby bloom species responded differently to nutrient input with time. Diatom species that appeared as early dominants were eventually replaced as the raphidophytes and dinoflagellates became predominant. Species competition was evident with some diatoms, such as *C. choctawhatcheeana,* maintaining a higher postbloom level of abundance even after the phosphorus loading was reduced. This could have been a response to reduced competition after other bloom species dropped out. Another possible long-term effect was that some species became more easily stimulated by lower concentrations of nutrients over the years of increased nutrient

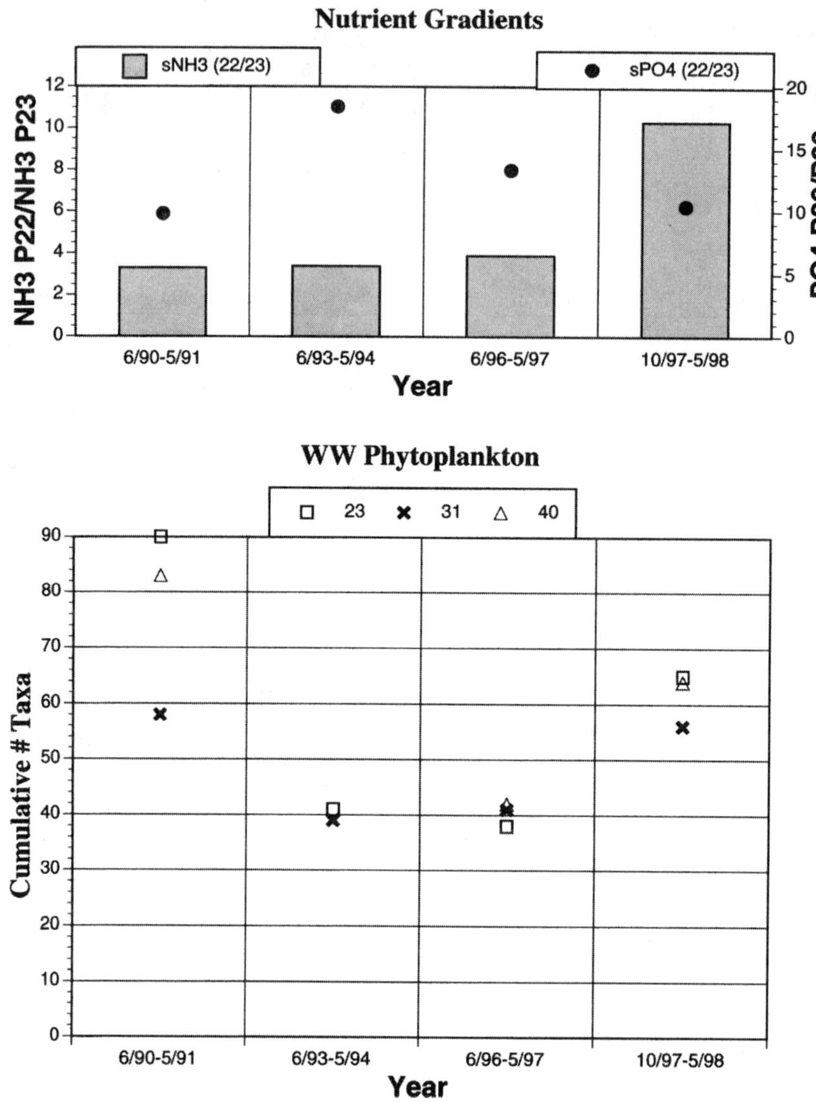

Figure 4.41 Annual averages of ammonia and orthophosphate gradients (station P22/station P23) plotted against cumulative numbers of whole-water phytoplankton species richness at stations P23, P31, and P40 in Perdido Bay during periods ranging as follows: 6/90–5/91; 6/93–5/94; 6/96–5/97; 10/97–5/98.

loading. It was almost as if the bay were "infected" by the plankton blooms so that, with time, remnants of the blooms remained in the bay, ready to take advantage of further increases in nutrient loading. It is thus possible that bloom species, such as *C. choctawhatcheeana* and *P. minimum*, were well established in the bay prior to increased nutrient loading, and remained in a more sensitive state to nutrient loading. If true, this means that the effects of the blooms could persist long beyond the cessation of the originally high levels of nutrient loading that caused them in the first place.

The state of the bay during periods of enhanced nutrient loading was an important factor in the origin of a bloom sequence. The original blooms occurred during the 1993–1994 drought that was accompanied by bay conditions that could enhance bloom formation. Low turnover rates, increased winter temperature and salinity, and increased water clarity could have been important variables in the generation of phytoplankton blooms (Livingston et al., 1997). The summer flood during 1994 effectively ended the previous bloom sequence, at least temporarily. The interannual drought–flood sequence was thus an important modifying factor in the response of the phytoplankton to nutrient loading. The state of the phytoplankton assemblages in the bay during periods of enhanced nutrient loading was another factor that could have contributed to the nature of the bloom successions. Each bloom species had a very different relationship to nutrient loading with obvious seasonal preferences for specific nutrients. The interrelationships of the various plankton species relative to the bloom sequences probably contributed to the generally inverse relationship of species richness to increased dominance of the bloom species. The relative (numerical) dominance of the different phytoplankton species proved to be a reliable indicator of community response. The relationships that could affect species interactions, such as toxic effects, predation, competition for light and nutrients, and crowding, probably changed seasonally and possibly interannually thereby altering the response of the bloom species to nutrient loading trends. This effect could have been enhanced by phytoplankton-induced water quality changes (i.e., nutrient concentration shifts, reduced dissolved oxygen, altered light transmission characteristics, and increased pH).

In summary, it appears that many factors influenced the observed occurrences of phytoplankton blooms in Perdido Bay over the 11-year study period. The timing and intensity of nutrient loading to the bay and nutrient concentrations in river–bay interfaces were closely associated with the bloom sequences. The influence of increased light penetration during a drought, both as a response to pulp mill modifications and interannual cycles of drought–flood conditions, also should be included in the evaluation of causative factors. There were distinct bloom successions that followed seasonal, species-specific patterns. Blooms originating in areas close to the mouth of Elevenmile Creek could have been modified by zooplankton predation with the result that peak bloom concentrations sometimes occurred in mid-bay areas downstream from the point of entry of the nutrients. Some blooms were baywide in distribution; others were specific to different parts of the bay. The main blooms occurred in the upper bay, which means that longitudinal salinity gradients could be a factor, especially to freshwater species such as blue-green algae. It is also likely that some plankton blooms needed altered nutrient gradients to be initiated, but, once formed, these blooms could have been sustained by nutrient loading from both anthropogenous and natural (river) sources as the plankton were transported down the bay. The extent of nutrient loading may be complementary to the concentration gradients of individual nutrients as the chief initiating factors for blooms. Thus, nutrient loading from the Perdido River system could have played an important role in the extent and duration of the plankton blooms. Any increase in the loading rates from the river could seriously exacerbate the bloom conditions in the bay. Although the nutrient gradients appeared to be involved in the some of the bloom successions, overall nutrient loading was still the primary controlling feature in the altered community structure of phytoplankton assemblages in Perdido Bay.

H. Phytoplankton Blooms and the Benthic Microalgae

The interaction of nutrient loading and nutrient dynamics in sediments of shallow estuaries has been well documented. Nutrient fluxes into and out of sediments are considered important to the productivity of both water column phytoplankton (Nixon, 1981; Hopkinson and Wetzel, 1982) and benthic algae (Nilsson et al., 1991; Rizzo et al., 1992; Cerco and Seitzinger, 1997). Rizzo et al. (1992) found that increased light attenuation decreased benthic algal activity, thus releasing more nutrients into the water column with resulting increases in phytoplankton production. In some cases, increased nutrient loading has led to increased benthic algal production (Sundback et al., 1991). Cerco and Seitzinger (1997), working in Indian River–Rehoboth Bay, Delaware, used field studies and modeling to show that benthic algae had a "major influence" on seasonal cycling of nitrogen and phosphorus. Nutrients were transferred via algal activity to sediments during late winter–spring periods of relatively high water column nutrient concentrations. The release of these nutrients through diagenesis during summer months of low water column nutrient concentrations enhanced phytoplankton production. The authors suggested that annual water column productivity was thus enhanced by the presence of benthic algae. The problems associated with these interactions include the varying relationships of light, nutrients, and temperature with complex assemblages of phytoplankton and benthic microalgae. Most of these studies indicated that there are interactions of water and sediment quality with seasonal cycles of phytoplankton and benthic microalgal activity. However, none of these studies included long-term, species-specific analyses of the interactions of the phytoplankton assemblages and microalgal species as a function of nutrient loading.

Benthic microbiota provide an important trophic link between water column and sediment primary productivity and the higher levels of associated food webs in estuarine systems (Federle et al., 1983a, b, 1986). Microalgae are an important component of estuarine sediments, and contribute to the trophic organization of the benthos (Livingston, 1984a). There are a number of studies concerning the qualitative composition of microphyte assemblages in coastal sediments. Admiraal (1984) provided a comprehensive review of sediment-inhabiting diatoms in estuaries. According to Admiraal (1984), most of the benthic microphytes are pennate diatoms, and are represented by both motile and immobile species. Light is often the primary limiting factor for the benthic microflora. Natural disturbance of the sediments often obscures vertical organization, and can be stimulatory to benthic productivity (Livingston, 1984a).

The relationship between benthic microphytes and water column phytoplankton has been established in the Apalachicola Bay system (Federle et al., 1983a, b, 1986). To evaluate the possible connection of these two forms of estuarine microphytes, a 20-month analysis was carried out in Perdido Bay (Figure 4.42). Monthly collections were taken of water quality, sediment nutrients,

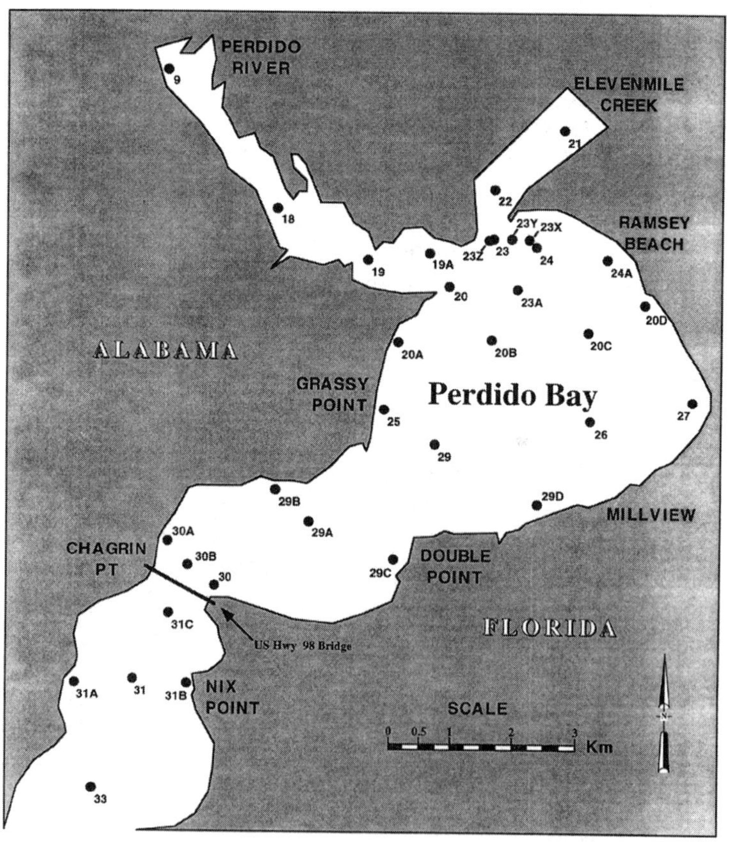

Figure 4.42 Map of Perdido Bay stations for the intensive 1993–1994 sampling program.

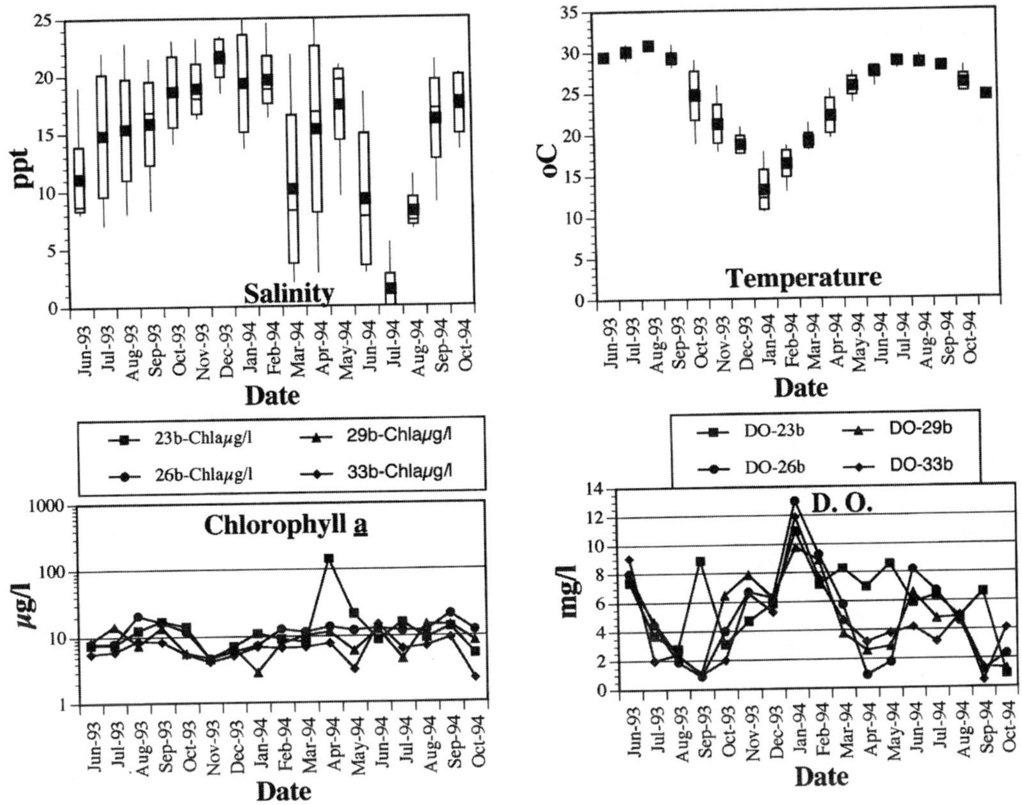

Figure 4.43 Surface salinity (ppt), water temperature (°C), chlorophyll *a* (μg l^{-1}), and dissolved oxygen (mg l^{-1}) in Perdido Bay at monthly intervals from June 1993 to October 1994.

water column phytoplankton (net and whole-water), benthic microphytes, water column phytoplankton found in the sediments (living and dead cells taken in core samples and processed to species), zooplankton, and infaunal macroinvertebrates. Fishes and invertebrates were taken quarterly. All biological collections were identified to species, counted, and weighed. The comprehensive data set was taken from March 1993 through October 1994; this happened to coincide with the initial, baywide plankton blooms of 1993–1994.

Salinity (Figure 4.43) increased during the 1993 drought to winter peaks. There was some reduction of salinity during spring 1994 that was followed by baywide salinity reductions in July 1994 following a major storm and increased river runoff. Drought conditions and high salinities were reestablished by the following fall. With the exception of a single peak of chlorophyll *a* during April 1994 at station P23, the chlorophyll *a* concentrations throughout the bay were low over the entire study period (Figure 4.43) despite massive phytoplankton blooms during this time (see Figure 4.35). Bottom dissolved oxygen reached relatively high concentrations throughout the upper and mid-bay areas during winter–spring periods of low temperature (Figure 4.43). Hypoxia during warm months was greatest at station P26 where some of the most intensive bloom activity was noted. Bottom dissolved oxygen increased in June 1994, a period of relatively low bloom activity. Hypoxia recurred following increased salinities and resumption of the drought in the fall. These data indicate that phytoplankton bloom activity was directly related to benthic (subhaloclinal) hypoxia. It also shows that, during bloom periods of low temperature, benthic dissolved oxygen was relatively high. Sediment total organic nitrogen (Figure 4.43) tended to be low during the bloom period (January–June, 1994). Sediment total phosphorus declined over this period to lows in June. Sediment total organic carbon peaked in April 1994 at the end of the bloom period.

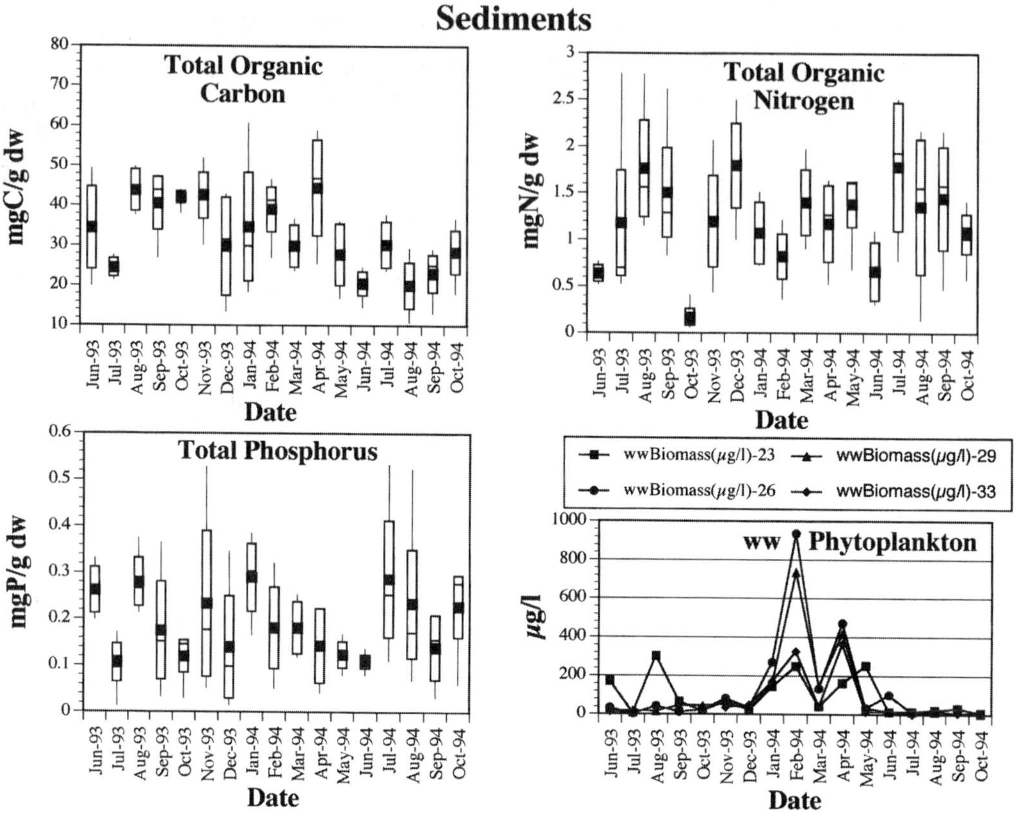

Figure 4.44 Sediment total organic carbon (mg C g^{-1} dw), sediment organic nitrogen (mg N g^{-1} dw), sediment total phosphorus (mg P g^{-1} dw), and whole-water phytoplankton biomass by station (μg l^{-1}) in Perdido Bay at monthly intervals from June 1993 to October 1994. All stations were used for this analysis. Sediment data are expressed as means with the 10th, 25th, 75th, and 90th percentiles of all stations.

The whole-water plankton biomass was highest at stations P26 and P29 during the bloom period (Figure 4.44). These biomass increases coincided with the blooms of *P. minimum* (Figure 4.45). The winter–spring blooms were characterized by a seasonal succession with *L. danicus* predominating in terms of percent biomass during January–February 1994. In February, there was joint dominance of this species with *P. minimum*, which was present as a subdominant at some stations throughout the winter months. This was followed by blooms of *M. throndsenii* (March), *C. choctawhatcheeana* (April), and *G. splendens* (May). The dinoflagellate *G. splendens* also dominated the plankton biomass during the summer of 1993. The whole-water phytoplankton dominance sequence followed that of the previously described net phytoplankton (Figure 4.35) with the exception of peak biomass of *P. minimum* during February 1994 and *G. splendens* in May 1994.

The living benthic microphyte assemblages in Perdido Bay (Tables 4.4 and 4.5) were dominated (numbers and biomass) by the species *Bacillaria paxillifer* and species of the genus *Navicula*. Other dominants included the genera *Pleurosigma* (biomass) and *Nitzschia* (numbers). There was no evidence of water column bloom species in the sediment microphytes. Living microphyte biomass in the sediments decreased throughout the bay from December 1993 through June 1994 (Figure 4.45). This decrease tended to be most pronounced at stations P26 and P29 (where blooms were most active) although the downward trend was evident throughout the bay. Species richness of the benthic microalgae (Figure 4.45) decreased at various stations from January through the

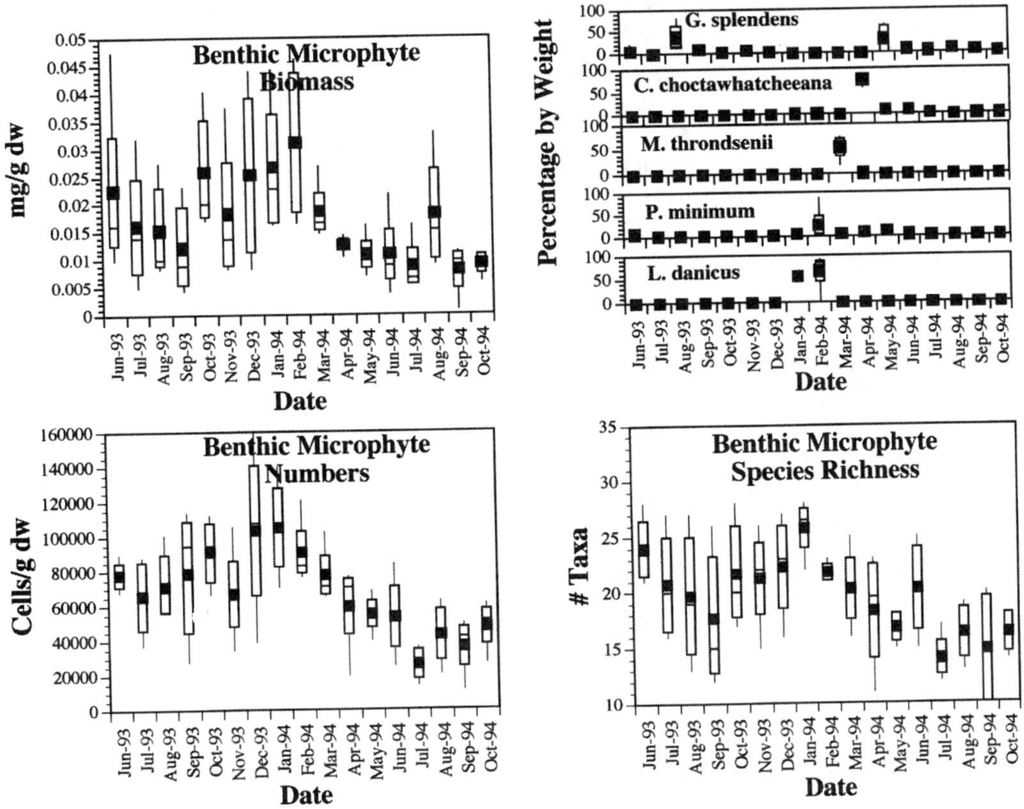

Figure 4.45 Living sediment microalgal numbers g^{-1} dw sediment, living sediment microalgal numbers of cells g^{-1} dw sediment, living sediment microalgal species richness, and whole-water phytoplankton percent dominance by weight of whole-water phytoplankton bloom species taken at monthly intervals from June 1993 to October 1994.

following summer. These trends tended to follow the water column phytoplankton bloom sequence in the bay during this period. The relatively steep decreases of numbers and species richness of the benthic microphytes coincided with the progression of water column blooms noted above. It is possible that the noted trends were simply part of a seasonal or interannual trend, and the noted correlation between the blooms and the reduced benthic microphyte community was spurious. A statistical analysis was run comparing the differences between grouped data from all stations from June–October 1993 and from June–October 1994. The difference of the mean numbers of cells g^{-1} dw sediment (1993, 79,159; 1994, 49,525) was significant ($p < 0.01$). The difference of the mean numbers of species (1993, 21.8; 1994, 16.8) was significant ($p < 0.01$). Mapping of the benthic microphytes in the sediments (Figure 4.46) indicate that they were least abundant in areas around the mouths of Elevenmile Creek and Bayou Marcus and in the western seagrass beds. The general increases with distance from the upper bay could indicate light limitation.

The benthic infauna in the bay during this period (see Figure 4.36) declined in numbers and species richness from October 1993 through June 1994. These declines were primarily due to a loss of the dominant polychaete worms (*Mediomastus ambiseta* and *Streblospio benedicti*). The bivalve mollusk *Rangia cuneata*, which should have been a primary biomass dominant during the 1993–1994 drought, was almost completely absent in unvegetated areas of the upper bay during the bloom period. These data are consistent with an hypothesized inhibition of the infauna that could have been related to changes in the trophic organization of the sediments during the 1994

Table 4.4 Summary of Numbers of Cells g^{-1} Sediment of Living Microalgae Taken by Coring at Stations in Perdido Bay from June 1993 to October 1994

Species	18	19	22	23	25	26	27	29	31	33	23A	24A	29A	29B	29C	31A	31B	Total
Navicula spp.	61,031	54,695	11,006	168,751	20,344	255,795	22,345	264,133	66,367	254,127	37,353	60,030	55,362	63,032	37,353	86,711	67,368	**1,585,803**
Nitzschia spp.	5,337	32,350	3,002	194,431	5,670	256,795	9,339	213,441	23,012	112,723	19,344	54,695	25,013	15,675	17,342	20,344	13,340	**1,021,849**
Bacillaria paxillifer	9,339	7,004	667	156,745	5,003	207,104	2,001	202,435	1,668	54,694	14,341	—	2,335	334	9,338	6,003	4,336	**683,345**
Diploneis spp.	3,335	16,342	1,001	40,354	8,338	66,367	6,003	54,695	13,007	48,024	5,670	6,337	21,344	26,347	16,009	16,676	6,670	**356,516**
Amphora spp.	5,670	5,670	1,001	45,690	6,004	41,354	2,668	56,362	13,341	105,386	3,336	8,004	11,673	10,673	4,336	16,342	14,674	**352,182**
Nitzschia sigmoidea	1,334	1,668	—	27,681	4,336	28,348	2,002	52,694	4,670	14,007	3,002	3,335	1,335	6,337	7,671	3,335	9,006	**170,761**
Nitzschia tryblionella	2,668	4,003	—	27,014	1,001	38,353	4,336	36,352	667	10,672	7,671	12,007	3,336	3,002	—	2,668	667	**154,416**
Tryblionella levidansis	—	—	—	12,006	—	83,375	—	42,021	—	4,002	—	—	—	—	—	—	—	**141,404**
Paralia sulcata	—	—	—	5,003	2,001	9,338	—	35,685	5,670	25,346	—	—	4,670	6,337	—	9,005	667	**103,721**
Diploneis elliptica	334	—	—	15,341	—	34,017	—	41,354	1,334	9,338	—	—	—	—	—	—	—	**101,718**
Diploneis bombus	—	—	—	10,005	—	36,685	—	40,020	—	11,339	334	—	—	—	—	—	—	**98,383**
Surirella fastuosa	—	—	—	3,002	334	12,006	—	54,361	3,335	17,342	—	—	2,001	1,001	—	334	2,668	**96,384**
Nitzschia dubia	2,335	2,335	—	7,338	—	29,682	2,335	22,012	667	1,334	3,336	15,008	4,669	1,668	—	—	1,001	**93,719**
Gyrosigma parkerii	—	667	—	6,670	2,002	8,004	334	21,344	2,335	8,671	—	1,668	3,002	8,672	14,008	4,002	7,338	**88,716**
Navicula yarrensis v. *americana*	—	—	—	1,334	—	8,004	—	21,011	4,002	19,343	—	—	—	5,003	334	3,002	3,002	**65,035**
Pleurosigma spp.	1,001	1,335	—	5,003	668	15,341	—	13,007	1,334	8,671	1,001	—	2,001	2,335	1,668	3,336	5,003	**61,704**
Cymatosira lorenziana	—	—	—	—	1,334	2,668	—	8,338	1,668	14,007	—	—	10,339	2,335	2,668	5,003	—	**48,360**
Surirella spp.	1,334	334	—	5,336	668	5,336	334	9,672	1,668	2,001	1,001	1,668	5,003	4,670	—	2,001	1,335	**42,361**
Thalassiosira spp.	—	334	—	2,668	—	3,002	—	10,005	1,668	11,339	334	667	1,001	1,668	1,001	1,334	2,335	**37,356**
Nitzschia panduriformis	—	334	—	2,668	—	2,668	—	4,002	—	23,345	—	—	—	1,001	—	1,668	—	**35,686**
Cocconeis spp.	9,339	667	668	2,001	1,334	667	3,669	1,668	3,336	—	2,001	1,001	667	1,334	—	—	4,336	**32,688**
Navicula yarrensis	—	—	—	4,002	—	6,337	—	7,337	3,002	3,335	—	—	334	1,001	668	1,668	1,334	**29,018**
Diploneis didyma	334	2,001	1,001	—	667	—	—	2,668	2,335	667	1,334	1,001	6,337	2,668	—	2,335	4,669	**28,017**
Undetermined centric diatoms	—	667	—	3,335	—	12,006	—	2,668	—	2,668	667	—	1,668	1,001	—	334	667	**25,681**
Gyrosigma spp.	—	—	—	2,668	—	1,334	—	3,002	1,001	12,673	—	—	1,334	334	334	2,335	—	**25,015**
Petrodictyon gemma	—	—	—	—	668	1,334	—	18,009	—	667	—	—	334	1,001	667	1,668	—	**24,348**
Cymatosira belgica	—	—	—	—	—	—	—	6,003	—	17,342	—	—	—	—	—	—	—	**23,345**
Melosira spp.	—	—	—	7,337	—	3,335	—	6,003	—	2,001	—	—	—	—	—	1,668	—	**20,344**
Campylodiscus echeneis	667	334	—	2,669	1,001	3,335	668	5,337	334	2,668	—	—	668	1,001	—	—	—	**18,682**
Entomoneis alata	—	—	—	2,001	—	5,670	—	6,003	—	2,001	667	1,001	—	—	—	334	—	**17,677**
Tabellaria fenestrata	5,336	3,335	—	1,334	—	2,668	—	1,334	—	—	2,668	334	—	334	—	—	—	**17,343**
Eunotia spp.	4,002	7,004	—	2,001	—	—	—	1,334	1,001	—	—	1,001	334	—	—	—	—	**16,677**
Nitzschia lorenziana	—	—	—	1,334	—	4,669	—	4,669	—	6,003	—	—	—	—	—	—	—	**16,675**
Eunotia pectinalis	667	4,003	668	6,670	—	667	—	3,335	—	—	—	—	—	—	—	—	—	**16,010**
Nitzschia gracilis	—	334	1,001	2,669	668	4,669	667	1,334	1,001	—	—	668	—	1,001	—	334	334	**14,680**
Pleurosigma elongatum	—	—	—	2,001	—	3,002	1,334	2,668	—	—	—	2,001	667	—	3,002	—	—	**14,675**

Achnanthes curvirostrum	—	—	—	5,336	—	1,334	—	4,669	—	2,668	—	—	—	—	—	—	—	**14,007**
Nitzschia circumsuta	—	—	—	1,001	—	3,669	334	4,669	1,668	667	—	—	334	1,001	—	—	—	**13,343**
Gyrosigma balticum	—	—	—	—	334	1,334	334	1,334	3,002	2,668	—	—	—	668	1,001	1,335	334	**12,344**
Gyrosigma sp. (s shaped)	—	—	—	1,334	—	6,003	—	4,002	—	667	—	—	—	—	—	—	—	**12,006**
Entomoneis spp.	—	334	—	3,002	334	2,001	334	2,335	—	667	—	—	—	334	—	—	667	**10,008**
Synedra spp.	1,334	1,001	—	667	334	2,001	—	1,334	—	—	2,001	667	—	—	667	—	—	**10,006**
Navicula lyra	—	—	—	—	—	2,001	—	3,002	2,002	1,334	—	—	334	334	—	—	—	**9,007**
Cocconeis scutellum	2,001	—	—	—	—	1,334	667	—	—	—	334	1,001	1,001	—	—	2,001	334	**8,673**
Frustulia rhomboides	1,334	1,001	—	2,001	—	1,668	—	1,668	—	—	—	—	—	334	—	—	—	**8,006**
Gyrosigma fasciola	—	—	—	1,334	—	—	—	—	—	6,670	—	—	—	—	—	—	—	**8,004**
Melosira dubia	—	—	—	—	—	3,335	—	4,669	—	—	—	—	—	—	—	—	—	**8,004**
Undetermined pennate diatoms	—	—	—	—	—	—	—	1,668	2,001	—	—	667	667	2,668	—	—	—	**7,671**
Cylindrotheca closterium	—	—	—	2,668	334	667	—	667	—	1,334	—	—	—	—	—	—	1,668	**7,338**
Nitzschia granulata	—	—	—	2,001	—	4,002	—	1,334	—	—	—	—	—	—	—	—	—	**7,337**
Caloneis acutiascula	334	334	—	334	334	2,001	—	1,668	668	—	334	—	—	—	334	334	—	**6,675**
Cocconeis placentula	—	—	—	667	—	2,668	—	3,335	—	—	—	—	—	—	—	—	—	**6,670**
Synedra tabulata	—	—	334	1,334	—	—	—	2,001	334	1,334	667	—	—	—	—	334	—	**6,338**
Eunotia pectinalis v. undulata	667	1,001	—	334	—	1,001	334	667	—	—	1,001	667	—	—	—	—	—	**5,672**
Actinocyclus spp.	—	—	—	—	—	—	—	3,002	—	2,001	—	—	—	—	334	—	—	**5,337**
Caloneis spp.	—	—	—	—	—	1,334	—	2,001	—	2,001	—	—	—	—	—	—	—	**5,336**
Gyrosigma macrum	—	—	—	—	334	—	—	334	334	1,334	—	—	—	334	334	667	1,001	**4,672**
Mastogloia spp.	—	—	—	667	—	—	—	—	—	2,001	—	—	—	—	—	2,001	—	**4,669**
Dictyocha fibula	—	—	—	1,334	667	—	—	1,001	—	—	—	—	334	334	—	334	—	**4,004**
Amphora arenaria	—	—	—	—	—	—	—	667	—	3,335	—	—	—	—	—	—	—	**4,002**
Gomphonema spp.	—	334	—	—	—	—	334	—	334	—	—	—	—	—	—	1,001	1,668	**3,671**
Nitzschia scalaris	334	—	—	667	—	1,334	—	1,334	—	—	—	—	—	—	—	—	—	**3,669**
Eunotia flexuosa v. eurycephala	—	—	—	667	334	334	—	1,001	—	—	334	667	—	—	—	3,337	—	**3,337**
Fragilaria spp.	—	—	—	667	—	—	—	2,001	—	667	—	—	—	—	—	—	—	**3,335**
Nitzschia hungarica	—	—	—	1,668	—	667	—	—	—	667	—	—	—	—	—	—	—	**3,002**
Actinoptychus undulatus	—	—	—	—	—	—	—	667	—	2,001	—	—	—	—	—	—	—	**2,668**
Eunotia serra	—	—	—	667	—	667	—	1,334	—	—	—	—	—	—	—	—	—	**2,668**
Fragilaria leptostauron	—	—	—	—	—	1,334	—	1,334	—	—	—	—	—	—	—	—	—	**2,668**
Navicula cf. confervacea	—	—	—	—	—	—	—	—	—	2,668	—	—	—	—	—	—	—	**2,668**
Pinnularia spp.	—	—	—	2,668	—	—	—	—	—	—	—	—	—	—	—	—	—	**2,668**
Aulacoseira granulata	—	—	—	—	2,001	—	334	—	—	—	—	—	—	—	—	—	—	**2,335**
Eunotia monodon	—	334	—	2,001	—	—	—	—	—	—	—	—	—	—	—	—	—	**2,335**
Navicula confervacea	—	2,335	—	—	—	—	—	—	—	—	—	—	—	—	—		—	**2,335**
Caloneis latiuscula	—	—	—	—	—	—	—	—	2,001	—	—	—	—	—	—	—	—	**2,001**

Table 4.4 (continued) Summary of Numbers of Cells g^{-1} Sediment of Living Microalgae Taken by Coring at Stations in Perdido Bay from June 1993 to October 1994

Species	18	19	22	23	25	26	27	29	31	33	23A	24A	29A	29B	29C	31A	31B	Total
Eunotia monodon v. constricta	—	—	—	1,334	—	667	—	—	—	—	—	—	—	—		—	—	**2,001**
Skeletonema costatum	—	—	—	—	—	—	—	—	—	2,001	—	—	—	—	—	—	—	**2,001**
Tropidoneis lepidoptera	—	—	—	1,334	—	667	—	—	—	—	—	—	—	—	—	—	—	**2,001**
Diploneis ovalis	—	—	—	—	—	—	—	—	—	—	1,668	—	—	—	—	—	—	**1,668**
Nitzschia punctata	—	—	—	—	—	—	—	—	—	1,334	—	—	334	—	—	—	—	**1,668**
Pinnularia acrosphaeria	—	—	334	—	—	1,334	—	—	—	—	—	—	—	—	—	—	—	**1,668**
Pinnularia viridis	334	—	—	667	—	—	—	667	—	—	—	—	—	—	—	—	—	**1,668**
Grammatophora marina	—	—	—	—	—	667	—	—	334	—	—	—	—	—	—	334	—	**1,335**
Achnanthes manifera	—	—	—	—	—	—	—	—	—	1,334	—	—	—	—	—	—	—	**1,334**
Amphora bigibba	—	—	—	—	—	667	—	—	—	667	—	—	—	—	—	—	—	**1,334**
Asterionella formosa	1,334	—	—	—	—	—	—	—	—	—	—	—	—	—	—	—	—	**1,334**
Asterionellopsis japonica	—	—	—	667	—	—	—	—	—	667	—	—	—	—	—	—	—	**1,334**
Diploneis fusca	—	—	—	—	—	667	—	—	—	667	—	—	—	—	—	—	—	**1,334**
Fallacia forcipata	—	—	—	667	—	667	—	—	—	—	—	—	—	—	—	—	—	**1,334**
Fragilaria brevistriata	—	—	—	—	—	1,334	—	—	—	—	—	—	—	—	—	—	—	**1,334**
Fragilaria pinnata	—	—	—	—	—	1,334	—	—	—	—	—	—	—	—	—	—	—	**1,334**
Melosira nummuloides	—	—	—	—	—	1,334	—	—	—	—	—	—	—	—	—	—	—	**1,334**
Navicula gracilis	—	—	—	—	—	—	—	667	—	667	—	—	—	—	—	—	—	**1,334**
Nitzschia littoralis	—	—	—	667	—	—	—	667	—	—	—	—	—	—	—	—	—	**1,334**
Opephora spp.	—	—	—	—	—	—	—	—	—	—	—	—	—	—	1,334	—	—	**1,334**
Psammodiscus nitidus	—	—	—	—	—	—	—	—	—	1,334	—	—	—	—	—	—	—	**1,334**
Rhopalodia gibberula	—	—	—	1,334	—	—	—	—	—	—	—	—	—	—	—	—	—	**1,334**
Surirella linearis	—	—	—	—	—	667	—	667	—	—	—	—	—	—	—	—	—	**1,334**
Tabellaria quadrisepta	—	—	—	—	—	667	—	667	—	—	—	—	—	—	—	—	—	**1,334**
Coscinodiscus spp.	—	—	—	—	—	—	—	334	—	—	334	—	—	—	—	—	334	**1,002**
Martyana martyi	—	334	—	—	334	—	—	—	—	—	334	—	—	—	—	—	—	**1,002**
Navicula hasta	—	—	—	—	—	334	334	—	—	—	—	—	334	—	—	—	—	**1,002**
Surirella ovata	—	334	—	334	—	—	—	—	—	—	—	—	334	—	—	—	—	**1,002**
Diploneis crabro	—	—	—	—	—	—	—	667	—	—	334	—	—	—	—	—	—	**1,001**
Melosira moniliformis	—	—	—	—	—	—	—	—	—	—	—	667	—	—	—	334	—	**1,001**
Navicula pygmaea	—	—	—	—	—	—	—	—	—	667	—	—	—	—	334	—	—	**1,001**
Nitzschia sigma	—	—	—	—	—	—	—	—	—	—	—	—	1,001	—	—	—	—	**1,001**
Pleurosigma angulatum	—	—	—	—	—	—	—	667	334	—	—	—	—	—	—	—	—	**1,001**
Trachyneis aspera	—	—	—	—	—	—	—	—	—	667	—	—	—	334	—	—	—	**1,001**
Actinella punctata	—	—	—	—	334	334	—	—	—	—	—	—	—	—	—	—	—	**668**

Nitzschia obtusa	—	—	334	—	—	—	—	—	—	—	—	—	—	334	—	—	—	**668**
Pinnularia gibba	—	—	—	—	—	—	—	—	—	—	334	334	—	—	—	—	—	**668**
Pinnularia major	334	—	—	—	—	334	—	—	—	—	—	—	—	—	—	—	—	**668**
Actinocyclus normanii v. *subsalsa*	—	—	—	—	—	—	—	—	—	667	—	—	—	—	—	—	—	**667**
Amphora binoides	—	—	—	—	—	—	—	—	—	667	—	—	—	—	—	—	—	**667**
Auliscus spp.	—	—	—	—	—	—	—	667	—	—	—	—	—	—	—	—	—	**667**
Caloneis silicula	—	—	—	—	—	667	—	—	—	—	—	—	—	—	—	—	—	**667**
Campylodiscus clypeus	—	—	—	—	—	—	—	—	—	667	—	—	—	—	—	—	—	**667**
Catacombus camschatica	—	—	—	—	—	667	—	—	—	—	—	—	—	—	—	—	—	**667**
Coscinodiscus argus	—	—	—	—	—	667	—	—	—	—	—	—	—	—	—	—	—	**667**
Coscinodiscus radiatus	—	—	—	—	—	—	—	—	—	667	—	—	—	—	—	—	—	**667**
Cyclotella spp.	—	—	—	—	—	—	—	667	—	—	—	—	—	—	—	—	—	**667**
Eunotia lunaris	—	—	—	—	—	—	—	667	—	—	—	—	—	—	—	—	—	**667**
Eupodiscus ocellatus	—	—	—	—	—	667	—	—	—	—	—	—	—	—	—	—	—	**667**
Grammatophora oceanica	—	—	—	—	—	—	—	—	—	667	—	—	—	—	—	—	—	**667**
Gymnodinium spp.	—	—	—	667	—	—	—	—	—	—	—	—	—	—	—	—	—	**667**
Hantzschia virgata	—	—	—	667	—	—	—	—	—	—	—	—	—	—	—	—	—	**667**
Mastogloia angulata	—	—	—	—	—	—	—	—	—	667	—	—	—	—	—	—	—	**667**
Melosira varians	—	—	—	—	—	—	—	667	—	—	—	—	—	—	—	—	—	**667**
Navicula americana	—	—	—	—	—	667	—	—	—	—	—	—	—	—	—	—	—	**667**
Navicula lyroides	—	—	—	—	—	—	—	667	—	—	—	—	—	—	—	—	—	**667**
Neidium iridis v. *amphigomphus*	—	—	—	—	—	667	—	—	—	—	—	—	—	—	—	—	—	**667**
Nitzschia insignis	—	—	—	—	—	—	—	—	—	667	—	—	—	—	—	—	—	**667**
Nitzschia lanceolata	—	—	—	—	—	—	—	—	—	667	—	—	—	—	—	—	—	**667**
Nitzschia longissima	—	—	—	—	—	—	—	—	—	667	—	—	—	—	—	—	—	**667**
Nitzschia pungens	—	—	—	—	—	—	—	667	—	—	—	—	—	—	—	—	—	**667**
Stauroneis phoenicenteron	—	—	—	667	—	—	—	—	—	—	—	—	—	—	—	—	—	**667**
Stenopterobia anceps	—	—	—	—	—	—	—	—	—	667	—	—	—	—	—	—	—	**667**
Synedra ulna	—	—	—	667	—	—	—	—	—	—	—	—	—	—	—	—	—	**667**
Thalassionema nitzschioides	—	—	—	—	—	—	—	—	—	—	—	—	—	—	—	667	—	**667**
Amphora obtusa	—	—	—	—	—	—	—	—	—	—	—	—	—	—	—	—	334	**334**
Auliscus pruinosus	—	—	—	—	—	—	—	—	334	—	—	—	—	—	—	—	—	**334**
Diploneis smithii	—	—	—	334	—	—	—	—	—	—	—	—	—	—	—	—	—	**334**
Eunotia curvata	334	—	—	—	—	—	—	—	—	—	—	—	—	—	—	—	—	**334**
Eunotia formica	—	—	—	—	—	—	—	334	—	—	—	—	—	—	—	—	—	**334**
Eunotia monodon v. *bidens*	—	334	—	—	—	—	—	—	—	—	—	—	—	—	—	—	—	**334**
Eupodiscus radiatus	—	—	—	—	—	—	—	—	—	—	334	—	—	—	—	—	—	**334**

Table 4.4 (continued) Summary of Numbers of Cells g^{-1} Sediment of Living Microalgae Taken by Coring at Stations in Perdido Bay from June 1993 to October 1994

Species	18	19	22	23	25	26	27	29	31	33	23A	24A	29A	29B	29C	31A	31B	Total
Gomphonema constrictum	—	—	—	—	—	—	334	—	—	—	—	—	—	—	—	—	—	334
Hemiaulus hauckii	—	—	—	—	—	334	—	—	—	—	—	—	—	—	—	—	—	334
Licmophora spp.	—	—	—	—	—	—	—	—	—	—	—	334	—	—	—	—	—	334
Nitzschia longissima v. reversa	—	—	—	334	—	—	—	—	—	—	—	—	—	—	—	—	—	334
Proboscia alata	—	—	—	—	—	—	—	—	—	—	—	—	—	—	—	334	—	334
Rhabdonema adriaticum	—	—	—	—	334	—	—	—	—	—	—	—	—	—	—	—	—	334
Rhaphoneis superba	—	—	—	—	—	—	—	—	—	—	—	—	334	—	—	—	—	334
Stauroneis anceps	—	—	—	—	—	—	—	334	—	—	—	—	—	—	—	—	—	334
Surirella febigeri	—	334	—	—	—	—	—	—	—	—	—	—	—	—	—	—	—	334
Surirella robusta	—	334	—	—	—	—	—	—	—	—	—	—	—	—	—	—	—	334
Terpsinöe musica	—	—	—	—	—	—	—	—	—	—	—	334	—	—	—	—	—	334
Thalassiothrix longissima	—	—	—	—	—	—	—	334	—	—	—	—	—	—	—	—	—	334
Triceratium reticulatum	—	—	—	—	—	—	—	—	334	—	—	—	—	—	—	—	—	334
Total	**122,402**	**153,759**	**21,017**	**820,421**	**68,048**	**1,250,634**	**61,374**	**1,340,019**	**172,102**	**857,095**	**112,069**	**175,763**	**170,432**	**176,438**	**120,737**	**202,114**	**157,090**	**5,981,508**

Data are also summed by species (across stations) and by date (across species).

Table 4.5 Summary of Biomass (µg) of Cells g^{-1} Sediment of Living Microalgae Taken in Perdido Bay from June 1993 to October 1994

Species	18	19	22	23	25	26	27	29	31	33	23A	24A	29A	29B	29C	31A	31B	TOTAL
Bacillaria paxillifer	0.003885	0.002920	0.003278	0.065215	0.002080	0.086170	0.000830	0.084175	0.000695	0.022750	0.005950	—	0.000970	0.000139	0.003890	0.002500	0.001810	**0.284257**
Navicula spp.	0.010370	0.009310	0.001870	0.028684	0.003455	0.043481	0.003800	0.039434	0.011300	0.043177	0.006350	0.010210	0.009420	0.010700	0.006360	0.014750	0.011450	**0.264121**
Pleurosigma spp.	0.003030	0.004040	—	0.015150	0.002020	0.046420	—	0.039370	0.004040	0.026260	0.003030	—	0.006050	0.007050	0.005050	0.010080	0.015220	**0.186810**
Petrodictyon gemma	—	—	—	—	0.004600	0.009200	—	0.124200	—	0.004600	—	—	0.002300	0.006900	0.004600	0.011500	—	**0.167900**
Gyrosigma balticum	—	—	—	—	0.003680	0.014700	0.003680	0.007350	0.033050	0.029400	—	—	—	0.007360	0.011000	0.014680	0.003680	**0.128580**
Nitzschia sigmoidea	0.000830	0.001041	—	0.017250	0.002706	0.017657	0.001250	0.032843	0.002920	0.008722	0.001870	0.002080	0.000833	0.003955	0.004786	0.002076	0.005610	**0.106429**
Nitzschia spp.	0.000373	0.002270	0.000211	0.013613	0.000397	0.017961	0.000654	0.012565	0.001610	0.007882	0.001355	0.003834	0.001750	0.001095	0.001217	0.001425	0.000934	**0.069146**
Entomoneis spp.	—	0.001590	—	0.014270	0.001590	0.009510	0.001590	0.011100	—	0.003170	—	—	—	0.001590	—	—	0.003170	**0.047580**
Entomoneis alata	—	—	—	0.003990	—	0.011290	—	0.011940	—	0.003990	0.001330	0.001990	—	—	—	0.000665	—	**0.035195**
Undetermined pennate diatoms	—	—	—	—	—	—	—	0.005900	0.007100	—	—	0.002370	0.0070	0.009450	—	—	—	**0.027190**
Paralia sulcata	—	—	—	0.001115	0.000446	0.002082	—	0.007958	0.001270	0.005652	—	—	0.001042	0.001412	—	0.002010	0.000149	**0.023135**
Frustulia rhomboides	0.003660	0.002750	—	0.005490	—	0.004575	—	0.004575	—	—	—	—	—	0.000915	—	—	—	**0.021965**
Thalassiosira spp.	—	0.000148	—	0.001180	—	0.001328	—	0.003834	0.000737	0.005005	0.000148	0.000295	0.000443	0.000735	0.000442	0.000590	0.001033	**0.015918**
Cymatosira lorenziana	—	—	—	—	0.000294	0.000587	—	0.001394	0.000367	0.003084	—	—	0.002277	0.000514	0.000585	0.001102	—	**0.010204**
Pinnularia viridis	0.001680	—	—	0.003360	—	—	—	0.003360	—	—	—	—	—	—	—	—	—	**0.008400**
Gyrosigma fasciola	—	—	—	0.001260	—	—	—	—	—	0.006300	—	—	—	—	—	—	—	**0.007560**
Tabellaria fenestrata	0.002260	0.001420	—	0.000566	—	0.001132	—	0.000566	—	—	0.001130	0.000142	—	0.000142	—	—	—	**0.007358**
Pleurosigma angulatum	—	—	—	—	—	—	—	0.003850	0.001930	—	—	—	—	—	—	—	—	**0.005780**
Nitzschia granulata	—	—	—	0.001664	—	0.003328	—	0.000554	—	—	—	—	—	—	—	—	—	**0.005546**
Proboscia alata	—	—	—	—	—	—	—	—	—	—	—	—	—	—	—	0.003960	—	**0.003960**
Eunotia pectinalis v. undulata	0.000346	0.000520	—	0.000173	—	0.000519	0.000173	0.000346	—	—	0.000520	0.000346	—	—	—	—	—	**0.002943**
Rhabdonema adriaticum	—	—	—	—	0.002170	—	—	—	—	—	—	—	—	—	—	—	—	**0.002170**
Eunotia pectinalis	0.000065	0.000389	0.000065	0.000647	—	0.000065	—	0.000324	—	—	—	—	—	—	—	—	—	**0.001553**
Nitzschia longissima	—	—	—	—	—	—	—	—	—	0.001480	—	—	—	—	—	—	—	**0.001480**
Cocconeis placentula	—	—	—	0.000136	—	0.000544	—	0.000680	—	—	—	—	—	—		—	—	**0.001360**
Actinoptychus undulatus	—	—	—	—	—	—	—	0.000000	—	0.001320	—	—	—	—		—	—	**0.001320**
Eunotia flexuosa v. eurycephala	—	—	—	0.000201	0.000101	0.000101	—	0.000302	—	—	0.000101	0.000201	—	—	—	—	—	**0.001007**
Cylindrotheca closterium	—	—	—	0.000302	0.000038	0.000075	—	0.000075	—	0.000151	—	—	—	—	—	—	0.000189	**0.000830**
Navicula confervacea	—	0.000409	—	—	—	—	—	—	—	—	—	—	—	—	—	—	—	**0.000409**
Synedra ulna	—	—	—	0.000396	—	—	—	—	—	—	—	—	—	—	—	—	—	**0.000396**

Table 4.5 (continued) Summary of Biomass (μg) of Cells g^{-1} Sediment of Living Microalgae Taken in Perdido Bay from June 1993 to October 1994

Species	18	19	22	23	25	26	27	29	31	33	23A	24A	29A	29B	29C	31A	31B	TOTAL
Gymnodinium spp.	—	—	—	0.000389	—	—	—	—	—	—	—	—	—	—	—	—	—	**0.000389**
Hemiaulus hauckii	—	—	—	—	—	0.000368	—	—	—	—	—	—	—	—	—	—	—	**0.000368**
Hantzschia virgata	—	—	—	0.000343	—	—	—	—	—	—	—	—	—	—	—	—	—	**0.000343**
Aulacoseira granulata	—	—	—	—	0.000252	—	0.000042	—	—	—	—	—	—	—	—	—	—	**0.000294**
Skeletonema costatum	—	—	—	—	—	—	—	—	—	0.000140	—	—	—	—	—	—	—	**0.000140**
Thalassionema nitzschioides	—	—	—	—	—	—	—	—	—	—	—	—	—	—	—	—	0.000120	**0.000120**
Nitzschia pungens	—	—	—	—	—	—	—	0.000077	—	—	—	—	—	—	—	—	—	**0.000077**
Cyclotella spp.	—	—	—	—	—	—	—	0.000048	—	—	—	—	—	—	—	—	—	**0.000048**
Achnanthes curvirostrum	—	—	—	0.000000	—	0.000000	—	0.000000	—	0.000000	—	—	—	—	—	—	0.000000	**0.000000**
Achnanthes manifera	—	—	—	—	—	—	—	—	—	0.000000	—	—	—	—	—	—	—	**0.000000**
Actinella punctata	—	—	—	—	0.000000	0.000000	—	—	—	—	—	—	—	—	—	—	—	**0.000000**
Actinocyclus normanii v. subsalsa	—	—	—	—	—	—	—	—	—	0.000000	—	—	—	—	—	—	—	**0.000000**
Actinocyclus spp.	—	—	—	—	—	—	—	0.000000	—	0.000000	—	—	—	—	—	—	—	**0.000000**
Amphora arenaria	—	—	—	—	—	—	—	0.000000	—	0.000000	—	—	—	—	—	—	—	**0.000000**
Amphora bigibba	—	—	—	—	—	0.000000	—	—	—	0.000000	—	—	—	—	—	—	—	**0.000000**
Amphora binoides	—	—	—	—	—	—	—	—	—	0.000000	—	—	—	—	—	—	—	**0.000000**
Amphora obtusa	—	—	—	—	—	—	—	—	—	—	—	—	—	—	—	—	—	**0.000000**
Amphora spp.	0.000000	0.000000	0.000000	0.000000	0.000000	0.000000	0.000000	0.000000	0.000000	0.000000	0.000000	0.000000	0.000000	0.000000	0.000000	0.000000	0.000000	**0.000000**
Asterionella formosa	0.000000	—	—	—	—	—	—	—	—	—	—	—	—	—	—			**0.000000**
Asterionellopsis japonica	—	—	—	0.000000	—	—	—	—	—	0.000000	—	—	—	—	—	—	—	**0.000000**
Auliscus pruinosus	—	—	—	—	—	—	—	—	0.000000	—	—	—	—	—	—	—	—	**0.000000**
Auliscus spp.	—	—	—	—	—	—	—	0.000000	—	—	—	—	—	—	—	—	—	**0.000000**
Caloneis acutiascula	0.000000	0.000000	—	0.000000	0.000000	0.000000	—	0.000000	0.000000	—	0.000000	—	—	—	0.000000	0.000000	—	**0.000000**
Caloneis latiuscula	—	—	—	—	—	—	—	—	0.000000	—	—	—	—	—	—	—	—	**0.000000**
Caloneis silicula	—	—	—	—	—	0.000000	—	—	—	—	—	—	—	—	—	—	—	**0.000000**
Caloneis spp.	—	—	—	—	—	0.000000	—	0.000000	—	0.000000	—	—	—	—	—	—	—	**0.000000**
Campylodiscus clypeus	—	—	—	—	—	—	—	—	—	0.000000	—	—	—	—	—	—	—	**0.000000**
Campylodiscus echeneis	0.000000	0.000000	—	0.000000	0.000000	0.000000	0.000000	0.000000	0.000000	0.000000	—	—	0.000000	0.000000	—	—	—	**0.000000**
Catacombus camschatica	—	—	—	—	—	0.000000	—	—	—	—	—	—	—	—	—	—	—	**0.000000**

Cocconeis scutellum	0.000000	—	—	—	—	0.000000	0.000000	—	—	—	0.000000	0.000000	0.000000	—	—	0.000000	0.000000	**0.000000**
Cocconeis spp.	0.000000	0.000000	0.000000	0.000000	0.000000	0.000000	0.000000	0.000000	0.000000	—	0.000000	0.000000	0.000000	0.000000	—	—	0.000000	**0.000000**
Coscinodiscus argus	—	—	—	—	—	0.000000	—	—	—	—	—	—	—	—	—	—	—	**0.000000**
Coscinodiscus radiatus	—	—	—	—	—	—	—	—	—	0.000000	—	—	—	—	—	—	—	**0.000000**
Coscinodiscus spp.	—	—	—	—	—	—	—	0.000000	—	—	0.000000	—	—	—	—	—	0.000000	**0.000000**
Cymatosira belgica	—	—	—	—	—	—	—	0.000000	—	0.000000	—	—	—	—	—	—	—	**0.000000**
Dictyocha fibula	—	—	—	0.000000	0.000000	—	—	0.000000	—	—	—	—	0.000000	0.000000	—	0.000000	—	**0.000000**
Diploneis bombus	—	—	—	0.000000	—	0.000000	—	0.000000	—	0.000000	0.000000	—	—	—	—	—	—	**0.000000**
Diploneis crabro	—	—	—	—	—	—	—	0.000000	—	—	0.000000	—	—	—	—	—	—	**0.000000**
Diploneis didyma	0.000000	0.000000	0.000000	—	0.000000	—	—	0.000000	0.000000	0.000000	0.000000	0.000000	0.000000	0.000000	—	0.000000	0.000000	**0.000000**
Diploneis elliptica	0.000000	—	—	0.000000	—	0.000000	—	0.000000	0.000000	0.000000	—	—	—	—	—	—	—	**0.000000**
Diploneis fusca	—	—	—	—	—	0.000000	—	—	—	0.000000	—	—	—	—	—	—	—	**0.000000**
Diploneis ovalis	—	—	—	—	—	—	—	—	—	—	0.000000	—	—	—	—	—	—	**0.000000**
Diploneis smithii	—	—	—	0.000000	—	—	—	—	—	—	—	—	—	—	—	—	—	**0.000000**
Diploneis spp.	0.000000	0.000000	0.000000	0.000000	0.000000	0.000000	0.000000	0.000000	0.000000	0.000000	0.000000	0.000000	0.000000	0.000000	0.000000	0.000000	0.000000	**0.000000**
Eunotia curvata	0.000000	—	—	—	—	—	—	—	—	—	—	—	—	—	—	—		**0.000000**
Eunotia formica	—	—	—	—	—	—	—	0.000000	—	—	—	—	—	—	—	—	—	**0.000000**
Eunotia lunaris	—	—	—	—	—	—	—	0.000000	—	—	—	—	—	—	—	—	—	**0.000000**
Eunotia monodon v. bidens	—	0.000000	—	—	—	—	—	—	—	—	—	—	—	—	—	—	—	**0.000000**
Eunotia monodon v. constricta	—	—	—	0.000000	—	0.000000	—	—	—	—	—	—	—	—	—	—	—	**0.000000**
Eunotia monodon	—	0.000000	—	0.000000	—	—	—	—	—	—	—	—	—	—	—	—	—	**0.000000**
Eunotia serra	—	—	—	0.000000	—	0.000000	—	0.000000	—	—	—	—	—	—	—	—	—	**0.000000**
Eunotia spp.	0.000000	0.000000	—	0.000000	—	—	—	0.000000	0.000000	—	—	0.000000	0.000000	—	—	—	—	**0.000000**
Eupodiscus ocellatus	—	—	—	—	—	0.000000	—	—	—	—	—	—	—	—	—	—	—	**0.000000**
Eupodiscus radiatus	—	—	—	—	—	—	—	—	—	—	0.000000	—	—	—	—	—	—	**0.000000**
Fallacia forcipata	—	—	—	0.000000	—	0.000000	—	—	—	—	—	—	—	—	—	—	—	**0.000000**
Fragilaria brevistriata	—	—	—	—	—	0.000000	—	—	—	—	—	—	—	—	—	—	—	**0.000000**
Fragilaria leptostauron	—	—	—	—	—	0.000000	—	0.000000	—	—	—	—	—	—	—	—	—	**0.000000**
Fragilaria pinnata	—	—	—	—	—	0.000000	—	—	—	—	—	—	—	—	—	—	—	**0.000000**
Fragilaria spp.	—	—	—	0.000000	—	—	—	0.000000	—	0.000000	—	—	—	—	—	—	—	**0.000000**
Gomphonema constrictum	—	—	—	—	—	—	0.000000	—	—	—	—	—	—	—	—	—	—	**0.000000**
Gomphonema spp.	—	0.000000	—	—	—	—	0.000000	—	0.000000	—	—	—	—	—	—	0.000000	0.000000	**0.000000**
Grammatophora marina	—	—	—	—	—	0.000000	—	—	0.000000	—	—	—	—	—	—	0.000000	—	**0.000000**
Grammatophora oceanica	—	—	—	—	—	—	—	—	—	0.000000	—	—	—	—	—	—	—	**0.000000**
Gyrosigma macrum	—	—	—	—	0.000000	—	—	0.000000	0.000000	0.000000	—	—	—	0.000000	0.000000	0.000000	0.000000	**0.000000**

Table 4.5 (continued) Summary of Biomass (μg) of Cells g^{-1} Sediment of Living Microalgae Taken in Perdido Bay from June 1993 to October 1994

Species	18	19	22	23	25	26	27	29	31	33	23A	24A	29A	29B	29C	31A	31B	TOTAL
Gyrosigma parkerii	—	0.000000	—	0.000000	0.000000	0.000000	0.000000	0.000000	0.000000	0.000000	—	0.000000	0.000000	0.000000	0.000000	0.000000	0.000000	**0.000000**
Gyrosigma sp. (s shaped)	—	—	—	0.000000	—	0.000000	—	0.000000	—	0.000000	0.000000	—	—	—	—	—	—	**0.000000**
Gyrosigma spp.	—	—	—	0.000000	—	0.000000	—	0.000000	0.000000	0.000000	—	—	0.000000	0.000000	0.000000	0.000000	—	**0.000000**
Licmophora spp.	—	—	—	—	—	—	—	—	—	—	—	0.000000	—	—	—	—	—	**0.000000**
Martyana martyi	—	0.000000	—	—	0.000000	—	—	—	—	—	0.000000	—	—	—	—	—	—	**0.000000**
Mastogloia angulata	—	—	—	—	—	—	—	—	—	0.000000	—	—	—	—	—	—	—	**0.000000**
Mastogloia spp.	—	—	—	0.000000	—	—	—	—	—	0.000000	—	—	—	—	—	0.000000	—	**0.000000**
Melosira dubia	—	—	—	—	—	0.000000	—	0.000000	—	—	—	—	—	—	—	—	—	**0.000000**
Melosira moniliformis	—	—	—	—	—	—	—	—	—	—	—	0.000000	—	—	—	0.000000	—	**0.000000**
Melosira nummuloides	—	—	—	—	—	0.000000	—	—	—	—	—	—	—	—	—	—	—	**0.000000**
Melosira spp.	—	—	—	0.000000	—	0.000000	—	0.000000	—	0.000000	—	—	—	—	—	0.000000	—	**0.000000**
Melosira varians	—	—	—	—	—	—	—	0.000000	—	—	—	—	—	—	—	—	—	**0.000000**
Navicula americana	—	—	—	—	—	0.000000	—	—	—	—	—	—	—	—	—	—	—	**0.000000**
Navicula cf. *confervacea*	—	—	—	—	—	—	—	—	—	0.000000	—	—	—	—	—	—	—	**0.000000**
Navicula gracilis	—	—	—	—	—	—	—	0.000000	—	0.000000	—	—	—	—	—	—	—	**0.000000**
Navicula hasta	—	—	—	—	—	0.000000	0.000000	—	—	—	—	—	0.000000	—	—	—	—	**0.000000**
Navicula lyra	—	—	—	—	—	0.000000	—	0.000000	0.000000	0.000000	—	—	0.000000	0.000000	—	—	—	**0.000000**
Navicula lyroides	—	—	—	—	—	—	—	0.000000	—	—	—	—	—	—	—	—	—	**0.000000**
Navicula pygmaea	—	—	—	—	—	—	—	—	—	0.000000	—	—	—	—	0.000000	—	—	**0.000000**
Navicula yarrensis v. *americana*	—	—	—	0.000000	—	0.000000	—	0.000000	0.000000	0.000000	—	—	—	0.000000	0.000000	0.000000	0.000000	**0.000000**
Navicula yarrensis	—	—	—	0.000000	—	0.000000	—	0.000000	0.000000	0.000000	—	—	0.000000	0.000000	0.000000	0.000000	0.000000	**0.000000**
Neidium iridis v. *amphigomphus*	—	—	—	—	—	0.000000	—	—	—	—	—	—	—	—	—	—	—	**0.000000**
Nitzschia circumsuta	—	—	—	0.000000	—	0.000000	0.000000	0.000000	0.000000	0.000000	—	—	0.000000	0.000000	—	—	—	**0.000000**
Nitzschia dubia	0.000000	0.000000	—	0.000000	—	0.000000	0.000000	0.000000	0.000000	0.000000	0.000000	0.000000	0.000000	0.000000	—	—	0.000000	**0.000000**
Nitzschia gracilis	—	0.000000	0.000000	0.000000	0.000000	0.000000	0.000000	0.000000	0.000000	—	—	0.000000	—	0.000000	—	0.000000	0.000000	**0.000000**
Nitzschia hungarica	—	—	—	0.000000	—	0.000000	—	—	—	0.000000	—	—	—	—	—	—	—	**0.000000**
Nitzschia insignis	—	—	—	—	—	—	—	—	—	0.000000	—	—	—	—	—	—	—	**0.000000**
Nitzschia lanceolata	—	—	—	—	—	—	—	—	—	0.000000	—	—	—	—	—	—	—	**0.000000**
Nitzschia littoralis	—	—	—	0.000000	—	—	—	0.000000	—	—	—	—	—	—	—	—	—	**0.000000**
Nitzschia longissima v. *reversa*	—	—	—	0.000000	—	—	—	—	—	—	—	—	—	—	—	—	—	**0.000000**
Nitzschia lorenziana	—	—	—	0.000000	—	0.000000	—	0.000000	—	0.000000	—	—	—	—	—	0.000000	—	**0.000000**
Nitzschia obtusa	—	—	0.000000	—	—	—	—	—	—	—	—	—	—	0.000000	—	—	—	**0.000000**

Nitzschia panduriformis	—	0.000000	—	0.000000	—	0.000000	—	0.000000	—	0.000000	—	—	—	0.000000	—	0.000000	—	**0.000000**
Nitzschia punctata	—	—	—	—	—	—	—	—	—	0.000000	—	—	0.000000	—	—	—	—	**0.000000**
Nitzschia scalaris	0.000000	—	—	0.000000	—	0.000000	—	0.000000	—	—	—	—	—	—	—	—	—	**0.000000**
Nitzschia sigma	—	—	—	—	—	—	—	—	—	—	—	—	0.000000	—	—	—	—	**0.000000**
Nitzschia tryblionella	0.000000	0.000000	—	0.000000	0.000000	0.000000	0.000000	0.000000	0.000000	0.000000	0.000000	0.000000	0.000000	0.000000	—	0.000000	0.000000	**0.000000**
Opephora spp.	—	—	—	—	—	—	—	—	—	—	—	—	—	—	0.000000	—	—	**0.000000**
Pinnularia acrosphaeria	—	—	0.000000	—	—	0.000000	—	—	—	—	—	—	—	—	—	—	—	**0.000000**
Pinnularia gibba	—	—	—	—	—	—	—	—	—	—	0.000000	0.000000	—	—	—	—	—	**0.000000**
Pinnularia major	0.000000	—	—	—	—	0.000000	—	—	—	—	—	—	—	—	—	—	—	**0.000000**
Pinnularia spp.	—	—	—	0.000000	—	—	—	—	—	—	—	—	—	—	—	—	—	**0.000000**
Pleurosigma elongatum	—	—	—	0.000000	—	0.000000	0.000000	0.000000	—	—	—	0.000000	0.000000	—	0.000000	—	—	**0.000000**
Psammodiscus nitidus	—	—	—	—	—	—	—	—	—	0.000000	—	—	—	—	—	—	—	**0.000000**
Rhaphoneis superba	—	—	—	—	—	—	—	—	—	—	—	—	0.000000	—	—	—	—	**0.000000**
Rhopalodia gibberula	—	—	—	0.000000	—	—	—	—	—	—	—	—	—	—	—	—	—	**0.000000**
Stauroneis anceps	—	—	—	—	—	—	—	0.000000	—	—	—	—	—	—	—	—	—	**0.000000**
Stauroneis phoenicenteron	—	—	—	0.000000	—	—	—	—	—	—	—	—	—	—	—	—	—	**0.000000**
Stenopterobia anceps	—	—	—	—	—	—	—	—	—	0.000000	—	—	—	—	—	—	—	**0.000000**
Surirella fastuosa	—	—	—	0.000000	0.000000	0.000000	—	0.000000	0.000000	0.000000	—	—	0.000000	0.000000	—	0.000000	0.000000	**0.000000**
Surirella febigeri	—	0.000000	—	—	—	—	—	—	—	—	—	—	—	—	—	—	—	**0.000000**
Surirella linearis	—	—	—	—	—	0.000000	—	0.000000	—	—	—	—	—	—	—	—	—	**0.000000**
Surirella ovata	—	0.000000	—	0.000000	—	—	—	—	—	—	—	—	0.000000	—	—	—	—	**0.000000**
Surirella robusta	—	0.000000	—	—	—	—	—	—	—	—	—	—	—	—	—	—	—	**0.000000**
Surirella spp.	0.000000	0.000000	—	0.000000	0.000000	0.000000	0.000000	0.000000	0.000000	0.000000	0.000000	0.000000	0.000000	0.000000	—	0.000000	0.000000	**0.000000**
Synedra spp.	0.000000	0.000000	—	0.000000	0.000000	0.000000	—	0.000000	—	—	0.000000	0.000000	—	—	0.000000	—	—	**0.000000**
Synedra tabulata	—	—	0.000000	0.000000	—	—	—	0.000000	0.000000	0.000000	0.000000	—	—	—	—	0.000000	—	**0.000000**
Tabellaria quadrisepta	—	—	—	—	—	0.000000	—	0.000000	—	—	—	—	—	—	—	—	—	**0.000000**
Terpsinöe musica	—	—	—	—	—	—	—	—	—	—	—	0.000000	—	—	—	—	—	**0.000000**
Thalassiothrix longissima	—	—	—	—	—	—	—	0.000000	—	—	—	—	—	—	—	—	—	**0.000000**
Trachyneis aspera	—	—	—	—	—	—	—	—	—	0.000000	—	—	—	0.000000	—	—	—	**0.000000**
Triceratium reticulatum	—	—	—	—	—	—	—	—	0.000000	—	—	—	—	—	—	—	—	**0.000000**
Tropidoneis lepidoptera	—	—	—	0.000000	—	0.000000	—	—	—	—	—	—	—	—	—	—	—	**0.000000**
Tryblionella levidansis	—	—	—	0.000000	—	0.000000	—	0.000000	—	0.000000	—	—	—	—	—	—	0.000000	**0.000000**
Undetermined centric diatoms	—	0.000000	—	0.000000	—	0.000000	—	0.000000	—	0.000000	0.000000	—	0.000000	0.000000	—	0.000000	0.000000	**0.000000**
Total	**0.026499**	**0.026807**	**0.002423**	**0.175393**	**0.023828**	**0.271094**	**0.012019**	**0.396820**	**0.065019**	**0.173083**	**0.021784**	**0.021468**	**0.027455**	**0.051957**	**0.037930**	**0.065458**	**0.043245**	**1.442282**

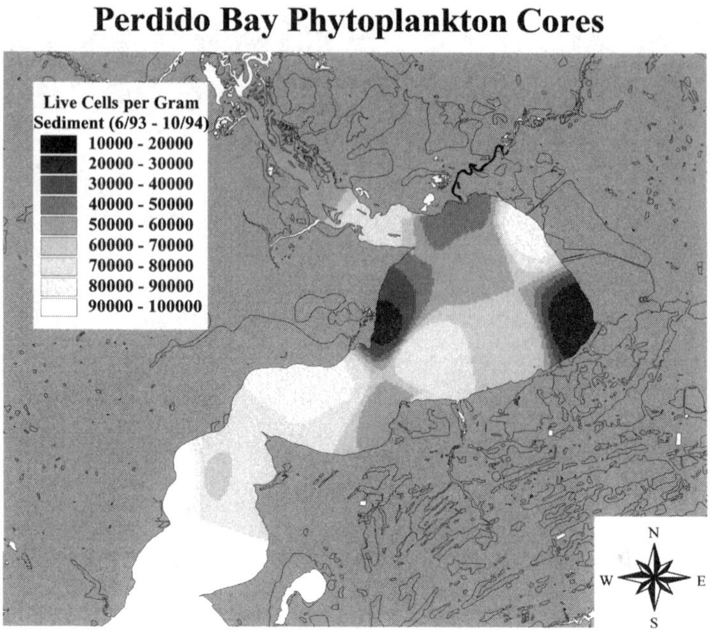

Figure 4.46 Map of the time-averaged data concerning sediment microalgal biomass (g^{-1} dw) taken in Perdido Bay at monthly intervals from June 1993 to October 1994.

phytoplankton blooms. The sediment microphyte data could provide an important link connecting the phytoplankton blooms with benthic infauna that represent the principal connection between the estuarine primary production and the upper levels of the food web (Livingston et al., 1997). It should be emphasized that such a link is based on correlative data taken over a relatively short period. However, there was a significant reduction of the benthic microphyte community in Perdido Bay that coincided with increased orthophosphate loading, a succession of phytoplankton blooms, and reductions of infaunal macroinvertebrate assemblages.

Although the processes associated with the postulated bloom effects (toxic substances, predation, competition) remain beyond the scope of this study, the response of the base of the trophic organization of bay sediments following the blooms of 1994 indicates a possible connection between the phytoplankton blooms, sediment microphytes, and infaunal macroinvertebrates. The causes of the observed changes in the sediment microflora could be due to various habitat changes, direct and/or indirect, that were associated with bloom species. Dinoflagellate resting stages produce toxins (Riegman, 1998). Increased organic loading from the water column to the sediments could alter microbial food webs, thus enhancing selection pressure on algal resting stages. This, in turn, could produce more toxic water column populations. Sediment changes during and after hypereutrophication could thus act as a selective factor for both benthic microphytes and water column plankton. Moderate eutrophication could produce increased useful productivity, whereas excessive nutrient loading could lead to favorable conditions for toxic species and/or inedible algae, thus leading to imbalances of the phytoplankton community, reductions of the benthic microbiota, and altered food webs via the infaunal macroinvertebrates. These hypotheses should be tested experimentally since, if there is a direct connection between the plankton blooms and the deterioration of the benthos, the alteration of phytoplankton associations due to changes in nutrient loading could lead to widespread reductions of secondary production in ways that are difficult to detect and currently remain outside of the scope of most research concerning sediment contamination.

I. Phytoplankton Blooms and Water Quality

1. Introduction

It is widely acknowledged that phytoplankton blooms are controlled to some degree by various physical, chemical, and biological factors that include water quality. Conversely, these blooms have a variety of effects on water and sediment quality. There are established water quality indicators that are commonly used to represent phytoplankton activities. Chlorophyll *a* is routinely considered to be an indicator of phytoplankton assemblages and is used to represent phytoplankton in descriptive studies and modeling. Particulate and dissolved organic carbon are also used as indicators of phytoplankton associations. Dissolved oxygen is considered to be a reliable indicator of phytoplankton activity. However, phytoplankton assemblages represent a very diverse and complex set of species that are not well defined in terms of natural history and are rarely studied over long periods as a community. Consequently, there is considerable confusion concerning how water and sediment quality affect phytoplankton distribution, and, conversely, how phytoplankton activities affect water and sediment quality. Complex feedback phenomena preclude simplistic generalizations.

Various studies have addressed the question of the effects of habitat on phytoplankton (Harding, 1994; Cloern, 1996; Hallegreaff and Fraga, 1997; Sin et al., 1999). Sin et al. (1999) found that river discharge regulated the magnitude, location, and timing of phytoplankton blooms in the York River estuary. This control was due to river flow impacts on residence time, nutrient loading, light transmission, and tidal mixing. Zooplankton grazing was not as important as resource limitation in controlling the phytoplankton dynamics. However, the long-term effects of river flow on species-specific changes in the phytoplankton community were not analyzed. Conditions prior to bloom outbreaks are not usually defined in such studies. This is typical of the literature concerning phytoplankton/habitat interactions. Consequently, few generalizations can be made concerning the origin and sustenance of a given bloom within the context of interannual changes of phytoplankton community dynamics. The complex life history cycles of various bloom species complicate generalizations. Another problem is that much of the published literature concerning individual bloom species is devoted to following such a population without taking supporting data concerning habitat variables. Quantitative estimates of changes in the phytoplankton species composition of the community as a whole are rarely included in the published literature. Consequently, questions concerning how blooms are initiated, how such blooms affect phytoplankton community structure, and the long-term effects of phytoplankton blooms on water and sediment quality have been largely ignored.

By analyzing 11 years of water quality data in the Perdido system relative to the quantitative and qualitative aspects of the phytoplankton distributions in Perdido Bay, it was possible to evaluate the long-term relationships of phytoplankton assemblages and water quality.

2. Long-Term Water Quality Trends

The long-term changes of water quality in Perdido Bay have been described above. These data are summarized here as annual averages (Figure 4.47). Surface water temperature increased throughout the bay during droughts (Figure 4.47a). Bottom salinities tended also to be higher during these periods. Bottom salinity in the bay varied inversely with distance from the Perdido River. Water color was highest in Elevenmile Creek during the first year of sampling; this was followed by a brief increase in 1994 that was followed by declines from 1995 through 1998 and a smaller peak in 1999. With the exception of Elevenmile Creek, which was affected by mill discharges, water color in Perdido Bay tended to be lowest during drought periods. The water color in the creek was significantly reduced following a paper mill color reduction program in 1990. Water

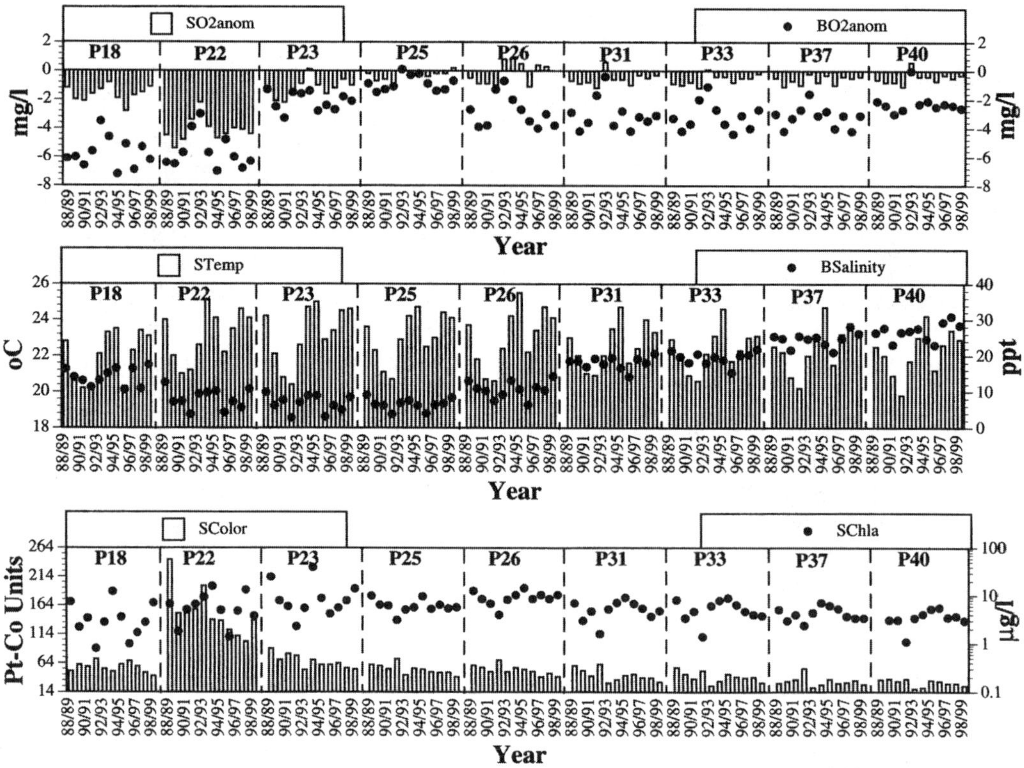

Figure 4.47 Annual averages of important water quality factors: (a) surface and bottom oxygen anomaly, surface temperature, bottom salinity, surface water color, surface chlorophyll *a*.

color throughout the bay followed interannual periods noted at the mouth of the Perdido River with peaks during high river flow years and reductions during the three drought periods. This is further evidence that disharge from Elevenmile Creek had a minimal influence on water color in the bay. Chlorophyll *a* concentrations (Figure 4.47a) were highest during periods of low river flow with peaks in 1993–1994. Although annual averages of chlorophyll *a* concentrations were generally low throughout the bay, peak values were noted at the mouth of Elevenmile Creek during 1989, 1994, and 1999 (e.g., drought years). Chlorophyll *a* was inversely associated with distance from the upper bay. In this way, chlorophyll gave general indications of areas of increased productivity with the highest concentrations found in upper bay areas during droughts. Overall, the most important factor concerning interannual changes of water temperature, salinity, water color, and chlorophyll *a* in Perdido Bay appeared to be the drought–flood cycle of the Perdido River.

There was a dichotomy of surface and bottom dissolved oxygen anomalies (oxygen concentrations standardized for solubility with temperature and salinity); surface values were usually higher than bottom values (Figure 4.47a). The smallest differences of surface and bottom oxygen anomalies occurred in the upper bay seagrass beds (station P25); the greatest such differences were found at the mouth of the Perdido River and Elevenmile Creek. Low DO anomalies in the deep hole at the mouth of the Perdido River (P18) was probably due to the effects of salinity stratification. The lowest average surface DO anomalies occurred in Elevenmile Creek (station P22) due to paper mill discharges of dissolved organic carbon and ammonia. The highest DO anomalies were noted at stations P23, P26, P31, and P40 from 1992 to 1994, a period of low river flow. Bottom oxygen anomalies tended to be higher in the lower bay (station P40), possibly reflecting the influx of highly oxygenated bottom water moving into the bay from the Gulf. The DO concentrations in any given part of the bay thus reflected local habitat conditions that included depth, the extent of salinity

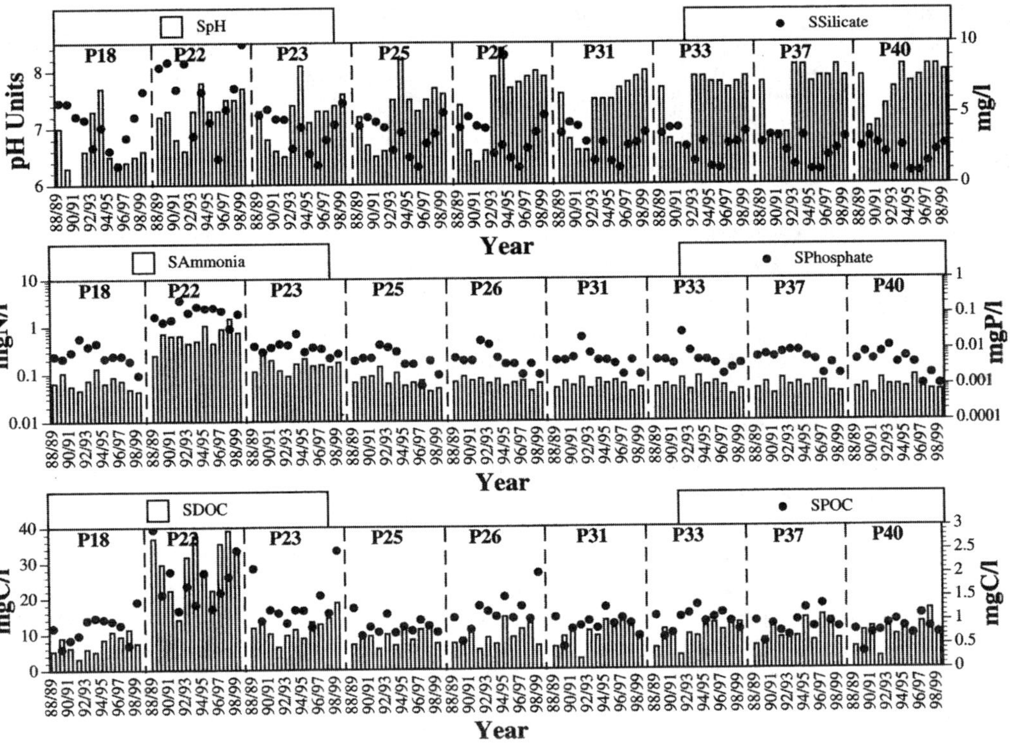

Figure 4.47(b) Surface pH, surface silica, surface ammonia, surface orthophosphate, surface dissolved organic carbon, surface particulate organic carbon) taken in the Perdido Bay system from October 1988 through September 1999.

stratification, the effects the pulp mill, the presence of seagrass beds, and the northward movement of bottom water from the Gulf. However, the highest such anomalies (surface and bottom) occurred during the initial phytoplankton blooms in 1993–1994. During the succession of blooms from diatom species to the larger forms of raphidophytes and dinoflagellates (i.e., increased phytoplankton biomass with time), there was a stepwise reduction of bottom oxygen anomalies from 1993 to 1999 at most stations. Although the drought could have had an indirect effect on this trend, the primary changes in reduced subhaloclinal oxygen anomaly could have been associated in time with the long-term bloom successions. The surface water oxygen anomalies were not affected by this trend with the possible exception of station P26 where there was a series of high surface oxygen anomalies from 1993 to 1997. The data thus indicate complex long-term changes in DO in the bay that were determined to some degree by drought–flood conditions, vertical distribution relative to the halocline, and long-term bloom successions.

Ammonia concentrations (Figure 4.47b) were highest in Elevenmile Creek (station P22). There were general increases of ammonia concentrations in the creek from 1995 through 1998. At various bay stations, average annual ammonia concentrations tended to be lowest during the last 3 years of sampling, although the trend to lower ammonia appeared to start in 1995. Phytoplankton blooms following increased orthophosphate loading could have accounted for the reduced concentrations of ammonia in the bay during periods of increased ammonia loading by the mill. Orthophosphate concentrations in Elevenmile Creek were generally higher than anywhere in the bay. Creek concentrations reached maximum levels from 1992 to 1997 with a decrease in 1998 and an increase in 1999. This trend followed the long-term changes of orthophosphate loading rates by the paper mill. Bay stations showed peak orthophosphate concentrations during 1992 with reduced concentrations from 1993 through 1999. The lowest such concentrations were usually during the last 2 to

3 years, and the pattern of orthophosphate in the bay during these years tended to be opposite that of Elevenmile Creek. The general trends of both nutrients in the bay indicated that, during the phytoplankton blooms, there were general reductions of ammonia and orthophosphate concentrations in Perdido Bay. The fact that reductions of orthophosphate occurred in Perdido Bay during periods of the highest loading of this nutrient indicate at least partial control of such concentrations by the plankton blooms. During the 1998 period of low orthophosphate loading, concentrations of this nutrient increased at most stations (i.e., reduced bloom activity), whereas ammonia concentrations were generally reduced (except P22 and P23). This is further evidence that the blooms affected the concentrations of these nutrients in complex ways and that nutrient concentrations in the bay were probably controlled by the phytoplankton in varying degrees over the 11-year period of study. However, nutrient concentrations in areas influenced by Elevenmile Creek (P22, P23) were often the controlling factors of phytoplankton bloom activity. The use of these nutrients as indicators of bloom activity was thus complicated by the complex interactions of nutrient loading and nutrient utilization by phytoplankton populations.

Average surface silica concentrations (Figure 4.47b) were highly variable in the upper bay; silica concentrations were inversely related to distance from the Perdido River mouth. There were precipitous reductions of silica from 1991–1992 through 1996 that were followed by increased silicate concentrations from 1997 through 1999. Silica concentrations were inversely related to pH, which increased throughout the bay in 1992–1993 at about the same time as the silicate decreases and DO increases. The pH levels remained high through 1999. There were sustained increases of pH throughout the bay during periods of increasing numbers and biomass of the phytoplankton. High pH appeared to be a relatively good indicator of the blooms. High DOC in Elevenmile Creek (Figure 4.47b) was consistent with mill discharge rates. Baywide DOC tended to increase during the blooms although this trend was tenuous. The POC concentrations tended to be highest in Elevenmile Creek. Although the long-term trends of POC were complex, there was a tendency for POC to peak during the drought periods.

These data indicate that long-term water quality changes in Perdido Bay reflected drought–flood sequences and the occurrence of species-specific successions of phytoplankton blooms over the 11-year period of observation. In most instances, the water quality trends were relatively complex so that simple and direct associations of individual factors with phytoplankton blooms and long-term, species-specific successions were not usually possible. There were both short- and long-term responses of factors such as dissolved oxygen and pH that indicated direct impacts due to increased phytoplankton activity associated with the blooms. According to these habitat changes, it is likely that some blooms started as early as 1992, a period when no phytoplankton data were taken. Nutrient relationships with these blooms were complicated and were affected by the blooms in ways that were highly variable in space and time.

3. Phytoplankton Blooms and Chlorophyll a

Chlorophyll *a* has long been used as an indicator of phytoplankton production. Most modeling efforts that are concerned with nutrients and estuarine productivity use direct transformations of chlorophyll into postulated phytoplankton biomass (i.e., organic carbon). The chlorophyll/phytoplankton biomass relationship was thought by some to be flawed. According to this view, there were too many sources of error due to species-specific differences of chlorophyll *a* concentrations among phytoplankton types. The extreme taxonomic complexity of phytoplankton assemblages together with varying levels of chlorophyll *a* in different species was also thought to complicate the relationship between this factor and phytoplankton biomass. However, various authors (Platt et al., 1970) qualified such criticism on the basis that since chlorophyll *a* is so widely used in oceanography as an index of phytoplankton standing crop, this factor can be used as an equivalent of algal biomass if the researcher is "aware of the variability involved." In recent years, the chlorophyll *a* phytoplankton relationship has proved to be somewhat more complicated. For example, there are differences in chlorophyll *a*, *b*, and *c* content in phytoplankton due to species-specific differences in cellular structure (i.e., diatoms and dinoflagellates; Hitchcock, 1982). Fully half of the dinoflagellates do not even have chlorophyll *a*, and many are phagotrophic (van den Hoek et al., 1995). There are major differences among species in the size and number of chloroplasts. These questions have led to renewed interest in the question of whether or not chlorophyll *a* can be used as a legitimate substitute for real phytoplankton data.

Within-species cell size of phytoplankton in Perdido Bay varied over a relatively wide range. For example, the range of cell diameter or length varied by 2.0 µm in *Cyclotella choctawhatcheeana* and *Miraltia throndsenii* (diatoms), 1.5 to 2.0 µm in *Merismopedia tenuissima* (blue-green alga) and some nanoflagellates, and 850 µm in *Rhizosolen*ia spp. (diatoms). Cell abundance data may be transformed into biomass by taking into account differences in cell size and shape. Thus, phytoplankton biomass can be estimated indirectly from cell counts after the determination of cell volume. Smayda (1978), however, discouraged the use of cell volume data calculated for the same taxonomic categories from other, similar localities. Because many diatom taxa exhibit considerable intraspecific variability of cell size, especially cell diameter and cell length, it follows that the cell volume and even the biomass of a given species may be somewhat variable. Therefore, our biomass transformations were calculated from our own observations of Perdido Bay plankton on a species-specific basis (see Appendix I).

A comparison of the chlorophyll *a* data with numbers and biomass of whole-water phytoplankton is given in Figure 4.48. The period 1993–1994 in Perdido Bay marked the beginning of consistently high orthophosphate loading from the pulp mill. The first phytoplankton blooms were noted in 1993–1994 (see Figure 4.35, Table 4.1). During this drought, high numbers of phytoplankton were taken in the Perdido system, marking the beginning of the long-term bloom sequences. Phytoplankton biomass increased during this period. There was increased chlorophyll *a* at this time at levels that were comparable to the increased chlorophyll *a* concentrations during the 1989 drought when no blooms were noted (Figure 4.35). During the period of generally increased orthophosphate loading (1994 to 1997), various bloom species prevailed in a series of seasonal and interannual progressions of diatoms, raphidophytes, and dinoflagellates. Pronounced changes in the phytoplankton community structure such as reduced species richness and altered dominance sequences accompanied the blooms. There were major changes in the balance of difference groups with the replacement of the diatoms and the golden brown algae by the raphidophytes, dinoflagellates, and the cryptophytes during later bloom years. During this period, there were significant increases in the numbers and biomass of the whole-water phytoplankton. The documented changes in the phytoplankton as well-defined seasonal and interannual bloom successions were accompanied by alterations in the habitat variables such as surface and bottom dissolved oxygen, pH, and silica. However, chlorophyll *a* did not reflect these major changes in the phytoplankton assemblages.

During the period from late 1997 through spring 1999, there was a general reduction of orthophosphate loading with periodic increases of ammonia loading from the mill (i.e., summer–fall, 1998). During this period, phytoplankton numbers and biomass declined, with some increases noted during summer periods of increased ammonia loading. Once again, relatively few of these changes were noted in the chlorophyll *a* data. Massive blooms throughout the bay (see Table 4.1) accompanied the resumption of orthophosphate loading in the spring and summer of 1999 with the reassertion of dominance by diatoms, raphidophytes, and dinoflagellates. Phytoplankton biomass reflected the increased dominance of the larger forms of dinoflagellates and raphidophytes. These changes were not fully represented by the chlorophyll *a* data; some blooms were associated with high chlorophyll *a*, but there was no real pattern to such correlations and statistical analyses of the data resulted in no significant associations of blooms and chlorophyll *a*. Thus, although chlorophyll *a* trends tended to reflect interannual sequences of drought–flood conditions, this variable gave no consistent indication of the quantitative or qualitative aspects of the long-term

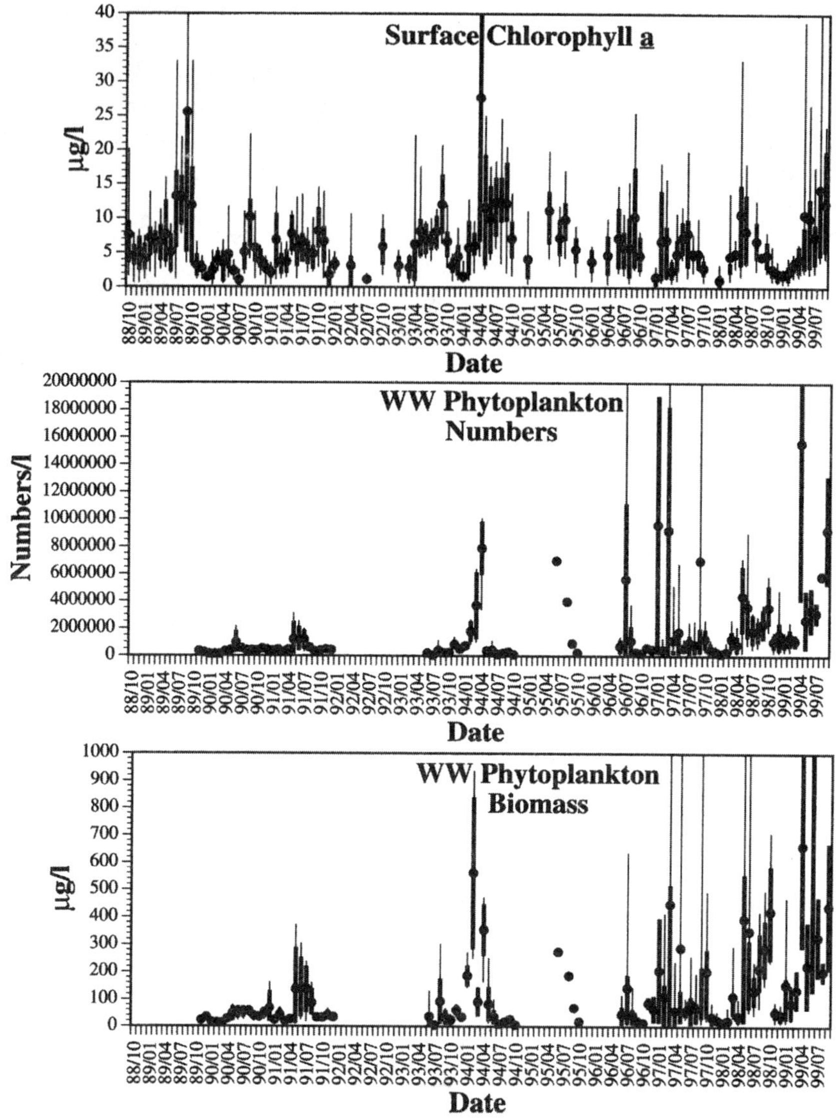

Figure 4.48 Surface chlorophyll *a* compared with biomass (μg l^{-1}) and abundance (numbers l^{-1}) of whole-water phytoplankton in Perdido Bay taken from 10/88 through 9/99. Data are expressed as means with the 10th, 25th, 75th, and 90th percentiles of all stations.

changes of the phytoplankton assemblages. There was a general correlation of chlorophyll with increased phytoplankton activity, but chlorophyll was not an adequate indicator of bloom activity.

The long-term chlorophyll *a* data were run against phytoplankton total numbers and biomass at stations P23, P31, and P40 in Perdido Bay (Figure 4.48). There were general associations of chlorophyll *a* with peak numbers and/or biomass, especially during interannual drought periods. However, the overall long-term trends of increasing phytoplankton numbers and biomass with time were not reflected in the chlorophyll *a* data. Regressions were run with phytoplankton biomass and phytoplankton total numbers as the dependent variables against the primary water quality factors by station by date over the 11-year period of study. Scattergrams of the long-term field data were examined and either logarithmic or square root transformations were made, where necessary, to approximate the best fit for a normalized distribution. Phytoplankton biomass at all stations was significantly associated with surface chlorophyll *a*; however, relatively little of the variation was accounted for ($r^2 = 0.189$, $p = 0.05$); phytoplankton numbers had a similar relationship to surface chlorophyll *a* ($r^2 = 0.190$, $p = 0.05$). When run by station (P23, P31, P40), only station P23 showed a significant relationship between these variables; again, the association was relatively weak ($r^2 = 0.167$, $p < 0.05$). Chlorophyll *a* was regressed against the primary habitat variables by station and biomass of the various bloom species. The primary association was found between chlorophyll *a* and the biomass of the following species: *C. choctawhatcheeana* ($r^2 = 0.286$, $p < 0.05$), *Miraltia throndsenii* ($r^2 = 0.241$, $p < 0.05$), *Chattonella subsalsa* ($r^2 = 0.160$, $p < 0.05$). There was a weak inverse relationship between chlorophyll *a* and bottom DO at station P31 ($r^2 = 0.11$, $p < 0.05$). Overall, with the possible exception of the biomass of two diatom bloom species, chlorophyll *a* was not closely associated with either the primary water quality variables or the main indices of phytoplankton biomass and numbers.

The reasons for the general lack of a strong correlation between chlorophyll *a* and the major microalgal indicators are not clear from this analysis, and further work along these lines is in progress. In river-dominated estuaries such as the Perdido, high DOC loading could affect the distribution of the microheterotrophic flagellates in what is essentially a heterotrophic system (D. Flemer, personal communication). The lack of strong connections between the chlorophyll *a* concentrations and counts and biomass estimates of the phytoplankton could be related in some way to the fact that Perdido Bay is not dominated by autotrophic processes. The relatively high flushing rates of the estuary could also be involved in the complex relationships of nutrients, chlorophyll, and phytoplankton assemblages, possibly as a sampling artifact. Light limitation, especially in the upper estuary, is another confounding feature of general associations among these variables. In any event, it is unlikely that chlorophyll *a* will represent the nonpigmented microheterotrophs as microflagellates or the larger, colorless dinoflagellates so that the emphasis on heterotrophic relationships could provide some understanding regarding these results. The poorly understood role of dinoflagellates in the trophic organization is a complicating factor.

The current widespread use of chlorophyll *a* as an indicator of phytoplankton numbers or biomass is thus somewhat problematical in river-dominated estuaries and the use of chlorophyll *a* as a quantitative indicator of phytoplankton assemblages could be compromised by the complex interrelationships of plankton populations in such systems. Chlorophyll *a* was significantly associated with phytoplankton biomass and numbers, but the relatively low r^2 values could be related to a combination of factors that could change relative to driving functions such as river flow and temperature. The Perdido studies indicated that chlorophyll *a* was not representative of the incidence of phytoplankton blooms (see Table 4.1), probably for similar reasons as those discussed above. The use of chlorophyll *a* as a direct substitute for phytoplankton in modeling in river-dominated estuaries is also problematical based on the comparisons of the chlorophyll distributions relative to direct observations of the changes in the phytoplankton assemblages in the Perdido Bay system over the 11-year period of study. There were significant correlations between chlorophyll *a* and phytoplankton numbers and biomass, but the high variance associated with these relationships would qualify the use of chlorophyll as a direct substitute for these indices of the phytoplankton community. The assumptions that underlie the use of chlorophyll *a* as a substitute for phytoplankton assemblages should be examined in other coastal and marine systems. It is possible that very different trophodynamic processes occur in offshore (blue-water) systems. In any case, the Perdido studies indicated that definition of phytoplankton blooms is best carried out with detailed, species-specific determinations of the phytoplankton assemblages. The use of chlorophyll *a* for such evaluations would have been misleading.

J. Long-Term Sediment Quality Trends in Perdido Bay

The analysis of long-term sediment quality data could provide a clue concerning a possible link between the initiation and long-term succession of phytoplankton blooms and associated adverse effects on the sediments of Perdido Bay. Trends of sediment total phosphorus in Perdido

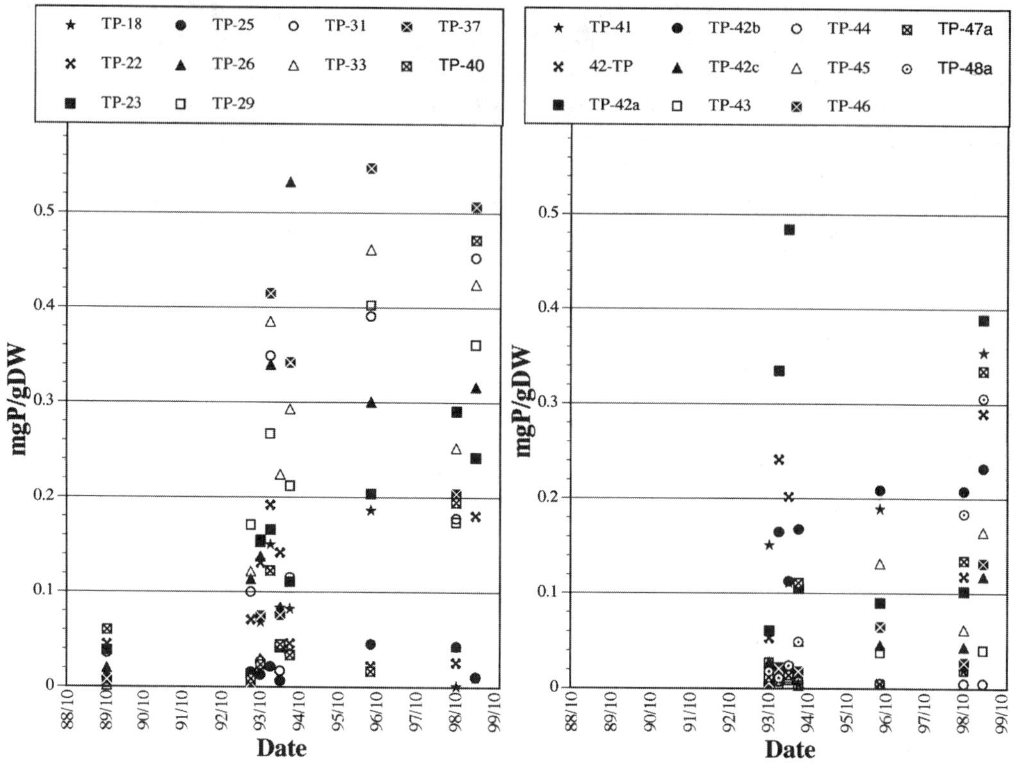

Figure 4.49(a) Long-term changes of sediment total phosphorus (TP) in Perdido Bay.

Bay are given in Figure 4.49a. Total phosphorus in sediments of the upper, mid and lower parts of the bay increased significantly between 1989 and 1993. These increases were evident everywhere except at the upper bay seagrass station and in the lower bay (P40). By 1996, there were further increases of sediment TP throughout the bay with the same exceptions as before. This time, however, concentrations in lower Elevenmile Creek were lower than before. By 1998, after reduced orthophosphate loading, there were general reductions of sediment TP in most parts of the bay (except station P23). The resumption of orthophosphate loading during spring 1999 was followed by increased TP in the sediments everywhere except the upper bay seagrass beds. These data indicate that increased orthophosphate loading from Elevenmile Creek was accompanied by baywide increases of total phosphorus in sediments. Because of the relative paucity of sediment nutrient data over the 11-year study period, only broad generalizations could be made regarding these trends. However, it appears that, with the exception of the upper bay seagrasses, sediments in upper, mid, and lower bay areas responded to nutrient loading to upper Perdido Bay.

Sediment total phosphorus in the lower bay (Figure 4.49a) was characterized by relatively low levels at various stations in 1993. However, in Wolf Bay (stations P42a, T42), there were high TP sediment concentrations at this time. By 1996, there was a general increase of TP in sediments of the lower bay with associated increases at stations P42b (Wolf Bay) and P41. These concentrations remained more or less the same in 1998. By spring 1999, there were again increases in sediment TP in Wolf Bay (P42, P42a, P42b) and in associated areas (P41). Areas receiving urban runoff (stations P47a and P48a) were also characterized by high sediment total phosphorus during 1999. These TP concentrations were comparable to those in the upper bay. The data indicate that postulated agricultural runoff in Wolf Bay and urban runoff in lower Perdido Bay could have been associated with increased concentrations of total phosphorus in sediments of bay receiving areas over the study period.

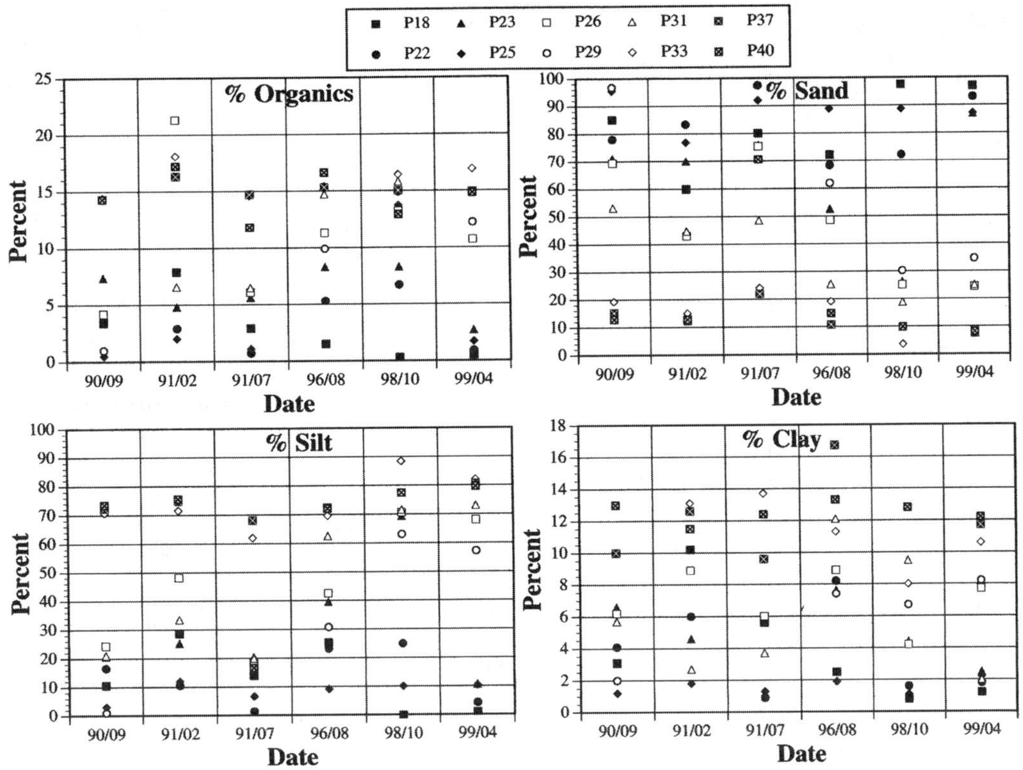

Figure 4.49(b) Long-term changes of sediment characteristics (%organics, %sand, %silt, %clay) in Perdido Bay.

Long-term trends of sediment quality characteristics in Perdido Bay are shown in Figure 4.49b. There was no observable change in clay percentages at most stations in Perdido Bay over the period of study. Percent clay concentrations increased from the upper to the lower bay with peaks at station P40. Percent sand was relatively high at stations P18 (river mouth), P23 (creek mouth), and P25 (seagrass beds). The deeper (depositional) areas in lower Perdido Bay (stations P33, P37, and P40) tended to have relatively high percent organics and high silt concentrations over the entire study period (Figure 4.49b). Sediments at stations P18 (the mouth of the Perdido River), P22 (Elevenmile Creek), and P25 (the seagrass beds) remained more or less constant in percent organics and percent silt concentrations over the study period. From 1990–1991 to 1996, percent organics and percent silt increased at stations P23, P26, P29, and P31. By October 1998, silt concentrations continued to increase at these stations in addition to increased silt at stations P33 and P40 in the lower bay. With the exception of decreased silt at station P23, these trends continued in 1999. Thus, silt concentrations increased at most stations in the upper, mid, and lower bay during the phytoplankton blooms from 1993 to 1999. These trends were evident regardless of season.

Long-term changes of sediment quality in Perdido Bay were generally associated with bloom frequency and distribution in areas where phytoplankton production was most intensive (Figure 4.49c). The relatively low frequency of sediment sampling restricted interpretation of results to a very general level. However, sediment quality, in terms of silt and organic content, was a reasonably good indicator of overall bloom occurrence in unvegetated, nondepositional areas of the bay. Even in depositional areas, there were some increases in silt concentrations with time. Since particulate loading into the bay did not follow these trends to any significant degree (see Figure 4.4), it is likely that the increased silt and organic matter noted in the sediments over the study period reflected the increased incidence of phytoplankton blooms.

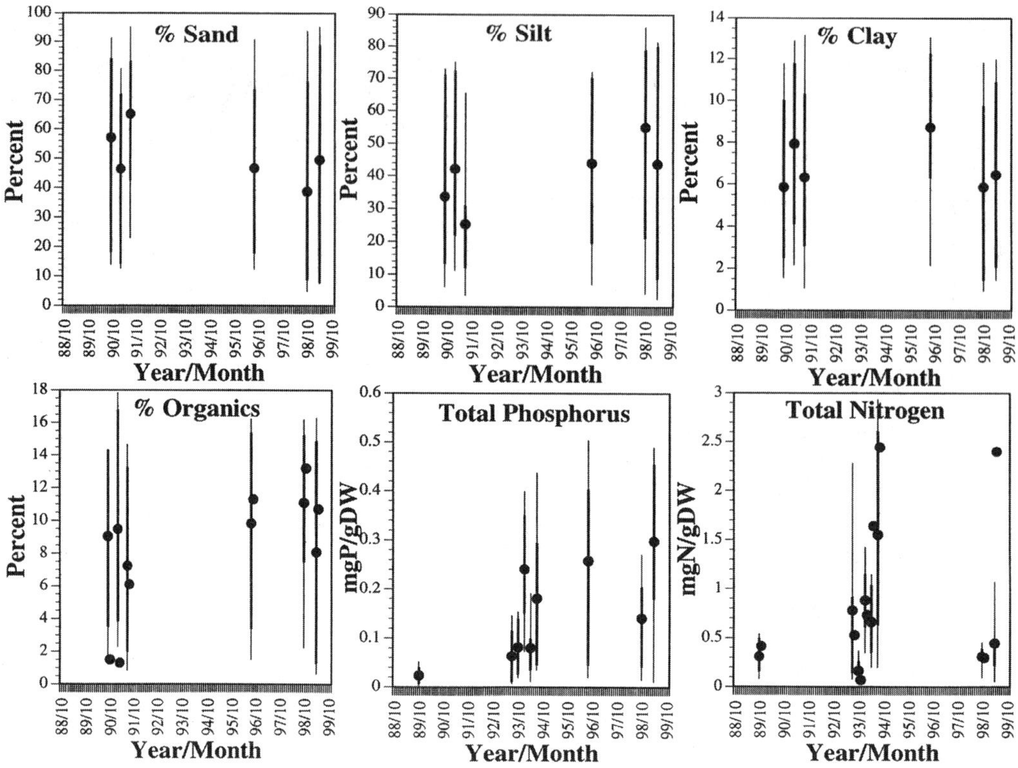

Figure 4.49(c) Mean values of %sand, %silt, %clay, %organics, TP, and total nitrogen in sediments taken in Perdido Bay (stations P18, P22, P23, P25, P26, P29, P31, P37, P40) over the 11-year study period.

The long-term sediment data provide evidence that sediment conditions (total phosphorus, particle size distribution, silt concentrations, percent organics) could have been affected by the occurrence of plankton blooms in Perdido Bay. These data are consistent with the results of the stable isotope analyses (Coffin and Cifuentes, 1999). The spatial and temporal pattern of accumulation of silt and organic matter in the sediments could provide a key to an understanding of the dynamic relationships between sediment quality and long-term changes of phytoplankton bloom activity in Perdido Bay.

K. Statistical Analyses of Phytoplankton Distributions

1. The 1993–1994 Bloom Period

A statistical analysis was run with the physical–chemical and biological data taken at 34 stations monthly over a 12-month bloom period from June 1993 through May 1994. Independent variables included depth, temperature, salinity, salinity stratification index, DO, oxygen anomaly, water color and turbidity, nutrients (nitrate, ammonia, orthophosphate, silica), POC, sulfide, chlorophyll *a*, and Secchi depths. Dependent biological variables included total numbers, species richness, and numbers of dominant species of the phytoplankton, zooplankton, and infaunal macroinvertebrate assemblages. All independent and dependent variables were transformed before use in a multiple linear regression analysis and principle component analysis. For each biological variable, a multiple regression analysis was performed to describe how it was related to the physical variables. The

Principal Component Analysis (PCA) was then used to reduce the number of physical–chemical variables into smaller sets of linear combinations of variables that accounted for most of the total variation of the original set. The relationships between the biological variables and the first few principal components were then established using a multiple linear regression model. Detailed explanations of the statistics used in this analysis are given in Appendix II.

Phytoplankton numbers were significantly ($p = 0.05$) and positively associated with the salinity stratification index (bottom salinity–surface salinity), surface color, and chlorophyll *a* and were negatively associated with orthophosphate. The positive association of chlorophyll with phytoplankton numbers represented a general trend of increased chlorophyll with phytoplankton abundance. The negative association of phytoplankton numbers with orthophosphate could reflect the uptake of this nutrient in areas surrounding the mouth of Elevenmile Creek where nutrient gradients from the creek to the bay were the highest in the entire system. This analysis indicates that phytoplankton were probably involved in the rapid reduction of orthophosphate at the mouth of the creek. Phytoplankton species richness was positively associated with surface color, surface orthophosphate, and surface POC. Species richness was negatively associated with the salinity stratification index. This result is compatible with the inverse relationship of bloom numbers, species richness, and orthophosphate concentrations. Zooplankton numbers were negatively associated with the stratification index and positively associated with surface color and surface DO. Zooplankton species richness was positively associated with surface color and negatively associated with chlorophyll *a*. Zooplankton species richness was inversely associated with phytoplankton numbers. Zooplankton numbers were also inversely associated with phytoplankton numbers in areas subject to the direct influence of flows from the Perdido River and Elevenmile Creek. The inverse relationship of phytoplankton and zooplankton numbers and the respective species richness indicators is consistent with the observed reduction of phytoplankton species richness during seasonal and interannual blooms of individual species.

The continued association of phytoplankton numbers with salinity stratification is not unusual. It is well known that estuarine phytoplankton thrive during stable (i.e., drought) conditions characterized by low turnover rates that accompany periodically reduced winter–spring river flows. The statistical association of drought conditions, salinity stratification, and nutrient loading from the mill during the 1993–1994 period was thus not unexpected. When such changes are placed within the context of the physical dimensions of the bay, the importance of the winter drought on the observed habitat trends and the associated changes in the phytoplankton/zooplankton assemblages of the bay becomes obvious. The association of low orthophosphate concentrations with increased numbers of phytoplankton was also consistent with the experimental data showing orthophosphate limitation on phytoplankton in upper Perdido Bay during winter–spring periods (Flemer et al., 1997), and the rapid uptake of orthophosphate in this area. Infaunal numbers were positively associated with bottom DO and were inversely associated with DO anomalies, surface chlorophyll *a*, surface nitrate, surface orthophosphate, and bottom ammonia. Infaunal species richness was positively associated with bottom DO and negatively associated with surface chlorophyll *a*, surface orthophosphate, bottom ammonia, and bottom sulfides.

The results of the PCA were similar to those of the regression analyses. Salinity stratification, surface nutrients, and chlorophyll *a* were generally associated with the phytoplankton and zooplankton distributions, and bottom DO and surface nutrients were associated with the distribution of infaunal macroinvertebrates in Perdido Bay. These analyses are consistent with the role of nutrient loading and the physical state of the bay in the initiation of the observed microalgal blooms. The statistically inverse relationship of zooplankton and infaunal macroinvertebrates with the phytoplankton blooms was also consistent with field observations. It should be emphasized that all of these relationships are based on correlations among the various dependent and independent variables, and are not necessarily indicative of cause-and-effect mechanisms regarding these variables.

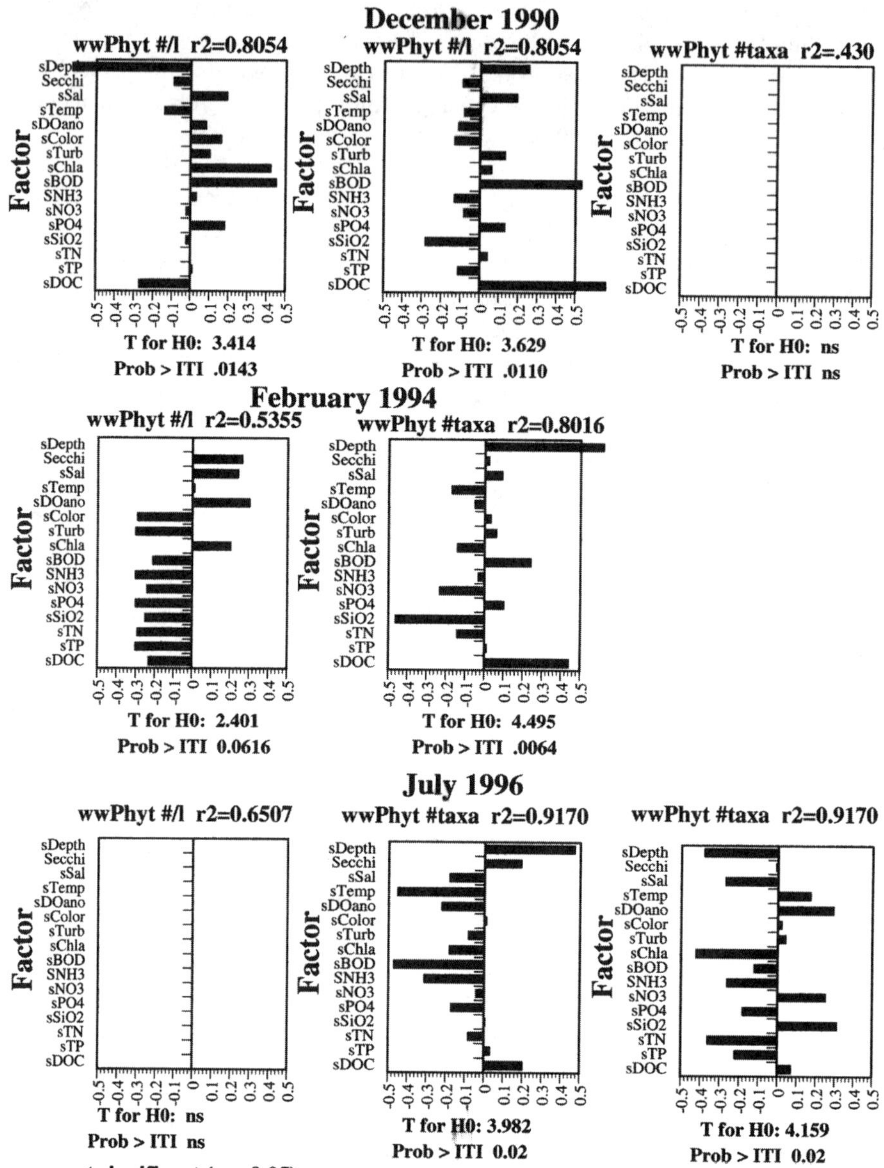

Figure 4.50 Results of PCA-regression analyses of net phytoplankton taken in December 1990 and February 1994 and whole-water phytoplankton taken in July 1996 (a), March 1998, and July 1998 (b). Dependent variables were phytoplankton cells l^{-1} and species richness by station.

2. Comparison of Long-Term Trends

A PCA/regression analysis was run with the primary habitat characteristics against whole-water phytoplankton numbers and species richness during prebloom (December 1990) and postbloom (February 1994, July 1996, March 1998, July 1998) periods (Figure 4.50). During the December 1990 prebloom period, phytoplankton cell numbers per liter were inversely proportional to depth (i.e., the highest numbers were in the upper bay) and were directly proportional to surface chlorophyll *a* and BOD. In this case, the chlorophyll *a* was generally associated with prebloom

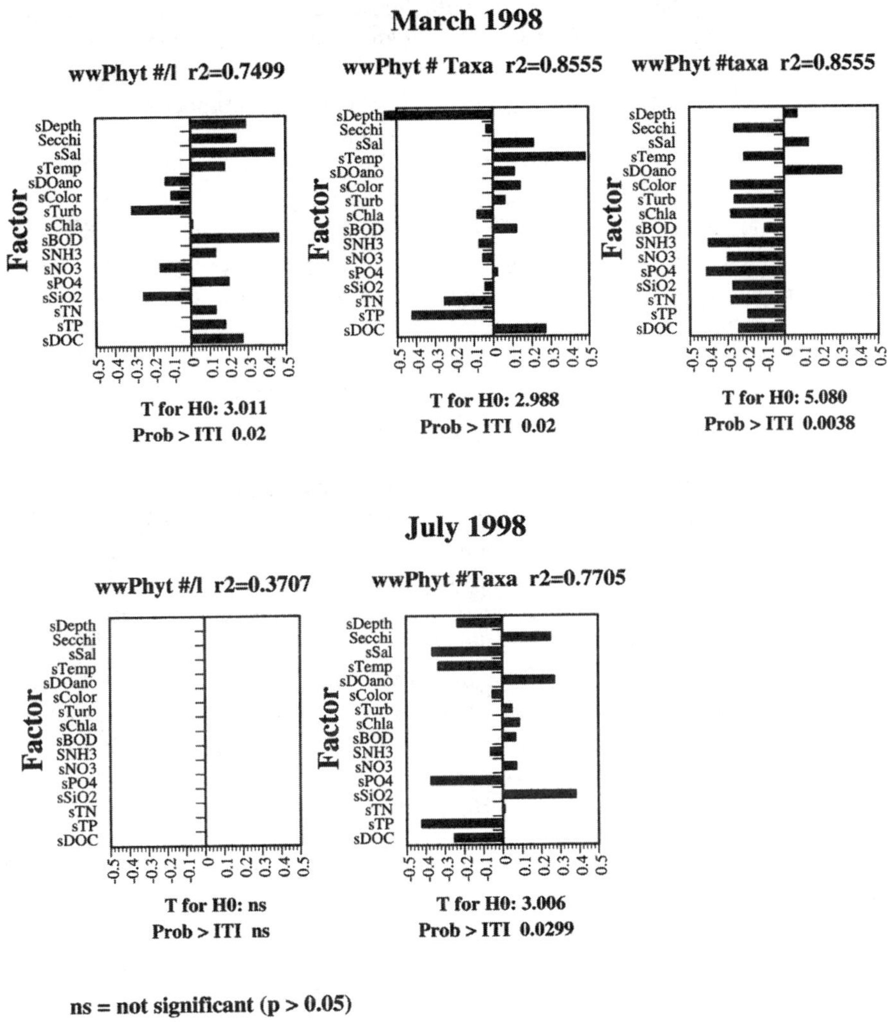

Figure 4.50 (continued)

phytoplankton numbers. However, in none of the subsequent analyses that were run during plankton blooms was chlorophyll *a* a significant factor associated with phytoplankton numbers. It is possible that under general conditions of natural nutrient loading and eutrophication, chlorophyll *a* is associated with phytoplankton numbers, but that during bloom sequences, it is not representative of the blooms. These results are consistent with the above observations that chlorophyll *a* concentrations were not uniformly high during blooms. Surface BOD and DOC were positively associated with phytoplankton abundance during the December 1990 period. No significant associations were noted for phytoplankton species richness during the prebloom period.

During the February 1994 bloom period, cell numbers per liter were inversely related to various nutrient concentrations although such relationships were only marginally significant. Phytoplankton abundance was also slightly associated with oxygen anomaly. Phytoplankton species richness during this initial bloom period was inversely related to silica and directly related to depth (i.e., higher species richness occurred in lower bay areas that were low in silica). Species richness was positively associated with DOC during the February 1994 blooms. In July 1996, there were no significant associations of phytoplankton abundance with specific habitat variables. Phytoplankton species richness at this time was inversely associated with surface BOD and water temperature and was positively associated with depth (i.e., higher phytoplankton species richness in the lower bay during warm periods). There was an indication that higher species richness was associated with reduced TN, chlorophyll *a*, and depth (i.e., reduced numbers of species in the upper bay during periods of increased phytoplankton activity). In March 1998, cell numbers per liter were strongly associated with increased BOD, increased salinity, and reduced turbidity. Phytoplankton species richness was inversely related to ammonia, orthophosphate, and total phosphorus (TP). In summer 1998, there was no significant relationship of phytoplankton numbers with any of the habitat variables. During this period of plankton blooms, phytoplankton species richness was inversely associated with orthophosphate and TP and was directly associated with silica. Thus, during the early years of relatively low nutrient loading, the surface ammonia and orthophosphate concentrations were not closely related to phytoplankton species richness. However, after increased nutrient loading of first orthophosphate and then both orthophosphate and ammonia, these nutrients were often inversely related to phytoplankton species richness.

A form of ordination called Canonical Correspondence Analysis (CCA) was used with the phytoplankton data as a direct analysis of various phytoplankton indices along gradients of key habitat variables. Ter Braak (1996) developed the mathematics of this method. Ter Braak and Smilauer (1998) provided the program. Preliminary PCA and regression analyses were run with the data to determine the most likely (significant) habitat factors to be run with the CCA analysis. The regressions were run with total numbers of cells per liter per station per date and species richness by station by date for net phytoplankton (December 1990, February 1994) and whole-water phytoplankton (July 1996, March 1998, and July 1998). Independent variables were similar to those run in the previous analysis of the 1993–1994 data. The above periods were chosen to allow a comparison of habitat-associated factors with phytoplankton assemblages before the blooms occurred (December 1990), during the initial primary blooms (February 1994), and during primary bloom periods before, during, and after the orthophosphate control program initiated by the mill (July 1996, March 1998, July 1998).

Results of the CCA analysis are shown in Figure 4.51. During December 1990, there were no real associations of the bloom species with the habitat indicators. Dominants such as *C. choctawhatcheeana* were found in upper and mid-bay areas, whereas *L. danicus* was found mainly in the lower bay. During the February 1994 bloom period, *L. danicus*, as the main bloom species, was closely associated with DOC and the inorganic nutrients. Station P22 was also associated with both ammonia and orthophosphate. During July 1996, the dominant bloom species (*H. akashiwo* and *Synedropsis* sp.) were associated with stations P18 and P23 in the upper bay, whereas *Gymnodium* spp. were found in the seagrass beds (station P25). During March 1998, both *Gymnodinium* sp. and *H. akashiwo* were associated with high concentrations of ammonia and orthophosphate at stations in the upper bay. Dissolved oxygen anomalies were inversely associated with these phytoplankton populations and with the inorganic nutrients. During the July 1998 blooms of *Synedropsis* sp., this species was closely associated with orthophosphate concentrations at upper and middle bay stations. The results of these analyses complement the previous PCA results and show the close associations of the orthophosphorus and ammonia concentrations with the various phytoplankton bloom species. With the possible exception of an association of *H. akashiwo* with chlorophyll *a* during the July 1996 blooms, chlorophyll *a* was not closely associated with most of the bloom species during the periods analyzed. These analyses are thus consistent with previous determinations that showed chlorophyll *a* to be significantly but weakly associated with phytoplankton distributions at any given time. However, chlorophyll *a* was not a reliable indicator of plankton blooms.

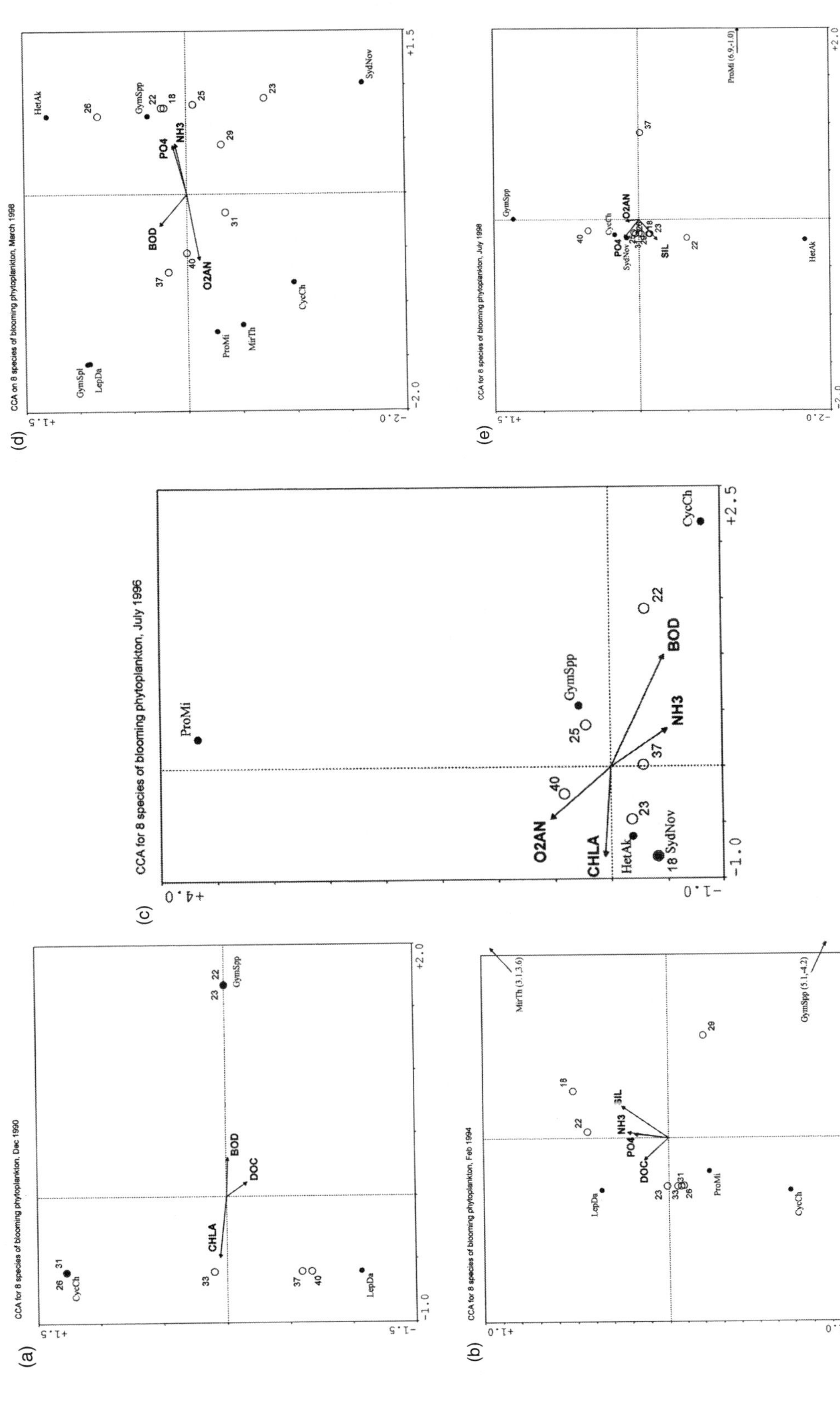

Figure 4.51 Results of the CCA analysis relating phytoplankton bloom sequences to statistically significant habitat factors during periods prior to and during increased nutrient loading to Perdido Bay: (a) December 1990; (b) February 1994; (c) July 1996; (d) March 1998; (e) July 1998.

CHAPTER 5

Effects of Blooms on Secondary Production

I. INTRODUCTION

There is considerable information concerning the toxicity of plankton bloom species and the adverse impacts of HABs on individual populations of invertebrates and fishes. Plankton blooms have been associated with effects on invertebrates (Buskey, 1997; Steidinger et al., 1998b), fishes (Shimada et al., 1983; Chang et al., 1990; Burkholder et al., 1992), and major fisheries (Smayda and Shimizu, 1993; Hallegreaff, 1995; Hallegreaff et al., 1995; Anderson and Garrison, 1997). Various authors (Anderson et al., 1997) have pointed out that anthropogenous sources of pollutants and nutrients are responsible for recent increases in both the frequency and severity of HABs. Riegman (1998) reviewed how the species composition of phytoplankton may be changed by the dynamic interactions among macronutrients. Much of this discussion, however, was conjectural, and the response of the phytoplankton as a community to anthropogenous alterations of nutrient loading to coastal systems remains largely unstudied. Likewise, relatively little information exists concerning system-level responses of the secondary producers to changes in the plankton communities.

According to Fryxell and Villac (1999), it is rare that data are available for events leading up to a given bloom. Sorokin et al. (1996), working in the Comacchio Lagoons (Ferrar, Italy), presented information showing reduced zooplankton and losses of benthic organisms and fishes as a response to an anthropogenous increase of nutrient loading and dominance of what was described as inedible cyanobacteria. In this case, adverse effects on higher trophic levels were mediated by changes in feeding patterns due to the altered state of the phytoplankton. However, the long-term, ecosystem-level cause-and-effect processes that were associated with the hypereutrophication remained largely unexplained.

Previous research indicates that the effects of plankton blooms are often cumulative and can occur in relatively complex ways (Fryxell and Villac, 1999). Toxic diatoms may have direct or indirect effects on grazers but it is more common that the primary adverse effect is on the consumer of the affected grazer. Harmful diatoms can cause mechanical injuries, especially to the gills of fishes. Substances associated with the blooms can also clog the gills. Blooms are frequently coupled with hypoxia or anoxia that, in turn, can cause additional changes in the producers and consumers. The early studies of Ryther (1954) concerning the association of nitrogen fertilization of Long Island Sound with the proliferation of chlorophytes and food web disruptions have not been followed up by detailed, ecosystem-level analyses of these relationships. Smayda (1997a) stated that published statements claiming bloom-caused ecological disruption are rarely based on ecological data; "Quantitative ecological studies to establish the *in situ* impacts of harmful algal species and their blooms should therefore be a high priority of future research" (Smayda, 1997a). This proposed line of research should include an evaluation of the effects of nutrient-related reorganization of the plankton communities on trophodynamic processes that can then lead to the various forms of adverse impacts at different levels of biological organization.

II. BAYWIDE TRENDS OF INVERTEBRATES AND FISHES

A. Introduction

Primary animal associations of shallow estuaries and coastal systems in the Gulf of Mexico include infaunal macroinvertebrates, epibenthic macroinvertebrates, and fishes (Sheridan, 1978; Laughlin, 1979; Sheridan, 1979; Sheridan and Livingston, 1979; 1983; Livingston, 1984a; 1991c). The infaunal macroinvertebrates represent the base of the macrotrophic food web (Livingston et al., 1997). Most invertebrate and fish populations in shallow coastal systems are benthic feeders (Livingston, 1984a), and the infauna, living in or on the sediments, represent the direct connection between primary production and the upper levels of the benthic food webs (Livingston et al., 1997). Interactions of these groups are mediated by competition and predator–prey relationships (Virnstein, 1977; Peterson, 1982; Whitlatch and Zajac, 1985). Thus, the sediment/water interface represents an important part of estuarine habitats, and sediment quality is integral to the overall level of secondary production in shallow coastal systems.

B. Community Interactions

When taken as baywide averages in response to mill-induced ammonia and orthophosphate loading and gradients in Perdido Bay (Figure 5.1), the relationships of ammonia and orthophosphate

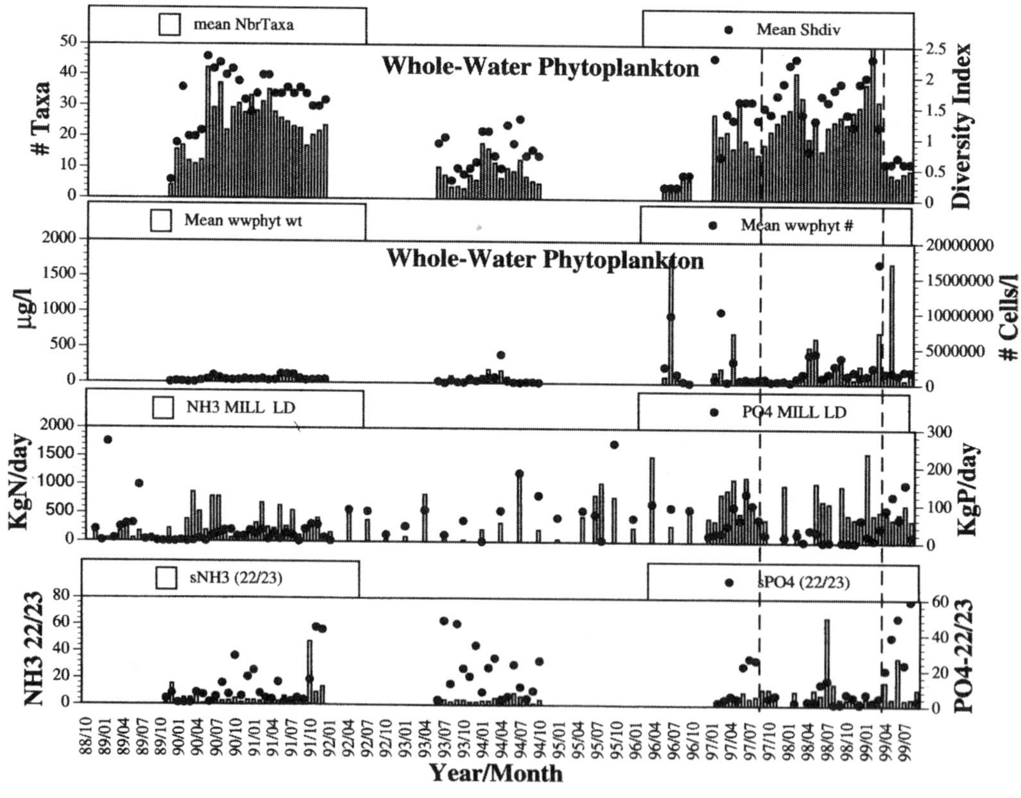

Figure 5.1 Mean species richness, Shannon diversity, numbers (# cells l^{-1}) and biomass (µg l^{-1}) of whole-water phytoplankton plotted against mill ammonia loading, mill orthophosphate loading, and nutrient ratios (ammonia and orthophosphate). Data were taken from stations P23, P31, and P40 in Perdido Bay from October 1988 through September 1999.

concentrations to whole-water phytoplankton associations are clarified. Using a Shannon diversity index of the phytoplankton community, the rapid recovery of the phytoplankton in response to the reduction of orthophosphate loading by the mill during late 1997–1998 becomes obvious. The rapid fall of this index in response to the summer 1998 ammonia loading by the mill and sharp reduction of this index after renewed orthophosphate loading in spring 1999 illustrate the rapid response of the sensitized phytoplankton bloom populations to renewed loading of ammonia and orthophosphate.

A statistical comparison of the physical–chemical habitat changes in 2-year intervals relative to the 1988 to 1990 period (considered a time of low nutrient loading) is shown in Table 5.1. Secchi depths in Elevenmile Creek (station P22) were significantly ($p = 0.05$) deeper from 1993 though 1999 than the first 2 years, a response to the mill's color removal program. Surface orthophosphate and surface chlorophyll *a* were significantly higher from 1993 to 1995 in Elevenmile Creek, whereas surface DO was significantly higher. Chlorophyll *a* was significantly lower in this area of the bay during the years 1990 to 1993; by 1993 to 1995, chlorophyll *a* was significantly higher, whereas there were significantly lower levels of chlorophyll *a* during the intensive bloom period (1995 to 1997). By the drought of 1997 to 1999, chlorophyll *a* levels were again higher in the bay. This represents statistical corroboration of the lack of a direct association between chlorophyll *a* and the long-term changes of bloom conditions in Perdido Bay.

Between 1997 and 1999, during some major ammonia loading from Elevenmile Creek, surface and bottom ammonia concentrations were significantly lower at station P23. At station P26, surface DO was significantly higher during 1993 to 1995 and 1997 to 1999, whereas surface ammonia was significantly lower at this station during 1997 to 1999. This pattern of reduced ammonia during increased ammonia loading was also evident at stations P31, P33, and P40. Thus, from 1997 to 1999 during periodically high ammonia loading, there were significantly lower concentrations of ammonia at most but not all stations in Perdido Bay. The initiation of orthophosphate loading during the 1993 to 1994 drought period was thus associated with initial increases of chlorophyll *a* that were then followed by subsequent decreases in chlorophyll *a* during the bloom years 1995 to 1997. At the same time, the significant reductions of ammonia concentrations in the bay during these periods of phytoplankton blooms could reflect increased utilization by phytoplankton as diatoms were replaced by raphidophytes and dinoflagellates and phytoplankton biomass increased accordingly. It is against this long-term history of seasonal and interannual changes in phytoplankton community structure and accompanying habitat changes that the major trends of secondary production and food web structure were analyzed.

Long-term changes of secondary producers (i.e., infaunal and epifaunal macroinvertebrates and fishes) in Perdido Bay are shown in Figure 5.2. These changes should be reviewed within the context of the drought–flood sequences. Infaunal numbers and cumulative (baywide) species richness peaked during the drought period from 1989–1990. From 1991 through 1992, there was a decrease in both indices that reached relatively low levels during the initial bloom years (1993–1994). There was a slight increase in numbers of infauna during the winter of 1997; however, there was no sign of infaunal recovery during the 1997–1998 period of reduced phytoplankton blooms. The baywide response of the epibenthic invertebrates was similar to that of the infauna. Peak levels of numbers and cumulative species richness occurred during 1989–1990 with reductions in both indices from 1991 to 1994. This was followed by reduced numbers of the invertebrate assemblages in Perdido Bay from 1994 to present with the exception of spring 1999, during which time relatively high numbers of invertebrates were taken. Fish assemblages showed a somewhat different long-term pattern. Peak fish cumulative species richness occurred during 1989 with high numbers during winter 1990. There were reductions of both indices from 1991–1992; this situation continued during the ensuing years. The lowest fish numbers were found during the 1998–1999 period. The changes of invertebrate and fish assemblages in Perdido Bay were thus associated with long-term cycling as opposed to the plankton responses to nutrient loading. This is consistent with the natural history attributes of these species (Livingston et al., 1997).

Table 5.1 Statistical Comparisons of Two 3-Year Averages of Station-Specific Habitat Variables Taken at Stations in Perdido Bay from 10/88 through 9/99

2-Year Means Station/Year Comparisons	Secchi	S-D.O.	S-chlor.a	S-NH3	S-PO4	B-D.O.	B-Chlor.a	B-NH3	B-PO4
18 (88-90)/18 (90/93)	1.04/0.89 WN	6.9/7.2 WN	2.8/3.4 WN	0.11/0.05 WN	0.006/0.017 WN	2.53/2.84 WN	6.19/3.94 WN	0.50/0.21 WN	0.008/0.013 WN
18 (88-90)/18 (93-95)	1.04/1.06 WN	6.9/7.5 WN	2.8/7.6 WN	0.11/0.10 WN	0.006/0.012 WN	2.53/2.33 WN	6.19/7.35 WN	0.50/0.30 WN	0.008/0.009 WN
18 (88-90)/18 (95-97)	1.04/0.96 WN	6.9/6.9 WN	2.8/1.9 WN	0.11/0.07 WN	0.006/0.006 WN	2.53/2.41 WN	6.19/4.80 WN	0.50/0.35 WN	0.008/0.012 WN
18 (88-90)/18 (97-99)	1.04/1.20 WN	6.85/7.85 WN	3.3/2.73 WN	.11/.04 WN	.006/.006 WN	2.53/3.34 WN	**6.19/3.80** **WS**	0.50/0.27 WN	0.008/0.006 WN
22 (88-90)/22 (90/93)	0.30/0.35 WN	3.6/5.2 WN	6.7/3.0 WN	0.52/0.45 WN	0.077/0.118 WN	2.65/4.14 WN	3.47/3.59 WN	0.33/0.29 WN	0.014/0.069 WN
22 (88-90)/22 (93-95)	**0.30/0.59** **WS**	3.6/4.3 WN	**6.7/10.6** **WS**	0.52/0.80 WN	**0.077/0.176** **WS**	**2.65/2.05** **WS**	3.47/7.14 WN	0.33/0.55 WN	0.014/0.068 WN
22 (88-90)/22 (95-97)	**0.30/0.56** **WS**	3.6/4.8 WN	6.7/4.0 WN	0.52/0.57 WN	0.077/0.129 WN	2.65/3.11 WN	3.47/4.56 WN	0.33/0.37 WN	0.014/0.018 WN
22 (88-90)/22 (97-99)	**0.30/0.53** **WS**	3.64/4.3 WN	4.2/4.2 WN	0.52/1.00 WN	0.07/0.06 WN	2.65/2.5 WN	3.47/2.5 WN	0.33/0.51 WN	0.014/0.011 WN
23 (88-90)/23 (90/93)	0.71/0.65 WN	6.4/6.6 WN	**8.8/4.4** **WS**	0.28/0.20 WN	0.007/0.014 WN	**6.0/4.7** **WS**	5.9/6.5 WN	0.26/0.11 WN	0.007/0.011 WN
23 (88-90)/23 (93-95)	0.71/0.87 WN	**6.4/8.1** **WS**	**8.8/24.0** **WS**	0.28/0.19 WN	**0.007/0.020** **WS**	6.0/6.6 WN	**5.9/26.8** **WS**	0.26/0.14 WN	0.007/0.013 WN
23 (88-90)/23 (95-97)	0.71/0.86 WN	6.4/7.3 WN	**8.8/4.3** **WS**	0.28/0.19 WN	0.007/0.014 WN	6.0/6.7 WN	5.9/5.7 WN	0.26/0.15 WN	0.007/0.009 WN
23 (88-90)/23 (97-99)	0.71/0.85 WN	6.4/7.7 WN	**8.8/4.8** **WS**	**0.28/0.11** **WS**	0.007/0.011 WN	6.0/6.3 WN	10.4/5.9 WS	**0.26/0.11** **WS**	0.007/0.007 WN
25 (88-90)/25 (90/93)	1.0/0.8 WN	8.1/8.0 WN	7.8/4.1 WN	0.10/0.11 WN	0.008/0.010 WN	7.4/7.7 WN	3.8/4.6 WN	0.11/0.11 WN	0.007/0.011 WN
25 (88-90)/25 (93-95)	1.0/0.9 WN	8.1/8.3 WN	7.8/7.1 WN	0.10/0.08 WN	0.008/0.007 WN	7.4/8.2 WN	3.8/7.6 WN	0.11/0.09 WN	0.007/0.007 WN
25 (88-90)/25 (95-97)	1.0/0.9 WN	8.1/8.7 WN	7.8/6.0 WN	0.10/0.07 WN	0.008/0.005 WN	7.4/7.6 WN	3.8/7.3 WN	0.11/0.08 WN	0.007/0.005 WN
25 (88-90)/25 (97-99)	1.0//0.96 WN	8.1/8.7 WN	7.8/6.0 WN	**0.10/0.043** **WS**	.008/.007 WN	7.4/7.6 WN	3.8/5.5 WN	0.11/0.06 WN	0.007/0.008 WN
26 (88-90)/26 (90/93)	0.9/0.8 WN	7.9/7.9 WN	12.4/5.2 WN	0.11/0.08 WN	0.006/0.013 WN	5.4/5.9 WN	10.3/7.2 WN	0.12/0.11 WN	0.008/0.015 WN
26 (88-90)/26 (93-95)	0.9/1.0 WN	**7.9/9.0** **WS**	12.4/11.8 WN	0.11/0.08 WN	0.006/0.006 WN	5.4/6.3 WN	10.3/10.5 WN	**0.12/0.08** **WS**	0.008/0.005 WN

26 (88-90)/26 (95-97)	0.9/1.0 WN	7.9/8.3 WN	12.4/7.3 WN	0.11/0.07 WN	0.006/0.005 WN	5.4/5.2 WN	10.3/4.9 WN	0.12/0.10 WN	0.008/0.005 WN
26 (88-90)/26 (97-99)	0.9/0.9 WN	**7.9/9.0** **WS**	12.4/6.7 WN	**0.11/0.04** **WS**	0.0006/0.007 WN	5.4/5.3 WN	10.3/8.7 WN	0.11/0.11 WN	0.008/0.006 WN
31 (88-90)/31 (90/93)	1.1/0.9 WN	**7.7/8.1** **WS**	3.6/4.2 WN	0.07/0.06 WN	0.007/0.013 WN	4.4/5.4 WN	4.4/6.1 WN	0.09/0.10 WN	0.006/0.019 WN
31 (88-90)/31 (93-95)	1.0/1.1 WN	7.7/7.7 WN	**3.6/7.9** **WS**	0.07/0.07 WN	0.007/0.006 WN	4.4/5.2 WN	**4.4/7.2** **WS**	0.09/0.08 WN	0.006/0.006 WN
31 (88-90)/31 (95-97)	1.1/1.1 WN	7.7/8.4 WN	3.6/4.9 WN	0.07 WN	0.007/0.005 WN	4.4/5.1 WN	4.4/4.8 WN	0.09/0.09 WN	0.006/0.006 WN
31 (88-90)/31 (97-99)	1.1/1.1 WN	7.7/8.4 WN	3.6/4.6 WN	**0.07/0.05** **WS**	0.007/0.007 WN	4.4/5.4 WN	4.4/4.8 WN	0.09/0.11 WN	0.006/0,006 WN
33 (88-90)/33 (90/93)	1.1/0.9 WN	7.5/7.9 WN	5.8/4.5 WN	0.06/0.06 WN	0.007/0.017 WN	4.3/5.4 WN	5.6/6.5 WN	2.1/1.8 WN	0.008/0.02 WN
33 (88-90)/33 (93-95)	1.1/1.2 WN	7.5/7.9 WN	5.8/8.0 WN	0.06/0.07 WN	0.007/0.006 WN	4.3/5.1 WN	5.6/6.2 WN	2.1/1.7 WN	0.008/0.006 WN
33 (88-90)/33 (95-97)	1.1/1.1 WN	**7.5/8.4** **WS**	5.8/5.4 WN	0.06/0.06 WN	0.007/0.005 WN	4.3/5.0 WN	5.6/5.1 WN	2.1/1.4 WN	0.008/0.005 WN
33 (88-90)/33 (97-99)	1.1/1.1 WN	**7.5/8.5** **WS**	5.8/4.0 WN	**0.06/0.04** **WS**	0.007/0.006 WN	4.3/4.7 WN	5.5/6.9 WN	**2.1/0.049** **WS**	0.008/0.005 WN
37 (88-90)/37 (90/93)	1.2/1.2 WN	7.7/7.8 WN	2.7/4.1 WN	0.06/0.05 WN	0.009/0.012 WN	4.8/4.9 WN	4.4/4.2 WN	0.09/0.10 WN	0.007/0.010 WN
37 (88-90)/37 (93-95)	1.2/1.3 WN	7.7/7.7 WN	**2.7/6.3** **WS**	0.06/0.06 WN	0.009/0.008 WN	4.8/5.2 WN	4.4/5.4 WN	0.09/0.11 WN	0.007/0.006 WN
37 (88-90)/37 (95-97)	1.2/1.4 WN	7.7/8.2 WN	2.7/4.0 WN	0.06/0.07 WN	0.009/0.005 WN	4.8/4.6 WN	4.4/4.1 WN	0.09/0.11 WN	0.007/0.005 WN
37 (88-90)/37 (97-99)	1.2/1.3 WN	7.7/8.1 WN	2.7/3.9 WN	**0.06/0.04** **WS**	0.009/0.007 WN	4.8/4.6 WN	0.09/0.12 WN	0.09/0.12 WN	0.007/0.006 WN
40 (88-90)/40 (90/93)	1.3/1.2 WN	7.9/8.0 WN	2.1/2.5 WN	0.06/0.05 WN	0.005/0.010 WN	5.6/5.5 WN	4.4/3.1 WN	0.08/0.08 WN	0.010/0.010 WN
40 (88-90)/40 (93-95)	1.3/1.3 WN	7.9/7.7 WN	**2.1/2.8** **WS**	0.06/0.05 WN	0.005/0.006 WN	5.6/5.7 WN	4.4/3.9 WN	0.08/0.11 WN	0.01/0.006 WN
40 (88-90)/40 (95-97)	1.3/1.5 WN	7/9/8.4 WN	2.1/2.1 WN	0.06/0.07 WN	0.005/0.005 WN	5.6/5.6 WN	4.4/3.1 WN	0.08/0.10 WN	0.01/0.005 WN
40 (88-90)/40 (97-99)	1.3/1.4 WN	7.9/8.1 WN	**2.1/3.8** **WS**	**0.6/0.4** **WS**	0.005/0.005 WN	5.6/5.5 WN	4.4/4.6 WN	0.08/0.11 WN	0.01/0.005 WN

Note: Data were grouped into two to 3-year periods for analysis. Data taken from 1990 through 1999 were compared with data taken during the 1988 to 1990 period of low phytoplankton bloom incidence. The statistical analysis was carried out using means testing methods listed in Appendix II.

WN = not significant (p = 0.05), WS = significant (p = 0.05), CV = RHO > critical value.

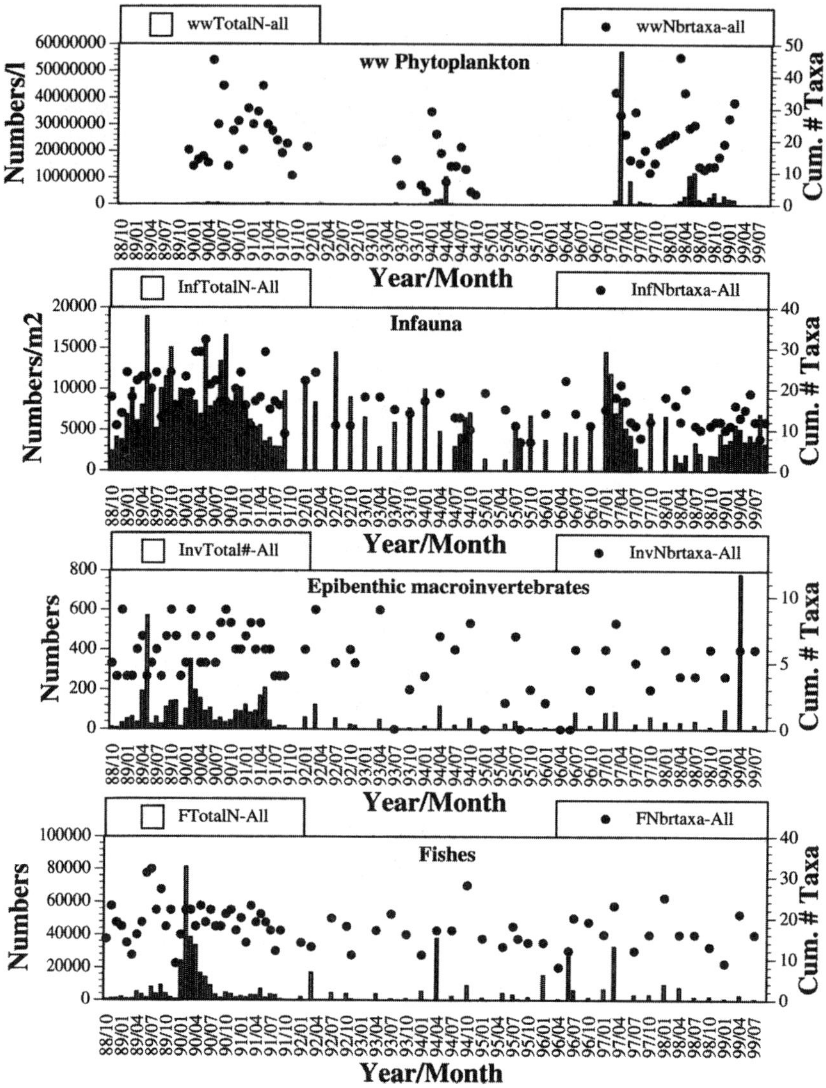

Figure 5.2 Whole-water phytoplankton average numbers and cumulative (all stations) species richness plotted against numbers and cumulative (all stations) species richness of infaunal macroinvertebrates, epibenthic macroinvertebrates, and fishes taken at stations in Perdido Bay from October 1988 through September 1999.

When viewed as deviations from the means of station-specific infaunal numbers and species richness (Figure 5.3), a somewhat different trend emerges. Infaunal numbers peaked during the 1989–1990 period. During the period 1993 to 1999, relatively low infaunal numbers were taken with occasional increases during summer 1996 and spring 1999. Infaunal species richness deviations from the mean were highest during 1989–1990; these high levels were followed by a steady decline such that this index was almost uniformly negative from mid-1992 through 1999. Deviations of invertebrate numbers were positive during disparate periods (winter 1992, spring 1993, winter 1997, winter 1999) with a general trend of negative deviations from 1992 through 1998. Invertebrate species richness deviations tended to be mostly negative during the phytoplankton bloom periods with periodic increases during winter 1997 and spring 1999. After strongly positive deviations during 1989–1990, deviations of fish numerical abundance were negative, with the exception of the period 1995–1996. Uniformly negative deviations were found during late 1998–1999. After

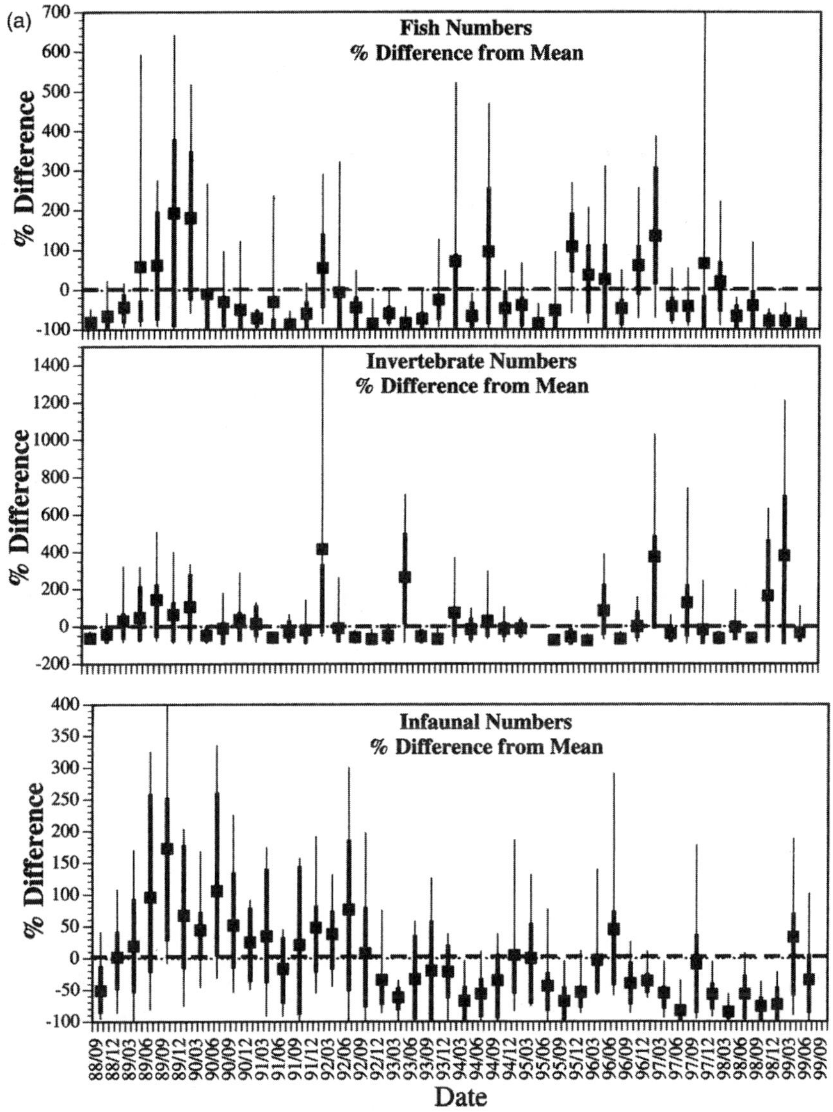

Figure 5.3 Numbers (a) and cumulative (all stations) (b) species richness of infaunal macroinvertebrates, epibenthic macroinvertebrates, and fishes plotted as deviations from the grand mean. Data were taken at stations in Perdido Bay from October 1988 through September 1999.

positive deviations of fish species richness during 1989–1990, there was a somewhat variable pattern of this index during ensuing years with a peak in 1996. The overall pattern was a general reduction of fish species richness from 1991 to present.

Station-specific changes taken as 2-year averages of numbers and species richness of the animal assemblages of Perdido Bay over the 11-year period are shown in Figure 5.4. Statistical comparisons of the 2-year trends of Perdido Bay animals is given in Table 5.2. Average infaunal numbers and species richness were uniformly high in the upper bay seagrass beds (station P25) over the entire 11-year period. These numbers were uniformly low at the mouth of the Perdido River (station P18) and in Elevenmile Creek (station P22) reflecting previously described habitat deterioration in these areas of the bay. Infaunal species richness was generally high at the mouth of Elevenmile Creek (station P23) from 1988 to 1993 with subsequent declines from 1993 to 1999. However, infaunal species richness from 1993 to 1995 was significantly lower ($p = 0.05$) at station P23 than that

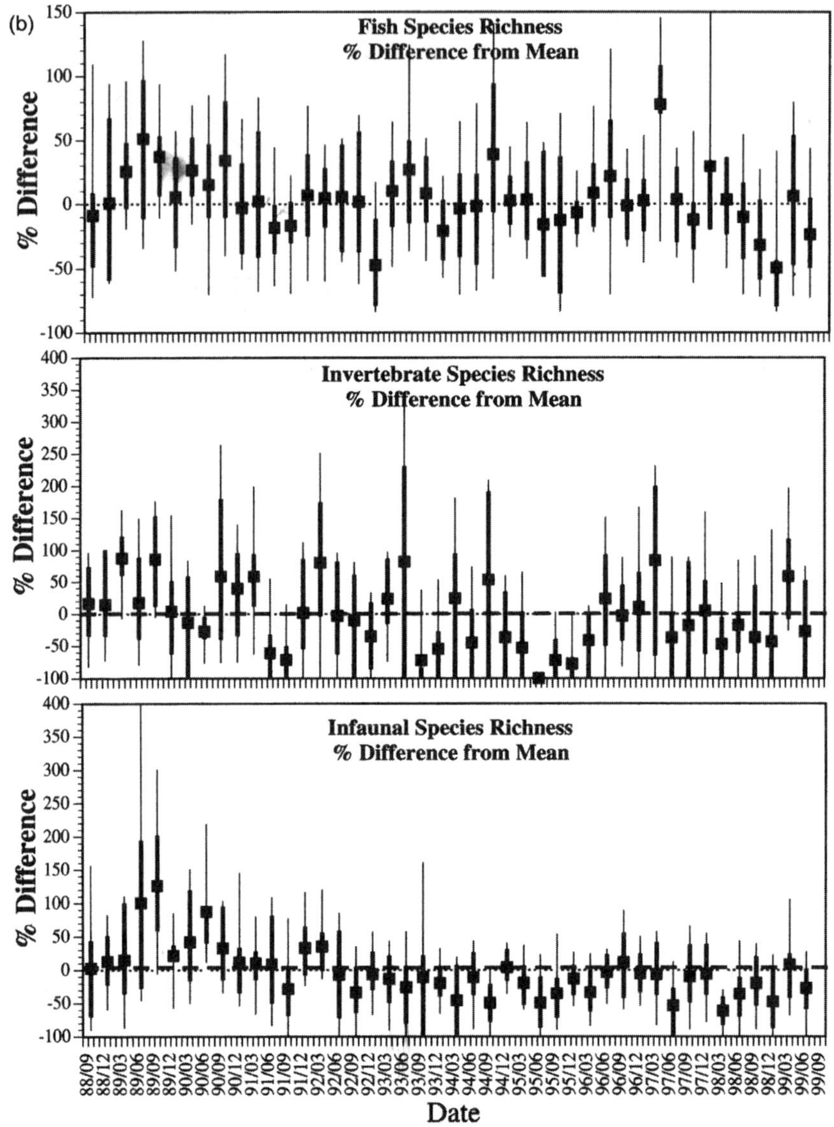

Figure 5.3 (continued)

during the 1990 to 1993 period. During the early years, bay infauna thus tended to be highest in the upper bay around Elevenmile Creek (P23) and the seagrass beds (P25). With time, infaunal numbers and species richness decreased at station P23 but remained relatively stable at station P25. During the initial years, there was increasing infaunal species richness from station P31 to P40; in the lower bay, this index was comparable to that of station P23. From areas in the upper to lower bay (P26 to P40), there were stepwise decreases of infaunal species richness from 1993 to 1999. Most such decreases were statistically significant (Table 5.2). Infaunal numbers were lowest at these stations; however, similar decreases with time were evident, many of which were statistically significant (Table 5.2). These data indicate that the overall deteriorating state of infaunal assemblages were not spatially uniform due to varying habitat conditions in the bay.

The trends of epibenthic macroinvertebrates (Figure 5.4) were similar to those of the infauna with certain exceptions. There were significant (Table 5.2) reductions of invertebrate numbers and

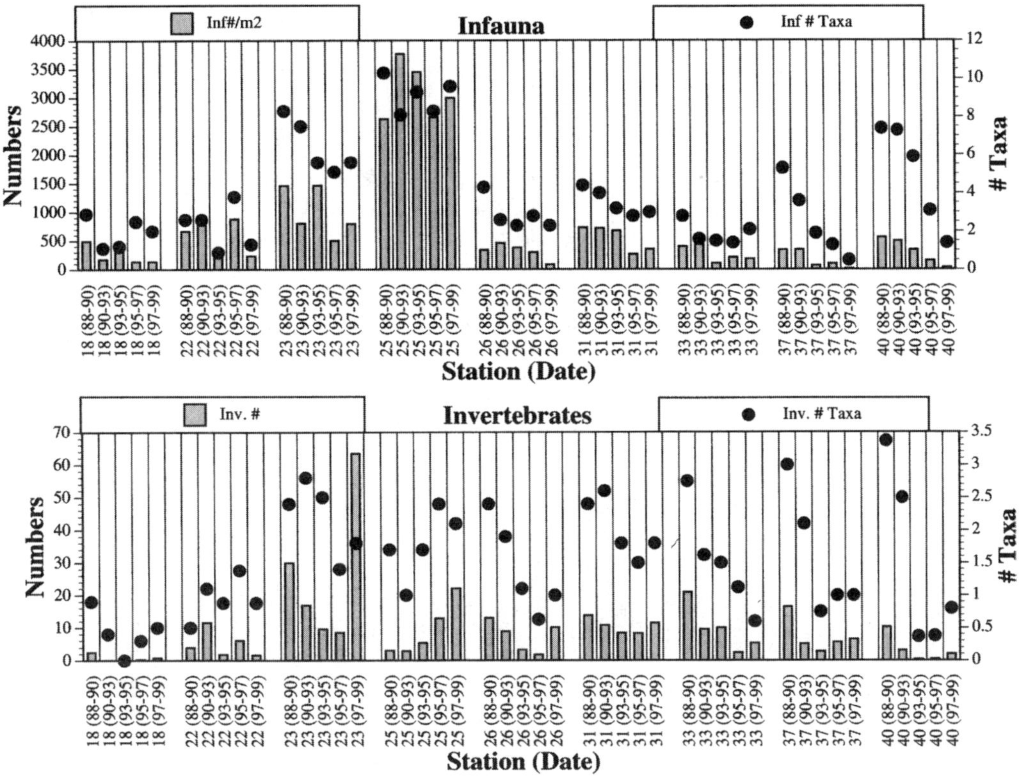

Figure 5.4 The 2-year averages of numbers and cumulative (all stations) species richness of infaunal macroinvertebrates, epibenthic macroinvertebrates, and fishes taken at stations in Perdido Bay from 10/88 through 9/99. Also shown are the 2-year averages of the FII biomass trends by station over the study period.

species richness at the mouth of the Perdido River during the bloom years. There was increased species richness with time in Elevenmile Creek (significantly so in 1997 to 1999). The significantly increased numbers of invertebrates at station P23 during the 1997 to 1999 period followed significantly lower values found during previous bloom periods. This trend could have reflected increased invertebrate numbers during spring 1999. However, invertebrate species richness remained significantly lower during these later periods. Although the invertebrate fauna in the seagrass beds (station P25) was not as well developed as the infauna (possibly a response to reduced predation pressure), there was a trend of increasing invertebrate numbers and species richness at station P25. As with the infauna, the early high (and increasing) invertebrate species richness in the upper bay from 1988 to 1990 fell during succeeding periods. These differences were statistically significant (Table 5.2). Temporal reductions of invertebrate community indices were more extreme with distance from the upper bay. These data thus show that major decreases of invertebrate numbers and species richness occurred over time from station P26 to station P40 with the trend of such losses starting in 1993 to 1995.

The infaunal and epibenthic macroinvertebrates are largely benthic species that are closely associated with bay sediments. These groups underwent prolonged declines in terms of numbers and species richness that were generally correlated with the trends of plankton blooms in Perdido Bay. Depositional areas in the lower bay appeared to be the most impaired such areas in terms of infaunal and epibenthic macroinvertebrate numbers and species richness as a result of the combined effects of salinity stratification, periodic hypoxia, and liquid mud. Prolonged exposure to plankton blooms appeared to exacerbate these impacts. Areas that were least affected by the blooms (i.e.,

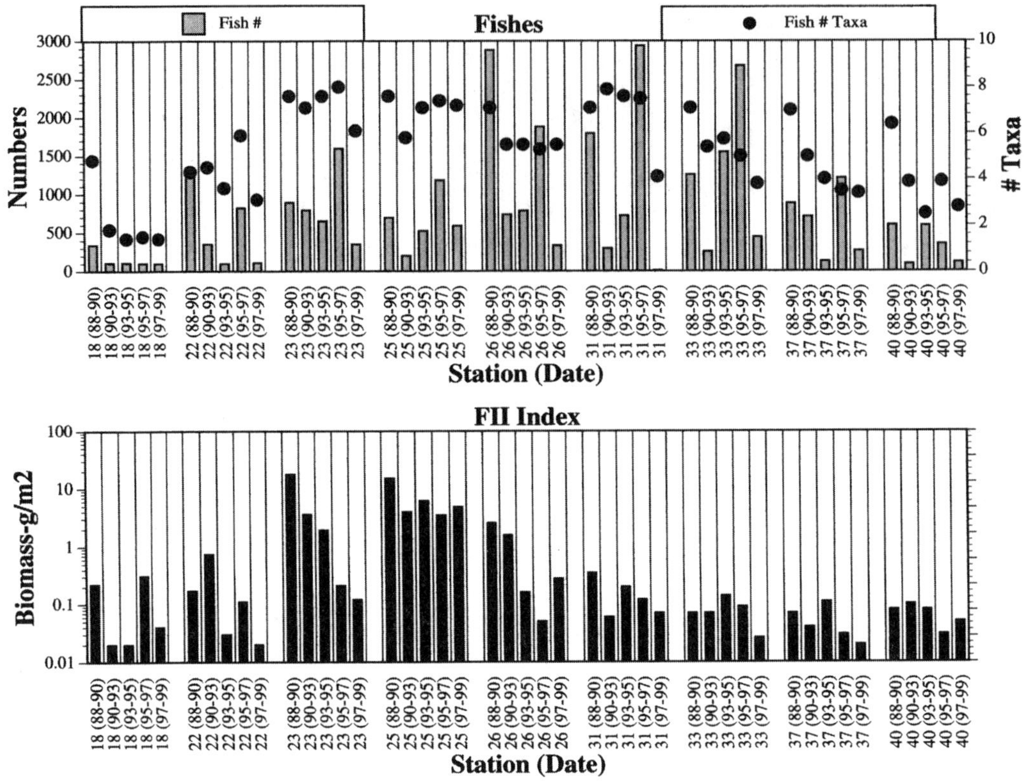

Figure 5.4 (continued)

upper bay seagrass beds) represent an ecological enclave in the upper bay during the years of sustained nutrient loading and phytoplankton blooms.

The trends of numerical abundance and species richness of the fishes were more complex than abundance and species richness of the invertebrates (Figure 5.4). At the mouth of the Perdido River (station P18), fish numerical abundance and species richness were significantly (Table 5.2) reduced during peak bloom periods. At station P23, fish species richness and numbers were relatively stable until the 1997 to 1999 period when there were significant reductions of these indices (Table 5.2). There was relatively little temporal change in the fish indices taken in the seagrass beds of the upper bay. Fish numbers were highest from station P26 to middle bay areas. These numbers declined during the period 1997 to 1999. In the lower bay, there were significant (p = 0.05) reductions of both fish numbers and fish species richness from 1995 to 1999. As with the infauna and epibenthic invertebrates, the debilitation of the fish assemblages was most evident in depositional areas of the lower bay. The spatial and temporal patterns of biological decline indicated that the sediments were somehow involved the deterioration of the Perdido Bay biota. These observations are consistent with the combination of effects of the prolonged plankton blooms superimposed over the trends of reduced water and sediment quality in the highly stratified depositional areas of the lower bay.

C. Trophic Organization

1. Research Approach

Trophic–dynamic aspects of aquatic systems (Lindeman, 1942) comprise the central theme of many recent ecological studies. The direct application of the trophodynamic concept to predictive

Table 5.2 Statistical Comparisons of Station-Specific 2-Year Averages of Numbers and Species Richness of Infaunal Macroinvertebrates, Epibenthic Macroinvertebrates, and Fishes Taken at Stations in Perdido Bay from 10/88 through 9/99

2-Year Means Station/Year Comparison	Infauna Numbers/m²	Infauna # Taxa	Invertebrate Numbers	Invertebrate # Taxa	Fish Numbers	Fish # Taxa	FII Total Biomass g/m²
18 (88-90)/18 (90/93)	496/176 WN	2.9/1.1 WN	**2.5/0.1 WS**	0.9/0.4 WN	**342/108 WS**	4.8/1.8 WN	**0.221/0.003 WS**
18 (88-90)/18 (93-95)	496/324 WN	2.9/1.2 WN	**2.5/0 WS**	**0.9/0.0 WS**	342/111 WN	4.8/1.4 WN	0.221/0.012 WN
18 (88-90)/18 (95-97)	496/137 WN	2.9/2.5 WN	**2.5/0.3 WS**	**0.9/0.3 WS**	**342/105 WS**	**4.8/1.5 WS**	0.221/0.308 WN
18 (88-90)/18 (97-99)	496/292 WN	2.9/2.0 WN	**2.5/0.8 WS**	0.9/0.5 WN	342/102 WN	**4.8/1.4 WS**	0.22/0.04 WN
22 (88-90)/22 (90/93)	667/876 WN	2.6/2.6 WN	3.9/11.6 WN	0.5/1.1 WN	1344/353 WN	4.3/4.5 WN	0.17/0.74 WN
22 (88-90)/22 (93-95)	**667/246 WS**	2.6/.88 WN	3.9/1.8 WN	.5/.88 WN	1344/100 WN	4.3/3.6 WN	0.17/0.03 WN
22 (88-90)/22 (95-97)	667/879 WN	2.6/3.8 WN	3.9/6.1 WN	0.5/1.38 WN	1344/823 WN	4.3/5.9 WN	0.17/0.11 WN
22 (88-90)/22 (97-99)	667/235 WN	2.6/1.3 WN	3.9/1.6 WN	**0.5/0.88 WS**	1344/111 WN	4.3/3.1 WN	0.17/0.02 WN
23 (88-90)/23 (90/93)	1463/804 WN	**8.3/4.5 WS**	30.0/16.9 WN	2.4/2.8 WN	892/794 WN	7.6/7.1 WN	17.8/3.61 WN
23 (88-90)/23 (93-95)	1463/1466 WN	**8.3/5.6 WS**	**30.0/9.6 WS**	2.42.5 WN	892/651 WN	7.6/7.3 WN	**17.8/1.93 WS**
23 (88-90)/23 (95-97)	**1463/503 WS**	8.3/5.1 WN	**30.0/8.5 WS**	**2.42/1.4 WS**	892/1597 WN	7.6/8 WN	**17.8/0.21 WS**
23 (88-90)/23 (97-99)	**1463/796 WS**	8.35/5.6 WN	**30.0/63.5 WN**	**2.42/1.87 WS**	**892/350 WS**	7.6/7.1 WN	**17.8/0.12 WS**
25 (88-90)/25 (90/93)	2628/3768 WN	10.3/8.1 WN	3.0/2.9 WN	1.7/1.0 WN	693/201 WN	7.6/5.8 WN	15.1/3.99 WN
25 (88-90)/25 (93-95)	2628/3452 WN	10.3/9.3 WN	3.0/5.4 WN	1.7/1.7 WN	693/523 WN	7.6/7.1 WN	15.1/6.21 WN
25 (88-90)/25 (95-97)	2628/2660 WN	10.3/8.3 WN	3.0/12.9 WN	1.7/2.4 WN	693/1180 WN	7.6/7.4 WN	15.1/3.51 WN
25 (88-90)/25 (97-99)	2626/1998 WN	10.3/9.6 WN	3.0/22.1 WN	1.7/1.1 WN	693/589 WN	7.6/7.2 WN	15.1/4.87 WN
26 (88-90)/26 (90/93)	538/461 WN	4.3/2.6 WN	13.1/8.9 WN	2.4/1.9 WN	2881/736 WN	7.1/5.6 WN	2.58/1.22 WN
26 (88-90)/26 (93-95)	538/381 WN	**4.3/2.3 WS**	**13.1/3.3 WS**	**2.4/1.1 WS**	2881/784 WN	7.1/5.5 WN	**2.58/0.16 WS**
26 (88-90)/26 (95-97)	**538/301 WS**	**4.3/2.8 WS**	**13.1/1.9 WS**	**2.4/0.63 WS**	2881/1881 WN	7.1/5.3 WN	**2.58/0.6 WS**
26 (88-90)/26 (97-99)	**538/84 WS**	**4.3/2.3 WS**	13.1/21.5 WN	**2.4/1.0 WS**	2881/333 WN	7.1/5.5 WN	**2.58/0.03 WS**
31 (88-90)/31 (90/93)	731/717 WN	4.4/4.0 WN	13.7/10.8 WN	2.4/2.6 WN	1792/296 WN	7.1/7.9 WN	0.35/0.06 WS
31 (88-90)/31 (93-95)	731/676 WN	4.4/3.2 WN	13.7/8.4 WN	2.4/1.8 WN	1792/721 WN	7.1/7.9 WN	0.35/0.20 WN
31 (88-90)/31 (95-97)	**731/264 WS**	4.4/2.8 WN	13.7/8.3 WN	**2.4/1.5 WS**	1792/2939 WN	7.1/7.5 WN	**0.35/0.12 WS**
31 (88-90)/31 (97-99)	**731/352 WS**	4.4/3.0 WN	13.7/11.5 WN	2.4/1.8 WN	1792/1203 WN	7.1/4.1 WN	0.35/0.37 WN
33 (88-90)/33 (90/93)	399/434 WN	2.8/1.6 WN	20.9/9.6 WN	2.75/1.63 WN	1252/256 WN	7.1/5.4 WN	0.066/0.066 WN
33 (88-90)/33 (93-95)	**400/106 WS**	**2.8/1.5 WS**	**20.9/10.0 WS**	**2.75/1.5 WS**	1252/1550 WN	7.1/5.8 WN	0.066/0.141 WN
33 (88-90)/33 (95-97)	400/209 WN	**2.8/1.4 WS**	**20.9/2.4 WS**	**2.75/1.12 WS**	1252/2675 WN	7.1/5.0 WN	0.066/0.085 WN
33 (88-90)/33 (97-99)	400/182 WN	2.8/2.1 WN	**20.9/5.3 WS**	**2.75/0.6 WS**	1252/444 WN	**7.1/3.8 WS**	0.066/0.0286 WN

Table 5.2 (continued) Statistical Comparisons of Station-Specific 2-Year Averages of Numbers and Species Richness of Infaunal Macroinvertebrates, Epibenthic Macroinvertebrates, and Fishes Taken at Stations in Perdido Bay from 10/88 through 9/99

2-Year Means Station/Year Comparison	Infauna Numbers/m²	Infauna # Taxa	Invertebrate Numbers	Invertebrate # Taxa	Fish Numbers	Fish # Taxa	FII Total Biomass g/m²
37 (88-90)/37 (90/93)	341/335 WN	5.3/3.6 WN	**16.5/5.1 WS**	3.0/2.1 WN	**881/129 WS**	7.0/5.0 WN	0.07/0.04 WN
37 (88-90)/37 (93-95)	**341/68 WS**	**5.3/1.9 WS**	**16.5/2.75 WS**	**3.0/0.75 WS**	881/709 WN	7.1/4.0 WN	0.07/0.11 WN
37 (88-90)/37 (95-97)	**341/102 WS**	**5.3/1.3 WS**	**16.5/5.6 WS**	**3.0/1.0 WS**	881/1207 WN	**7.0/3.5 WS**	**0.07/0.029 WS**
37 (88-90)/37 (97-99)	**341/16 WS**	**5.3/0.5 WS**	**16.5/6.5 WS**	**3/1.5 WS**	**881/285 WS**	**7.1/3.4 WS**	**0.07/0.02 WS**
40 (88-90)/40 (90/93)	560/498 WN	7.4/7.3 WN	10.9/7.1 WN	3.4/2.5 WN	601/144 WN	6.4-4.0 WN	0.08/0.10 WN
40 (88-90)/40 (93-95)	560/343 WN	7.4/5.9 WN	**10.9/0.4 WS**	**3.4/0.4 WS**	601.242 WN	**6.4/2.3 WS**	0.08/0.08 WN
40 (88-90)/40 (95-97)	**560/156 (WS)**	**7.4/3.1 WS**	**10.9/0.5 WS**	**3.4/0.4 WS**	601/353 WN	6.4/3.9 WN	**0.08/0.03 WS**
40 (88-90)/40 (97-99)	**560/34 WS**	**7.4/1.4 WS**	**10.9/2.0 WS**	**3.4/0.8 WS**	601/117 WN	6.4/2.8 WN	**0.08/0.05 WS**

Note: Data were grouped into 2- to 3-year periods for analysis. Data taken from 1990 through 1999 were compared with data taken during the 1988 to 1990 period of low phytoplankton bloom incidence. The statistical analysis was carried out using means testing methods listed in Appendix II.

WN = not significant (p = 0.05), WS = significant (p = 0.05), CV = RHO > critical value.

models is, however, difficult. Empirical demonstrations of aquatic food webs (Odum and Heald, 1972) can lead to successful applications (Odum et al., 1982). However, most paradigms of trophic organization of complex systems such as estuaries are based on the species as the central unit of trophic interactions. Such designation is not consistent with detailed information concerning species-specific ontogenetic feeding changes of many coastal invertebrates and fishes (Leber, 1983; 1985; Livingston, 1982; 1984b; 1988). The extreme complexity of aquatic food webs has been both demonstrated and ignored by aquatic ecologists. Generalizations concerning how to define the different food classes can be difficult even within a given developmental stage of a single species much less as the cumulative result of myriod ontogenetic progressions among hundreds of species that are prevalent in coastal areas.

Livingston et al. (1997) went from very detailed trophic studies to more generalized trophic models based on specific criteria for the definition of feeding classifications. By crossing traditional taxonomic lines with trophic definitions and by quantifying significant portions of the estuarine food webs over prolonged periods, the authors were able to compare inclusive trophic designations that crossed superficial taxonomic boundaries. The results of this reorganization of long-term field data into a functional trophic hierarchy could then be used as an index of change that allowed realistic comparisons of disparate estuarine and coastal systems. The specific criteria for the detailed delineation of trophic organization proved to be less important than the consistency of the data on which the trophic models were based. The details of the trophic organization could then be applied to more general models of classification (i.e., reorganization into realistic trophic levels). The obverse approach (i.e., going from generalized models based on unqualified assumptions to specific comparisons) that is so prevalent in much of the theoretical scientific literature on trophic organization remains relativistic and without field verification.

There was a dichotomy of responses of the trophic elements in Apalachicola Bay (Livingston et al., 1997). The herbivores and omnivores were directly linked to physical and chemical controls that were implicitly associated with the primary productivity of the estuary. The Apalachicola River, which had an important impact on the main habitat features of the Apalachicola system (Livingston, 1984a; 1991c; 1997d; Livingston et al., 2000) was directly linked to primary production of the estuary and, consequently, was an important factor in the response of these trophic components. The various levels of carnivory, on the other hand, were associated more closely with biological factors (i.e., the distribution of food in the form of herbivores and omnivores); none of the carnivore groups was related directly to specific habitat components. This work suggested that river flow and primary production were mainly associated with changes in the lower trophic levels and that the carnivores were associated primarily with these trophic components. Physical variables such as temperature and salinity were involved as indirect controls of the biological interactions at the higher trophic levels of the estuary through their influence on the primary production and herbivore response to such production. The natural history of each species was an important factor concerning both seasonal and interannual changes in both the physical habitat and the available food in the estuary. More recent analyses in other bay systems (Livingston, 1999) indicated that the entire trophic organization of river-dominated estuaries was related both in space and time to the qualitative and quantitative features of nutrient loading relative to the physiographic features of the system in question.

There is little doubt that trophodynamic aspects of population and community progressions in space and time are central to an understanding of the generation of secondary production in coastal areas (Livingston, 1987c; Livingston et al., 1997). However, the extensive scientific literature concerning soft-sediment biological organization in coastal systems tends to be based on issues that are usually tangential to estuarine trophodynamic processes (Peterson, 1992; Livingston et al., 1997). Most studies are so limited in space and time that fundamental coastal trophodynamic processes remain undetermined, overgeneralized, or, worse, misrepresented. Interactions between biotic and abiotic aspects of the trophic organization in estuarine systems thus remain inadequately defined even though it is generally recognized that trophodynamic processes are the central unifying feature among most aquatic systems. Few studies take into account spatial variability, interannual changes of the physical–chemical environment, or the highly variable nature of determinants of community organization in coastal systems (Dayton et al., 1992). Even less well developed is the relationship between the most important single aspect of estuarine and coastal primary production, the phytoplankton community, and the resultant food web. It should be intuitively apparent that changes in the quality and quantity of the primary sources of organic matter could affect associated food webs. Yet, there remains a void in the scientific literature both with respect to the structural components of phytoplankton communities and to the interactions of temporal changes of such components with associated food webs and secondary production in general. Much of the literature concerning the effects of plankton blooms on invertebrates and fishes is related to immediate effects on individual populations without regard to more fundamental, long-term changes in the system.

The primary database for the seven estuarine and coastal systems of the 30-year sampling program was transformed into a trophodynamic entity. All biological data (as biomass$\cdot m^{-2} \cdot mo^{-1}$ of the infauna, epibenthic macroinvertebrates, and fishes) were transformed from species-specific information into a matrix based on the trophic classification. The trophic organization was developed from empirical data concerning ontogenetic feeding stages of the species found in these systems (see Figure 2.1). Ontogenetic feeding units were determined from a series of detailed stomach content analyses carried out with epibenthic invertebrates and fishes in the various estuaries of the northeastern Gulf region (Sheridan, 1978; 1979; Laughlin, 1979; Sheridan and Livingston, 1979; 1983; Livingston, 1980; 1982; 1984b; unpublished data; Stoner and Livingston, 1980; 1984; Laughlin and Livingston, 1982; Stoner, 1982; Clements and Livingston, 1983; 1984; Leber, 1983; 1985). Based on the long-term stomach content data for each size class, the invertebrates and fishes were reorganized into their trophic ontogenetic units using cluster analyses. Infaunal macroinvertebrates were also organized by feeding preference based on a review of the scientific literature (Livingston, unpublished data). Livingston et al. (1997) gave a detailed listing of the trophic organization by growth stage by species. The field data taken over the 31-year sampling period in coastal regions of the northeastern Gulf were then reordered into trophic levels so that monthly changes in the overall trophic organization of the respective bays could be determined over the various system-specific study periods. Feeding patterns were derived from trophic studies in the respective estuaries and bay systems during the long-term studies of the upper Gulf of Mexico. It was assumed that feeding habits at this level of detail did not change over the period of observation; this assumption was based on previous, long-term analyses of species-specific fish feeding habits that remained stable in a given area over a 6- to 7-year period (Livingston, 1980).

The long-term (11-year) field data from Perdido Bay were transformed into individual feeding units of the infauna, epibenthic macroinvertebrates, and fishes taken during the long-term field program. Station-specific data were used and these data were also averaged (g m^{-2}) so that station location was not a factor in the some of the analyses. The data were summed across all taxonomic lines (e.g., herbivorous infauna were pooled with herbivorous epibenthic macroinvertebrates, herbivorous fishes, etc.). Final trophic levels included herbivores (feeding on phytoplankton and benthic algae), omnivores (feeding on detritus and various combinations of plant and animal matter), primary carnivores (feeding on herbivores and detritivorous animals), secondary carnivores (feeding on primary carnivores and omnivores), and tertiary carnivores (feeding on primary and secondary carnivores and omnivores). More detailed transformations of the field data were made (i.e., planktivorous herbivores, macrophyte-feeding herbivores, etc.). All data were given as ash-free dry mass$\cdot m^{-2}\cdot mo^{-1}$ or as percent ash-free dry mass$\cdot m^{-2}\cdot mo^{-1}$. Unless otherwise indicated, the data were presented as monthly, quarterly, or annual means of different combinations of stations. In this way, the long-term database of the collections of infauna, epibenthic macroinvertebrates, and fishes was reorganized into a quantitative and detailed trophic matrix based not on species designations (Livingston, 1988; Livingston et al., 1997) but on the complex ontogenetic feeding stages of the various organisms in the Perdido estuary. The so-called FII (fish–infauna–invertebrate) indices thus represented a reorganization of all biological collections into a biomass-based system that defined, quantitatively, the entire trophic organization of a square meter of water column/sediment. Changes of these indices at stations throughout Perdido Bay were determined from October 1988 through September 1999. This approach was also applied to the data from the other coastal systems in the database.

2. *Long-Term Trophodynamic Trends*

The FII total biomass analysis (Figure 5.4) of the Perdido Bay system allowed a generalized view of the station-specific trends in the bay over the 11-year study period. The river and creek mouth stations (P18, P22) were represented by a relatively depauperate fauna that is consistent with the deteriorated habitats in these parts of the bay. The highly productive stations in the upper bay (P23, P26, and P25) had divergent, long-term trends of FII with unvegetated stations showing serious reductions with time relative to the more stable grass-bed area (P25). The salinity stratified middle and lower bay stations had generally low FII biomass with biomass reductions during the period 1995 to 1999. Detailed, station-specific, annual fluctuations of the infaunal, invertebrate, and fish biomass totals were compared with the annual FII biomass values at the different levels of trophic organization (Figure 5.5). Each station showed a distinctive pattern of temporal deterioration with a general trend of reduced infaunal biomass in the bay from 1995–1996 to 1999. The infaunal macroinvertebrates were generally depauperate in the highly stratified hole at the mouth of the Perdido River (P18). The generally high biomass of the infauna in Perdido Bay at stations P23 and P25 (1988 to 1995) was associated with herbivores that were predominant in upper bay areas. There was a general decline of infaunal biomass throughout the bay. Temporal declines of infauna in the most productive areas (P23, P25) began in 1996 and 1998, respectively. There were progressive declines of infauna in the upper bay that continued to low levels in 1999. In the lower bay, the declines were even more extreme, with reductions to negligible infaunal biomass during the last 2 to 3 years of sampling. The data indicate that during periods of the early blooms (i.e., dominated by diatoms), there were no substantial declines. It was during the period 1997 to 1999

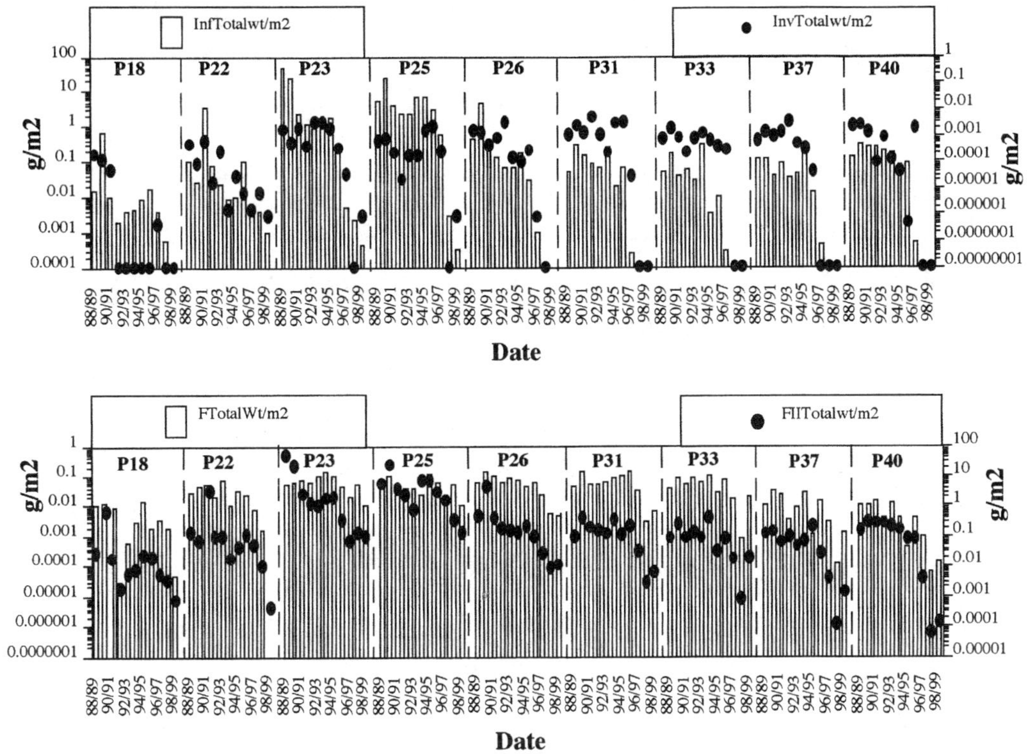

Figure 5.5 Total biomass m^{-2} for infauna (inf), epibenthic macroinvertebrates (inv), fishes (F), and FII indices (FII). Data are averaged over 1-year periods from October 1988 through September 1999.

(i.e., ascendancy of the raphidophytes and dinoflagellates) that the precipitous declines began throughout the bay. Areas already stressed by habitat deterioration were characterized by further declines. Seagrass areas in the upper bay were buffered from immediate impacts, but eventually were subject to serious declines of infaunal biomass.

The Perdido Bay invertebrates followed a similar pattern to that of the infauna (Figure 5.5). Again, there were biomass declines that followed station-specific patterns with major reductions occurring usually during the period 1996 to 1999. In general, the invertebrate biomass, while one to two orders of magnitude less than that of the infauna, was closely associated with infaunal biomass. The distinctive station-specific patterns of biomass decline was probably due to the interaction of existing habitat conditions, sediment quality changes, and the close trophic associations between these two groups of organisms (Livingston et al., 1997). Invertebrates were virtually nonexistent at station P18 during 6 of the 11 years of sampling. There was a progressive decline of invertebrates in Elevenmile Creek (P22) that started in 1995. The invertebrates were more evenly distributed throughout the bay than the infauna during the early years of sampling. At most stations, the sharp invertebrate declines occurred during the period from 1997 to 1999.

The fishes had less well defined decreases that tended to be most pronounced during the last 2 years of sampling. The fishes tended to be less affected by the blooms than the infauna or the epibenthic macroinvertebrates. It is possible that this had something to do with the fact that most of the fishes had less direct contact with the sediments. However, many of the fishes in Perdido Bay feed at the sediment/water interface, and are therefore subject to indirect habitat changes through altered changes in the trophic organization.

The benthic herbivores represent an important link between the primary producers and the upper levels of the food web (Livingston et al., 1997). In areas characterized by water quality problems (P22), high salinity stratification (P18), and the presence of liquid mud (P31, P33, P37, and P40),

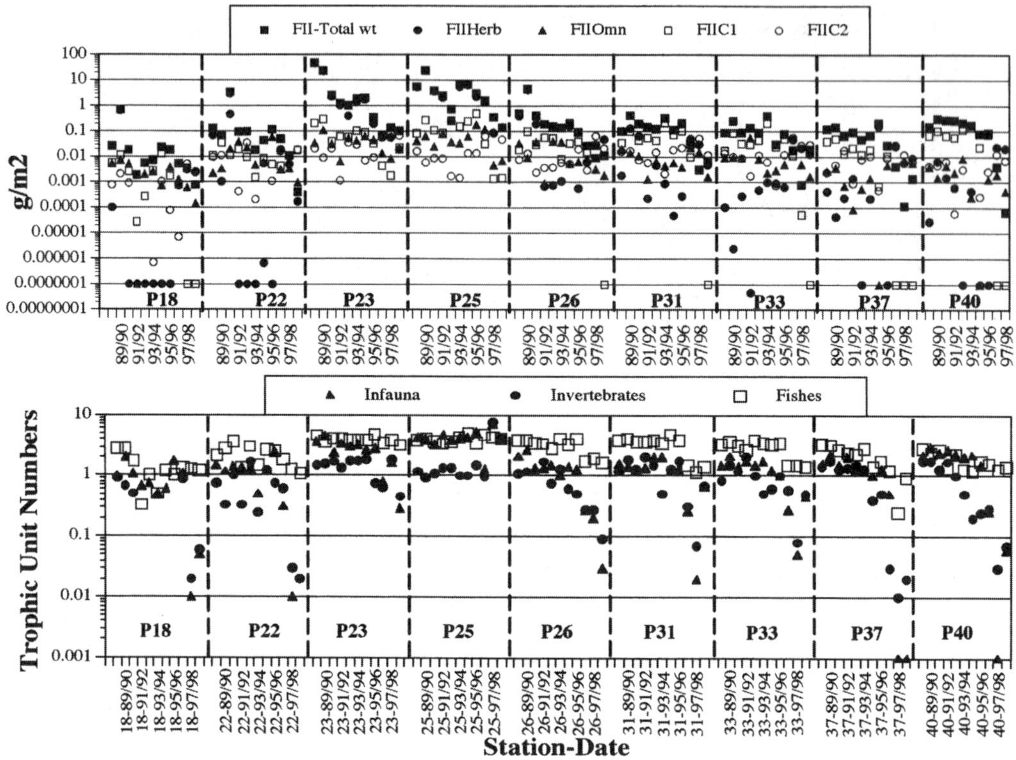

Figure 5.6 Total biomass m^{-2} for all herbivores (FIIHerb), omnivores (FIIOmn), C1 carnivores (FIIC1), C2 carnivores (FIIC2), and C3 carnivores (FIIC3). Data are averaged over 1-year periods from October 1988 through September 1999. Annual averages of numbers of trophic units broken down by animal type (infauna, epibenthic macroinvertebrates, fishes) and taken at stations in Perdido Bay from October 1988 through September 1999.

there was a much-reduced herbivore component, an indication of a disjunct trophic response to sediment deterioration (Figure 5.6). Herbivores were most concentrated in the upper bay at stations P23 and P25 and, to a lesser extent, P26 (Figure 5.6). The primary carnivores (C1) became increasingly dominant in lower parts of the bay. Secondary carnivores (C2) appeared to be least affected by temporal changes over the period of study. Top predators (C3) were largely absent at the mouth of the Perdido River (P18) and in the lower bay.

One of the important aspects of the food web structure in Perdido Bay is the organization of the trophic units (numbers of trophic units by group; Livingston et al., 1997) that represent one aspect of the trophic diversity of the estuary. Trophic units are the individual feeding groups differentiated along the lines of the ontogenetic development of the individual species. The distribution of the trophic units in Perdido Bay over the study period (Figure 5.6) indicated a general decline of trophic diversity across taxonomic lines at most stations. With the exception of the upper bay seagrass beds (station P25), there were systematic reductions of this index throughout the bay for the various animal groups with time. Infaunal and epibenthic macroinvertebrates had more precipitous declines of this index than the fishes. The least such impacts were at station P23 where there were declines of infaunal and epibenthic macroinvertebrate trophic richness during the 1995 to 1997 period with temporary increases during 1997–1998 (low nutrient loading). The lowest levels of trophic unit richness occurred during 1998–1999. Fish trophic richness appeared to be unaffected at this station. Trophic diversity was not affected in the seagrass beds (P25). Middle to lower bay areas (P26 to P40) showed pronounced declines of this index from 1994 to 1999 for the infaunal and epibenthic macroinvertebrates. Declines of fish trophic diversity appeared to occur somewhat later and were less severe than those of the invertebrates groups. Overall, the lowest levels of trophic diversity tended to be reached during the last 2 to 3 years of sampling in most areas of the bay. This coincided with the increasing dominance of the raphidophytes and dinoflagellates in Perdido Bay during this period.

Among the infaunal and epibenthic macroinvertebrates, the downward trend of FII biomass was most evident during the last 2 to 3 years of the survey (Figure 5.5). The general decreases of infaunal and invertebrate biomass and the deterioration of the basic food web structure in Perdido Bay tended to occur about 2 years later than declines of the species richness indices (Figure 5.4) of these groups. The station-specific trends need to be interpreted within the varying levels of habitat deterioration due to causes other than the plankton blooms. Station P18 was located in a deep depression at the end of the Perdido River which was characterized as highly stratified and periodically affected by liquid mud and hypoxia at depth. Station P22 is in Elevenmile Creek that was adversely affected by pulp mill effluents (hypoxia, high ammonia concentrations). There was a general decline of FII biomass from the upper to the lower bay (Figure 5.4). The lower bay, represented as depositional areas with relatively high salinity stratification, associated liquid mud deposits, and periodic hypoxia, deteriorated rapidly from 1997 to 1999. The highest biomass in the system was noted at station P23 (1988 to 1990) and station P25 (throughout the 11-year period). The general decreases of FII biomass were statistically significant ($p = 0.05$) at stations P23, P26, P31, P37, and P40 (Table 5.2).

Overall, these data indicated that there were serious disruptions of the food web throughout most of Perdido Bay, especially during periods associated with the later phases of the phytoplankton blooms. The most rapid deterioration occurred during periods of increasing dominance of the raphidophyte and dinoflagellate bloom species. From a trophodynamic perspective, the primary disruption to the baywide food web structure occurred from 1996 through 1999, a time of diverse phytoplankton blooms with some recovery due to the moratorium of orthophosphate loading from the mill in late 1997–1998. The annual life history cycles of many of the dominant fishes and invertebrates could have been involved in the general lack of response of these organisms to the more rapid recovery of the phytoplankton assemblages. Trends of the trophic units did reflect some recovery during the period of reduced orthophosphate loading, an indication of potential sensitivity of this index to changes in trophic successions. The brief return of the bay that was characterized by high numbers of shrimp and blue crabs in the upper bay during spring 1999 was probably halted by the resumption of orthophosphate loading and the return of massive plankton blooms in the bay. This brief recovery of some invertebrate groups was not reflected in the overall trophic organization of the bay during 1999, however, and was reflected in only a partial return of overall FII biomass and a brief increase of trophic unit numbers in the upper bay during 1998–1999 (Figures 5.4 and 5.6).

It is necessary to analyze the more detailed trophic changes that occurred in Perdido Bay over the 11-year study period (Figure 5.7). The primary form of herbivory among the infaunal macroinvertebrates was plankton-feeding herbivory (InfHP) (Figure 5.7a). During prebloom years, most of the planktivorous infauna were found at stations P23, P25, and P26, which is consistent with the general distributions of the phytoplankton during the early years of sampling. This is typical of river-dominated estuaries that are periodically orthophosphate limited (Livingston et al., 1997). There was a decline of such organisms at station P23 during 1991 that could be reflection of the effects of the drought on the trophic organization of the bay (Livingston et al., 1997). There was a more precipitous decline of planktivorous infauna during the period 1995–1996 with the lowest levels of reached during the last 3 years of sampling. In the seagrass beds (P25), such decline was delayed, with a reduction of the plankton-feeding infaunal herbivores occurring during 1996–1997. This decline was not as precipitous as that occurring at station P23. Farther down the bay (P26), there was an even steeper decline of this group of organisms that started during 1996–1997. In the middle to lower bay, there was a generally low level of plankton-feeding infaunal herbivores that remained more or less stable over the study period at station P31 and actually increased at stations

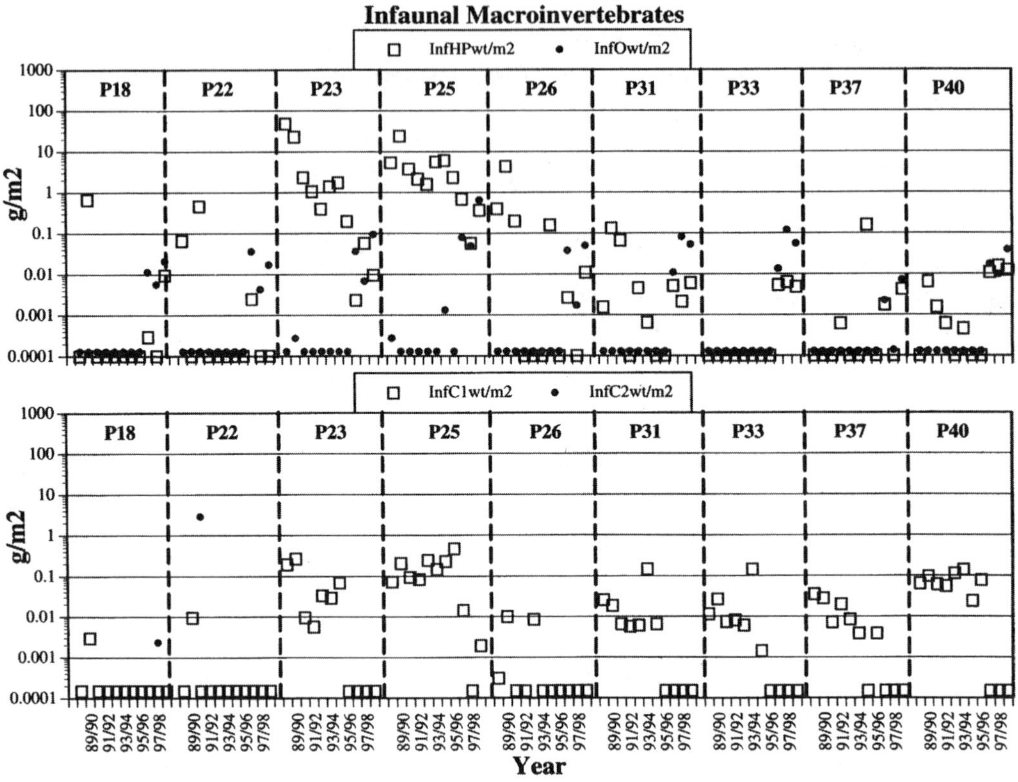

Figure 5.7 Total biomass m^{-2} for trophic classifications of infauna, invertebrates, and fishes taken in Perdido Bay from October 1988 to September 1999 and averaged annually. (a) Infaunal planktivores (InfHP), infaunal omnivores (InfO), and infaunal primary carnivores (InfC1), and infaunal secondary (InfC2) carnivores.

P33, P37, and P40 during the last 3 years of sampling. The omnivorous infauna (InfO) tended to increase during the last 3 years at most stations. These increases of infaunal omnivore biomass cooccurred with the declines of the herbivorous infauna at some stations that could indicate some form of competitive interaction. The data thus indicate that disruption of one of the primary trophic groupings (i.e., planktivorous herbivores; Livingston et al., 1997; InfHP) accounted for some of the main alterations of infaunal biomass with time. The plankton-feeding infauna include various bivalve mollusks (*Brachidontes exustus, Mactra fragilis, Polymesoda caroliniana, Rangia cuneata*) that were prevalent in upper Perdido Bay during the early years of sampling but disappeared after the main plankton blooms began.

The infaunal C1 predator biomass (InfC1), a group prevalent at all stations except P18 and P22, also fell precipitously throughout the bay during the 1997 to 1999 period (Figure 5.7a). The primary carnivores of the infaunal associations tended to decrease during the last 4 years at station P23 and, to a lesser degree, during the last 3 years at station 25. The C1 carnivores were largely absent from stations P18, P22, and P26 by the third year of the study. The C1 carnivores, in relatively low abundance in middle to lower bay areas, completely disappeared from these areas during the last 4 to 5 years of the study. These data show that the infaunal associations throughout the bay were affected by existing conditions with major disruptions in areas with diminished habitat (P18, P22, P31, P33, P37, and P40). There were shifts in the trophic organization led by reductions of infaunal planktivores that cooccurred in various ways with the incidence of water column phytoplankton blooms and the rising dominance of the raphidophytes and the dinoflagellates. The C1 infaunal

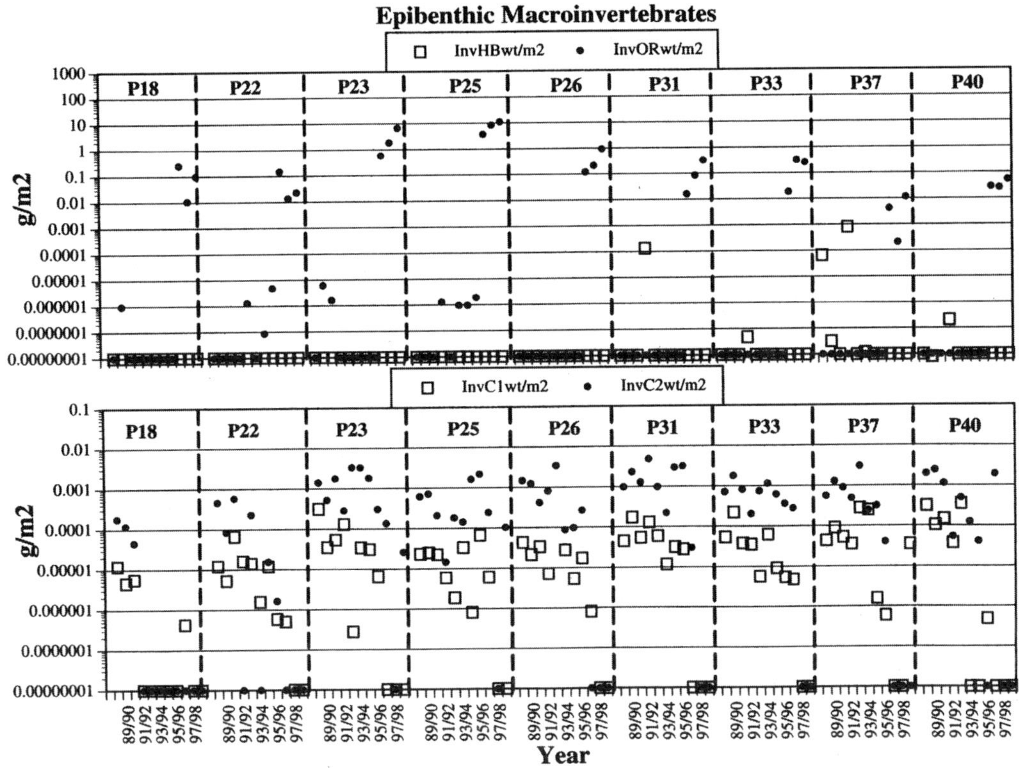

Figure 5.7(b) Invertebrate herbivores (InvHB), invertebrate omnivores (InvOR), invertebrate primary carnivores (InvC1), and invertebrate secondary carnivores (InvC2).

predators followed established patterns of declines of their primary sources of food, both before and after the phytoplankton bloom periods.

Detailed changes of the trophic structure of the epibenthic macroinvertebrates in Perdido Bay are given in Figure 5.7b. Invertebrate herbivores (InvHB), feeding mostly on benthic plants, were largely absent. The general omnivores (InvOR), feeding mainly on detritus and plant matter, tended to follow the pattern of the infaunal macroinvertebrate omnivores in space and time. There was a general increase in omnivores throughout the bay during the last 3 years of sampling. This paralleled the collapse of the infaunal planktivorous herbivores and could have been a result of altered competitive pressures. The invertebrate primary carnivores (InvC1; at least 50% of the diet was herbivores) and secondary carnivores (InvC2; feeding mainly on primary carnivores) were widely distributed throughout the bay. The primary carnivores included young blue crabs (*Callinectes sapidus*) and young-of-the-year penaeid shrimp (*Penaeus* spp). Both trophic groups (C1, C2 carnivores) dropped out at the mouth of the Perdido River by the fourth year and the primary carnivores dropped out of the lower bay (P40) by 1994. There was a general collapse of the invertebrate primary and secondary carnivores throughout the bay during the last 2 to 3 years of the survey.

The early loss of infaunal herbivores, usually dominant in a trophically balanced system, added substance to the likelihood that the observed trends of this group were somehow related to changes in the benthic microflora which accompanied the initiation of the water column phytoplankton blooms in 1993–1994. The trophospecific reductions of secondary production in the depositional areas usually preceded those in the shallower upper bay areas (P23) which, in turn, tended to precede those in the seagrass beds of the upper bay. The sequence of the progressive losses of this trophic grouping gives credence to the hypothesis that the sediments were a focal point of the

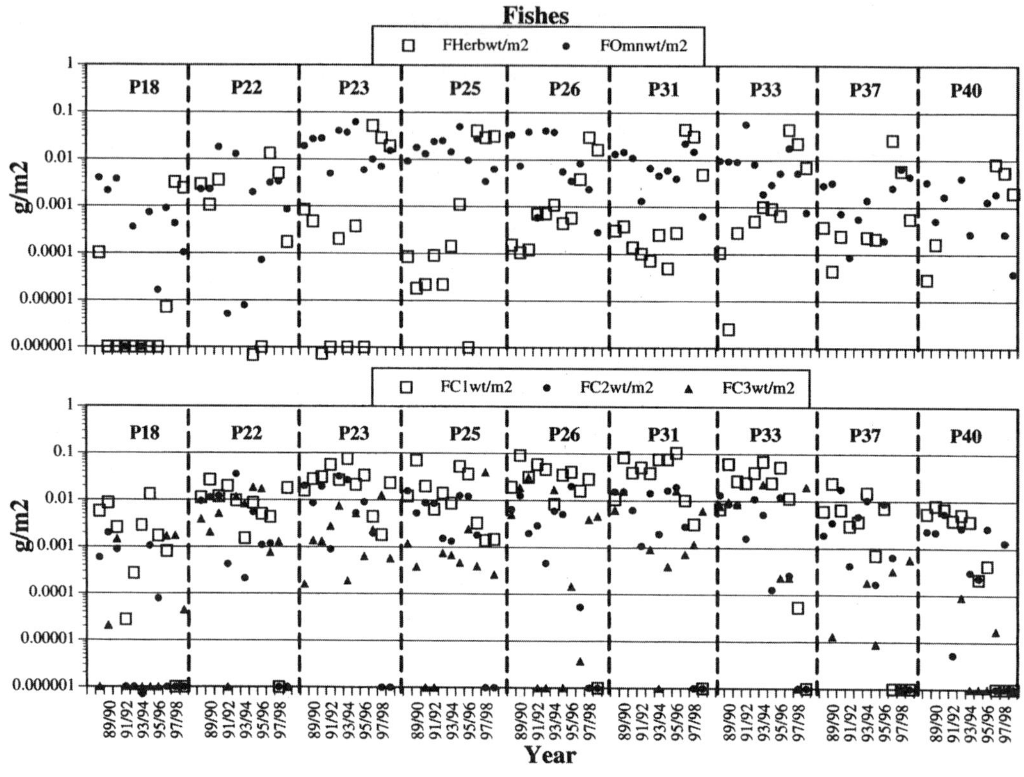

Figure 5.7(c) Fish herbivores (FHerb), fish omnivores (FOmn), fish primary carnivores (FC1), fish secondary carnivores (FC2), and fish tertiary (FC3) carnivores.

observed deterioration of the trophic organization of the bay during the period of intensive phytoplankton blooms. The progressive changes of the invertebrate biota that tended to follow alterations of the phytoplankton communities associated with the phytoplankton blooms also supported the hypothesis that the loss of the benthic microflora during the blooms could have played a role in the overall deterioration of the bay.

The trophic response of the fishes (Figure 5.7c) was more complex than that of the infaunal and epibenthic macroinvertebrates. Fish assemblages in the upper and middle parts of the bay were dominated by the omnivores (FOmn) and primary carnivores during the early years of sampling. The omnivorous fishes tended to decrease at most stations through time. Coincident with the reductions of the omnivores, there was an increase in herbivorous fishes during the last 3 years of sampling. This represented yet another indicator of changes in the phytoplankton communities during this period. The fish primary carnivores (FC1) tended to undergo major reductions at most stations during the last 2 to 3 years of the project, especially in middle to lower bay areas. The C1 predators in the bay included the early-to-middle-stage sciaenids (*Leiostomus xanthurus*, *Micropogonias undulatus*) and young-of-the-year pinfish (*Lagodon rhomboides*). These fish stages were primarily bottom feeders that specialized in foraging upon sediment infaunal and epibenthic invertebrate populations. The secondary carnivores (FC2) tended to drop out of the bay almost entirely by the last 2 years of the field analyses. The larger fish predators (FC3) in the bay had no clear trend through time, although it should be noted that this group was not well sampled by the methods of collection. The fish trophic organization changed in a significant way in Perdido Bay during periods characterized by the intensive phytoplankton blooms of the last 3 years (1997 to 1999). These changes were consistent with reductions of herbivorous infaunal and epifaunal invertebrates that provide the primary source of food for the carnivorous fishes and with the increase of phytoplankton biomass in Perdido Bay during the last 3 years of sampling.

The timing and extent of the changes of the Perdido Bay food webs during the early years were consistent with previous studies concerning the timed responses of different trophic levels to changes in state variables such as river flow (Livingston et al., 1997). The Perdido Bay results showed that seasonal and interannual successions of plankton blooms, occurring as species-specific sequential seasonal trends and interannual successions, were eventually associated with major changes of the trophic organization of the bay through time. The pattern of the deterioration indicated sediment changes that were somehow associated with the plankton blooms as the most likely explanation, although such interpretation remains circumstantial and the mechanism involved (effects of algal toxins, disruption through competition and/or predation, etc.) awaits experimental verification. It is most likely that the observed response of the trophic organization of Perdido Bay was related to a combination of factors rather than any single effect. The reduction of the important planktivorous infauna that cooccurred with the bloom successions indicates a possible link with the postulated adverse response of the sedimentary microflora to the plankton blooms. These changes could also have been exacerbated by the increased benthic hypoxia that accompanied the blooms, but seasonal hypoxia was not the primary factor in the deterioration of the bay.

In any case, the qualitative and quantitative changes of the phytoplankton communities of Perdido Bay that accompanied the nutrient-stimulated plankton blooms were closely associated with the long-term pattern of significant reductions of secondary production through habitat-specific changes in the trophic organization of the bay. Despite the general increases of plankton biomass through time, the disjunct trophic response precluded successive increases in secondary production. These changes could only be interpreted within the context of natural drought-flood sequences, existing habitat deterioration such as the effects of pulp mill effluents on Elevenmile Creek, and changes in subhaloclinal aquatic and sedimentary habitat due to the salinity stratification that was associated with the dredged opening of Perdido Pass. The complexity of the interannual bloom sequences was a key element in the overall trophic response of Perdido Bay to anthropogenous nutrient loading through time.

CHAPTER 6

Comparison of Gulf Coastal Systems

It cannot be assumed that the eutrophication process that operates in a given estuary will work in precisely the same manner as that in other such systems. The differences of specific combinations of habitat characteristics, river flow, and nutrient loading patterns, and physiographic features such as depth distributions preclude easy generalization. The composition and distribution of species-specific coastal populations, from microalgae to fishes, varies widely from estuary to estuary so that generalizations based on such populations are difficult to make with respect to the effects of eutrophication. There is still no scientific consensus concerning whether or not bloom generation and the effects of HABs on secondary production follow similar pathways among different coastal systems. In some ways, the study of eutrophication processes in a single estuary is limited by the specificity of local ecological conditions. Therefore, generalization of eutrophication processes should be based on comparisons of disparate coastal areas.

The difficulties involved in such comparisons include the general lack of consistent databases in different coastal systems and the high level of variation of species composition from estuary to estuary. The development of a robust set of phytoplankton community indices is necessary if coastal areas are to be compared. Given the relatively limited application of current bloom definitions, there should be a more consistent definition of how relative dominance by bloom species is related to the community as a whole. By transforming species-specific data of invertebrates and fishes into universal trophic equivalents, the problem of using questionable indices based on individual (species-specific) populations is eliminated. By using the entire phytoplankton community as a subject for comparison, the usual problems associated with what is often anecdotal data concerning plankton blooms becomes more manageable. Therefore, the use of the trophic organization of phytoplankton community indices and interchangeable energy units of the trophic organization provides a universal basis for the comparison of different estuaries regardless of system-specific habitat components and species distributions.

I. HABITAT CONDITIONS

Habitat conditions in river–estuaries along the northeastern Gulf are determined by various factors. These features have been outlined in Chapters 2 and 3. The barrier estuaries of the Florida Panhandle are dominated by a series of rivers from the Perdido to the Apalachicola. The western Panhandle estuaries tend to be characterized by low turbidity and high color relative to the Choctawhatchee and Apalachicola systems. Wind is a major factor in the determination of water circulation and salinity in most of the shallow Gulf coastal estuaries. In areas where river flow is a factor, frictional effects on the saltwater regimes are combined with salt-wedge circulation and geostrophic forces to affect the estuarine salinity distribution which, in turn, is closely associated with important water quality factors. The physical dimensions of a given river–estuarine basin are thus important in the determination of the distribution of water and sediment quality. The size, depth, and location of passes of each system represent important determinants of key habitat conditions. These drainage characteristics are important in the determination of salinity relationships in the respective study areas. Together with freshwater influxes, basin characteristics, and associated geostrophic forces, the complex interactions of these factors determine spatial/temporal water and sediment quality characteristics of the Gulf coastal estuaries of the northeastern Gulf of Mexico.

II. SALINITY STRATIFICATION AND HABITAT DETERIORATION

A. Stratification, Dissolved Oxygen, and Sediment Condition

A series of 2-year databases was taken from the long-term field analyses of the various coastal systems; data were chosen based on comparable levels of river flow conditions in the long-term sequences of flow rates in each of the chosen drainage systems. A combined database of the various physical–chemical variables and stratification indices was constructed and regressions of stratification indices were run with bottom dissolved oxygen (DO). A detailed account of methods used for this analysis is given by Livingston (1997a). The classification process for salinity and DO is given in Appendix I.

Salinity stratification indices and mean depths for the study areas are shown in Figure 6.1. These indices are categorized as follows:

Highly stratified (HS): salinity difference (bottom – surface) ≥10 ppt.
Partially stratified, strong (PSS): salinity difference ≥5 and <10 ppt.
Partially stratified, weak (PSW): salinity difference ≥2 and <5 ppt.
Vertically homogeneous (VH): salinity difference <2 ppt.

Dissolved oxygen indices were defined as follows:

Anoxic to severely hypoxic (ASH): DO ≤ 2 mg l^{-1}
Hypoxic (HYP): DO > 2 and ≤ 4 mg l^{-1}
Biologically "neutral" (BN): DO > 4 mg l^{-1}

Salinity stratification of the Econfina and Fenholloway river–estuaries was maximal in the lower parts of the drainage areas (stations E3, F5; E6, F6; E7, F9) (Figure 6.1). Offshore areas in both systems were not strongly stratified. The overall level of stratification was somewhat less in the Econfina estuary than that shown in the Fenholloway system. The relatively higher stratification of the Fenholloway estuary could have been due to greater depths relative to the reference sites. Dissolved oxygen (Figure 6.2) was low in the Fenholloway River–estuary throughout the water column. Areas characterized by high stratification and the presence of pulp mill effluents (F4, F5, F6) had moderate to severe hypoxic conditions in surface and bottom waters of the Fenholloway estuary relative to the Econfina system. Increased DOC and ammonia from pulp mill effluents were responsible for these conditions in the Fenholloway system (Livingston, 1993a; Livingston et al., 1998b). In offshore areas, mean DO levels in the Fenholloway drainage were slightly lower than DO concentrations in the Econfina estuary. However, in most instances, DO minima were comparable in the two offshore systems. Hypoxic conditions (HYP) at depth in the estuaries were directly associated with mean depth ($r^2 = 0.402$, p = 0.0003), mean bottom salinity ($r^2 = 0.425$, $p = 0.0002$) and vertical stratification indices ($r^2 = 0.225$, $p = 0.01$), whereas hypoxic conditions were inversely related to mean surface salinity ($r^2 = 0.441$, $p = 0.0001$).

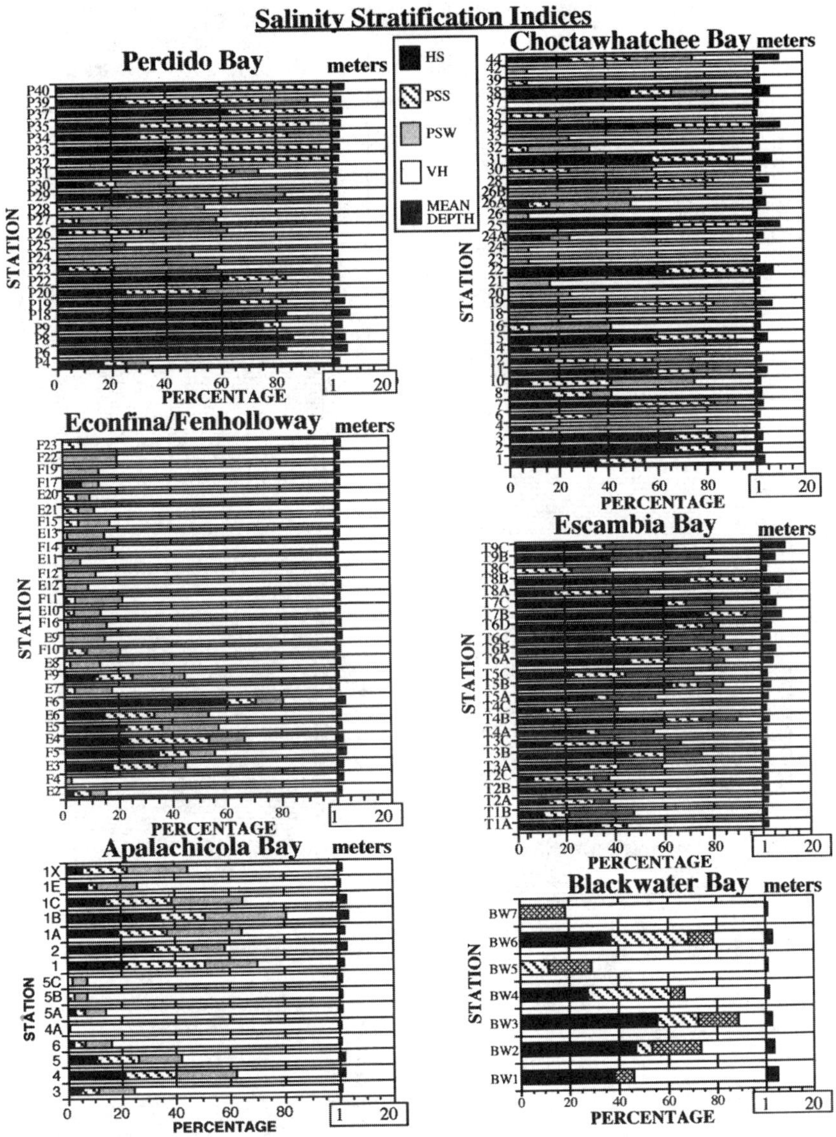

Figure 6.1 Salinity stratification indices and station-specific depths for the various study areas averaged over 24 month periods. Stratification designations include highly stratified (HS) (salinity difference [bottom – surface] ≥ 10 ppt); Partially stratified, strong (PSS) (salinity difference ≥ 5 and < 10 ppt); Partially stratified, weak (PSW) (salinity difference ≥ 2 and < 5 ppt); Vertically homogeneous (VH) (salinity difference < 2 ppt).

The Apalachicola system is relatively shallow compared with other estuaries along the northeast Gulf Coast (Figure 6.1). Depth appears to be an extremely important factor in eutrophication processes. Early studies (Dawson, 1955; Gorsline, 1963) did not mention salinity stratification in the Apalachicola Bay system. However, more recent analyses indicate that the estuary is periodically stratified (Figure 6.1) (Livingston, 1997a; Livingston et al., 2000). In Apalachicola Bay, residual surface flows are directed primarily to Sike's Cut and West Pass (Livingston et al., 2000). The highest level of stratification in the bay occurred at the artificial Sike's Cut (station 1b). At West Pass (station 1a), tidal variability was predominant. Bottom salinities here were often less than those in the vicinity of Sike's Cut, even though the depth of West Pass was usually greater than

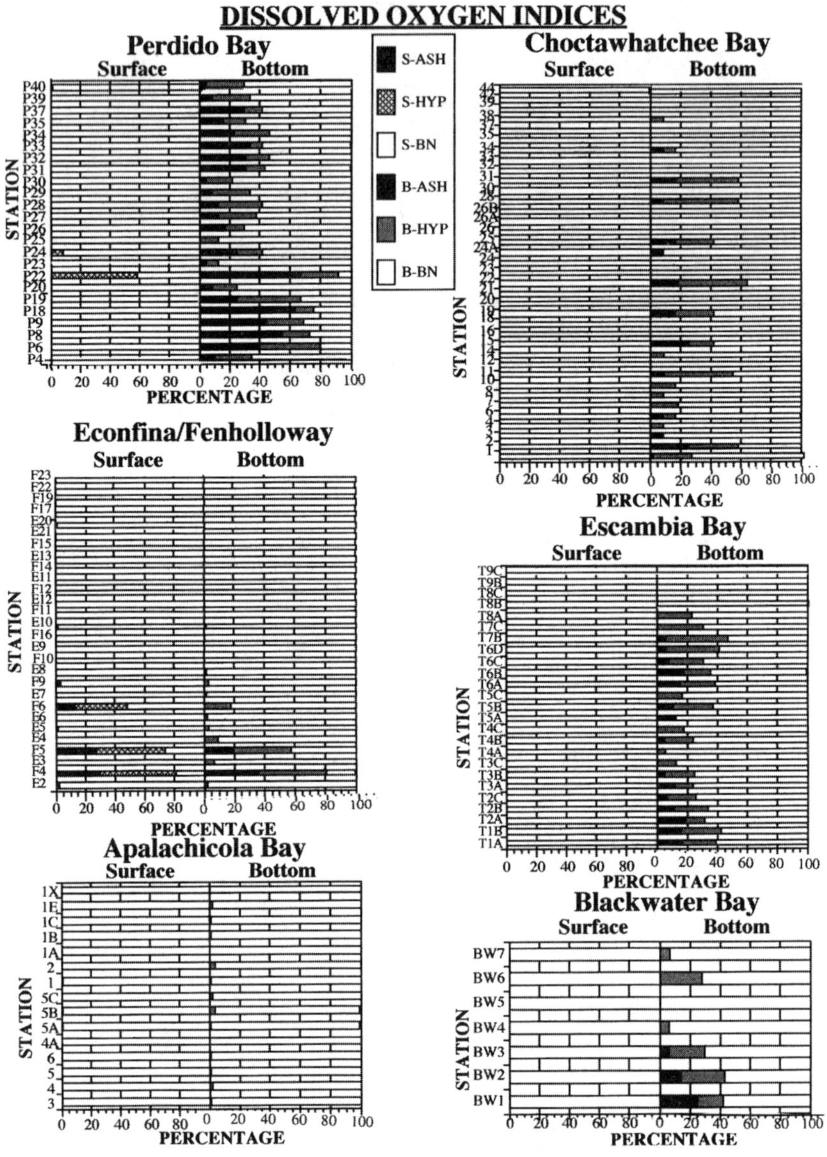

Figure 6.2 Surface and bottom dissolved oxygen indices for stations in the various study areas averaged over 24-month periods. Index designations include anoxic to severely hypoxic (ASH) (DO ≤ 2 mg l^{-1}); hypoxic (HYP) (DO > 2 and ≤ 4 mg l^{-1}); biologically "neutral" (BN) (DO > 4 mg l^{-1}).

that of Sike's Cut by more than 1 m. The artificial pass opens onto the primary basin of Apalachicola Bay, whereas West Pass has restricted access due to oyster bar development (Livingston et al., 2000). Average bottom current velocities were highest at the passes and in the eastern region of the bay, with the highest average bottom velocities at Sike's Cut. This situation appears to have an important influence on the bottom current structure of Apalachicola Bay due to the density differentials brought about by the heavier Gulf water penetrating through the cut. Stratification was relatively less well developed in western sections of the bay approaching Indian Pass where tidal and subtidal variation was prominent. The lowest levels of stratification were noted in East Bay (stations 4a, 5a, 5b, 5c). These areas are notable for having shallow depths and generally low salinities.

During periods of high river flow, there were increased salinity gradients in all directions, and the Apalachicola estuary was generally stratified with the range of salinity gradients dependent on season and location. Using a combination of regression and analysis of variance with the long-term database, Livingston (1984b) showed that surface to bottom salinity-by-month interactions were significant at the dredged Sike's Cut and in the Intracoastal Waterway at the mouth of the river. Thus, stratification was significant over prolonged periods in two areas that were affected by dredging in the bay. The combination of river flow, depth (dredging), and the entrances to the Gulf (including Sike's Cut) were important variables in the long-term trends of stratification in the bay. There was a relatively low incidence of hypoxic conditions in the Apalachicola system (Figure 6.2). The fact that the deepest and most highly stratified areas were found at the highly active river mouth and at the passes that were subject to highly oxygenated Gulf waters could, in part, account for this finding.

The relatively shallow Apalachicola system, although periodically stratified, was one of most well mixed estuaries along the northeastern Gulf. This factor allowed increased utilization of mineralized nutrients from the bottom by the phytoplankton with periodic (i.e., drought) conditions enhancing production of benthic microalgae. By comparison, the deeper, highly stratified Choctawhatchee system had relatively little such interchanges between bottom and surface areas. Choctawhatchee Bay has shallow shelf areas with relatively steep slopes that eventually level out to average depths of 3 m in eastern sections and 8 m in western sections of the central portions of the estuary. Areas along the periphery of the bay were generally less than 2 m deep. The deepest parts of the bay were positively correlated with high stratification (Figure 6.1) (HS, $r^2 = 0.335$, $p = 0.001$; PSS, $r^2 = 0.465$, $p = 0.01$). Extreme hypoxia (ASH) (Figure 6.2) was highly correlated, ($r^2 = 0.605$, $p = 0.001$) with highly stratified conditions in deeper areas of the bay. There was a general decrease of such correlation with decreasing stratification (PSS, $r^2 = 0.322$, $p = 0.001$; PSW, $r^2 = -0.294$, $p = 0.01$; VH, $r^2 = -0.553$, $p = 0.001$). Hypoxic conditions (HYP) showed a similar trend (HS, $r^2 = 0.451$, $p = 0.01$; PSS, $r^2 = 0.702$, $p = 0.01$; PSW, $r^2 = -0.384$, $p = 0.01$; VH, $r^2 = -0.552$, $p = 0.01$). Areas characterized by DO concentrations exceeding 4.0 mg l^{-1} showed an opposite trend as a function of the stratification indices (HS, $r^2 = -0.562$, $p = 0.001$; PSS, $r^2 = -0.672$, $p = 0.0001$; PSW, $r^2 = 0.407$, $p = 0.01$; VH, $r^2 = 0.627$, $p = 0.001$).

A regression of subhalocline hypoxic conditions (HYP) with mean depths in the Choctawhatchee system showed a relatively high r^2 value ($r^2 = 0.636$, $p = 0.003$) as did the relationship of hypoxia with high levels of stratification ($r^2 = 0.562$; $p = 0.0001$). Depth was a leading variable in the distribution of hypoxia in the bay, with salinity stratification as a necessary prerequisite for deteriorating water quality conditions. There was thus a direct association of depth, salinity stratification, and the extent of hypoxia in Choctawhatchee Bay.

The primary difference between the Choctawhatchee system and Apalachicola Bay was the shallow depths in the Apalachicola system and the lack of hypoxia despite stratification at the passes and in areas directly receiving Apalachicola River flow. However, other factors were associated with these trends. Bayou stations located in urban areas of the Choctawhatchee system also had high levels of hypoxia with or without stratification indicating adverse effects due to urban runoff (Livingston, 1997a). Dead-end canals surrounded by high levels of urbanization and marina development in the bay (Old Pass Lagoon) showed the lowest DO at depth in the entire bay. In nearby areas where there was more exchange with the open Gulf, hypoxic conditions at depth were minimal.

The present state of eutrophication in both Apalachicola Bay and Choctawhatchee Bay remains unknown as there have been no ecosystem-level studies for the past 15 years in either system. Recent anecdotal data concerning the Choctawhatchee estuary indicate the generation of plankton blooms in western parts of the system due to urban runoff (see above). The effects of the red tide blooms remain undetermined. In the Apalachicola system, the effects on the bay of a malfunctioning sewage treatment plant in the city of Apalachicola remain undocumented.

High-salinity water from the Gulf moves up Pensacola Bay and into the deeper parts of Escambia Bay and East Bay (Livingston, 1999). As in most river-dominated estuaries, salinity variation was high in the upper and midportions of Escambia Bay: in both the Blackwater and Escambia systems, river flow dominated the physical habitat. The lowest salinities were noted during winter–spring periods of high river flow. Peak salinities occurred during fall periods of relatively low river flows. North–south gradients of salinity were evident in both systems. Salinity stratification was maximal in deeper areas of the lower and midportions of Escambia Bay (Figure 6.1). High salinity stratification in Blackwater Bay/East Bay occurred in deeper parts of both systems.

Surface DO was highest in upper Escambia Bay (Livingston, 1999). The lowest DO (Figure 6.2) occurred in areas of the highest salinity stratification indices, usually at depth (i.e., below the halocline). Regressions of DO with other physical–chemical variables indicated that surface DO was significantly (but weakly) associated with surface salinity (i.e., near the mouth of the Escambia River). Bottom DO was significantly (inversely) associated with bottom temperature and the salinity stratification index. The depth/hypoxia relationship occurred only in upper parts of both the Blackwater and Escambia systems. In Pensacola Bay, deeper areas were not associated with hypoxic conditions at depth due to active exchanges of bottom water with the Gulf. In Blackwater Bay, bottom oxygen concentrations were highest in the upper bay and deeper parts of East Bay. Surface oxygen increased with distance from the river mouth. This situation was similar to that in Escambia Bay, where the highest oxygen was noted in the middle bay areas. Temporal relationships between bottom DO concentrations and salinity stratification indices indicated that bottom DO in the Pensacola Bay system was controlled by salinity stratification in the upper and middle parts of Escambia Bay and the Blackwater/East Bay system.

Spatial/temporal trends of salinity in Perdido Bay followed seasonal and interannual patterns of freshwater runoff. Winter to early spring periods represent the low-salinity period, whereas fall months of drought usually are characterized by the highest salinities of the year. Salinity stratification of the estuary essentially creates two very different habitats that are depth related (Figure 6.1). The estuarine portions of the Perdido River and Elevenmile Creek showed the highest levels of salinity stratification with such stratification evident in over 80% of the samples. These areas represent the highest levels of stratification of all of the bays surveyed. Most areas of the upper bay (stations P24 to P30) were characterized by low levels of vertical stratification. These were the shallowest stations in the bay. Overall, mean depth in Perdido Bay was highly correlated with high salinity stratification (HS) ($r^2 = 0.688$, $p = 0.001$) and was inversely correlated with vertical homogeneity (VH) ($r^2 = -0.468$, $p = 0.01$).

The analysis of the relationships of depth, salinity stratification, and DO distribution in Perdido Bay was carried out with data taken before the initiation of the plankton blooms. With the exception of the lower end of Elevenmile Creek (station P22), shallow waters above the halocline in Perdido Bay were characterized by DO concentrations above 4 mg l^{-1} (Figure 6.2). The highest levels of severe to moderate hypoxia at depth were located in Elevenmile Creek which was affected by high DOC (leading to high BOD) and ammonia released from the pulp mill located at the head of the creek (Livingston, unpublished data). High levels of severe to moderate hypoxia were also found in estuarine portions of the Perdido River system. Relatively low levels of hypoxia occurred at depth in upper portions of the estuary. In the lower estuary, subhalocline hypoxic conditions were found in just over 40% of the samples taken.

The distribution of hypoxia in Perdido Bay was comparable to DO conditions found in deeper portions of Choctawhatchee Bay. Extreme bottom water hypoxia in Perdido Bay was highly correlated with high levels of stratification (HS, $r^2 = 0.749$, $p = 0.001$). There were inverse correlations of bottom hypoxia with decreasing levels of stratification (PSS, $r^2 = -0.210$, $p = 0.01$; PSW, $r^2 = 0.521$, $p = 0.01$; VH, $r^2 = -0.418$, $p = 0.003$). Vertically homogeneous salinity conditions showed an opposite trend (HS, $r^2 = -0.787$, $p = 0.01$; PSS, $r^2 = 0.275$, $p = 0.001$; PSW, $r^2 = 0.521$, $p = 0.001$; VH, $r^2 = 0.406$, $p = 0.01$). Regressions of severe bottom hypoxia in Perdido Bay showed relatively strong associations with mean depth ($r^2 = 0.255$, $p = 0.01$) and high salinity stratification

($r^2 = 0.562$, $p = 0.0001$). There was a negative association of severe hypoxia with surface salinity ($r^2 = 0.248$, $p = 0.007$), and hypoxia was positively associated with depth ($r^2 = 0.255$, $p = 0.01$). These trends were similar to those noted in the Choctawhatchee Bay system. Both estuaries were subjected to increased salinity stratification by the creation of artificial passes to the Gulf that are currently maintained by the U.S. Army Corps of Engineers.

Detailed studies by Niedoroda (1992) and modeling by Gallagher (see Livingston, 1993a) indicated that the physiography of Perdido Bay, together with seasonal and interannual fluctuations of the Perdido River, determined the stratification potential and bottom DO regime of the estuary. Flushing rates of the bay played an important role in the generation of plankton blooms, which, in turn, were associated with increased surface DO and reduced bottom DO during warm periods. Saltwater intrusion into lower Elevenmile Creek was found at distances of 2 to 3 km from the mouth of the creek. The level of salinity stratification was quite significant in the creek, with considerable differences between surface and bottom salinities, thereby limiting mixing of the surface and bottom layers. The modeling analysis indicated that the decrease of surface layer DO ranged from about 6.0 mg l^{-1} at mile 2.5 to between 3.0 and 4.0 mg l^{-1} at the mouth. This decrease was due primarily to the oxidation of the paper mill's wastewater carbonaceous BOD and ammonia. The lower layer DO levels of 1.0 to 2.0 mg l^{-1} were due almost entirely to the effect of Perdido Bay water quality and sediment oxygen demand as a consequence of stratification. However, the long-term studies in Perdido Bay indicated that salinity stratification exacerbated the effects of hypereutrophication through increased subhaloclinal hypoxia as phytoplankton biomass increased throughout the bay.

The history recorded in long cores sampled in Perdido Bay sediments (Brush, 1991) indicated changes that were possibly related to non-point sources such as timbering, agriculture, and municipal development, but very little if any changes could be specifically related to the paper mill activity in Elevenmile Creek. The historical data also indicated that salinity increases in the Perdido estuary were a sharp deviation from the natural (i.e., freshwater) conditions of the estuary in the recent past (i.e., prior to the opening of Perdido Pass).

The modeling data, combined with the descriptive field analyses and the field/laboratory experiments, indicated that hypoxia in Perdido Bay prior to the plankton blooms was the product of highly complex chemical interactions that were mediated in large part by the hydrographic characteristics of the estuary. The establishment of a strong halocline was a central feature of the development of hypoxia in Perdido Bay. A series of screening models (Klein, 1991) were run to determine the importance of specific state variables in the hydrodynamics of Perdido Bay. The Klein models showed that meteorological events in the Gulf region were responsible for transport conditions that were consistent with the movement of nutrients and organic matter up the bay from postulated sources in the lower bay. These analyses all indicated that the opening and continued maintenance of Perdido Pass had extensive, adverse effects on habitats in deeper parts of Perdido Bay.

Based on rates of river flow and the physiography of the various alluvial river estuaries in the northeastern Gulf, Niedoroda (1999) found that circulation and mixing in the Escambia Bay system was intermediate between the more "vigorous" Apalachicola Bay system and the more restricted Choctawhatchee and Perdido Bay systems. Depth and stratification were highest in Choctawhatchee Bay; reduced wind mixing and high stratification led to similar reductions of mixing in the Perdido system. Salinity stratification was closely associated with depth (usually occurring in depths >2 m); this stratification was closely associated with subhaloclinal hypoxia in both systems. The lack of vertical mixing also stabilized the sediments in these systems, leading to formation of liquid mud, which represented an impaired habitat for bottom-dwelling organisms. The combination of periodic, warm-season hypoxia and liquid mud formation resulted in poor benthic habitat in the Choctawhatchee and Perdido estuaries, a situation that was exacerbated by the radical alteration of the inshore salinity regimes by dredged passes to the high salinity Gulf waters (Livingston, 1997a, b). Thus, the dredged openings of the lower parts of both Choctawhatchee Bay and Perdido Bay impaired useful benthic production in areas with depths >2 m which includes considerable portions of these systems. A similar adverse impact on oyster production due to increased predation in high-salinity bottom waters entering through the artificial Sike's Cut in Apalachicola Bay has been documented (Livingston et al., 2000). The plankton blooms in Perdido Bay led to exacerbation of existing depletion of subhaloclinal DO, and there were indications of additional adverse effects on sediment quality through as yet unknown changes due to these blooms. However, the impacts of the plankton blooms on Perdido Bay should be placed within the context of an already damaged system due to dredging activities in the lower bay.

III. NUTRIENT LOADING AND NUTRIENT CONCENTRATIONS

A. Introduction

A review was made of the long-term river flow conditions of the subject river–estuarine systems of the northeastern Gulf of Mexico (Figure 2.1). The U.S. Geological Survey (Tallahassee, FL, M. Franklin, personal communication) provided river flow data. River flows were then calculated for the various systems based on the entry at the head of the respective estuaries. River flows and nutrient concentration/loading data were compiled over a series of 2-year periods. An attempt was made to choose periods that were close to nondrought conditions in each system. Nutrient data for Apalachee Bay, Pensacola Bay (Blackwater Bay, Escambia Bay), Choctawhatchee Bay, Apalachicola Bay, and Perdido Bay were taken by Livingston (unpublished data).

B. Flow Rates and Nutrient Loading/Concentrations

A comparison of river flows and nutrient loading characteristics was carried out in the subject bay systems. Comparative flow rates are shown in Figure 6.3. The Apalachicola River system had the greatest flow rates of all the systems analyzed. Escambia and Choctawhatchee River flows were comparable. The Perdido River (including the Alabama Blackwater River and the Styx River), the Yellow River, and the Florida Blackwater River were roughly comparable. In most cases, peak flows occurred during winter–spring periods, and seasonal drought periods were usually noted during fall periods. The Fenholloway River, the Econfina River, and Elevenmile Creek were also comparable at considerably lower flow rates than the alluvial rivers.

Nutrient loading and nutrient concentrations were determined for the respective rivers. Data for Elevenmile Creek were calculated from January 1990 through December 1991, a period when nutrient loading to Perdido Bay was incorporated into the bay food webs without evidence of phytoplankton imbalances. Nutrient loading and nutrient concentrations were also calculated from January 1994 through December 1995, a period when increased orthophosphate loading from the mill was associated with increased phytoplankton blooms and changes in the food webs and secondary production of the bay.

Orthophosphate loading was highest in the largest alluvial rivers; the Apalachicola, Choctawhatchee, and Escambia systems (Figure 6.4). This loading was also high in the Fenholloway system and in Elevenmile Creek (1994–1995). The Econfina system was comparable to that of the Perdido, Florida Blackwater, and Yellow Rivers, and Elevenmile Creek (1990–1991). The orthophosphate concentration data (Figure 6.4) indicated comparability among the Econfina, Choctawhatchee, and Elevenmile Creek (1990–1991) systems at one level and the Escambia, Apalachicola, Perdido, Blackwater, and Yellow River systems at a somewhat lower level. Orthophosphate concentrations in the Fenholloway River and Elevenmile Creek (1994–1995) were one to two orders of magnitude higher than those found in the other systems. These comparisons indicate that nutrient loading was higher in the alluvial rivers than the nutrient-enriched Fenholloway River and Elevenmile Creek after increased orthophosphate loading (1994–1995). However, these small streams

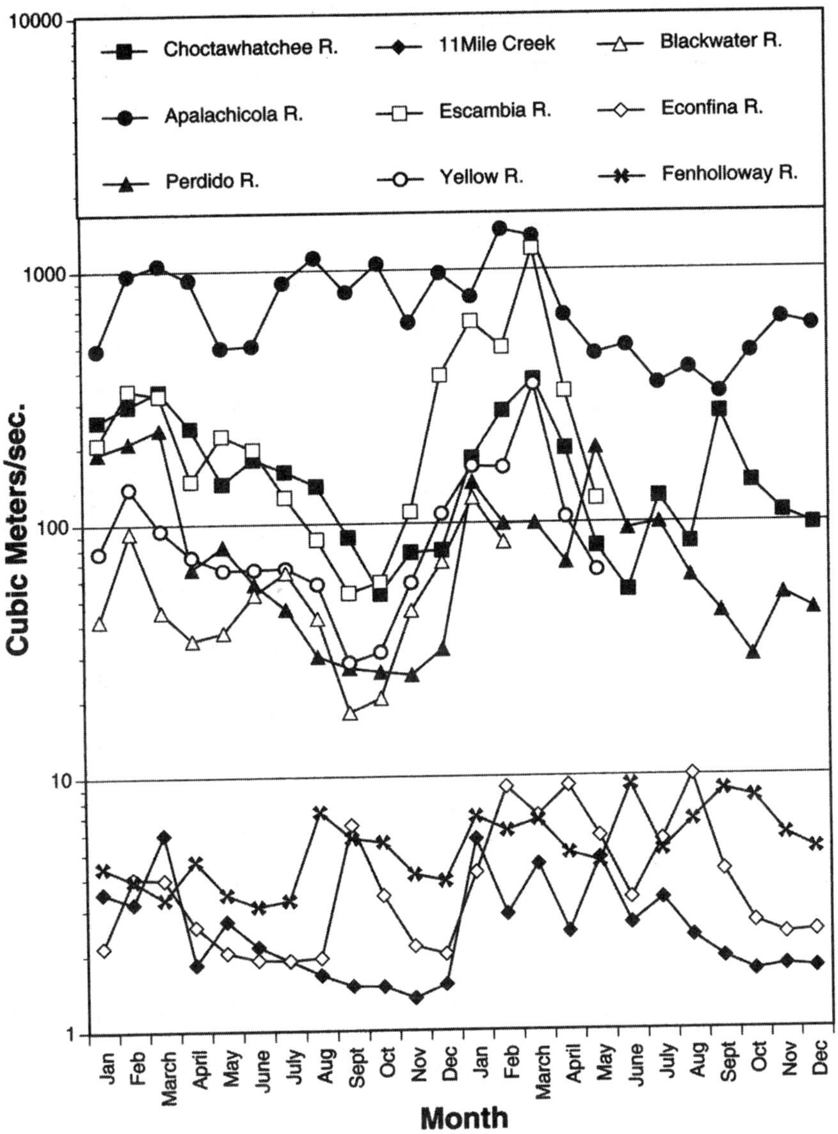

Figure 6.3 River flows ($m^3 s^{-1}$) for the Apalachicola River (January 1994 to December 1995), the Choctawhatchee River (January 1987 to December 1988), the Escambia River (January 1997 to May 1998), the Perdido River (January 1994 to December 1995), the Blackwater River (January 1997 to May 1998), the Yellow River (January 1997 to May 1998), the Econfina River (January 1992 to December 1993), the Fenholloway River (January 1992 to December 1993), and Elevenmile Creek (January 1994 to December 1995).

had significantly higher orthophosphate concentrations than the alluvial rivers. Total phosphorus loadings were comparable among the various rivers despite the flow disparities among these systems. Total phosphorus concentrations were highest in the Fenholloway River and Elevenmile Creek (1994–1995) (Livingston, 1999). Total phosphorus data tended to follow the orthophosphate trends although there was somewhat less difference between the 1990–1991 and 1994–1995 Elevenmile Creek TP concentration levels than those shown with the orthophosphate data.

Ammonia loading (Figure 6.4) was comparable among the various systems with the exception of the Econfina, which had relatively low ammonia loading. Ammonia concentrations (Figure 6.4) were an order of magnitude higher in the Fenholloway River and Elevenmile Creek (1990–1991;

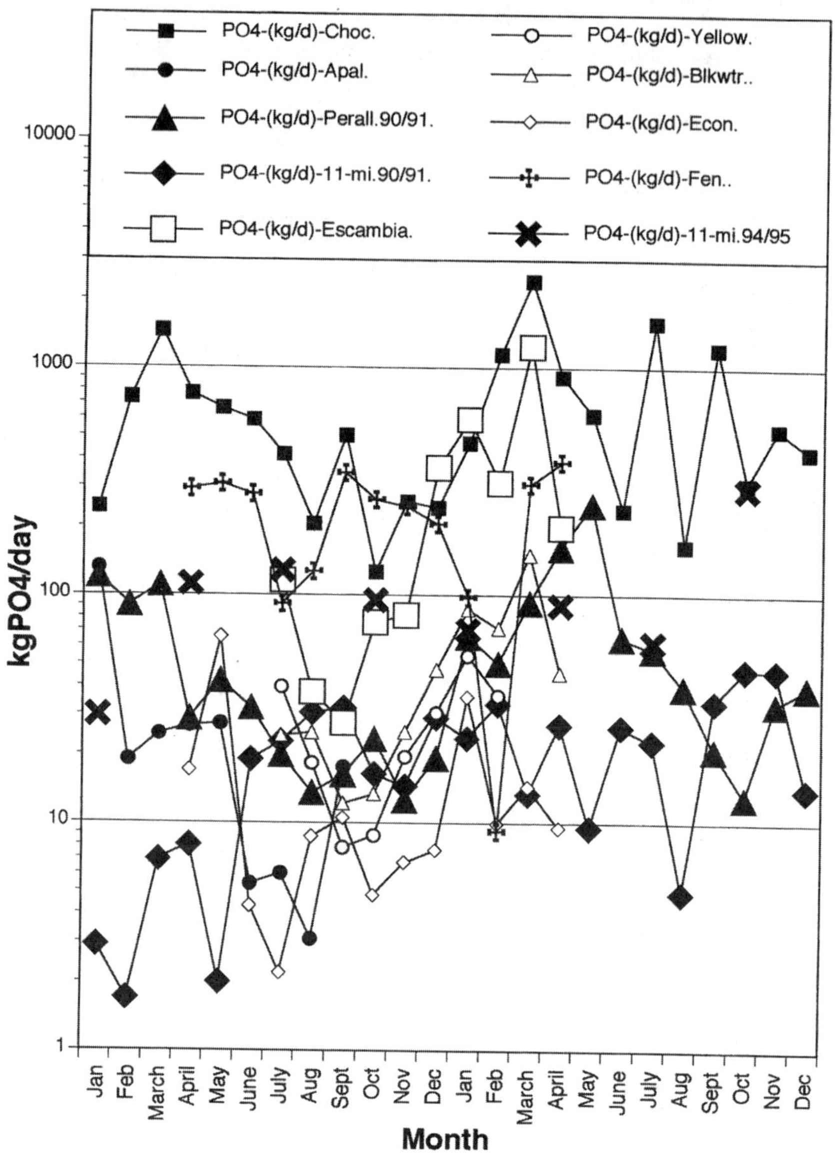

Figure 6.4 Nutrient loading (kg d^{-1}) and nutrient concentrations (mg l^{-1}) of orthophosphate and ammonia for the Apalachicola River (January 1990 to September 1990), the Choctawhatchee River (January 1987 to December 1988), the Escambia River (August 1997 to April 1998), the Perdido River (January 1994 to December 1995), the Blackwater River (July 1997 to February 1998), the Yellow River (July 1997 to February 1998), the Econfina River (April 1992 to April 1993), the Fenholloway River (April 1992 to April 1993), and Elevenmile Creek (January 1994 to December 1995).

1994–1995) than the other rivers. Thus, as with orthophosphate, ammonia loading in the small, nutrient-enriched streams was comparable to that of the alluvial rivers, whereas ammonia concentrations in the Fenholloway River and Elevenmile Creek were much higher than those in the alluvial rivers. Nitrite/nitrate loading (Livingston, 1999) appeared to be directly related to river flow characteristics in most of the systems tested. Somewhat higher nitrate/nitrite concentrations occurred in Elevenmile Creek (1990–1991; 1994–1995); the lowest such measurements were found in the Econfina system.

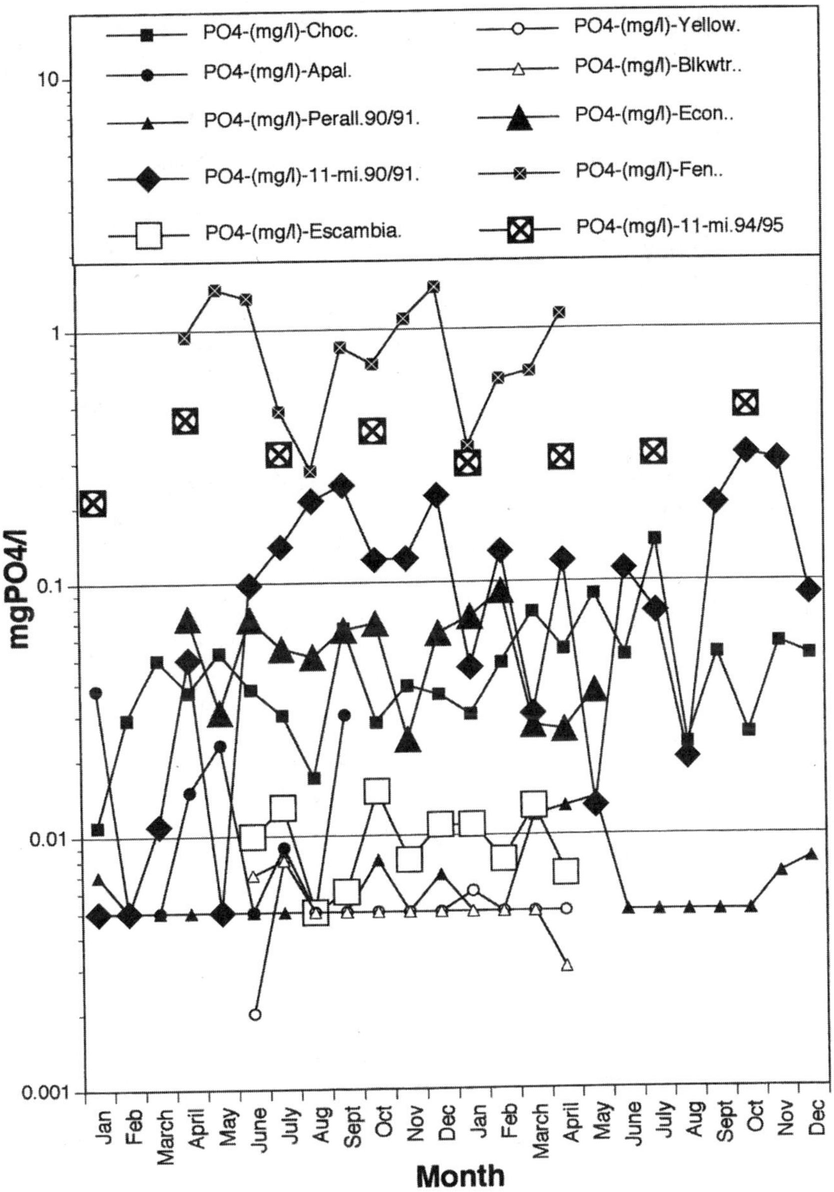

Figure 6.4 (continued)

The nutrient loading and nutrient concentration data indicate that there were differences between absolute nutrient loading and nutrient concentration. Nutrient loading can be high as a function of higher river flow, whereas concentrations of nutrients vary primarily as a function of flow differences and anthropogenous sources such as pulp mill discharges, agricultural discharges, and urban runoff. The question concerning the differentiation of operable factors that initiate blooms (i.e., loading vs. concentration) is complex. However, overall loading is the key factor although such loading should be evaluated within the context of the flow characteristics of the delivering medium, the resultant nutrient concentrations, and the assimilative capacity of the receiving system.

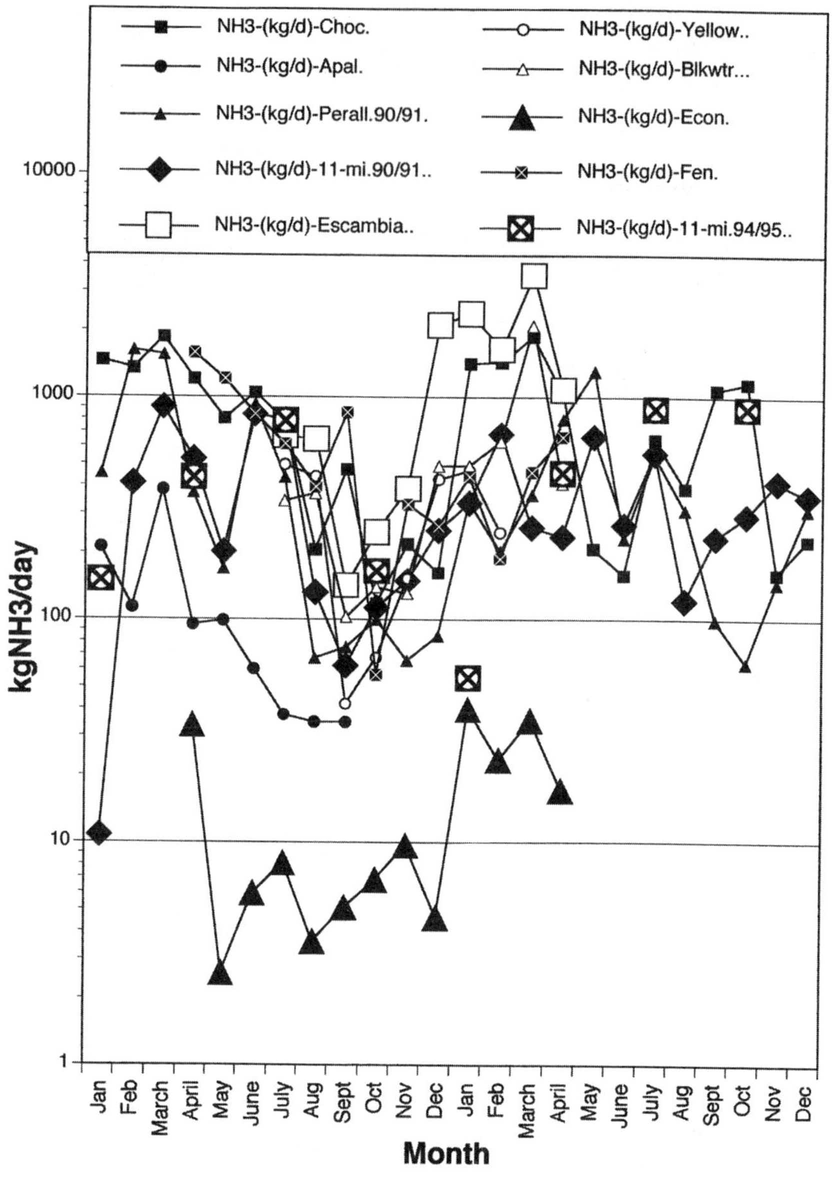

Figure 6.4 (continued)

A comparison between the relatively unpolluted Apalachicola system and the Perdido system indicates certain similarities. Iverson et al. (1997) found that dissolved silicate did not limit phytoplankton production in the largely mesotrophic Apalachicola Bay (Fulmer, 1997). This was the case in the Perdido system prior to the blooms (Livingston, 1992; 1997b, c; Flemer et al., 1997; Livingston et al., 1998a). The possible effects of silica limitation after the diatom blooms could have contributed to the long-term bloom succession and the replacement of many of the diatom species with raphidophytes, dinoflagellates, and cryptophytes. In the Apalachicola system, orthophosphate availability limited phytoplankton during both low- and high-salinity winter periods and during the summer at stations with low salinity. Nitrogen, on the other hand, was limiting during summer periods of moderate to high salinity in the Apalachicola estuary (Iverson et al., 1997).

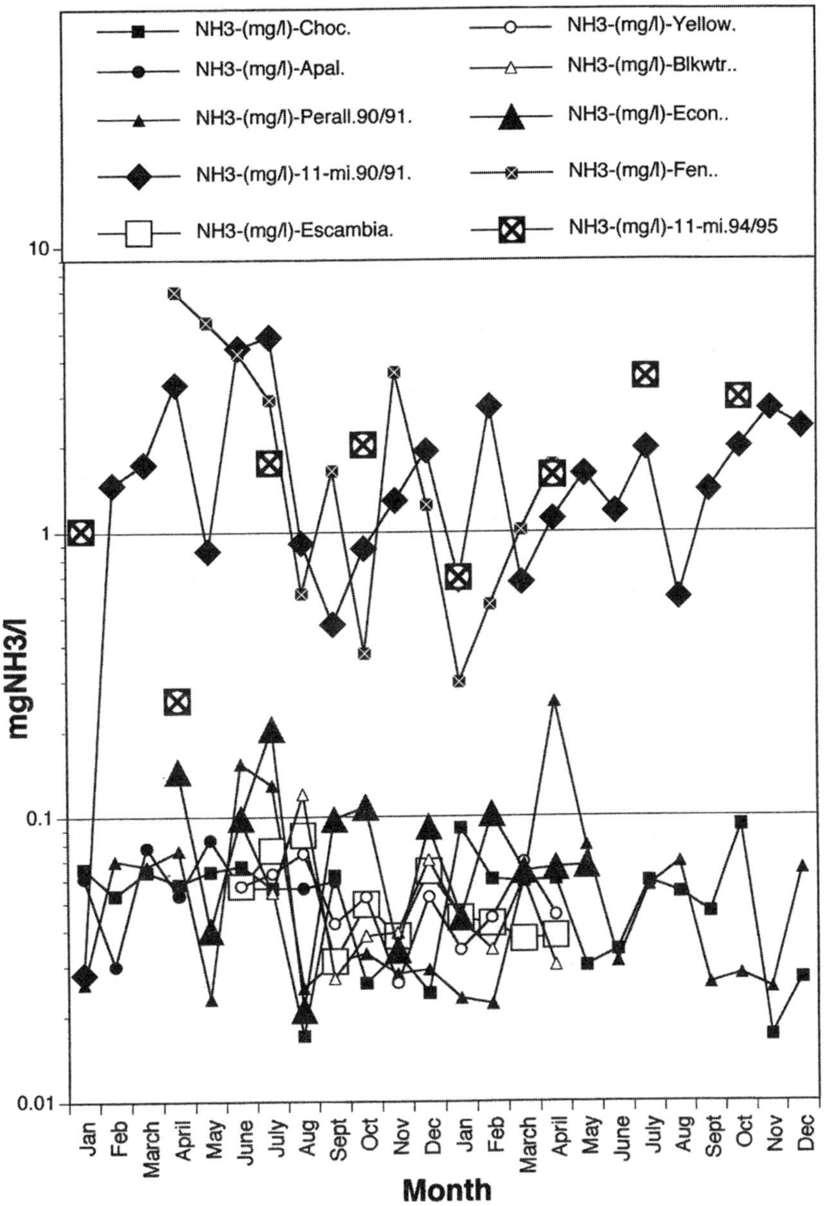

Figure 6.4 (continued)

Light and temperature limitation was highest during winter–spring periods, thus limiting primary production during this time (Iverson et al., 1997). High chlorophyll *a* levels during winter periods were attributed to low zooplankton grazing during the cooler months (Iverson et al., 1997). These conditions prevailed in the Perdido system during periods of moderate nutrient loading (1989 to 1991). Nitrogen input to primary production was limited by the relatively high flushing rates in the Apalachicola system. Flow rates affected the development of nutrient limitation in the Apalachicola system with nutrient limitation highest during low-flow summer periods. This also appears to have been the case in Perdido Bay.

Iverson et al. (1997) noted that there had been no notable increase in chlorophyll *a* concentrations in Apalachicola Bay over the past two decades despite increases in nitrogen loading due to increased basin deposition of this nutrient. However, no phytoplankton data were taken in the Apalachicola system. The potential effects of nutrient loading from a malfunctioning sewage treatment plant were not tested. Perdido Bay, on the other hand, was subject to complex changes of both nutrient loading and nutrient concentrations due to ammonia and orthophosphate discharges from the pulp mill on Elevenmile Creek. The effects of the discharge of these nutrients appeared to be related to specific temporal patterns relative to the nutrient requirements of individual plankton species and to the complex reactions of these species to combinations of these nutrients.

Nutrient loading in the Apalachicola River system remained relatively high without apparent hypereutrophication in the bay whereas relatively lower nutrient loading in the Perdido system with higher concentrations of reactive inorganic phosphorus and nitrogen compounds in Elevenmile Creek ended in years of plankton blooms. This contrast strengthens the hypothesis that high nutrient concentrations, as a product of high nutrient loading into a creek with relatively low flows, could have been an important factor in the initiation of plankton blooms in Perdido Bay. Nutrient loading per se was probably not the only factor in the bloom process. The drought–flood periodicity of river input is an important factor in the overall pattern of eutrophication in these systems. The added nutrient loading from the Perdido River could also have been important in sustaining the plankton blooms once they were initiated by the high nutrient concentrations from Elevenmile Creek. Thus, the complex relationships of flow rates, nutrient loading, and the resulting nutrient concentrations appeared to be important in the initiation and sustenance of plankton blooms in the Perdido estuary.

IV. WATER QUALITY

Salinity tended to be lowest in the larger river-dominated estuaries during winter periods. There were longitudinal salinity gradients in the alluvial systems during all seasons. These salinity gradients were comparatively less well developed in Apalachee Bay where low salinity occurred only in inshore areas. Water temperature in the various coastal systems had relatively high ranges due to seasonal variation. Ammonia and orthophosphate concentrations (Figure 6.5) were highest at the mouths of Elevenmile Creek (stations P21, P22, P23) and the Fenholloway River (F09). There were also relatively high orthophosphate concentrations in inshore parts of the Fenholloway system (F14). Orthophosphate concentrations were generally near detection limits (0.002 to 0.005 mg l^{-1}) in most of the systems although periodic increases appeared in some of the estuaries during winter. With the exception of the creeks affected by pulp mill effluents, ammonia concentrations tended to be relatively uniform in the various systems during all seasons shown. There were some high ammonia concentrations in the Apalachicola system during winter flooding. Ammonia concentrations were relatively high at the mouths of the Fenholloway River and Elevenmile Creek.

Secchi depths (Figure 6.5) were lowest in Elevenmile Creek during 1989 and at the mouth of the Fenholloway River. Secchi depths were also relatively low in the Apalachicola Bay system. There was a general increase in Secchi depths with distance from the mouths of the rivers in Escambia–Pensacola, Choctawhatchee, Apalachicola, and Perdido Bay systems. Silica concentrations in all of the bays tended to decrease with distance from river mouths. The highest silica concentrations were in areas associated with Elevenmile Creek, whereas the lowest such concentrations were in offshore areas of Apalachee Bay during spring and summer periods. Water color (Figure 6.5) was highest in Elevenmile Creek (1989) and at the mouth of the Fenholloway River system. Water color tended to be higher in the big alluvial systems during winter periods of increased river flow. The lowest color levels were in offshore parts of the Apalachee Bay.

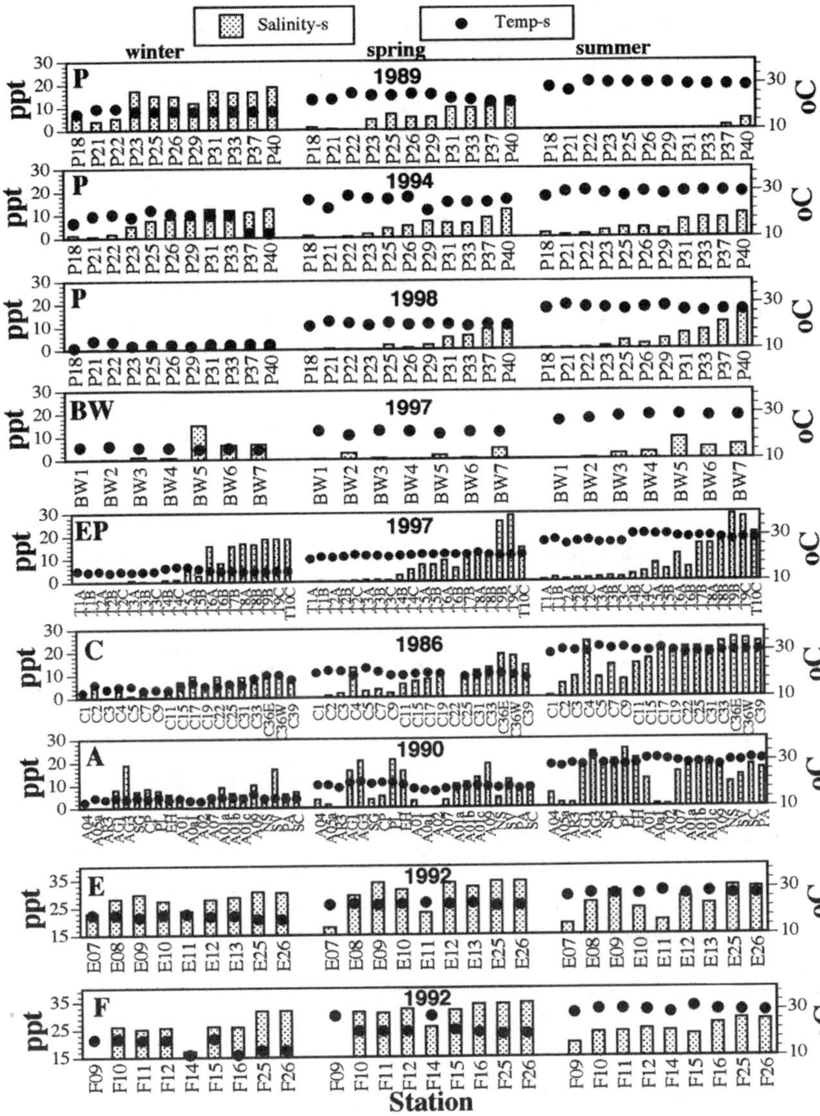

Figure 6.5 Water quality distributions of surface salinity (ppt), surface temperature (°C), orthophosphate (mg P l^{-1}), ammonia (mg N l^{-1}), bottom dissolved oxygen (mg l^{-1}), surface color (Pt-Co units), surface chlorophyll *a* (μg l^{-1}), surface ratio turbidity (NTU), surface silica (mg l^{-1}), Secchi depths (m), and total organic carbon (mg C l^{-1}) taken in the following bay systems: Perdido Bay (P, 1989, 1994, 1998), Blackwater Bay (BW, 1997), Escambia/Pensacola Bay (EP, 1992), Choctawhatchee Bay (C, 1986), Apalachicola Bay (A, 1990), the Econfina River–estuary (E, 1992), and the Fenholloway River–estuary (F, 1992).

Chlorophyll *a* concentrations tended to be relatively low in all systems with certain exceptions. The highest chlorophyll concentrations were found in the upper Perdido system during the spring of 1994. Periodic high chlorophyll *a* concentrations were also found in upper parts of the Escambia system during the summer. Moderately high chlorophyll *a* concentrations were noted in outer bay areas of the Apalachicola system during winter months and in inshore parts of the Econfina (E11) and Fenholloway (F14) systems during summer. Moderately high chlorophyll *a* was also found in marina areas of the Choctawhatchee Bay system (C36W) during summer months. There was a general trend of higher chlorophyll *a* concentrations in the upper parts of the various bay systems.

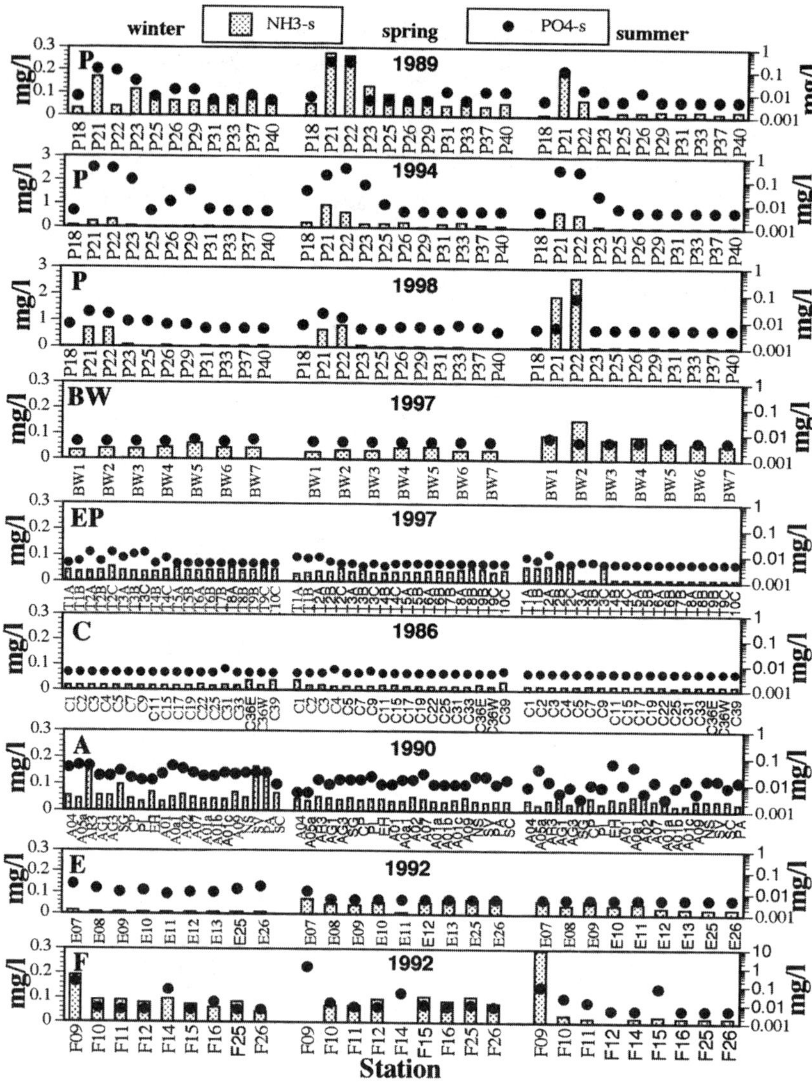

Figure 6.5 (continued)

Turbidity was relatively high in upper Apalachicola Bay, upper Escambia Bay, and upper Choctawhatchee Bay during winter months of high river flow. The TOC concentrations (Figure 6.5) were highest in Elevenmile Creek (station P21). The TOC concentrations were relatively high in most of the systems during winter months and tended to be lowest in offshore parts of Apalachee Bay.

The water quality data tended to reflect several factors. Inorganic nutrient gradients were highest in the small creeks that were affected by pulp mill effluents. These areas were also highest in TOC. The big alluvial systems tended to have periodically high turbidity and color that reflected the effects of winter–spring flooding. With the possible exception of upper Perdido Bay during spring 1994 and upper Escambia Bay during summer 1997, there were no indications of hypereutrophication or phytoplankton blooms in any of the systems based on chlorophyll *a* concentrations. In fact, the chlorophyll *a* concentrations, while higher in the upper parts of most systems, were relatively low overall. Chlorophyll *a* was a general indicator of phytoplankton activity, but was not

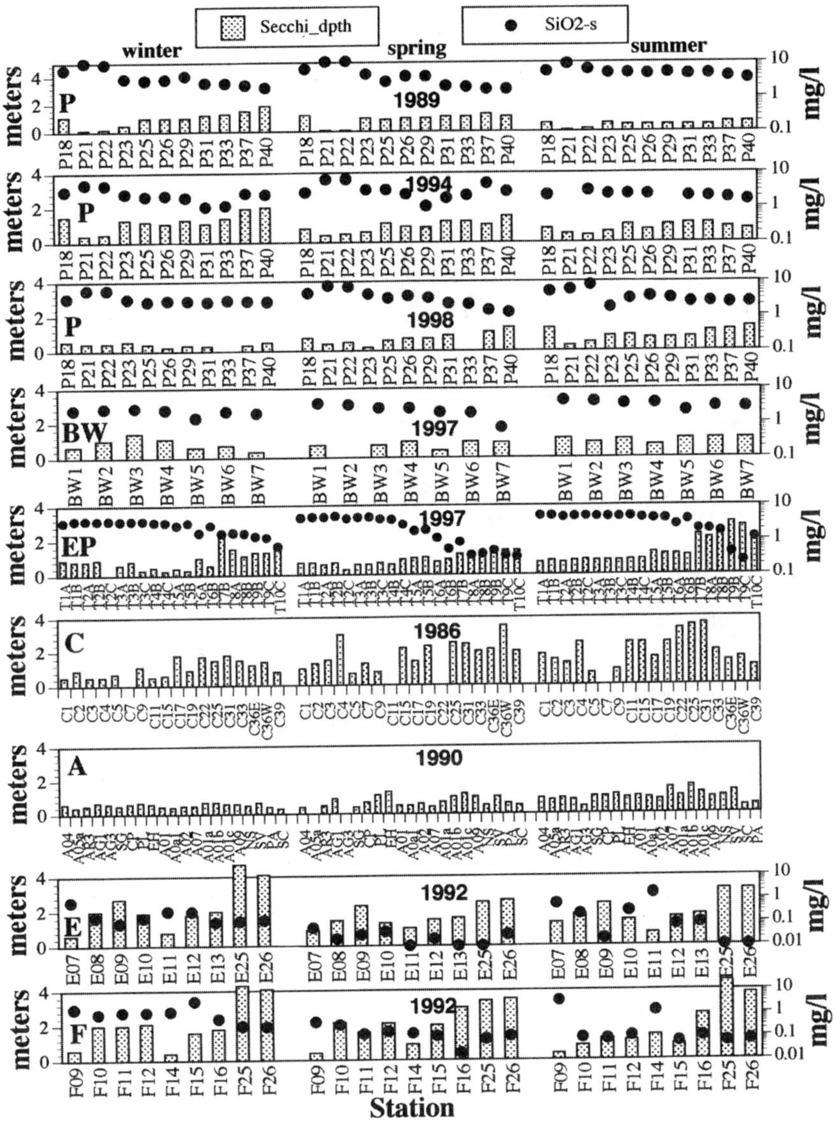

Figure 6.5 (continued)

an indicator of blooms. Overall, the water quality factors from the various estuarine and coastal areas of the upper Gulf of Mexico indicated relatively uniform conditions with river runoff and the presence of pulp mill effluents as the primary controlling features.

V. SEDIMENT COMPARISONS

Sediments were taken with coring devices at fixed stations. Variables determined included particle size distribution (phi units), percent organics, and sand/silt/clay fractions. The sediments were analyzed according to methods described by Mahoney and Livingston (1982) as defined by Galehouse (1971) (Appendix I).

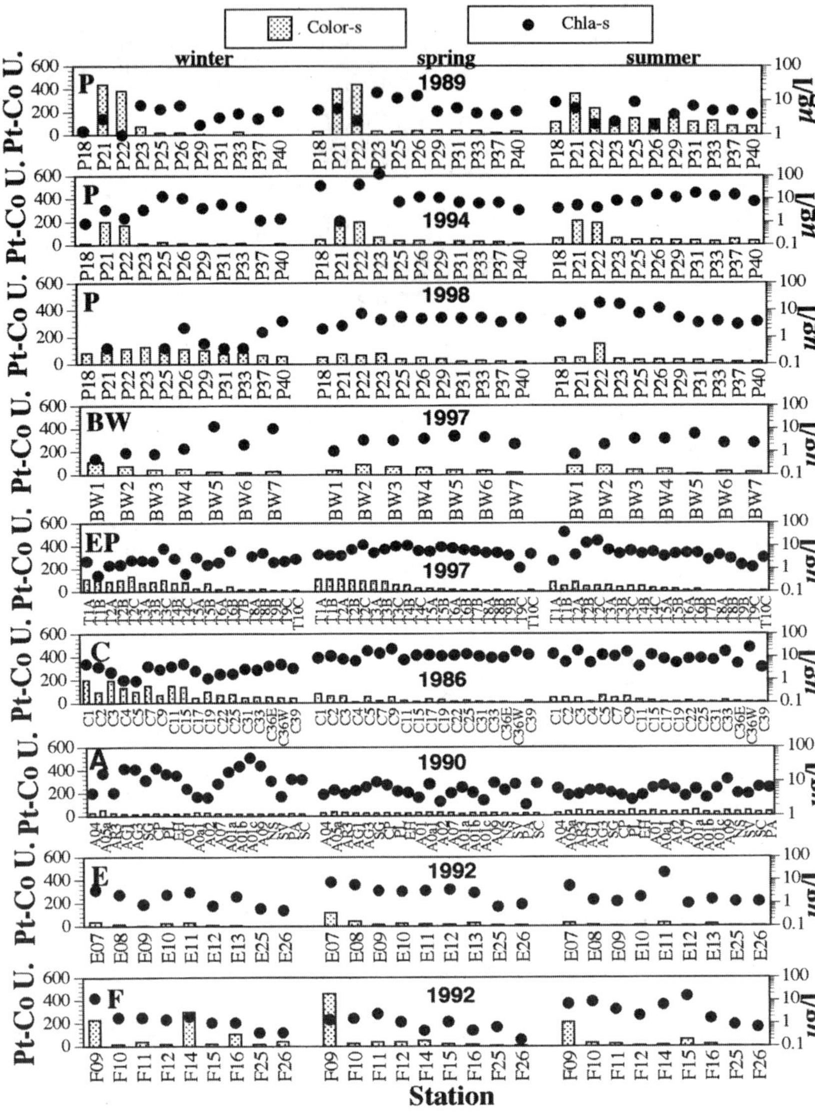

Figure 6.5 (continued)

Results of a comparative sediment analysis are given in Figures 6.6 and 6.7. The distribution of sediment quality in the Escambia–Pensacola Bay system was typical of that found in the Perdido and Choctawhatchee estuaries. Core data indicated that the overall distribution of sediments in the Pensacola system was related to a combination of depth distribution and proximity to the Escambia and Blackwater/Yellow River mouths. There were relatively coarse sediment particles (high sand concentrations) in shallow fringing areas of Escambia Bay and Blackwater–East Bay. In the deeper, more central, parts of the respective bays, a north–south gradient of finer particles existed with the highest concentrations of fine particles found in lower Escambia Bay and lower East Bay. Sand distribution was highest in the shallow, fringing areas of upper Escambia Bay and upper Blackwater–East Bay. The highest silt concentrations occurred in midportions of Escambia Bay and in lower parts of East Bay. Clay was most highly concentrated in lower Escambia Bay. Both silt and clay concentrations were relatively high in upper Pensacola Bay and in lower parts of Escambia

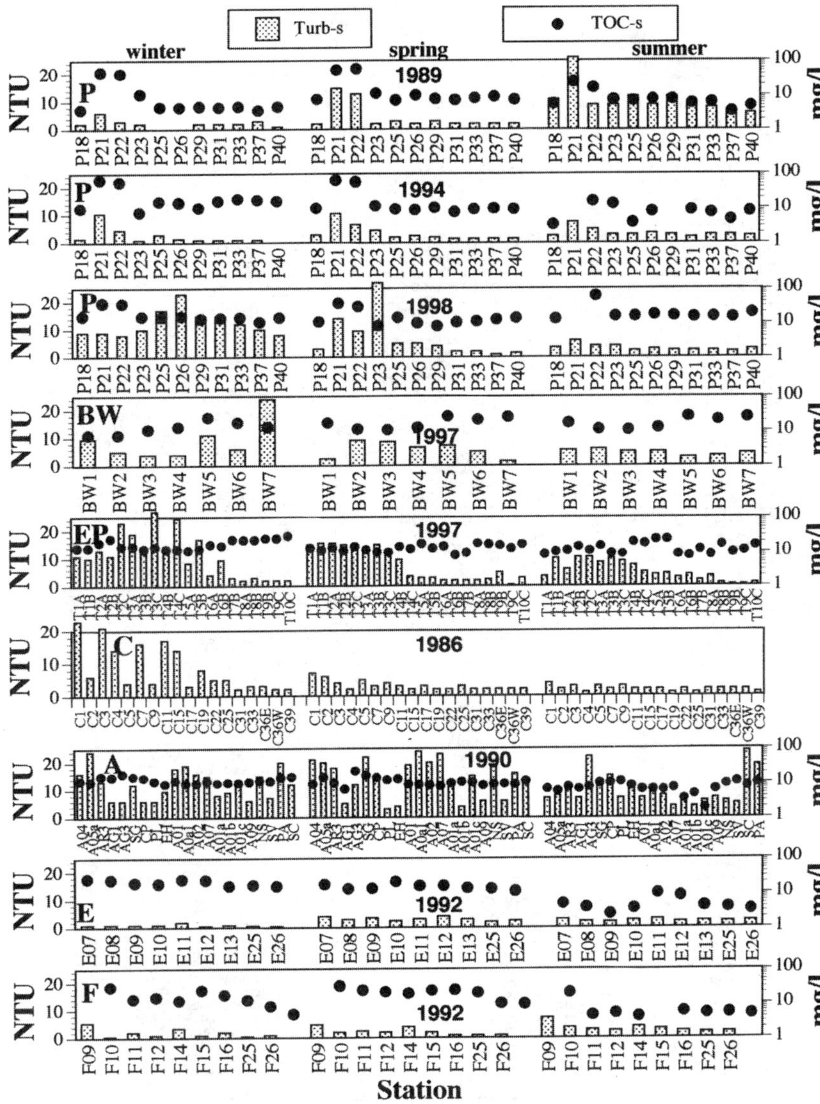

Figure 6.5 (continued)

Bay. A diving survey in Pensacola Bay indicated a thin veneer of silty sediments on top of a sandy base. Areas in deeper sections of Pensacola Bay had higher concentrations of fine particles. Silt concentration distribution suggested that storm water runoff from the City of Pensacola affected sediment composition in lower Escambia Bay and upper Pensacola Bay. The relatively high percentage of organic matter observed in sediments at these stations in the western part of the lower bay supported this hypothesis.

In the Apalachicola estuary, upland stations in East Bay (near extensive salt marshes, 5b; or in *Vallisneria americana* beds, 4a) had the highest silt-clay fractions with correspondingly high organic concentrations (Figure 6.6). In the Choctawhatchee Bay system, deeper areas characterized by high salinity stratification had relatively high silt fractions and high concentrations of organic matter. Relatively well developed deposits of liquid mud characterized these areas. In Perdido Bay, sediment composition was directly related to depth and salinity stratification. As in the Escambia and

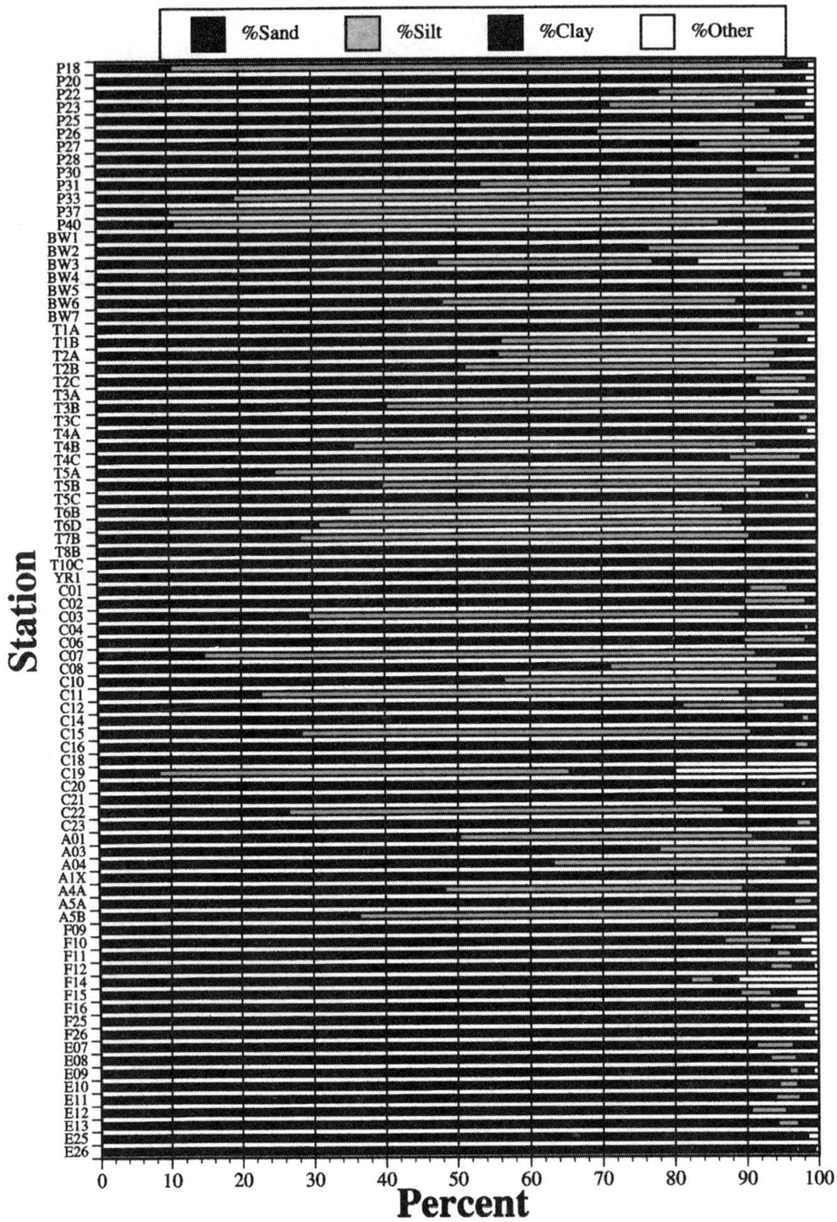

Figure 6.6 Sediment analyses (% sand, % silt, % clay, % other) in bay systems of the northeastern Gulf of Mexico.

Choctawhatchee estuaries, the bottom of the shallow periphery of the bay was characterized by sandy substrates. The shelf graded down a slope to deeper parts of the bay which were characterized as silty with extensive deposits of liquid mud. In general, bay areas characterized by well-developed, stable haloclines at depths exceeding 2 m usually had extensive deposits of liquid mud that were characterized by high silt-clay fractions and relatively high amounts of organic matter. The Apalachee Bay estuaries had uniformly sandy conditions with relatively large particle size and low amounts of organic matter. Inshore Apalachee Bay stations had somewhat higher silt–clay fractions, with the Fenholloway estuary having higher organic compositions than cognate stations in the Econfina estuary.

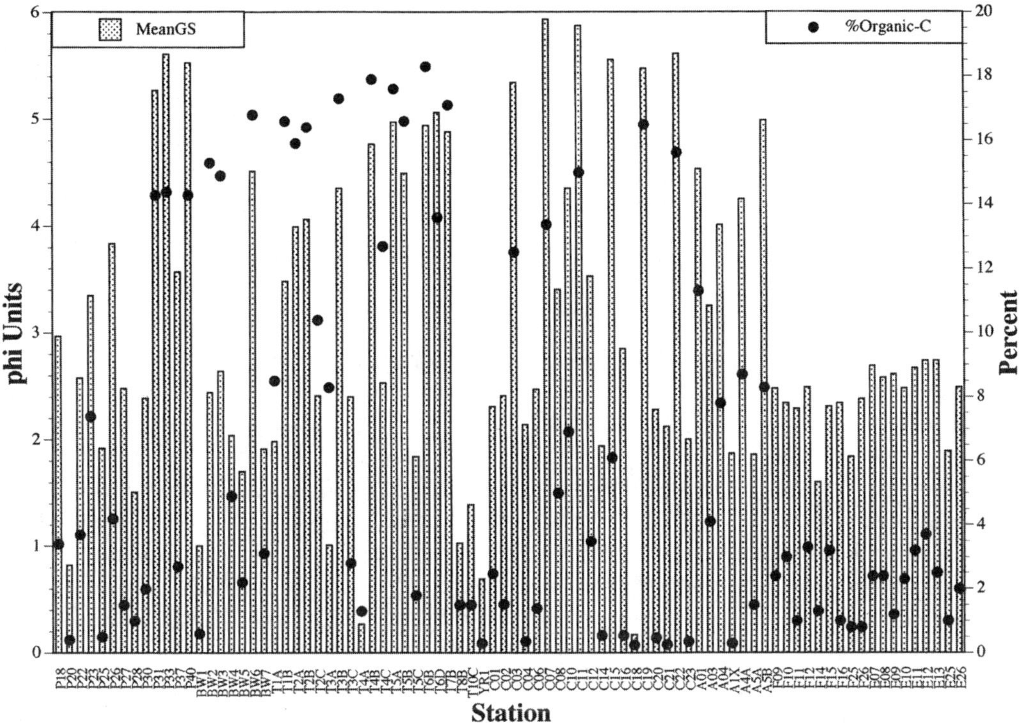

Figure 6.7 Sediment grain sizes (phi units) and percent organics in bay systems of the northeastern Gulf of Mexico.

At the head of Choctawhatchee Bay, sediments at station C01 in the delta area had fine-grained brown sand. At stations farther down the bay (C07 to C19), the bottom sediments were somewhat variable with station C07 having the highest phi of all stations and station C18 having the lowest. There was a fine flocculent layer above the sandy substrate in upper bay areas. There was a distinct salt wedge throughout the deeper parts of the bay below the sandy shelf in a manner very much like that described in Perdido Bay. In general appearance and consistency, the liquid mud in Choctawhatchee Bay was almost identical to that in Perdido Bay. In Apalachicola Bay, where most areas are less than 2 m in depth, no such liquid mud was found. The silt–clay matrix in upper parts of the bay were part of the solid sediment foundation. Areas characterized by high-salinity stratification in Apalachicola Bay were in high-energy passes where water movement precluded the formation of nephlos. Overall, liquid mud occurred only in areas deeper than 2 m where there was pronounced salinity stratification and limited current action. Reduced water column turbulence below the halocline was necessary for the suspension of fine particulates in a liquid matrix above the solid sediment bottom. The development of the liquid mud in the Perdido and Choctawhatchee systems appeared to be directly associated with artificial openings in the barrier islands that bordered the Gulf of Mexico.

Benthic deposits can have an important effect on the oxygen content of natural waters. Present studies estimate that sediment oxygen demand (SOD) can comprise up to 50% of the total oxygen depletion for a given body of water. SOD encompasses two separate processes: the biological respiration of all living organisms; and the chemical oxidation of reduced substances in the sediment such as divalent iron, manganese, and sulfate. A diver-deployed Benthic Lander (Rowe and Boland, 1991) was used in the determination of SOD across the sediment/water interface (Appendix I). Duplicate samples were taken after the instrument was grounded in sediments at a given station. Dissolved oxygen probes (one for each of the duplicate units) were attached to YSI DO meters that were monitored at 2- to 5-minute intervals. Sampling stations included areas that were adjacent to the major sources of freshwater input (including Elevenmile Creek, the Perdido River, the Choctawhatchee River, and the Apalachicola River) to the respective bays. The data indicated that SOD levels in the Apalachicola, Choctawhatchee, and Perdido estuaries were generally similar. During summer periods, SOD in the Choctawhatchee system was somewhat higher than that in the Perdido system. However, there was no substantial difference in the SOD noted in the Perdido and Choctawhatchee systems. A comparison of SOD in areas that had similar salinity regimes in the Apalachicola, Choctawhatchee, and Perdido estuaries indicated no significant differences among the three systems.

VI. PHYTOPLANKTON ORGANIZATION

A. Introduction

There is very little scientific literature concerning the effects of nutrification on phytoplankton communities. Howarth et al. (2000) point out that increased loading of nitrogen and phosphorus may be associated with reductions of silica that, in turn, limits diatom growth leading to changes in phytoplankton community composition (Jorgensen and Richeardson, 1996). This can affect the size distribution as well as the species-specific content of the plankton assemblages. Changes from diatoms to flagellates can lead to food web changes that can then lead to hypothetical alterations of fish populations (Greve and Parsons, 1977). Radach et al. (1990) found that with increases of the nitrogen:silica and phosphorus:silica ratios over a 23-year period, diatoms decreased and flagellates increased along the German coast. Shifts of nitrogen:phosphorus ratios increased summer bloom incidence of the haptophyte *Phaeocystis* (Riegman et al., 1992) which was associated with phosphorus limitation. Dale et al. (1999), studying long-term dinoflagellate cyst abundance in sediments in Oslofjord (Norway), found that, during periods of nutrient and organic enrichment, the abundance of cysts doubled with distinct increases in one species (*Lingulodinium machaerophorum*). Riegman (1998) indicated that phytoplankton species composition may be changed by the dynamic interactions among macronutrients. However, as in the relatively few studies listed above, detailed, quantitative, long-term phytoplankton community changes due to altered nutrient loading have not been studied in a systematic fashion. This lack of such information is remarkable considering the importance of phytoplankton to the issue of eutrophication.

The Perdido Bay study indicated that the structure of the phytoplankton communities in the upper bay was determined in large part by nutrient loading and nutrient concentration gradients at the end of Elevenmile Creek. Increased orthophosphate loading during 1993–1994 was associated with phytoplankton blooms of various diatoms, general increases of numerical and biomass percentages of bloom species, and associated reductions of plankton species richness. Interannual sequences favored increased representation by raphidophytes, dinoflagellates, and cryptophytes, and these changes of the plankton communities were associated with reductions of numbers, biomass, and species richness of the infaunal and epibenthic macroinvertebrates and fishes. The plankton blooms could have affected the bay food webs through reductions of sedimentary microphyte assemblages that, in turn, led to the observed loss of planktivorous infaunal macroinvertebrates. This herbivore component was largely eliminated during later bloom successions leading to a general collapse of the entire food web in areas of the bay affected by the blooms. Reduced secondary production was usually most pronounced in depositional areas that were already low in numerical abundance of benthic invertebrates and fishes due to periodic hypoxia and associated development of liquid mud as a result of salinity stratification. Thus, hypereutrophication in the form of plankton blooms and associated changes in the phytoplankton communities of Perdido Bay were associated with the progressive deterioration of secondary production of Perdido Bay through processes that altered the food web structure.

The use of specific phytoplankton community indices and associated food web distortions found in the Perdido system could be used to evaluate the general ecological status of other Gulf estuaries. Accordingly, a comparative analysis was carried out to evaluate the status of the various estuarine systems in the study area.

B. Phytoplankton Communities in the Study Areas

1. Wolf Bay

Wolf Bay, adjacent to the lower Perdido estuary (see Figure 2.2a), has been affected by nutrient loading from postulated agricultural sources for years (Livingston et al., 1998a). A land-use characterization of the Perdido system (Patrick, 1991) indicated that "according to agreements with EPA, the Wolf Bay drainage basin and Gulf Shores are not included in the Perdido Bay study." The reason for the removal of key parts of the Perdido drainage system from a land-use analysis was not given. A recent review of agricultural activities in Baldwin County, Alabama (A. Niedoroda, personal communication) indicated that the production of cotton in this area increased by a factor of 26 between 1988 and 1995, and that comparable shifts in the agricultural activities in this region have continued to the present time. Possible changes related to the altered use of fertilizers in this area of the bay due to shifts in agricultural activities are thus directly relevant to the observed deterioration of Wolf Bay.

This bay was sampled for water quality quarterly from October 1993 through September 1999 as part of a comprehensive analysis of the Perdido drainage system. Between July 1996 and July 1997, a series of five quarterly analyses was conducted to determine baywide distributions of water and sediment quality and infaunal macroinvertebrates in the Perdido system (including Wolf Bay). Whole-water phytoplankton samples were taken monthly in Wolf Bay from August 1998 through September 1999. Ammonia and orthophosphate concentrations were periodically high in Wolf Bay (Figure 6.8). Chlorophyll *a* was also occasionally high in Wolf Bay, with the highest concentrations usually occurring during periods of high orthophosphate concentrations in the upper bay (Figure 6.8). Sediment analyses of the entire Perdido Bay system (Figure 6.9) indicated high silt concentrations in middle to lower bay depositional areas (P31, P33, P37, and P40), and in Wolf Bay (P42b, 42a, and 42). Diving surveys indicated that liquid mud was common in Wolf Bay. The distribution of liquid mud was associated with spatial patterns of high silt concentrations in the sediments. The highest total phosphorus concentrations occurred at stations P26, P29, P31, P33, and P37 in Perdido Bay and in parts of Wolf Bay (Figure 4.11). The sediment composition of Wolf Bay closely resembled that in middle to lower Perdido Bay (Figure 6.9). Thus, water and sediment quality characteristics in Wolf Bay indicated that this part of the system was subjected to nutrient loading and habitat response to such loading in a fashion similar to that noted in upper Perdido Bay. In this case, the source appeared to be non-point agricultural activities in the Wolf Bay drainage basin (Livingston et al., 1998a).

The phytoplankton biota of Wolf Bay (Table 6.1) was dominated by bloom indicators (Cryptophytes) and/or bloom species (*Gymnodium* spp., *Heterosigma akashiwo)*, with well-established populations of other bloom species (*Prorocentrum minimum*, *Cyclotella choctawhatcheeana*). Peaks of *Prorocentrum* occurred in January 1999. The diatom *C. choctawhatcheeana* peaked during April 1999, and *Heterosigma* peaked during summer 1999. The seasonal sequence of bloom dominants followed that of upper Perdido Bay. The near absence of the chrysophytes was further indication that Wolf Bay was dominated by the same bloom species progressions and alterations in the plankton community structure that were found in upper Perdido Bay during increased nutrient loading from Elevenmile Creek. A comparison of Wolf Bay whole-water phytoplankton species richness (station P42b) with this index in the upper bay (Station P23) during periods before and during plankton blooms (Figure 6.10) indicated that Wolf Bay phytoplankton species richness was even lower than that in the upper bay during the 1993–1994 blooms. The relatively high phytoplankton species

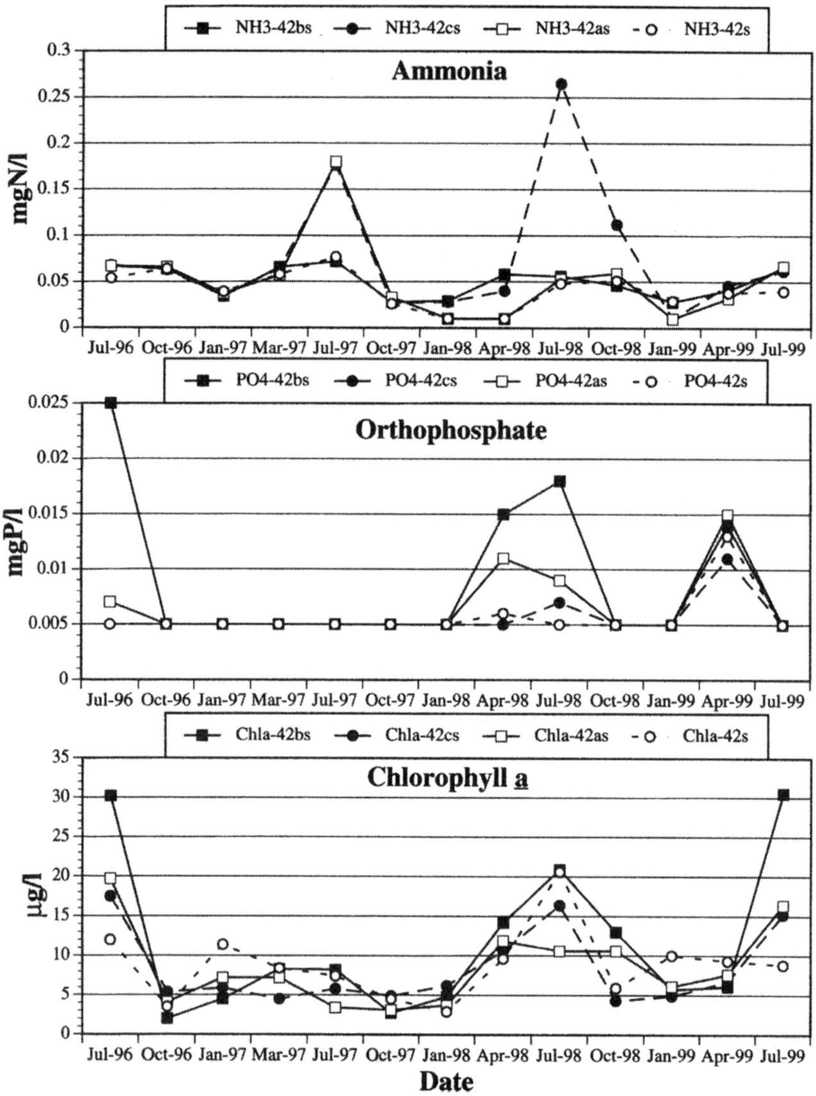

Figure 6.8 Analyses of ammonia (mg N l^{-1}), orthophosphate (mg P l^{-1}), and chlorophyll *a* (µg l^{-1}) at stations in Wolf Bay at quarterly intervals from July 1996 through July 1999.

richness during the nonbloom years (1990–1991) at station P23 was matched only by the period of 1998–1999 at this station when orthophosphate loading from the mill was low. With resumption of the nutrient loading into Elevenmile Creek in spring 1999, phytoplankton species richness again was reduced to levels comparable to those in Wolf Bay during a period characterized by high nutrient concentrations in the upper Wolf Bay drainage. These data give further indication that Wolf Bay was seriously damaged by nutrient loading in a manner similar to that described for upper Perdido Bay.

Water quality and sediment data in Wolf Bay showed signs of deterioration similar to that in the main stem of the bay during and after the initiation of the plankton blooms. High silt concentrations and the presence of liquid mud were similar to conditions in bloom-affected areas of the middle and lower Perdido Bay (Figure 6.11). Infaunal macroinvertebrates are considered to be good indicators of sediment/water quality in coastal systems. Results of the infaunal macroinvertebrate analyses are given in Figure 6.11. Extremely low infaunal numbers and/or species richness were

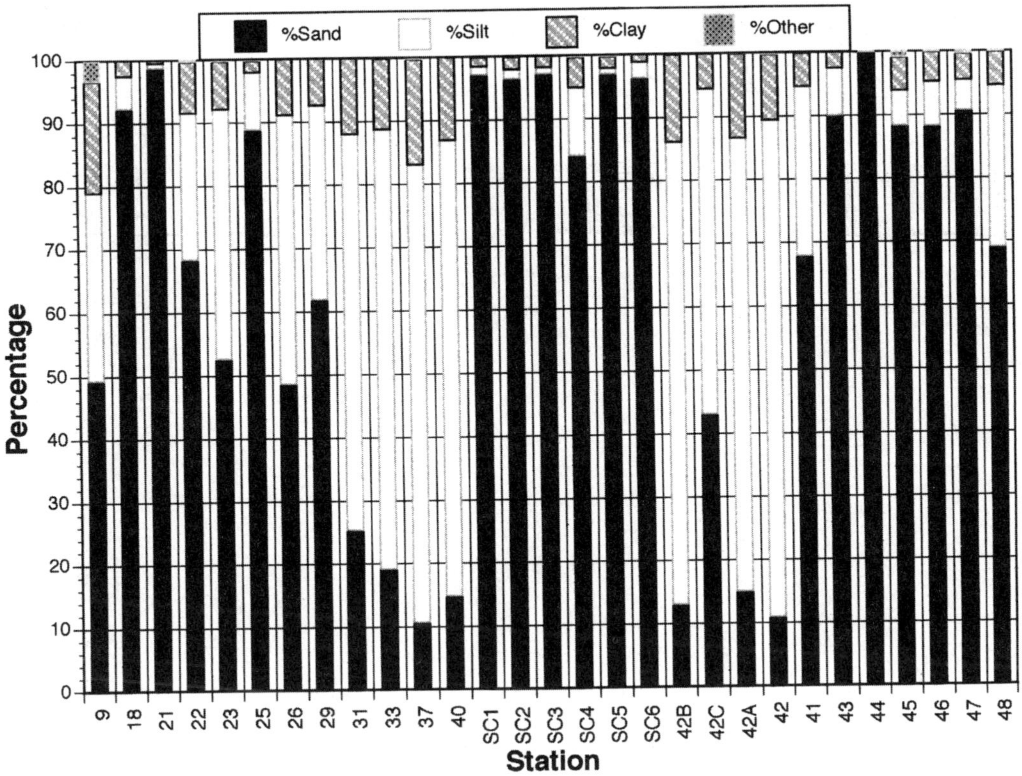

Figure 6.9 Sediment analyses (percent sand, percent silt, percent clay, percent other) in Perdido Bay and Wolf Bay during 1997.

noted at the mouth of the Perdido River (P18), in lower Elevenmile Creek (P22), in eastern parts of upper Perdido Bay (P26, P29), in lower Perdido Bay (P31, P33, P37, P40, and P41), in Wolf Bay (P42A, P42) and in the lower bay east and south of Ono Island (P48). These areas coincide with (1) deep holes at the mouth of the Perdido River characterized by reduced water and sediment quality, (2) eastern areas of the upper bay affected by phytoplankton blooms, (3) depositional areas of the lower bay characterized by salinity stratification, liquid mud, and periodic hypoxia, (4) Areas of Wolf Bay affected by non-point nutrient pollution, and (5) lower bay areas affected by urban runoff. These data illustrate the variety of impacts on Perdido Bay due to different combinations of anthropogenous activities that included dredging and a combination of point and non-point sources of nutrient loading.

The overall distribution of the infauna in Perdido Bay corresponded generally to a combination of depth, salinity stratification, and water/sediment quality. Areas characterized by low infaunal community indices coincided with high silt fractions in the sediments and the presence of liquid mud. Deeper (depositional) areas, characterized by high stratification and high concentrations of silt and liquid mud (middle to lower bay stations), had relatively low infaunal abundance and species richness. Although Wolf Bay and the bay south of Ono Island were not highly stratified, the infaunal indices were generally low in these areas as well. Only high silt concentrations of the sediments were associated with the trends of degradation of the infaunal biota in these lower bay areas. The connection of phytoplankton blooms and corresponding deterioration of the sediments emerges as a reasonable explanation for the almost total lack of benthic infaunal macroinvertebrates in Wolf Bay. Regression analyses of the data (Table 6.2) tested the postulated associations of high nutrient concentrations, altered phytoplankton community indices through increased representation by bloom species, deterioration of the benthic habitat, and losses of benthic secondary production in

Table 6.1 Whole-Water Phytoplankton Taken in Wolf Bay (station P42b) Monthly (except 9/98) from August 1998 through September 1999

Species	98/08	98/10	98/11	98/12	99/01	99/02	99/03	99/04	99/05	99/06	99/07	99/09	TOTAL	%
*Undetermined cryptophytes**	2,200,117	772,474	3,082,563	702,491	1,113,722	5,675,376	886,778	8,595,741	348,580	1,318,337	2,276,098	593,852	**27,566,128**	**49.0**
Undetermined nannoflagellates	671,166	570,524	339,915	253,937	1,134,383	1,310,831	802,767	1,061,466	541,198	691,827	849,788	447,222	**8,675,023**	**15.4**
Cryptomonas spp.	45,989	600,517	37,324	11,997	161,960	2,867,025	529,400	93,345	66,650	2,000	205,282	27,327	**4,648,815**	**8.3**
Gymnodinium spp.*	1,037,074	57,986	23,328	6,665	67,317	49,340	45,339	104,013	373,907	238,607	1,657,586	183,288	**3,844,449**	**6.8**
*Heterosigma akashiwo**	179,289	151,296	46,655	13,330	281,930	217,361	100,013	22,670	130,634	73,982	623,178	22,661	**1,862,998**	**3.3**
Chlamydomonas spp.	0	215,280	0	0	36,658	1,062,800	28,004	16,002	5,332	667	1,333	0	**1,366,076**	**2.4**
Chrysochromulina spp.	1,333	7,998	8,665	3,333	192,619	258,699	18,669	268,034	100,642	69,316	85,979	32,659	**1,047,945**	**1.9**
Pyramimonas spp.	11,997	13,330	4,666	0	202,616	166,688	32,004	6,668	272,599	80,647	111,972	34,658	**937,844**	**1.7**
*Prorocentrum minimum**	33,325	0	1,333	0	523,203	13,335	2,667	0	5,332	9,998	136,633	7,332	**733,157**	**1.3**
*Cyclotella choctawhatcheeana**	48,655	11,997	2,000	42,656	105,307	8,001	78,677	145,352	111,306	29,993	68,650	41,990	**694,583**	**1.2**
Undetermined oxytoxoid flagellate	0	0	642,506	8,665	0	0	0	29,337	0	2,000	0	0	**682,508**	**1.2**
Chaetoceros wighami	0	0	0	6,665	69,316	14,669	350,711	73,343	14,663	0	0	0	**529,367**	**0.9**
Chaetoceros subtilis	0	0	0	0	13,997	0	469,392	1,334	6,665	0	0	0	**491,388**	**0.9**
Amphidinium spp.	0	667	3,333	1,333	352,579	10,668	2,667	18,669	11,997	51,321	31,326	5,999	**490,558**	**0.9**
Dictyosphaerium ehrenbergianum	0	431,892	0	0	0	0	0	0	0	0	0	0	**431,892**	**0.8**
*Chattonella cf. subsalsa**	213,947	0	0	0	0	0	0	1,334	19,995	52,654	6,665	0	**294,595**	**0.5**
Skeletonema costatum	0	0	0	0	99,309	1,334	174,689	0	0	0	0	0	**275,332**	**0.5**
Gymnodinium cf. splendens	30,659	5,332	13,997	10,664	3,333	53,340	16,002	8,001	0	11,997	29,326	3,333	**185,984**	**0.3**
Undetermined pennate diatoms	2,666	53,987	4,666	3,333	9,998	6,668	48,006	9,335	5,999	3,333	4,666	2,000	**154,655**	**0.3**
Gyrodinium spp.	0	0	0	0	0	0	0	0	49,988	29,993	58,652	0	**138,633**	**0.2**
Pseudoscourfieldia spp.	9,331	0	0	0	1,333	0	2,667	0	0	9,998	87,312	8,665	**119,305**	**0.2**
Eutreptiella spp.	33,325	0	0	0	667	40,005	2,667	0	11,997	2,666	15,330	667	**107,323**	**0.2**
Undetermined dinoflagellates	25,327	0	667	3,999	6,665	5,334	6,668	4,001	6,665	11,997	29,993	2,666	**103,981**	**0.2**
Miraltia sp.	0	19,329	0	0	2,666	4,001	58,674	0	2,000	2,000	2,000	0	**90,670**	**0.2**

Dactyliosolen fragilissimus	0	0	0	0	57,319	0	12,002	0	0	0	0	0	**69,321**	**0.1**
Chaetoceros spp.	0	0	0	0	17,996	2,667	18,669	24,003	2,666	0	0	0	**66,001**	**0.1**
Ankistrodesmus spp.	0	64,651	0	667	667	0	0	0	0	0	0	0	**65,984**	**0.1**
*Miraltia throndsenii**	0	0	0	0	1,333	40,005	14,669	0	0	667	6,665	0	**63,339**	**0.1**
Nitzschia reversa	0	667	0	0	2,666	1,334	13,335	5,334	2,000	8,665	11,997	3,333	**49,330**	**0.1**
Cerataulina pelagica	0	0	0	0	35,325	0	9,335	0	0	0	0	0	**44,660**	**0.1**
Nitzschia spp.	667	20,562	667	0	1,333	0	17,336	0	667	0	0	0	**41,331**	**0.1**
Scenedesmus spp.	0	30,359	0	0	0	0	10,668	0	0	0	0	0	**41,327**	**0.1**
Gymnodinium sanguineum	12,664	0	0	0	0	2,667	0	1,334	3,999	4,666	9,331	0	**34,661**	**0.1**
Cyclotella spp.	1,333	11,331	0	10,664	2,000	1,334	0	0	1,333	667	1,333	0	**29,995**	**0.1**
Euglena spp.	667	3,333	0	0	0	6,668	1,334	1,334	0	0	15,330	0	**28,665**	**0.1**
Spermatozopsis exsultans	0	27,993	0	0	0	0	0	0	0	0	0	0	**27,993**	**0.0**
Undetermined nannococcoids	0	1[illegible],997	0	0	0	0	0	6,668	0	0	0	0	**18,665**	**0.0**
Ceratium hircus	0	0	0	0	0	0	0	0	5,332	11,331	0	0	**16,663**	**0.0**
Selenastrum spp.	0	16,663	0	0	0	0	0	0	0	0	0	0	**16,663**	**0.0**
Synedropsis sp. *novum**	0	7,332	0	0	0	0	5,334	2,667	0	0	0	0	**15,333**	**0.0**
Mallomonas spp.	0	[illegible]3,997	0	0	0	0	0	1,334	0	0	0	0	**15,331**	**0.0**
Prorocentrum micans	0	0	0	0	0	0	0	1,334	3,999	7,998	667	0	**13,998**	**0.0**
*Leptocylindrus danicus**	0	0	0	0	0	0	2,667	8,001	1,333	0	0	0	**12,001**	**0.0**
Amphora spp.	0	0	0	0	1,333	0	1,334	2,667	2,000	1,333	1,333	0	**10,000**	**0.0**
Ankistrodesmus sp.	0	8,665	0	0	0	0	0	0	0	0	0	0	**8,665**	**0.0**
Synedra spp.	0	6,665	0	0	0	1,334	0	0	0	0	0	0	**7,999**	**0.0**
Thalassionema spp.	0	0	0	0	0	4,001	0	0	2,666	0	0	667	**7,334**	**0.0**
Chaetoceros simplex	0	0	0	0	7,332	0	0	0	0	0	0	0	**7,332**	**0.0**
Undetermined curved cell w/ long bristle	0	5,999	0	0	0	0	0	0	0	0	0	0	**5,999**	**0.0**
Dinobryon bavaricum	0	1,333	0	0	0	0	4,001	0	0	0	0	0	**5,334**	**0.0**
Navicula spp.	0	667	0	0	0	0	4,001	0	0	0	667	0	**5,334**	**0.0**
Oocystis spp.	0	0	0	0	0	0	5,334	0	0	0	0	0	**5,334**	**0.0**
Crucigenia tetrpedia	0	5,332	0	0	0	0	0	0	0	0	0	0	**5,332**	**0.0**
Thalassiosira spp.	0	0	0	0	0	0	0	4,001	667	0	0	0	**4,668**	**0.0**
Achnanthes spp.	1,333	667	0	0	667	0	0	0	667	0	667	667	**4,666**	**0.0**

Table 6.1 (continued) Whole-Water Phytoplankton Taken in Wolf Bay (station P42b) Monthly (except 9/98) from August 1998 through September 1999

Species	98/08	98/10	98/11	98/12	99/01	99/02	99/03	99/04	99/05	99/06	99/07	99/09	TOTAL	%
Synedra ulna	0	0	0	0	667	0	1,334	0	0	0	0	1,333	**3,334**	**0.0**
Anabaena spp.	0	3,333	0	0	0	0	0	0	0	0	0	0	**3,333**	**0.0**
Pseudonitzschia spp.	0	0	0	0	0	0	2,667	0	0	0	0	0	**2,667**	**0.0**
Asterionellopsis glacialis	0	0	0	2,000	667	0	0	0	0	0	0	0	**2,667**	**0.0**
Stylochrysalis spp.	0	2,666	0	0	0	0	0	0	0	0	0	0	**2,666**	**0.0**
Cymbella spp.	667	0	0	0	0	0	1,334	0	0	0	0	0	**2,001**	**0.0**
Chaetoceros affinis	0	0	0	0	2,000	0	0	0	0	0	0	0	**2,000**	**0.0**
Phacus spp.	0	1,333	0	0	0	0	0	0	0	0	667	0	**2,000**	**0.0**
Coscinodiscus spp.	0	0	0	0	0	0	0	1,334	0	0	0	0	**1,334**	**0.0**
Eunotia spp.	0	0	0	0	0	0	1,334	0	0	0	0	0	**1,334**	**0.0**
Gyrosigma sp. *(w/stauros)*	0	0	0	0	0	0	0	1,334	0	0	0	0	**1,334**	**0.0**
Pleurosigma spp.	0	0	0	0	0	0	1,334	0	0	0	0	0	**1,334**	**0.0**
Scenedesmus dimorphus	0	0	0	1,333	0	0	0	0	0	0	0	0	**1,333**	**0.0**
Synedra cf. delicatissima	0	1,333	0	0	0	0	0	0	0	0	0	0	**1,333**	**0.0**
Closterium spp.	0	0	0	0	667	0	0	0	0	0	0	0	**667**	**0.0**
Cosmarium spp.	0	667	0	0	0	0	0	0	0	0	0	0	**667**	**0.0**
Diploneis spp.	0	0	0	0	0	0	0	0	667	0	0	0	**667**	**0.0**
Entomoneis spp.	0	0	0	0	667	0	0	0	0	0	0	0	**667**	**0.0**
Golenkinia spp.	0	667	0	0	0	0	0	0	0	0	0	0	**667**	**0.0**
Gonatozygon spp.	0	0	0	0	0	0	0	0	0	667	0	0	**667**	**0.0**
Protoperidinium spp.	0	0	0	0	0	0	0	0	667	0	0	0	**667**	**0.0**
Staurastrum spp.	0	667	0	0	0	0	0	0	0	0	0	0	**667**	**0.0**
Total	**4,561,529**	**3,161,881**	**4,212,284**	**1,083,731**	**4,512,211**	**11,825,484**	**3,784,481**	**10,519,988**	**2,114,809**	**2,729,322**	**6,330,422**	**1,420,316**	**56,256,455**	**100.0**

Note: Data are presented as cells l^{-1} with bloom indicators and bloom species marked by asterisks.

* Indicators of Phytoplankton Blooms

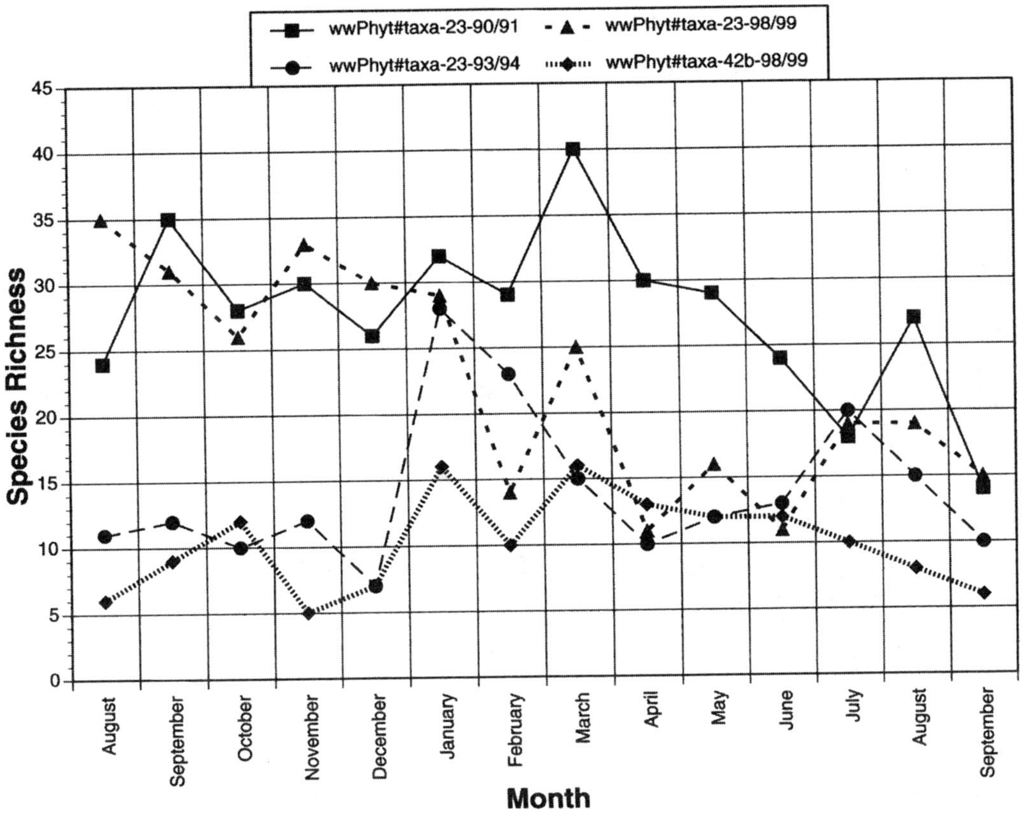

Figure 6.10 Comparison of whole-water phytoplankton species richness at station P23 in upper Perdido Bay (1990–1991; 1993–1994; 1998–1999) and station P42b in Wolf Bay (1998–1999). Data are shown monthly for the respective time periods.

the form of infaunal macroinvertebrates. Infaunal numerical abundance was positively associated with various forms of nutrients and was negatively associated with salinity stratification and silt content of the sediments. Infaunal species richness was positively associated with sandy sediments and negatively associated with bottom salinity, silt content of the sediments, and fine particles in sediments. Overall, nutrients and sediment characteristics were closely associated with the distribution of numbers and species richness of the infaunal macroinvertebrates in Perdido Bay.

The Wolf Bay data indicate that nutrient loading from non-point sources such as agricultural activities can lead to changes in the receiving estuary that are similar to those associated with point-source loading. These changes can be related to processes associated with alterations of the phytoplankton assemblages that include increased dominance by noted bloom species and low species richness. Overall, by September 1999, the combination of dredging at the mouth of the estuary and nutrient loading from multiple sources accounted for widespread habitat deterioration and major reductions of secondary production throughout much of the Perdido Bay system.

2. Escambia Bay

The Escambia Bay system has been subjected to continuous anthropogenous stress for decades (Livingston, 1997a). Submerged aquatic vegetation, formerly abundant in this system, has been largely absent since the mid-1970s (Collard, 1991a, b). Formerly productive oyster bars are no longer commercially viable, although a small oyster fishery still exists. Some of the most extensive fish kills in the scientifically documented literature occurred in Escambia Bay during the late 1960s

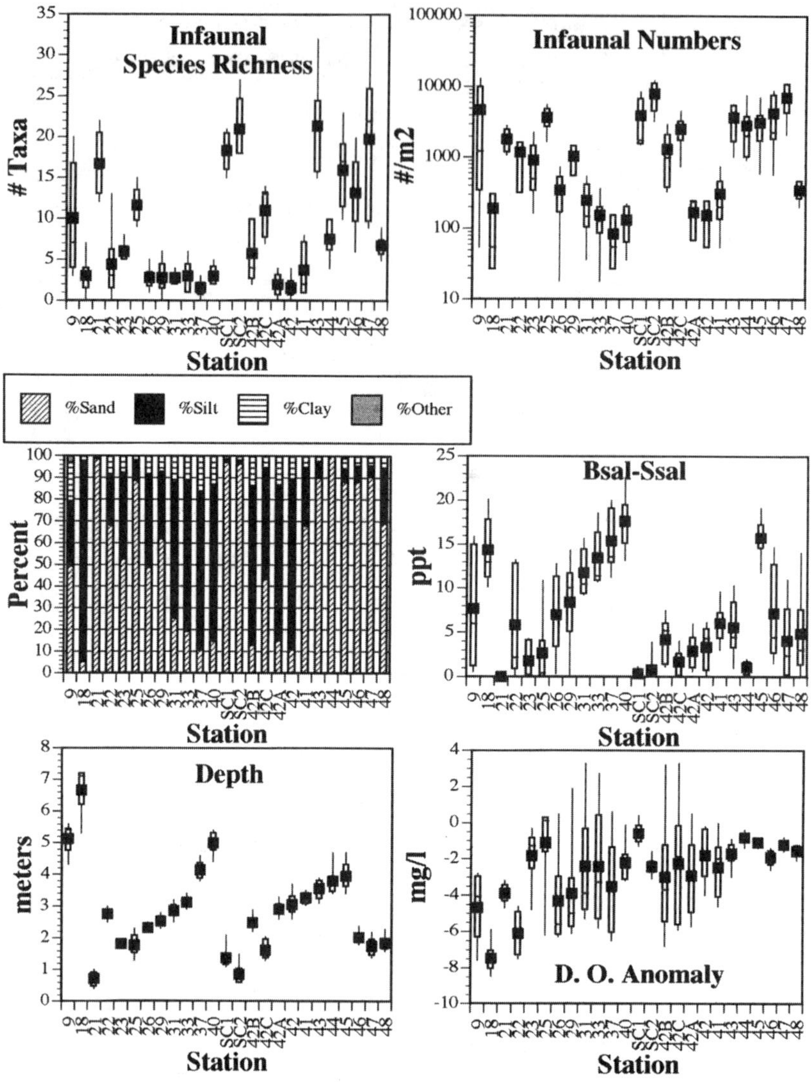

Figure 6.11 Averages of five quarterly collections of infaunal macroinvertebrate showing numbers m^{-2} and species richness in Perdido Bay from July 1996 through July 1997. Also shown are salinity stratification indices (Bsal-Ssal), depth (m), and bottom DO anomalies (mg l^{-1}). Data are shown as means with the 10th, 25th, 75th, and 90th percentiles. Also shown are the results of sediment analyses taken during this period.

and early 1970s. Livingston (1997a) found that the fish kills were due to the effects of hyper-eutrophication from point- and non-point-source nutrient loading and habitat alterations due to dredging. Dredging activities enhanced the effects of cultural eutrophication by inhibiting tidal and wind-produced exchanges of bay water. An additional physical alteration in the form of a bridge restricted turnover rates in the upper bay. The fish kills were the final consequence of a distorted DO regime, which, in turn, was influenced by the sequential interaction of temperature, precipitation, altered water circulation patterns, nutrient-enhanced primary production, and a disjunct food web organization (Livingston, 1997a). The role of phytoplankton blooms in the deterioration of Escambia Bay was never determined.

Detailed, ecosystem-level analyses (similar to those carried out in the Perdido system) were performed in the Blackwater and Escambia–Pensacola Bay systems during 1997–1998 (Livingston,

Table 6.2 Regression Analysis of Infaunal Macroinvertebrates Collected from Perdido Bay in July and October 1996

INFAUNA (7/96) p values, r2 values

	SAL-ppt	BSAL-SSAL	Cnd-µmho/cm	Temp-°C	pH	Sed. % Org.
NUMBERS	—	—	—	—	—	p = 0.03, -0.21
SPECIES RICHNESS	P = 0.03, -0.12	P = 0.0008,-0.28,	—	P = 0.01, -0.41	—	p = 0.01, -0.21
	DO-mg/l	**O2anom-mg/l**	**Color-NCASI**	**Turb-NTU**	**Ratio Turb-NTU**	**Sed. % Sand**
NUMBERS	P = 0.01, 0.20	—	P = 0.0001, 0.37	P = 0.0022, 0.24	P = 0.0002, 0.33	p = 0,01, 0.29
SPECIES RICHNESS	P = 0.005, 0.23	P = 0.03, 0.14	P = 0.012, 0.17	—	—	p = 0.002, 0.55
	BOD-mg/l	**Alk-mg/l**	**TSS-mg/l**	**POM-mg/l**	**PIM-mg/l**	**Sed. % Silt**
NUMBERS	P = 0.0001, 0.61	—	—	—	—	p = 0.01, -0.28
SPECIES RICHNESS	—	—	—	—	—	p = 0.001, -0.31
	Si-mgSiO2/l	**S--, mgS/l**	**NH3-mgN/l**	**NO2-mgN/l**	**NO3-mgN/l**	**Sed. % Clay**
NUMBERS	P = 0.001, 0.27	P = 0.0001, 0.12	P = 0.0001, 0.38	P = 0.0001, 0.67	P = 0.0052, 0.20	—
SPECIES RICHNESS	P = 0.0253, 0.14	P = 0.0092, 0.18	—	P = 0.007, 0.19	—	p = 0.01, -0.19
	PON-mgN/l	**DON-mgN/l**	**TON-mgN/l**	**TN-mgN/l**	**PO4-mgP/l**	**Sed. Phi Units**
NUMBERS	—	P = 0.0001, 0.42	P = 0.0001, 0.62	P = 0.0001, 0.69	P = 0.0001, 0.43	p = 0.01, -0.26
SPECIES RICHNESS	—	P = 0.0208, 0.15	—	P = 0.0362, 0.13	—	p = 0.002, -0.56
	POP-mgP/l	**DOP-mgP/l**	**TOP-mgP/l**	**TDP-mgP/l**	**TP-mgP/l**	**Chla-µg/l**
NUMBERS	P = 0.0001, 0.52	—	P = 0.0024, 0.24	P = 0.0001, 0.72	P = 0.0001, 0.49	—
SPECIES RICHNESS	P = 0.0377, 0.12	—	—	—	—	P = 0.047, 0.11
	POC-mgC/l	**DOC-mgC/l**	**TOC-mgC/l**	**IDC-mgC/l**	**TDC-mgC/l**	**Chlb-µg/l**
NUMBERS	P = 0.0103, 0.41	P = 0.0001, 0.50	P = 0.0001, 0.51	P = 0.0042, 0.45	P = 0.0001, 0.42	—
SPECIES RICHNESS	—	P = 0.0313, 0.13	P = 0.037, 0.12	—	—	—

Table 6.2 (continued) Regression Analysis of Infaunal Macroinvertebrates Collected from Perdido Bay in July and October 1996

	INFAUNA-10/96 p values, r2 values					
	SAL-ppt	**BSAL-SSAL**	**Cnd-µmho/cm**	**Temp-°C**	**pH**	**Sed. % Org.**
NUMBERS	P = 0.0013, -0.26	P = 0.047, -0.18	P = 0.0328, -0.13	—	P = 0.0001, -0.57	p = 0.04, -0.20
SPECIES RICHNESS	P = 0.0001, -0.41	P = 0.0018, -0.25	—	—	P = 0.0001, -0.41	p = 0.001, -0.33
	DO-mg/l	**O2anom-mg/l**	**Color-NCASI**	**Turb-NTU**	**Ratio Turb-NTU**	**Sed. % Sand**
NUMBERS	—	—	—	P = 0.0001, 0.38	P = 0.0001, 0.34	p = 0.01, 0.30
SPECIES RICHNESS	P = 0.0023, 0.24	P = 0.0444, 0.11	P = 0.0287, 0.12	—	—	p = 0.0003, 0.40
	BOD-mg/l	**Alk-mg/l**	**TSS-mg/l**	**POM-mg/l**	**PIM-mg/l**	**Sed. % Silt**
NUMBERS	P = 0.0001, 0.69	—	—	—	—	p = 0.01, 0.30
SPECIES RICHNESS	—	—	—	—	—	P = 0.0002, -0.41
	Si-mgSiO2/l	**S--, mgS/l**	**NH3-mgN/l**	**NO2-mgN/l**	**NO3-mgN/l**	**Sed. % Clay**
NUMBERS	P = 0.0001, 0.61	P = 0.0002, 0.22	P = 0.0001, 0.78	P = 0.0001, 0.79	P = 0.0001, 0.32	—
SPECIES RICHNESS	P = 0.001, 0.27	—	—	P = 0.0329, 0.12	—	p = 0.003, -0.27
	PON-mgN/l	**DON-mgN/l**	**TON-mgN/l**	**TN-mgN/l**	**PO4-mgP/l**	**Sed. Phi Units**
NUMBERS	—	P = 0.0001, 0.66	P = 0.0001, 0.63	P = 0.0001, 0.86	P = 0.0001, 0.73	—
SPECIES RICHNESS	P = 0.0211, 0.14	P = 0.0037, 0.22	0.0097, 0.18	P = 0.0064, 0.19	—	p = 0.001, -0.59
	POP-mgP/l	**DOP-mgP/l**	**TOP-mgP/l**	**TDP-mgP/l**	**TP-mgP/l**	**Chla-µg/l**
NUMBERS	P = 0.0381, 0.12	—	—	P = 0.0001, 0.65	P = 0.0001, 0.62	—
SPECIES RICHNESS	—	—	—	—	—	P = 0.0004, -0.31
	POC-mgC/l	**DOC-mgC/l**	**TOC-mgC/l**	**IDC-mgC/l**	**TDC-mgC/l**	**Chlb-µg/l**
NUMBERS	—	P = 0.0001, 0.40	P = 0.0001, 0.37	P = 0.0001, 0.50	P = 0.0001, 0.35	—
SPECIES RICHNESS	—	—	—	—	—	—

Note: Dependent variables include numbers m^{-2} and species richness. Independent variables include bottom water quality and sediment quality features. Statistical methods used are listed in Appendix II.

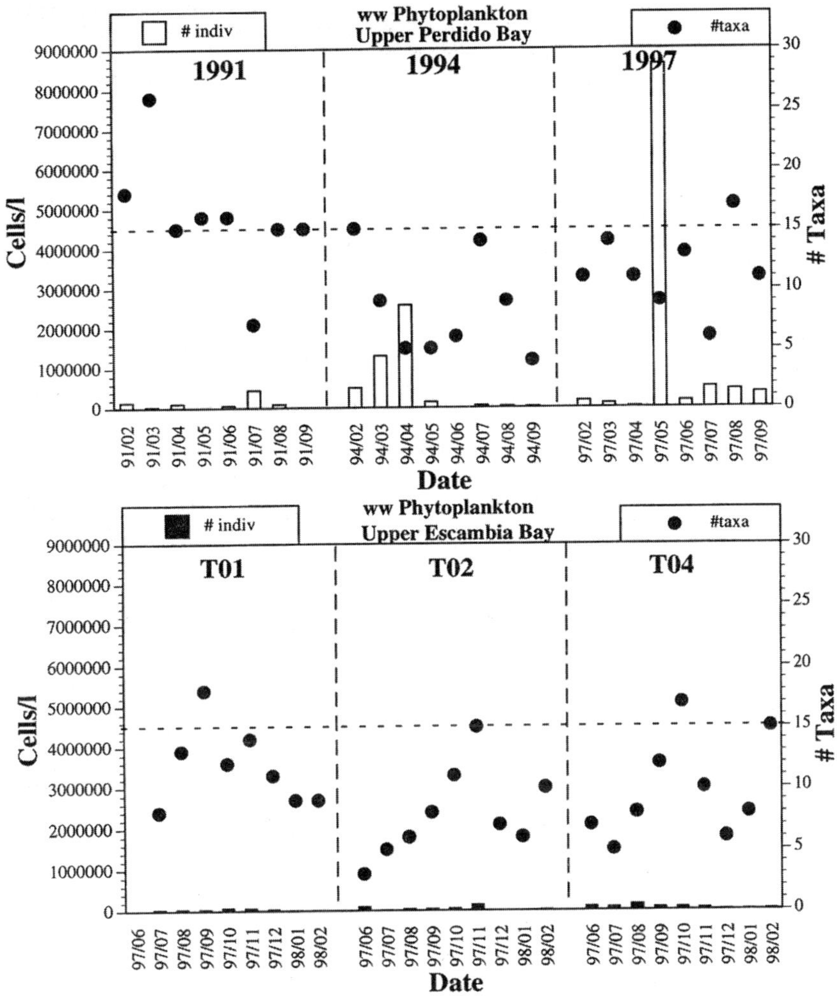

Figure 6.12 Comparison of whole-water phytoplankton numbers (cells l^{-1}) and species richness in upper Perdido Bay (station P23; 1991, 1994, 1997) with phytoplankton numbers (cells l^{-1}) and species richness in Escambia Bay (stations T01, T02, T04; 1997–1998).

1999). This included analysis of the phytoplankton communities in Escambia Bay. A comparison was made of phytoplankton indices of upper Escambia Bay with those of upper Perdido Bay (Figure 6.12). The Perdido data span the pre- and postbloom periods. Numbers of phytoplankton in the Perdido system during 1991 were comparable to those in Escambia Bay. Phytoplankton numbers were low in Escambia Bay compared with the numerical phytoplankton abundance in Perdido Bay during the bloom periods. Phytoplankton species richness in Escambia Bay was generally lower than the prebloom Perdido Bay index; the Escambia Bay species richness levels resembled those noted during bloom periods in Perdido Bay.

A comparison was made of the qualitative composition of the phytoplankton assemblages and the relative dominance of individual bloom species in the various study areas (Table 6.3). The number of dominant and subdominant phytoplankton species in Escambia Bay above the 0.1% level was relatively low (Table 6.3a). Generally low cumulative species richness in the upper bay and higher cumulative species richness in the lower bay characterized the Escambia Bay phytoplankton assemblages. Even though population dominance was low, the most numerous populations were bloom species (*C. choctawhatcheeana, P. minimum, H. akashiwo*). The phytoplankton community of

Table 6.3 List of Phytoplankton Numbers and Relative Dominance in Escambia Bay (May 1997 through March 1998), Apalachee Bay (Econfina and Fenholloway estuaries; April 1992 to March 1993), and Choctawhatchee Bay (September 1985 to August 1986), Perdido Bay (June 1990 to May 1991; June 1993 to May 1994; June 1996 to May 1997; June 1997 to May 1998). a. Escambia Bay.

Species	T1A	T3B	T4B	T5B	T6B	%
*Cyclotella choctawhatcheeana**	74759	96404	369131	573595	593740	13.48
*Prorocentrum minimum**	2997	125542	36963	20815	8159	1.54
*Heterosigma akashiwo**	6993	65601	2998	333	30137	0.84
Cylindrotheca closterium	25975	14986	10157	19314	9325	0.63
Dactyliosolen fragilissimus	1665	666	1166	14819	45455	0.5
Skeletonema costatum	0	1332	5328	14486	35965	0.45
Bacillaria paxillifer	667	666	11156	25808	9324	0.38
Cryptomonas sp.	38628	0	0	0	0	0.31
Amphidinium cf. vigrense	0	0	500	26307	7826	0.27
Undetermined pennate diatoms	3330	4830	6995	7993	9158	0.26
Nitzschia reversa	5994	0	2498	18150	999	0.22
*Gymnodinium cf. splendens**	6827	1832	1665	14320	1499	0.21
*Miraltia throndsenii**	3164	2498	5162	2831	10656	0.19
Chaetoceros spp.	7493	1665	1166	5162	8658	0.19
Undetermined centric diatoms	1998	2997	3664	5662	8159	0.18
Caloneis sp.	0	0	0	0	22311	0.18
Thalassionema nitzschioides	1998	1499	4996	6328	6994	0.17
Cerataulina pelagica	1499	500	1166	5495	10990	0.16
Nitzschia sigmoidea	0	0	666	17150	1665	0.15
Chaetoceros compressus	0	0	999	11988	5162	0.14
Navicula spp. (rhombic)	0	0	0	0	16317	0.13
Leptocylindrus danicus	167	0	1499	5662	7493	0.12
Chaetoceros laciniosus	0	0	0	999	11489	0.1
Total Numbers	**1,439,244**	**1,760,096**	**2,996,702**	**3,109,271**	**3,208,165**	**12,663,330**
Cumulative # Species	**40**	**41**	**67**	**101**	**90**	

Note: Data were averaged and shown for all species with relative abundance equal to or greater than 0.01% of the total in the Perdido system, and 0.10% for the other estuaries (where total numbers of species could not be fully represented). Species richness indices represent the cumulative number of phytoplankton species taken over each sampling period. Bloom species are marked by asterisks.

Escambia Bay appears to reflect a system that has previously suffered extreme hypereutrophication and has not fully recovered. Further, this system could be susceptible to future increases in nutrient loading, which is highly likely due to increasing levels of untreated storm water runoff, sewage input, and other forms of point-source nutrient loading.

Niedoroda (1999) compared the physical attributes of upper Perdido Bay and upper Escambia Bay. The data indicated that the physical dimensions of these bays are similar. Both systems are microtidal, although tidal action is greater in Escambia Bay as a result of differences in tidal ranges and tidal prisms. Perdido Bay is more sheltered and is poorly aligned for wind-mixing, whereas the broader Escambia system is more favorably aligned with prevailing wind action. Consequently, Escambia Bay is better mixed with less-time-stable haloclines than the Perdido estuary. Flow rates in the Escambia River (mean annual flow, 231 m^3 sec^{-1}) are greater than those in the Perdido River system (mean annual flow, 88 m^3 sec^{-1}). Based on increased flow rates into a comparable upper bay area, the flushing rates in the Escambia system are greater than those in the Perdido estuary with significantly shorter residence times (Niedoroda, 1999). During drought periods, residence time becomes an important factor in the creation of blooms through increased concentrations of nutrients and more stable conditions in terms of turnover rates and concentration of plankton.

The data from Escambia Bay support the hypothesis that bays that have been affected by excessive nutrient loading and resultant phytoplankton blooms remain infected by the various bloom

Table 6.3 (continued) b. Econfina system.

Species	E06	E07	E08	E09	E25	E10	E11	%
Thalassionema nitzschioides	38295	54446	353147	29138	10323	120214	65268	12.05
Cyclotella choctawhatcheana	48,413	76,931	122,567	48,776	28,720	114,455	44,598	8.7
Chaetoceros compressus	1665	4329	41625	16650	81585	304362	6327	8.2
Skeletonema sp. 1	7326	425574	1332	0	666	0	4995	7.9
Nitzschia section pseudonitzschia sp.	666	1665	10989	264735	11489	39960	1665	5.95
Cylindrotheca closterium	87246	67766	23477	7826	3497	8159	61272	4.66
Minutocellus scriptus	20979	30636	16650	1332	1332	58941	94239	4.02
Rhizosolenia fragilissima	333	8658	134199	8658	25974	30636	5661	3.84
Chaetoceros diversus	6327	15651	62937	5328	0	81585	21978	3.48
Cocconeis scutellum	21312	14320	35465	32135	3830	59275	19315	3.33
Bacteriastrum hyalinum	0	8658	59940	6993	833	69930	5328	2.72
Thalassiosira sp. 26	6993	7992	18648	666	2331	64269	35964	2.46
Asterionella japonica	999	10656	65435	1332	8991	41126	333	2.31
Nitzschia pungens	999	333	77256	5328	26973	999	333	2.02
Chaetoceros gracilis	333	0	0	0	333	89910	0	1.63
Chaetoceros sp. 3	0	0	88911	0	0	0	0	1.6
Amphidinium sp. 4	0	0	78255	0	0	0	0	1.41
Rhizosolenia stolterfothii	26640	3996	18981	3330	5329	6993	8159	1.32
Chaetoceros perpusillus	0	0	33633	0	0	38295	0	1.29
Rhizosolenia setigera	27972	5662	8658	3663	6660	10657	6660	1.26
Skeletonema costatum	30969	15651	4995	1998	1998	8658	2664	1.2
Amphiprora alata	17316	16484	333	333	0	333	30969	1.18
Nitzschia sp. 32	3996	0	25974	3996	666	20646	5661	1.09
Achnanthes manifera	7326	12322	7660	3996	1998	8492	13154	0.99
Rhizosolenia alata f. gracillima	0	0	0	45621	0	0	0	0.82
Leptocylindrus danicus	333	6827	5328	2997	10989	1998	12321	0.73
Gonyaulax sp.	31635	1665	999	333	3330	0	333	0.69
Johannesbaptistica sp.	0	0	1332	5328	30636	0	0	0.67
Chaetoceros brevis	0	0	8658	666	4329	23310	0	0.66
Melosira nummuloides	31635	666	0	0	0	0	0	0.58
Cerataulina pelagica	0	0	20646	666	0	9657	0	0.56
Synedra tabulata	1998	2832	6328	6327	500	11988	667	0.55
Minidiscus sp.	1332	9990	0	0	0	0	18648	0.54
Johannesbaptistica pellucida	999	3330	2331	8658	4995	6660	666	0.5
Rhizosolenia alata	0	0	9324	1665	7327	4662	1998	0.45
Chaetoceros coarctatus	333	666	4329	3663	2331	11822	333	0.42
Prorocentrum micans	999	1665	3830	2498	3996	5329	1998	0.36
Prorocentrum minimum	999	999	4329	2331	333	8325	1665	0.34
Paralia sulcata	5994	1499	3996	2664	833	999	1665	0.32
Cocconeis disculus	2664	1998	1998	999	5828	1665	2332	0.31
Cymatosira lorenziana	3663	3663	3664	1832	999	2997	333	0.31
Chaetoceros cf. fragilis	0	16650	0	0	0	0	0	0.3
Chaetoceros affinis	0	999	6327	333	0	8658	0	0.29
Chaetoceros subsecundus	0	7992	5328	0	0	0	1998	0.28
Hemiaulus hauckii	0	333	333	10656	0	666	1332	0.24
Grammatophora marina	3996	2664	1499	2332	833	1499	333	0.24
Dictyocha speculum	0	0	3996	1665	999	6327	0	0.23
Bacillaria paxillifer	6993	1499	0	0	0	2664	0	0.2
Bellorochea malleus	666	5994	0	0	0	3663	0	0.19
Chaetoceros constrictus	0	0	9324	0	0	999	0	0.19
Rhizosolenia imbricata	0	0	333	666	6328	1665	0	0.16

Table 6.3b (continued)

Species	E06	E07	E08	E09	E25	E10	E11	%
Chaetoceros dichaeta	0	0	0	0	8991	0	0	0.16
Ceratium hircus	0	999	2997	1665	500	2331	333	0.16
Miraltia throndsenii	1665	1665	999	0	0	999	3330	0.16
Chaetoceros lorenzianus	0	0	2997	0	0	4995	0	0.14
Chaetoceros subtilis	0	333	1998	0	0	2664	2997	0.14
Chaetoceros didymus	0	0	999	0	666	4662	333	0.12
Actinoptychus undulatus	999	2997	167	0	333	1665	333	0.12
Cyclotella sp.	1998	3330	0	0	0	0	0	0.1
Total Numbers	**501626**	**888797**	**1433428**	**567096**	**331260**	**1340905**	**505808**	**5568920**
Cumulative # Species	**76**	**73**	**76**	**71**	**64**	**82**	**63**	

Note: Data were averaged and shown for all species with relative abundance equal to or greater than 0.01% of the total in the Perdido system, and 0.10% for the other estuaries (where total numbers of species could not be fully represented). Species richness indices represent the cumulative number of phytoplankton species taken over each sampling period. Bloom species are marked by asterisks.

species, especially when exposed to continued nutrient loading from urban runoff. Future increases of anthropogenous nutrient loading could be accompanied by sudden and continuing changes of the phytoplankton community of the upper bay that would include increased dominance by bloom species and continued reduction of phytoplankton species richness. The deteriorated state of the sediments and secondary production in upper Escambia Bay (Livingston, 1999) could be associated with the observed changes of the phytoplankton communities in these areas. In any case, Escambia Bay fits the profile of a system that has been seriously affected by nutrient loading and the effects of phytoplankton blooms. The phytoplankton community structure of upper Escambia Bay, together with sediment characteristics and a reduced level of secondary production, fit the model that characterized Wolf Bay and postbloom Perdido Bay.

3. Apalachee Bay

Phytoplankton assemblages in the unpolluted Econfina system (Table 6.3b) were characterized by low dominance, the general absence of bloom species, and relatively high cumulative species richness. The number of species above 0.01% of the total numbers was so high that the full complement of phytoplankton species could not be represented in Table 6.3b. Low numbers of cells per liter were also typical of what is an oligotrophic offshore system that is low in nutrient concentrations. There are no alluvial rivers that drain into Apalachee Bay. In the Fenholloway system, the two inshore stations were characterized by low cumulative species richness relative to other parts of the offshore area (Table 6.3c). In both the Econfina and Fenholloway systems, *C. choctawhatcheeana* was a major dominant. However, this species did not appear in bloom concentrations and there were no noxious bloom species among the dominants in either estuary.

Apalachee Bay presents a very different set of conditions than the alluvial systems to the west. There is no barrier-enclosed bay or alluvial river. Salinities in the nearshore waters remain relatively high, whereas most of the estuaries to the west are enclosed within barrier systems with correspondingly low salinities. Nutrient loading in the form of ammonia and orthophosphate in the polluted Fenholloway River is comparable to that in Elevenmile Creek. Both of these small rivers receive relatively high volumes of nutrient-enriched pulp mill effluents that exceed their assimilative capacities. The loading rates of ammonia and orthophosphate and the concentration gradients at the entry points to the respective estuaries are comparable. However, color levels in the Fenholloway estuary are much higher than those in the Perdido system. Consequently, light becomes limiting to plant life in the nearshore waters of the Fenholloway estuary (Livingston, 1993a; Livingston

Table 6.3 (continued) c. Fenholloway system.

Species	F06	F09	F10	F16	F25	F11	F14	%
Skeletonema costatum	6660	19980	1051947	46573	1998	475524	958041	43.45
*Cyclotella choctawhatcheana**	24,517	38,784	147,041	54,720	24,285	129,211	122,565	9.18
Nitzschia section pseudonitzschia sp.	0	4329	34299	18066	42458	384615	1332	8.23
Cylindrotheca closterium	17316	151515	52614	19481	5329	43124	63603	5.99
Rhizosolenia stolterfothii	0	0	24143	162671	33633	6494	333	3.86
Minutocellus scriptus	47619	55944	13653	4664	1332	14652	88578	3.84
Thalassionema nitzschioides	1998	7659	31968	23810	29804	18648	16817	2.22
Bacteriastrum hyalinum	371	0	76923	4774	1998	45621	666	2.21
Johannesbaptistica sp.	0	0	92241	999	0	35964	333	2.2
Rhizosolenia setigera	6993	10989	14985	2997	3830	64103	7326	1.89
Chaetoceros lorenzianus	0	0	0	0	2664	91242	0	1.59
Gymnodinium sp.*	0	0	0	85914	0	0	0	1.46
Cocconeis scutellum	12063	9657	15485	10989	1166	12821	10989	1.24
Chaetoceros compressus	0	0	0	0	36963	31635	0	1.16
Asterionella japonica	999	4995	10323	18871	1998	15984	1332	0.92
Rhizosolenia alata f. gracillima	0	0	17649	35964	0	0	0	0.91
Bacillaria paxillifer	999	4662	13320	10656	999	5661	14319	0.86
Skeletonema sp. 1	4995	15984	13986	666	0	0	4329	0.68
Nitzschia sp. 32	3996	6327	3996	2083	4995	3663	13653	0.66
Thalassiosira sp. 26	0	8658	11988	0	0	999	16650	0.65
Navicula sp. 46	0	0	0	0	0	30969	6660	0.64
Rhizosolenia fragilissima	333	9657	11988	2110	6327	999	1998	0.57
Bellorochea malleus	333	7992	2331	7659	0	6660	7659	0.55
*Leptocylindrus danicus**	0	5994	5328	7326	2165	4496	6327	0.54
Merismopedia thermalis	20901	4662	999	0	0	0	0	0.45
Cymatosira lorenziana	666	999	1665	15569	0	6660	999	0.45
Chaetoceros diversus	0	0	4329	7992	1665	9324	2997	0.45
Chaetoceros didymus	0	0	0	4662	18648	666	1665	0.44
Paralia sulcata	0	666	3663	14652	1166	2997	2331	0.43
Cocconeis disculus	1332	2664	1665	4995	7660	2664	2997	0.41
Prorocentrum micans	0	0	11655	3860	2165	1998	1665	0.36
Amphiprora-like sp.	4329	3663	1998	666	0	8159	2331	0.36
Melosira nummuloides	19314	0	0	0	0	0	0	0.33
*Prorocentrum minimum**	666	999	2331	999	0	666	13653	0.33
Cerataulina pelagica	0	666	13653	0	0	2331	333	0.29
Johannesbaptistica pellucida	3663	1665	1332	3663	3331	1332	1832	0.29
Synedra tabulata	666	666	2165	1998	167	4829	4995	0.26
Merismopedia punctata	12987	999	0	0	0	167	333	0.25
Chaetoceros affinis	0	0	0	6993	666	6660	0	0.24
Achnanthes manifera	666	1332	1998	2165	2166	2664	2331	0.23
Chaetoceros cf. fragilis	0	0	0	0	0	11988	0	0.2
Rhizosolenia alata	0	4995	0	1611	4329	333	333	0.2
Gyrosigma fasciola	0	5328	999	167	0	666	2664	0.17
Chaetoceros coarctatus	0	0	0	7992	0	1332	0	0.16
Chaetoceros subtilis	0	4995	333	0	0	2664	0	0.14
Bellarochea-like diatom	7992	0	0	0	0	0	0	0.14
Dictyocha speculum	0	0	333	1998	4329	666	0	0.12
Streptotheca thamensis	666	666	1665	1332	0	1665	1332	0.12
Rhizosolenia delicatula	0	0	0	666	5661	0	666	0.12
Chaetoceros vixvisibilis	0	0	0	0	0	6993	0	0.12
Epiphytic *Navicula* sp.	0	0	6993	0	0	0	0	0.12
Grammatophora marina	333	0	833	1665	333	2331	1332	0.12

Table 6.3c (continued)

Species	F06	F09	F10	F16	F25	F11	F14	%
Biddulphia mobiliensis	0	1665	1332	1998	0	1499	0	0.11
Amphiprora alata	999	999	0	0	0	0	3996	0.1
Corethron cryophyllum	666	0	333	333	1665	2664	0	0.1
Total Numbers	**255317**	**412920**	**1607727**	**597684**	**264912**	**1414755**	**1340494**	**5893809**
Cumulative # Species	**54**	**51**	**71**	**75**	**60**	**72**	**62**	

Note: Data were averaged and shown for all species with relative abundance equal to or greater than 0.01% of the total in the Perdido system, and 0.10% for the other estuaries (where total numbers of species could not be fully represented). Species richness indices represent the cumulative number of phytoplankton species taken over each sampling period. Bloom species are marked by asterisks.

et al., 1998b). In addition, nitrogen (not phosphorus) is limiting to phytoplankton in the Econfina estuary (Livingston, unpublished data). Phytoplankton in inshore Econfina estuarine areas were more abundant than those in cognate areas of the Fenholloway system. Estuarine and inshore parts of the Fenholloway system were low in phytoplankton species richness compared with Econfina cognate stations (Table 6.3c). However, parts of the Fenholloway offshore areas had higher numbers of phytoplankton than comparable Econfina areas. This was a consequence of the effects of increased nutrient loading from the Fenholloway system, higher nutrient concentrations in nearshore areas, and increasing light penetration in offshore areas due to dilution of river runoff by clear, oligotrophic Gulf waters (Livingston et al., 1998b).

The question remains: Why are there no phytoplankton blooms off the Fenholloway estuary where nutrient loading from the Fenholloway River is comparable to that in Elevenmile Creek as it enters Perdido Bay? The reason for this could be found in the distribution of water color and resulting inhibition of light penetration relative to the high nutrient gradients in the Fenholloway estuary. The high nutrient gradients in Gulf receiving areas off the Fenholloway River–estuary occur within areas of low light transmission (Stations F09 and F10; Figure 6.13). The data (Livingston et al., 1998b) show that by the time such nutrients reach offshore areas of increased light transmission (Figure 6.14), the ammonia and orthophosphate concentrations are too dilute to stimulate phytoplankton blooms. The increase in offshore, nonbloom, phytoplankton species supports this hypothesis. This is contrary to the situation in Perdido Bay where light penetration in the upper bay is not substantially influenced by mill effluents. Nutrient gradients occur in areas with enough light penetration to allow bloom formation. The presence of relatively high quantities of compounds such as the tannins and/or ammonia in the Fenholloway estuary (Livingston et al., 1998b) could also have an inhibitory effect on the phytoplankton populations in inshore areas of the Fenholloway drainage (Livingston, unpublished data). However, it is likely that phytoplankton suppression in the Fenholloway estuary was due primarily to the reduced light penetration.

4. *Choctawhatchee Bay*

Net phytoplankton taken over a 12-month period in the Choctawhatchee Bay system are shown in Table 6.3d. With the exception of the polluted marina in Old Pass Lagoon, there were no signs of hypereutrophication in the Choctawhatchee system as a whole. Overall, phytoplankton production was relatively low in Choctawhatchee Bay in the mid-1980s. The high cumulative species richness indices were due, in part, to the method of collection since net samples are selective for relatively large plankton that are easily identified. However, there was no sign of dominance by bloom species in most areas of the bay. It is possible that sequestering of nutrients in the main stem (i.e., deeper, salinity-stratified areas) could have been involved in the relatively low level of phytoplankton production in large parts of the Choctawhatchee system. However, as noted above, some of the polluted bayous and the hypereutrophicated Old Pass Lagoon were indicative of

Table 6.3 (continued) d. Choctawhatchee Bay.

Species	C03	C07	C11	C15	C19	C22	C25	C31	C34	C36E	C38	%
Skeletonema costatum	12502	19490	0	29679	0	0	0	28	0	2075530	34	13.42
Chaetoceros socialis	9323	39842	26515	31590	27555	20830	20526	71813	241960	103300	129807	8.58
Coscinodiscus centralis	113216	110226	13955	138333	24531	18335	18721	33969	45505	43648	55772	5.98
Chaetoceros decipiens	25964	59297	6150	46782	38728	15423	11446	35564	92848	115059	114666	5.25
Chaetoceros didymus	39252	53999	10239	55799	28992	27354	28495	36021	30131	62880	27787	4.16
*Cyclotella choctawhatcheana**	13626	15654	5912	31684	14428	7915	19099	19448	11021	73040	23788	3.33
*Leptocylindrus danicus**	1444	14619	0	12899	1460	6860	209	4802	179	83250	1598	3.17
Pyrocystis sp. 1	8374	15048	2665	5459	5863	16404	14563	23313	26162	11050	108237	2.99
Chaetoceros brevis	35633	37682	3045	76802	7514	7489	5609	24070	48838	69908	14106	2.9
Striatella unipunctata	1271	1576	462	371	0	2124	2188	16610	47436	19167	81510	2.51
Ceratium tripos	496	4507	1121	20034	10197	15127	8397	22641	21336	15523	50504	2.46
Rhizosolenia stolterfothii	2124	10455	236	30193	511	21605	5230	5021	36517	45030	42614	2.22
Falcula hyalina	3219	11890	505	98355	3699	7141	5970	13304	4342	128610	3415	2.14
Rhabdonema adriaticum	4496	3341	1098	3002	3234	5824	5056	12805	5268	27494	46520	2.08
Thalassionema nitzschioides	12617	10626	3550	3019	1141	3717	516	5294	514	162823	39568	2.05
Coscinodiscus granii	24907	22804	11036	16287	10222	9344	6198	16717	4364	11643	33392	1.98
Cocconeis scutellum	6084	4988	57	2463	966	1850	3887	21203	52989	107565	5787	1.71
Ceratium hircus	7629	11639	2469	23264	2725	6050	3517	10740	59104	21648	21946	1.38
Chaetoceros coarctatus	73467	2840	1782	3962	17916	4510	19053	8890	21888	12041	3446	1.36
Ceratium fusus	245	4525	350	31088	4032	5117	4970	5012	16372	6591	31697	1.34
Chaetoceros lorenzianus	1080	6170	398	0	7457	5504	1840	9561	2160	30954	36476	1.1
Chaetoceros laciniosus	2911	2163	942	5038	11731	5101	5463	16320	30460	28818	4491	1.09
Dinophysis caudata v. pedunculata	0	245	0	5550	0	370	0	3910	1260	3303	56923	1.03
Cyclotella striata	19372	27313	404	57129	2495	4037	2288	1332	2445	21125	2223	1
Ceratium trichoceros	3257	1019	305	3527	3068	6511	5831	12852	7570	4387	13667	1
Bacteriastrum hyalinum	7095	3906	369	2860	11752	3251	5500	9303	22856	18796	3183	0.87
Paralia sulcata	4650	3644	128	118	2049	681	110	8335	774	40705	10967	0.84
Chaetoceros didymus v. protuberar s	12065	3330	5662	3780	7510	0	7428	9460	4403	27339	4400	0.81
Rhizosolenia delicatula	0	0	0	0	0	45	0	0	0	119176	2822	0.76
Chaetoceros affinis	0	440	0	368	820	699	0	1586	129	60980	5303	0.67
Thalassiothrix frauenfeldii	690	860	0	419	1520	1754	122	6986	805	23310	7250	0.66
Cyclotella meneghiniana	5241	2237	2340	2708	1455	530	5127	3099	1548	44753	5547	0.66
Ceratium furca	130	118	231	17	2615	1247	1630	4320	5905	180	23462	0.64
Rhizosolenia imbricata	0	0	0	61	2530	728	310	382	2295	18307	22532	0.62

Table 6.3d (continued)

Species	C03	C07	C11	C15	C19	C22	C25	C31	C34	C36E	C38	%
Chaetoceros lauderi	1228	696	485	489	2240	1278	680	2935	13800	14890	30448	0.6
Chaetoceros densus	0	0	1927	1075	5458	0	514	4630	1857	1548	40650	0.59
Chaetoceros eibonii	3080	2098	0	33000	0	2123	689	0	0	24100	770	0.44
Rhizosolenia alata	112	0	0	2171	486	48	0	0	37660	3244	9479	0.42
Dinophysis caudata	0	0	0	0	1656	600	2051	4090	32933	13864	1174	0.4
Dictyocha fibula	3972	1417	1627	233	2270	3495	2221	1396	877	9450	883	0.38
Ceratium tripos v. atlanticum	0	0	0	345	0	0	1210	1385	1033	205	36913	0.37
Thalassiosira eccentrica	15839	22852	0	14223	845	0	0	0	0	0	260	0.35
Rhizosolenia setigera	1580	1530	0	545	0	0	510	0	0	9206	24610	0.34
Grammatophora marina	977	667	186	540	680	807	884	36	14415	6936	2218	0.34
Dinophysis caudata f. acutiformis	0	38	0	963	0	1671	0	3322	0	3878	21295	0.33
Prorocentrum micans	0	0	0	1516	1260	400	2383	0	384	12363	24100	0.32
Nitzschia acicularis v. closterioides	1819	3000	0	44401	0	0	0	0	0	0	0	0.31
Bacteriastrum varians	9475	4650	0	5537	0	67	0	0	0	29123	9	0.31
Nitzschia longissima	592	161	170	754	430	0	0	0	158	3082	18826	0.31
Chaetoceros gracilis	585	276	0	42	0	0	0	0	1612	4827	14808	0.29
Hemidiscus cuneiformis	0	0	0	0	0	0	0	0	0	0	29400	0.28
Actinocyclus ehrenbergii	1346	2263	791	929	2153	1232	746	799	2551	1104	2507	0.28
Chaetoceros messanensis	17796	7102	0	5244	3510	1122	0	0	820	3150	65	0.28
Synedra ulna	21308	14666	156	3202	0	0	0	0	0	0	0	0.27
Stephanopyxis turris	0	0	0	0	0	0	0	232	690	108	22283	0.27
Cymatosira lorenziana	214	0	0	0	0	0	0	81	4446	35111	322	0.26
Amphiprora gigantea	2213	2618	75	1910	0	754	0	44	0	18452	7889	0.26
Thalassiosira oestrupii	3100	5351	84	16751	1732	892	1240	282	868	648	673	0.26
Bacillaria paxillifer	6334	5134	164	27646	406	0	0	179	0	0	0	0.25
Chaetoceros pelagicus	0	0	0	258	2530	525	327	580	104	30467	0	0.24
Guinardia flaccida	0	395	0	133	1155	515	1030	0	0	32419	670	0.24

Opephora martyi	675	254	0	1264	166	343	2001	2500	814	3139	3176	0.24
Nitzschia sigma	1352	978	32	0	3727	1175	911	7512	59	639	45	0.24
Opephora schwartzii	71	186	799	673	2330	1861	1957	8253	532	3655	339	0.22
Synedra tabulata	707	7555	1328	1184	369	227	634	99	0	338	2618	0.21
Navicula cf. ammophila	57	142	0	29800	0	0	0	0	0	0	0	0.19
Bacillaria paradoxa	1583	2842	1007	378	1120	133	126	582	0	6098	443	0.18
Delphineis surirella	278	145	154	1600	492	169	0	0	282	20661	2526	0.17
Cyclotella sp.	1180	0	3820	525	0	565	1230	0	0	1690	2090	0.17
Prorocentrum gracile	0	0	0	0	0	2665	0	0	0	2424	1885	0.17
Coscinodiscus oculus-iridis	0	156	0	51	89	8	0	0	80	27	12100	0.16
Synedra acus	21118	2653	0	92	0	0	0	0	0	0	0	0.15
Eunotia pectinalis	3570	3027	366	280	0	0	0	29	0	0	0	0.15
Nitzschia lorenziana	1040	2232	1210	1010	0	970	0	0	13000	0	0	0.15
Prorocentrum compressum	0	0	0	0	0	840	0	0	239	21900	17	0.15
Chaetoceros debile	0	0	0	0	539	0	0	0	21600	0	0	0.15
Rhizosolenia fragilissima	508	1800	0	8920	0	0	0	18	0	11400	0	0.14
Rhizosolenia hebetata	0	0	0	0	0	1770	42	0	1247	0	5005	0.14
Thalassiothrix mediterranea	0	0	0	0	8	0	0	0	90	0	15100	0.13
Fragilaria sp.	0	0	298	0	0	0	0	0	18	20000	0	0.13
Striatella interrupta	0	0	0	0	0	0	0	795	0	0	4171	0.12
Nitzschia longa	1091	0	0	0	1120	0	0	191	0	13290	407	0.11
Synedra undulata	0	0	0	0	0	0	0	119	0	0	1190	0.11
Chaetoceros pseudocurvisetus	0	0	0	124	765	33	0	0	120	14967	0	0.1
Total Numbers	**638552**	**657403**	**128962**	**1007153**	**312786**	**279285**	**256496**	**552920**	**1028510**	**4113583**	**1438008**	**15989411**
Cumulative # Species	**216**	**182**	**87**	**157**	**115**	**132**	**93**	**146**	**134**	**191**	**186**	

Note: Data were averaged and shown for all species with relative abundance equal to or greater than 0.01% of the total in the Perdido system, and 0.10% for the other estuaries (where total numbers of species could not be fully represented). Species richness indices represent the cumulative number of phytoplankton species taken over each sampling period Bloom species are marked by asterisks.

Table 6.3 (continued) e. Perdido Bay

Species (6/90-5/91)	23	31	40	%
*Cyclotella choctawhatcheeana**	79658	238840	188581	2.57
Leucocryptos marina	77590	68020	40000	0.94
Cryptomonas acuta	65400	10400	11400	0.44
*Prorocentrum minimum**	2542	53197	27849	0.42
Plagioselmis sp. A (*sensu Thron*)	26400	34000	0	0.31
Amphidinium amphidinioides	12400	31400	2600	0.24
Cryptomonas acerosa	26515	11000	8600	0.23
Plagioselmis sp. B (*sensu Thron*)	29400	7800	0	0.19
Pedinomonas mikron	20900	2509	10800	0.17
Pedinomonas sp.	0	34000	0	0.17
*Gymnodinium splendens**	23046	7365	3288	0.17
Cyclotella sp.	4811	11427	10137	0.13
Anabaena macrosporum	5462	17094	0	0.11
Synedra fasciolata	11533	7701	1059	0.1
Bacillaria paxillifer	15500	907	300	0.08
Skeletonema costatum	0	4200	9600	0.07
Synedra fasciolata	2083	3084	6013	0.06
Nitzschia sigmoidea	10600	0	200	0.05
Dictyocha fibula	1000	4980	4382	0.05
Pseudoscourfieldia sp.	989	800	6979	0.04
*Gymnodinium cf. splendens**	4200	396	2800	0.04
Dinobryon sertularia	4906	666	0	0.03
Synedra fasciolata	0	0	5400	0.03
Calycomonas ovalis	1982	0	3000	0.03
Pachysphaera marshalliae	0	0	4400	0.02
*Leptocylindrus danicus**	0	792	2600	0.02
Ankistrodesmus falcatus	0	3000	0	0.02
Rhizosolenia fragilissima	1400	800	800	0.02
Rhizosolenia delicatula	0	0	2946	0.01
Mallomonas cf. schwemmeki	2039	888	0	0.01
Scenedesmus quadricauda	1451	0	1200	0.01
Entomoneis alata	1800	200	400	0.01
Pyramimonas cf. adriatica	0	0	2400	0.01
Calycomonas gracilis	0	800	1400	0.01
Chaetoceros subtilis	0	0	2200	0.01
Thalassionema nitzschioides	400	435	1344	0.01
Spermatozopsis sp.	933	1110	0	0.01
Amphidinium musicolum	0	200	1800	0.01
Pyramimonas cf. propulsar	0	0	2000	0.01
Frustulia rhomboides	1906	0	0	0.01
Prorocentrum micans	400	400	1097	0.01
Paralia sulcata	800	200	755	0.01
Cymatosira lorenziana	566	0	1189	0.01
Eunotia pectinalis	1744	0	0	0.01
Cyclotella sp. 7 (9μ)	1000	618	0	0.01
Asterionellopsis japonica	0	400	1200	0.01
Cyclotella striata	600	600	300	0.01
Dinobryon divergens	133	1332	0	0.01
Diploneis smithii	1400	0	0	0.01
Nitzschia sp. 6	1400	0	0	0.01
Petrodictyon gemma	1400	0	0	0.01
Nitzschia panduriformis	200	588	557	0.01
Nitzschia longissima	0	1022	200	0.01
Cerataulina pelagica	400	0	800	0.01
Falcula-like sp.	600	600	0	0.01

Table 6.3e (continued)

Species (6/90-5/91)	23	31	40	%
Synedra fasciolata	0	0	1200	0.01
Nitzschia pungens	0	392	738	0.01
Dinophysis caudata	0	0	1100	0.01
Pyramimonas cf. grossii	0	0	1000	0.01
Pyramimonas cf. nephroides	0	0	1000	0.01
Total Numbers	**6938960**	**8288827**	**4480658**	**19708445**
Cumulative # Species	**90**	**58**	**83**	

Note: Data were averaged and shown for all species with relative abundance equal to or greater than 0.01% of the total in the Perdido system, and 0.10% for the other estuaries (where total numbers of species could not be fully represented). Species richness indices represent the cumulative number of phytoplankton species taken over each sampling period. Bloom species are marked by asterisks.

potential problems due to anthropogenous nutrient loading. It is possible that increased nutrient loading has led to phytoplankton blooms, expanded losses of habitat, and fish and marine mammal kills.

5. Perdido Bay

Phytoplankton species associations in Perdido Bay during the period June 1990 to May 1991 were characterized by low relative dominance, low numbers of cells per liter, and relatively high cumulative numbers of species (Table 6.3e). There was a relatively high number of species above the 0.01% level. The presence of two bloom species (*C. choctawhatcheeana, P. minimum*) in the numerically dominant species could be associated with temporary periods of high orthophosphate loading although there were no signs of blooms during this period. By 1993–1994, the phytoplankton composition had changed, with high dominance by five bloom species, relatively high numbers of cells per liter, low cumulative species richness, and relatively few species above the 0.01% level (Table 6.3e). Bloom species included *C. choctawhatcheeana, M. throndsenii, L. danicus, P. minimum*, and *G. cf. splendens*. By 1996–1997, cumulative phytoplankton species richness was even lower, with high relative dominance of a biota characterized by a different set of bloom species that included *M. throndsenii, H. akashiwo, Synedropsis* sp., *C. choctawhatcheeana, P. minimum*, and *G. splendens* (Table 6.3e). Once again, there were relatively few species above the 0.01% level. During 1997–1998, increased cumulative species richness, and a reversion to dominance characterized the Perdido Bay phytoplankton biota by *C. choctawhatcheeana* (Table 6.3e). Species numbers above the 0.01% level had increased. This partial recovery could have been associated with reduced orthophosphate loading by the mill during this period. However, numbers of phytoplankton were still high and other bloom species (*H. akashiwo, Synedropsis* sp.) were still dominant, probably as a result of high ammonia loading during summer 1998. The numbers of species included above the 0.01% level had increased during the 1997–1998 period relative to the previous bloom years. Thus, these data indicated a partial recovery during the period of a return to low orthophosphate loading.

C. Phytoplankton Indices

The comparison of the phytoplankton communities in the different bays indicates that indices such as species richness and relative abundance of bloom species are reasonable indicators of the effects of nutrient loading in a given estuary. A general comparison of the 12-month cumulative

Table 6.3e (continued)

Species (6/93-5/94)	23	26	29	33	%
*Cyclotella choctawhatcheeana**	2557667	8912999	8862501	7931666	57.59
*Miraltia throndsenii**	1380335	4858001	5762666	442667	25.35
*Leptocylindrus danicus**	236001	2359334	1857001	837668	10.78
*Prorocentrum minimum**	457670	153003	148836	79668	1.71
Gymnodinium sp.*	174669	152302	188470	91001	1.24
Cerataulina pelagica	38668	215334	56334	67334	0.77
Cryptomonas sp.	250333	0	0	0	0.51
Chaetoceros simplex v. calcitrans	3000	20667	113001	38001	0.36
Pseudonitzschia pseudodelicatissi	4335	78667	34667	42667	0.33
*Gymnodinium cf. splendens**	95669	20668	6836	334	0.25
Chaetoceros subtilis	4667	26668	37334	27667	0.2
Rhizosolenia fragilissima	0	2667	18334	57667	0.16
Chaetoceros affinis	1000	25334	15667	18001	0.12
Skeletonema costatum	0	26668	10667	20334	0.12
Chaetoceros simplex v. calcitrans	334	27334	8334	11667	0.1
Cylindrotheca closterium	8734	7303	8901	4334	0.06
Bacillaria paxillifer	20002	3667	4669	0	0.06
Chaetoceros fragilis	0	13668	8667	667	0.05
Dictyocha fibula	1334	1334	667	334	0.01
Thalassionema nitzschioides	1000	7335	4668	4668	0.04
Bacteriastrum hyalinum	334	4334	6000	3334	0.03
Asterionellopsis japonica	1000	6334	668	5001	0.03
Eucampia cornuta	2334	4001	3668	1667	0.02
Prorocentrum micans	0	667	1333	6668	0.02
Pachysphaera pelagica	2334	1667	3002	1001	0.02
Scenedesmus quadricauda	134	134	267	0	0
Gyrosigma sp. (s shaped)	667	2667	2334	0	0.01
Pachysphaera marshalliae	0	2000	667	3000	0.01
Dictyocha spiculum	2667	334	2000	0	0.01
Total Numbers	**5251961**	**16947831**	**17177699**	**9702356**	**49079836**
Cumulative # Species	**41**	**52**	**47**	**39**	

Note: Data were averaged and shown for all species with relative abundance equal to or greater than 0.01% of the total in the Perdido system, and 0.10% for the other estuaries (where total numbers of species could not be fully represented). Species richness indices represent the cumulative number of phytoplankton species taken over each sampling period. Bloom species are marked by asterisks.

phytoplankton species richness indices is shown in Figure 6.15. Inshore Fenholloway stations have relatively low phytoplankton species richness compared with Econfina cognates. Upper Escambia Bay (stations T1a, T3b), characterized by a history of hypereutrophication, periodic high phytoplankton production in the upper bay, and existing low sediment quality (Livingston, 1999) represents an impaired phytoplankton biota that has undergone partial recovery but is still characterized by relatively high bloom species representation. In the Perdido system (Figure 6.15), the relatively high cumulative species richness in the early years deteriorated into reduced numbers of phytoplankton species during bloom periods. Following a period of low orthophosphate loading from the mill, there was, according to this index, a partial recovery that was qualified by the response of the phytoplankton to summer 1998 ammonia loading. An annual cumulative species richness of 60 appears to be a reasonable guideline for a healthy phytoplankton assemblage in these Gulf estuaries. In each case, the phytoplankton community structure, represented by species richness, was indicative of the water quality and nutrient loading characteristics of the receiving estuaries. The phytoplankton species richness of the different coastal systems was also closely associated

Table 6.3e (continued)

Species (6/96-5/97)	18	21	22	23	23A	25	26	29	31	37	40	SC1	SC2	%
*Miraltia throndsenii**	5328	0	2664	23643	5332	12740268	10829704	21922662	11195237	67816329	18598467	0	0	59.04
*Heterosigma akashiwo**	65103	1000	2498	31930222	4692160	11868371	2167349	255972	36984	196402	0	92741	276224	21.28
Synedropsis sp. *Novum**	16817	167	12999	11793136	5332	29989	20662	1058660	1332	23996187	77648	2484847	49451	16.31
*Cyclotella choctawhatcheeana**	13987	1398	10323	94614	171957	398834	331059	290784	205476	342818	215175	248418	465368	1.15
*Prorocentrum minimum**	0	0	666	48618	9331	48953	666003	133866	416584	22813	8658	333	333	0.56
Cryptomonas sp.	0	0	0	848808	0	0	0	0	0	0	0	0	0	0.35
Skeletonema costatum	11655	0	1998	28638	0	25308	96237	30636	87914	97910	141858	0	0	0.22
Chaetoceros subtilis	0	0	0	1665	0	11655	71928	34965	109897	96408	83583	0	0	0.17
*Gymnodinium simplex**	3164	833	2833	22489	19995	17664	17318	8326	26307	92123	9657	118881	50949	0.16
Spermatozopsis exsultans	1333	21313	124432	247216	0	0	0	0	0	1667	0	0	0	0.16
Chaetoceros simplex	0	0	999	2331	0	8325	36297	23310	60273	91242	49617	0	0	0.11
Gymnodinium alexandrium-like*	0	0	0	0	0	0	0	0	0	0	0	0	0	0
Chlamydomonas sp. *w/2 pyrenoids i*	2165	57276	35497	14665	0	0	0	0	0	0	0	0	0	0.05
Nitzschia reversa	0	0	333	27640	0	23310	21313	6663	8993	5828	2997	0	333	0.04
*Gymnodinium cf. splendens**	0	0	0	3663	3999	7996	27972	9324	14319	3831	8330	0	0	0.03
Chaetoceros affinis	0	0	0	0	0	666	0	0	11988	27473	24642	0	0	0.03
Merismopedia tenuissima	0	2997	55611	0	0	0	0	0	0	0	0	0	0	0.02
Cylindrotheca closterium	1166	0	333	333	0	2997	999	1665	4330	10493	27324	833	3164	0.02
Chaetoceros sp. in mucilage	0	0	0	0	0	0	0	0	0	0	44289	0	0	0.02
Bacillaria paxillifer	0	3491	666	1998	0	21645	333	1998	1998	1499	666	333	0	0.02
Merismopedia punctata	0	0	30969	7992	0	0	0	0	0	0	0	0	0	0.02
Anabaena sch. v. recta f. rotundi	16983	0	0	0	0	0	0	0	0	0	0	0	0	0.01
Dinobryon divergens	13986	0	0	4666	5332	0	999	0	0	0	0	0	0	0.01

Table 6.3e (continued)

Species (6/96-5/97)	18	21	22	23	23A	25	26	29	31	37	40	SC1	SC2	%
Synedra fasciolata	9825	1332	167	2833	0	3333	1999	1999	333	0	0	0	0	0.01
Chaetoceros compressus	0	0	0	0	0	0	0	0	18648	0	11655	0	0	0.01
Dictyosphaerium pulchellum	5994	10989	7992	0	0	0	0	0	0	0	0	0	0	0.01
Cerataulina pelagica	0	0	0	333	0	0	1332	666	3996	9657	9990	0	0	0.01
Asterionellopsis glacialis	0	0	0	0	0	0	333	0	4329	7659	6327	0	0	0.01
Chaetoceros diversus	0	0	666	0	0	0	0	0	9657	1332	4329	0	0	0.01
Dinobryon sertularia	9657	0	2664	0	0	0	0	0	0	0	0	0	0	0.01
Pterosperma sp.	0	0	0	13328	0	0	0	0	0	0	0	0	0	0.01
Total Numbers	**204971**	**181324**	**350788**	**45219293**	**4914771**	**25222968**	**14301829**	**23785492**	**12243572**	**92848317**	**19361176**	**2948052**	**867801**	**242450354**
Cumulative # Species	**40**	**33**	**40**	**38**	**10**	**36**	**27**	**25**	**41**	**49**	**42**	**14**	**24**	

Note: Data were averaged and shown for all species with relative abundance equal to or greater than 0.01% of the total in the Perdido system, and 0.10% for the other estuaries (where total numbers of species could not be fully represented). Species richness indices represent the cumulative number of phytoplankton species taken over each sampling period. Bloom species are marked by asterisks.

Table 6.3e (continued)

Species (6/97-5/98)	18	22	23	25	26	29	31	37	40	%
*Cyclotella choctawhatcheeana**	15318	2331	834503	4833944	2185490	3969387	7970854	7695861	3269196	56.33
*Heterosigma akashiwo**	142191	43623	596742	1489569	3734084	864229	122753	27972	22811	12.89
Synedropsis sp.*	4995	1998	56280	1085639	1383286	808264	540495	100733	41626	7.36
Cryptomonas sp.	111222	0	2637140	0	0	0	0	0	0	5.03
Merismopedia tenuissima	0	1070595	0	0	0	0	0	0	0	1.96
Dictyosphaerium pulchellum	35298	666	56620	253080	173160	427239	99900	2664	0	1.92
Pedinophora sp.	0	1015080	0	0	0	0	0	0	0	1.86
Spermatozopsis exsultans	4995	751785	23644	45621	50641	41625	14652	2664	833	1.71
Chaetoceros subtilis	0	0	0	3330	999	0	47952	445225	399624	1.64
Synedra fasciolata	27639	1666	16985	249084	209795	319014	61272	333	0	1.62
Pandorina morum	0	459877	2664	0	0	0	0	0	0	0.85
Dinobryon sertularia	18981	999	6660	99567	60273	131868	7659	333	0	0.6
*Miraltia throndsenii**	0	0	0	1333	333	333	49952	153513	87080	0.54
Synedra cf. delicatissima	999	3330	0	79259	70263	85581	14985	333	1332	0.47
*Gymnodinium cf. splendens**	8991	0	19481	26307	46626	39975	27308	14486	3864	0.34
Skeletonema costatum	0	0	12322	7326	1332	0	31968	56943	68765	0.33
Dinobryon divergens	38295	2331	61605	17316	8658	7659	2331	0	0	0.25
Ankistrodesmus sp.	6327	0	0	46620	24642	45288	666	2664	666	0.23
Dinobryon bavaricum	4995	999	667	19314	31302	42291	666	0	0	0.18
*Leptocylindrus danicus**	666	1665	0	0	0	0	0	41792	48020	0.17
Chrysochromulina sp.	0	0	88412	0	0	0	0	0	0	0.16
Rhizosolenia longiseta	8159	333	56279	1666	6328	666	333	0	0	0.14
Pedinophora sp.	0	70263	0	0	0	0	0	0	0	0.13
Gymnodinium sp.*	0	0	0	0	0	0	0	63603	0	0.12
Nitzschia reversa	0	0	333	11655	500	6827	16317	14153	10656	0.11
Bacillaria paxillifer	666	333	10994	17316	9324	10324	5661	333	666	0.1
Prorocentrum minimum	0	667	167	999	167	500	8659	22317	15984	0.09
Brachymonas-like sp.	0	48655	0	0	0	0	0	0	0	0.09
Dictyosphaerium ehrenbergianum	15318	0	0	13320	2664	0	7992	0	6660	0.08
Cyclotella striata	0	0	0	2997	3663	3663	8658	12987	12321	0.08
Mallomonas sp.	1998	0	17316	12987	2000	7992	0	0	0	0.08
Aulacoseira granulata	19314	0	18994	0	0	0	0	0	666	0.07
Chaetoceros socialis	0	0	0	0	0	0	0	2997	34965	0.07
Gonium pectorale	333	23992	10656	0	0	0	0	0	0	0.06

Table 6.3e (continued)

Species (6/97-5/98)	18	22	23	25	26	29	31	37	40	%
Ankistrodesmus spiralis	9324	333	0	7332	1665	2500	2000	11322	0	0.06
Entomoneis-like sp.	0	0	33483	0	0	0	0	0	0	0.06
Nitzschia palea	0	0	31302	333	0	0	0	0	0	0.06
Ankistrodesmus falcatus	14486	8996	2331	1332	333	1998	999	666	0	0.06
Stylochrysalis cf. libera	0	30636	0	0	0	0	0	0	0	0.06
Asterionellopsis glacialis	0	0	0	666	666	666	3331	6327	11256	0.04
Navicula spicula	0	0	0	333	0	0	0	6660	15318	0.04
Aulacoseira sp.	1665	12654	0	0	1665	2997	1665	0	0	0.04
Scenedesmus dimorphus	0	4662	8325	1332	1332	1332	0	1332	0	0.03
Selenastrum sp.	0	0	333	2997	2000	2664	1332	2997	999	0.02
Closterium sp.	1665	2664	0	2331	666	1998	3663	333	0	0.02
Cylindrotheca closterium	0	666	2332	1333	0	0	1499	3165	3997	0.02
Prorocentrum compressum	0	0	0	0	0	333	0	3998	8093	0.02
Crucigenia tetrpedia	0	2664	0	1332	0	0	1332	3996	1332	0.02
Chaetoceros decipiens	0	0	0	0	0	0	0	1332	8658	0.02
Nitzschia acicularis	0	333	8997	0	0	0	0	0	0	0.02
Stylochrysalis sp.	0	9324	0	0	0	0	0	0	0	0.02
Synedra parasitica	0	0	0	0	9324	0	0	0	0	0.02
Dactyliosolen fragilissimus	0	0	0	0	0	0	0	5328	2531	0.01
Prorocentrum micans	333	0	0	0	0	0	0	3666	3198	0.01
Crucigenia quadrata	0	2664	0	1332	1332	1332	0	0	0	0.01
Leptocylindrus minimus	0	0	0	0	0	0	0	333	6200	0.01
Nitzschia sigmoidea	0	0	2332	1665	999	333	999	167	0	0.01

Phacus longicauda	0	5999	0	0	0	0	0	0	0	0.01
Synedra fasciolata	333	0	1333	2331	0	333	1666	0	0	0.01
Dictyocha spiculum	0	0	0	0	0	667	0	4662	500	0.01
Chroococcus sp.	0	0	0	5328	0	0	0	0	0	0.01
Pediastrum duplex	0	0	5328	0	0	0	0	0	0	0.01
Teleaulax sp.	0	0	4829	0	0	0	0	0	0	0.01
Diploneis elliptica	0	0	4666	0	0	0	0	0	0	0.01
Treubaria setigerum	1665	0	0	0	666	1998	333	0	0	0.01
Acanthoceras magdeburgerse	2664	0	666	0	333	333	0	0	333	0.01
Selenastrum sp.	666	1332	0	666	1332	0	333	0	0	0.01
Pinnularia spp.	1665	333	333	500	333	666	333	0	0	0.01
Eunotia pectinalis	333	666	0	1665	333	666	0	0	0	0.01
Quadrigula chodatii	0	0	3663	0	0	0	0	0	0	0.01
Gyrosigma sp. *(w/stauros)*	0	0	0	2000	0	667	0	666	0	0.01
Crucigenia fenestrata	3330	0	0	0	0	0	0	0	0	0.01
Cyclotella sp.	333	0	0	666	167	0	1999	0	0	0.01
Elakatothrix gelatinosa	0	0	0	666	0	1332	999	0	0	0.01
Semiorbis sp.	1332	0	0	1665	0	0	0	0	0	0.01
Synedra acus	0	0	2997	0	0	0	0	0	0	0.01
Ceratium hircus	0	0	0	0	0	0	0	1333	1499	0.01
Amphidinium sp.	0	0	0	0	0	0	999	1665	167	0.01
Grand Total	**529639**	**3637395**	**4663037**	**8367543**	**8034504**	**6842701**	**9196550**	**8983760**	**4380897**	**54636026**
Cumulative # Species	**63**	**45**	**65**	**64**	**54**	**51**	**56**	**58**	**65**	

Note: Data were averaged and shown for all species with relative abundance equal to or greater than 0.01% of the total in the Perdido system, and 0.10% for the other estuaries (where total numbers of species could not be fully represented). Species richness indices represent the cumulative number of phytoplankton species taken over each sampling period. Bloom species are marked by asterisks.

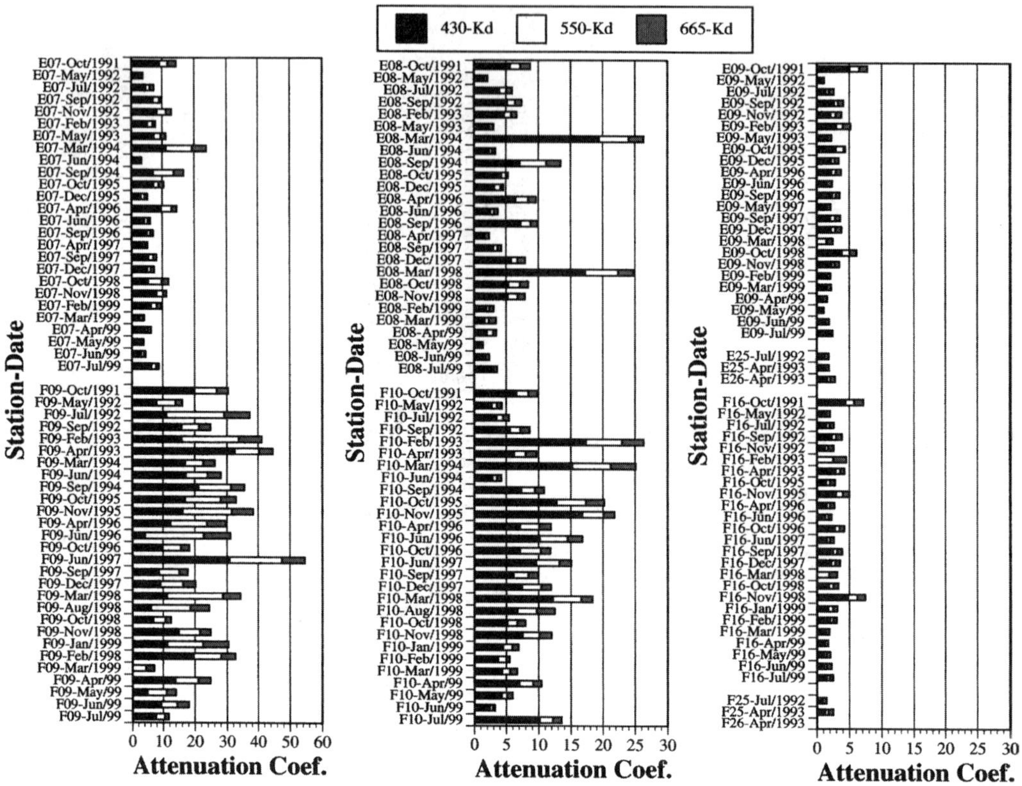

Figure 6.13 Comparison of attenuation coefficients (Kd) at three wavelengths (430, 550, 665 nm) of light in cognate inshore waters of the Econfina and Fenholloway estuaries. Data were taken monthly to quarterly from October 1991 through July 1999.

with the relative abundance of known bloom species. Thus, high relative dominance of bloom species is another reasonable indicator of nutrient-enriched systems with impaired phytoplankton assemblages.

VII. FOOD WEB RESPONSE TO PLANKTON ORGANIZATION

A. Introduction

The extreme complexity of aquatic food webs has been documented in many ways. The trophic indices on which this section is based were used to define generalized trophic models from detailed stomach-content analyses. By crossing traditional taxonomic lines with reconstituted trophic definitions, and by quantifying significant portions of the estuarine food webs in a series of river-dominated bays and coastal systems over prolonged periods, it became possible to use trophic relationships as a common coin of comparison. It should be emphasized that no system can be understood within itself; only through comparison with other systems can such knowledge be achieved. Most such comparisons are often hampered by the lack of uniformity of sampling methods at one end, and the limited scope (spatial and temporal) of the given study on the other. Not the least of such problems is the fact that few studies include detailed community analyses of the most fundamental part of the trophic system, the benthic and water column microalgae.

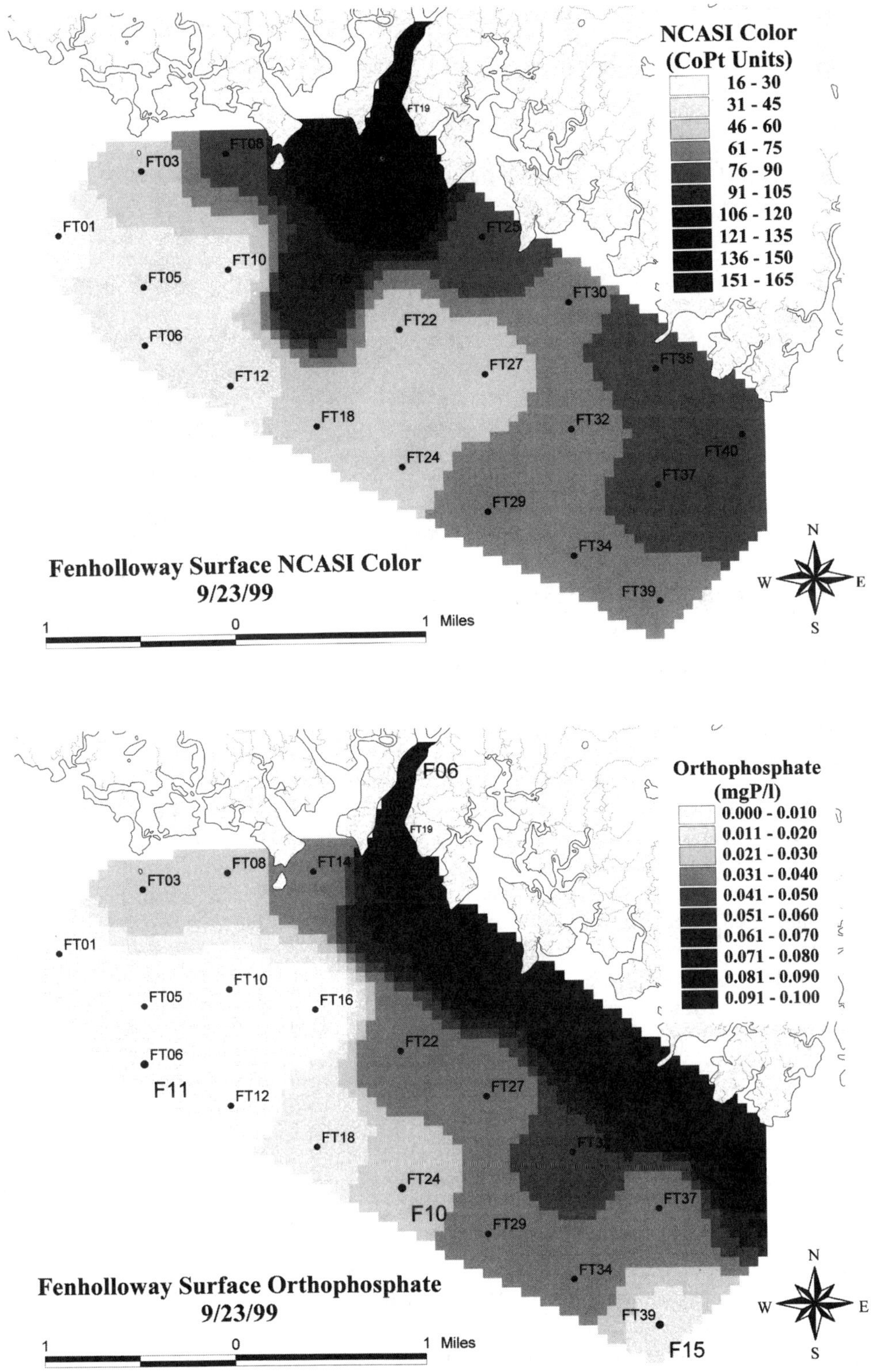

Figure 6.14 Spatial distribution of water color (Pt-Co units) (a), orthophosphate (mg P l^{-1}) (b), and ammonia (mg N l^{-1}) (c) in the Fenholloway estuary during September 1999.

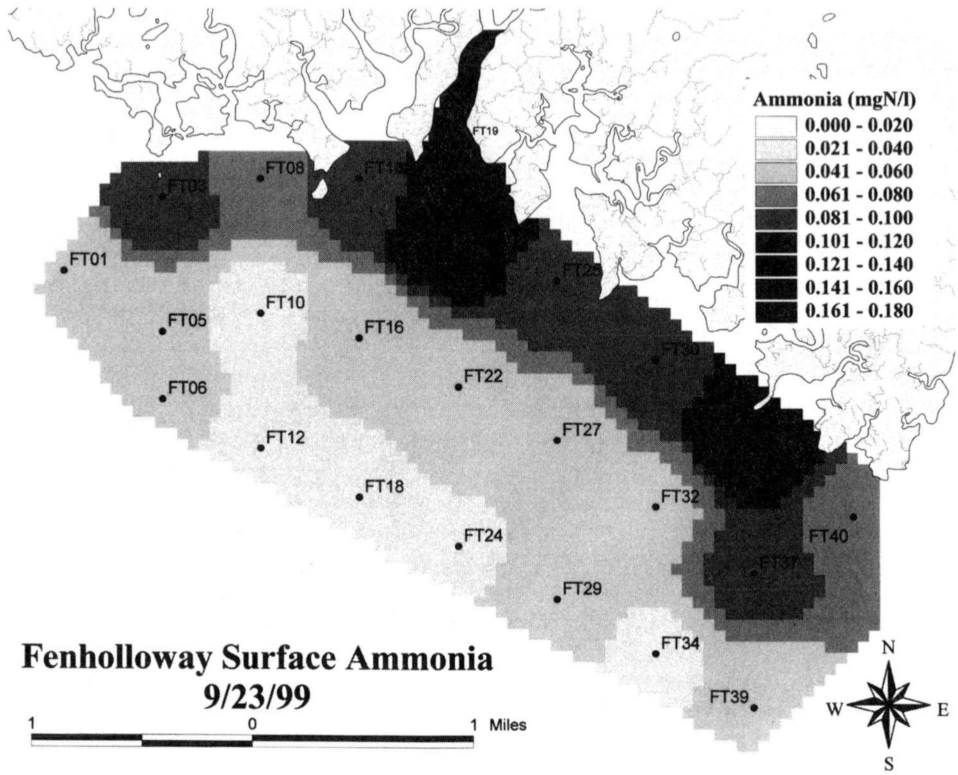

Figure 6.14 (continued)

There are direct connections between sediment quality and changes in phytoplankton and zooplankton associations (Bloesch and Burgi, 1989). Experiments concerning the response of infauna to nutrient additions (Posey et al., 1999) indicated that there was increased benthic microalgal production with increased nutrients and reduced infauna but there was no significant interaction among these factors. Benthic microalgae may be enhanced through increases of phosphorus and nitrogen (Dauer et al., 1982; Howarth, 1988; Becker, 1996) with possible increases of certain benthic grazers (Dauer et al., 1982; Posey et al., 1999). Benthic microalgae are directly connected to estuarine food webs through infaunal macroinvertebrates (Freeman, 1989). Moderate levels of phytoplankton production can also increase food for suspension feeders (Livingston, 1982) with possible indirect increases of detrital food (Christensen and Kanneworff, 1985). Increased abundance or growth of infaunal macroinvertebrates due to increased nutrient loading may be due to various processes that can be complicated by the effects of predation (reduced predation, increased infauna, reduced microphytes) although no such effects were found in estuarine areas (Posey et al., 1999). In some ways, the lack of detailed (species-specific) data concerning the benthic microphytes qualify the results of most of these studies. However, there is ample evidence of the close interaction between benthic microphytes and infaunal macroinvertebrates.

The complex interactions between nutrient cycling and resultant food web responses complicate any evaluation of the effects of anthropogenous nutrient loading to an estuary. Micheli (1999), in a review of marine mesocosm experiments, found that although nutrient additions generally enhanced phytoplankton biomass, the complexity of marine food webs precluded the usually simple assumptions on which many theoretical models depend. Qualitative aspects of primary production (i.e., the edibility of phytoplankton) controls the herbivore component; increased nutrient loading often leads to increased phytoplankton productivity and decreased utilization of phytoplankton biomass by higher trophic levels. Open estuarine systems, with advection of nutrients and individual

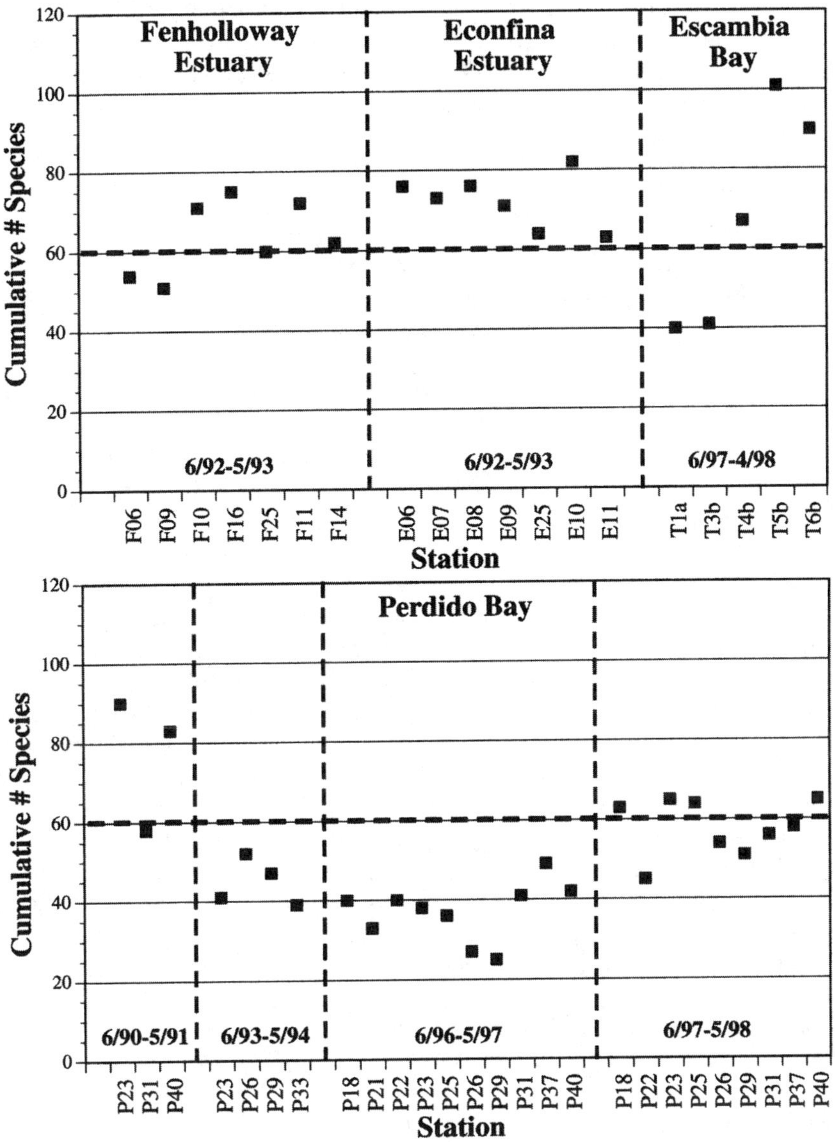

Figure 6.15 Comparison of cumulative whole-water phytoplankton species richness in the Econfina and Fenholloway estuaries, Escambia Bay, and Perdido Bay over 12-month intervals.

organisms from the inshore system, often are characterized by the uncoupling of the various trophic levels due to the migratory nature of the primary dominants. This represents a basic difference between the response of lacustrine and estuarine/coastal systems to anthropogenous nutrient loading.

The connection between benthic infauna and water column and benthic microphytes has been well established. Some studies (Netherlands Institute of Ecology, 1999) found that benthic invertebrates in the Bay of Aiguillon (Atlantic coast of France) depend mainly on benthic diatoms and to some extent phytoplankton for food. Mallin and Paerl (1994) noted that phytoplankton community structure in coastal systems is important to higher trophic levels in various ways. Phytoplankton are grazed by zooplankton accounting for approximately 38 to 45% of the daily phytoplankton productivity. The presence of noxious or otherwise inedible phytoplankton will not only decrease the trophic efficiency, but also could lead to possible changes in sediment quality through associated changes in the phytoplankton/zooplankton assemblages. The transfer rates of the integration of phytoplankton biomass into the estuarine food webs also depend on seasonal changes in water quality and species-specific responses of migrating populations on a seasonal basis (Mallin and Paerl, 1994). Such changes are closely associated with the quantitative and qualitative aspects of the phytoplankton assemblages.

B. Existing Conditions

The database for Apalachee Bay includes over 20 years of research that was started in 1971. The Econfina River–estuary was used as an unpolluted blackwater reference site for the Fenholloway system that was affected by pulp mill effluents. Databases for Apalachicola Bay and Perdido Bay represent long-term efforts (11 to 20 years) whereas those for the Pensacola Bay system and Choctawhatchee Bay represent short-term (18 to 36 months) databases. The Apalachicola Bay system is considered to be one of the last major alluvial river–estuaries that remains relatively unpolluted, with an undammed, intact floodplain and natural river flows (Livingston, 1997d). This system is still characterized by natural assemblages of freshwater and estuarine wetlands (Livingston, 1984b; Livingston et al., 1997a). During the 1970s and 1980s, it was considered to be one of the most productive river–estuaries along the Gulf Coast (Livingston, 1984a).

The Choctawhatchee River–estuary has been damaged by a series of anthropogenous activities that include dredging operations and unrestricted urban runoff. Because of physical changes associated with the opening of East Pass, the Choctawhatchee system was in a relatively unproductive state during the mid-1980s (Livingston, 1986a, b; 1989). The Pensacola Bay system, with most of its useful production in upper Escambia Bay, remains in a state of limited recovery from extreme hypereutrophication in the 1960s and 1970s. There is a continuing adverse response to ongoing pollution and nutrient loading from both point and non-point sources. Parts of lower Escambia Bay and Pensacola Bay have been severely affected by a broad range of toxic agents from urban runoff. The Perdido system, also affected adversely by the opening of a barrier island pass with the resulting stratification, benthic hypoxia, and liquid mud, has undergone serious changes in phytoplankton dynamics due to nutrient loading from a pulp mill in the upper bay and from non-point agricultural and urban sources in the lower bay. The results of nutrient loading to the Perdido system include establishment of a series of plankton bloom species, altered phytoplankton dominance hierarchies, associated deterioration of sediment quality, exacerbation of existing subhaloclinal hypoxia, and the deterioration of the estuarine food webs with severe reductions of secondary production throughout the bay.

C. Food Web Comparisons

Environmental well-being of a given aquatic system may be measured by the response of the food web structure to natural and anthropogenous sources of eutrophication. This process involves complex interactions of many species that follow relatively stereotypic seasonal cycles (Livingston, 1976; 1984a; 1986a; 1990; 1991b; 1991c; 1992; Livingston et al., 1974; 1997). In terms of frequency of occurrence in upper northeastern Gulf estuaries, infaunal macroinvertebrate assemblages are represented by various species: *Mediomastus ambiseta* (below-surface deposit feeder and detritivorous omnivore), *Streblospio benedicti* (above-surface deposit feeder and detritivorous omnivore), *Hobsonia florida* (above-surface deposit feeder and detritivorous omnivore), *Grandidierella bonnieroides* (surface deposit feeder and general omnivore), and *Parandalia americana* (predator, primary carnivore). Larger forms of infaunal macroinvertebrates include plankton-feeding herbivores such as *Mactra fragilis* and *Rangia cuneata*. Dominant epibenthic macroinvertebrates that are often found in these coastal areas include palaemonetid shrimp (*Palaemonetes* spp., detritivorous omnivores), xanthid crabs (*Rhithropanopeus harrisi*, primary carnivores), blue crabs (*Callinectes*

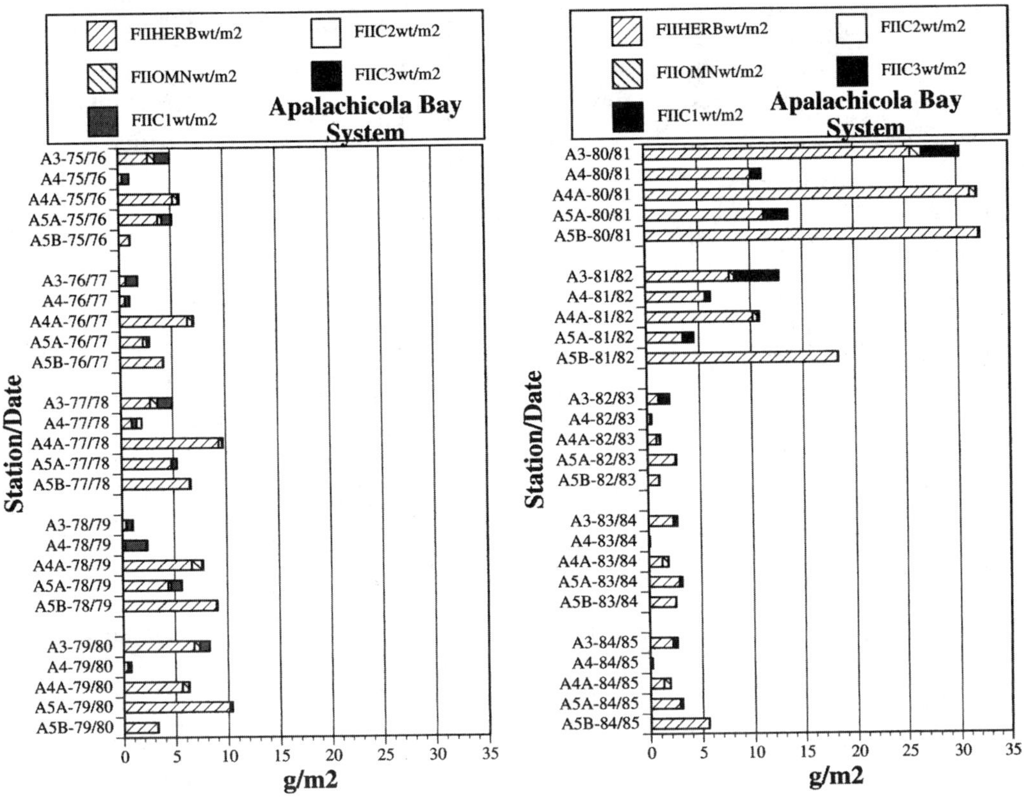

Figure 6.16 Total biomass (g m^{-2}) of FII indices by station in Apalachicola Bay as annual averages from 1975 to 1985. The FII indices include herbivores, omnivores, C1 carnivores, C2 carnivores, and C3 carnivores.

sapidus, primary carnivores at <30 mm; secondary carnivores at >30 mm), and penaeid shrimp (*Penaeus setiferus*, *P. duorarum*, and *P. aztecus*, primary carnivores at <25 mm; secondary carnivores at >25 mm). Most of these invertebrate species are browsers, grazers, or seize-and-bite predators.

Dominant fishes in upper Gulf estuaries include plankton-feeding primary carnivores such as bay anchovies (*Anchoa mitchilli*) and benthic feeding primary carnivores such as spot (*Leiostomus xanthurus*), hogchokers (*Trinectes maculatus*), young Atlantic croakers (*Micropogonias undulatus*, <70 mm), and silver perch (*Bairdiella chrysoura*, 21 to 60 mm). Secondary carnivores among the dominant fishes included larger croakers (>70 mm), Gulf flounder (*Paralichthys albigutta*), sand seatrout (*Cynoscion arenarius*), and juvenile spotted seatrout (*C. nebulosus*). Tertiary carnivores include the larger spotted seatrout (*C. nebulosus*), southern flounder (*P. lethostigma*), and sharks of various species. With the exception of the bay anchovies, all of the above species live and feed at or near the sediment/water interface with most of the trophic organization of the bay dependent on interactions among bottom-living infaunal and epibenthic macroinvertebrates and fishes.

Annual averages of the fish–invertebrate–infauna (FII) indices for the Apalachicola Bay system are presented in Figure 6.16. The FII indices in the Apalachicola system indicate a highly productive bay, with balanced distributions of the five primary trophic levels (Livingston et al., 1997). Interannual variation of the secondary production of the Apalachicola system is evident. Within limited interannual ranges of moderate to high freshwater flows of the Apalachicola River, there was little change in the trophic organization of the receiving estuary between 1975 and 1980. The physical instability of the estuary due to seasonal river flow fluctuations was actually a major component

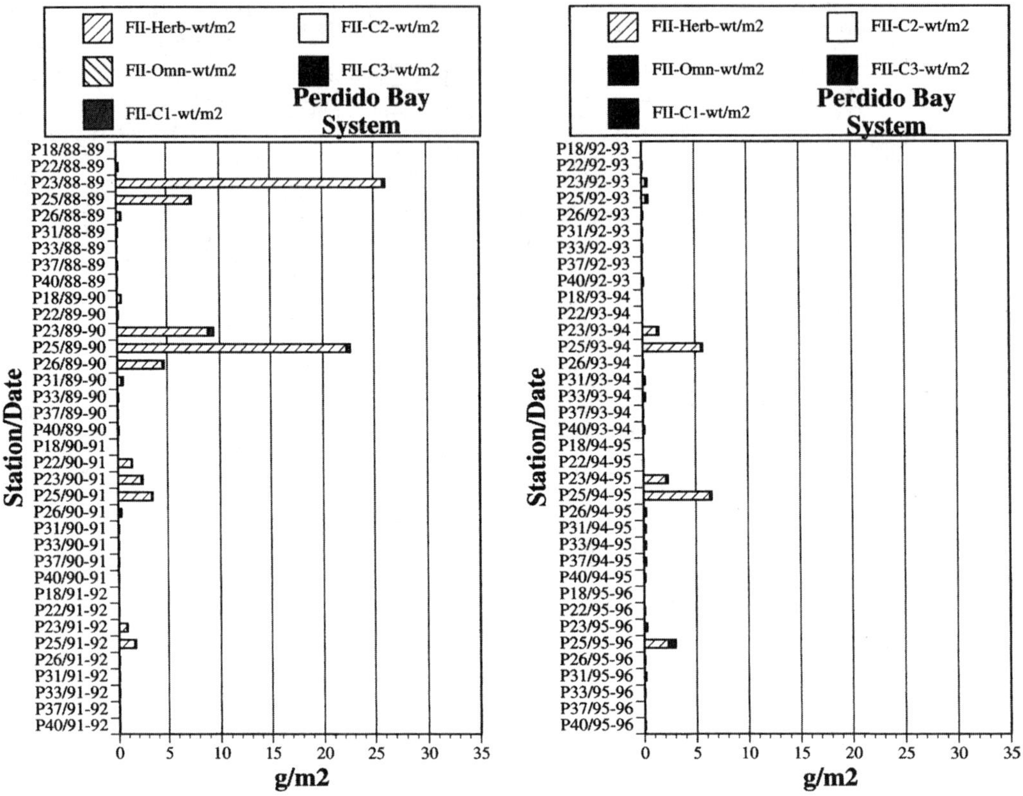

Figure 6.17 Total biomass (gm^{-2}) of FII indices by station in Perdido Bay as annual averages from 1988–1999. The FII indices include herbivores, omnivores, C1 carnivores, C2 carnivores and C3 carnivores.

in the continuation of a biologically stable estuarine system (Livingston et al., 1997). However, when a specific threshold of freshwater reduction was reached during a prolonged natural drought (1980 to 1982), clarification of the normally turbid and highly colored river–estuarine system led to rapid changes in the pattern of primary (microphyte) production. These changes were associated with major changes in the trophic structure of the system. Increased light penetration due to the cessation of river flow was an important factor in the temporal response of bay productivity and herbivore/omnivore abundance. The reduction of nutrient loading during the 1980 to 1982 drought period was postulated as a major cause of the loss of productivity of the river-dominated estuary both during and after the drought (Livingston et al., 1997). Recovery of predrought productivity, with resumption of increased river flows, was, likewise, a long-term event that appeared to be under way during increased river flow years from 1983 to 1985.

In the Perdido Bay system, the period 1988 to 1991 was marked by high secondary production in the upper bay (Figure 6.17). The highest FII production, dominated by planktivorous herbivores, was found just off Elevenmile Creek and in the seagrass beds of the upper bay. The period of relatively high secondary production (comparable to that in the Apalachicola system) was characterized by a prolonged drought. Secondary production levels were relatively high in marked contrast to those observed at stations in depositional areas down the main stem of the bay. The habitat of these stations of low secondary production was characterized by strong haloclines, periodic (i.e., during warm months) hypoxia at depth, and the continuous presence of liquid mud. As shown before, liquid mud appeared to be an important factor in the reduction of secondary benthic production through elimination of infaunal herbivore assemblages. The period from 1991 to 1993 was characterized by marked reductions of secondary production in the upper bay. This pattern

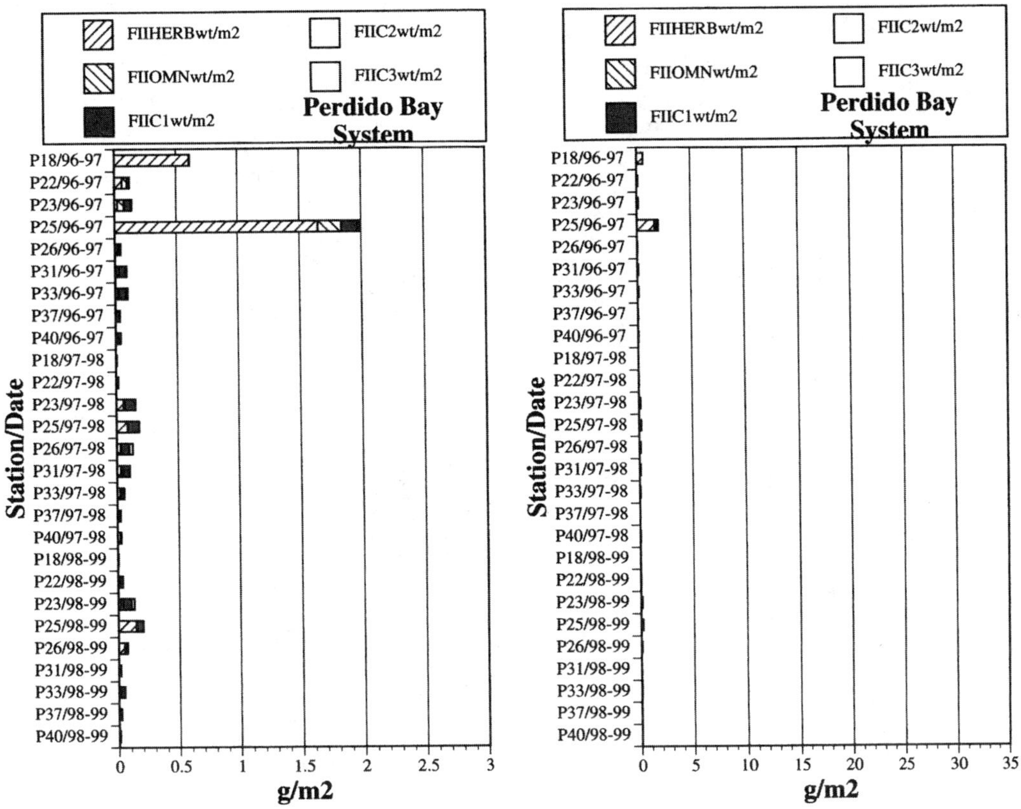

Figure 6.17 (continued)

resembled that in the period following the drought in the Apalachicola system, and could have been due to similar trends of primary production as a result of nutrient limitation following drought conditions. It is also possible that increased nutrient loading during this period was having an effect on the system. By 1993 to 1995, there was a slight increase in FII production, especially at station P25 during the initial phytoplankton blooms in the bay. However, by 1995–1996, FII production was generally limited to the upper bay seagrass beds. From 1997 to 1999, FII production was extremely low throughout the study area. The loss of FII biomass in Perdido Bay over this period did not follow the long-term pattern exhibited during the drought–flood period of 1988 to 1991 in the bay. The extreme reduction of secondary production was generally associated with the initiation and succession of phytoplankton blooms, especially in the later years.

The Choctawhatchee Bay system currently has one of the least well developed fringing (emergent) vegetation zones of the various Gulf estuaries in the region (see Figure 2.2b). This vegetation is limited to a few peripheral areas bordering several bayous and the river delta. In addition, the input of high-salinity water into the relatively deep inshore bay system has led to strong salinity stratification in large portions of the estuary with subhaloclinal hypoxia at depth during warm periods. It has also led to the establishment of a more-or-less permanent layer of liquid mud that has been associated with broad areas of relatively low secondary production. Statistical analyses indicated that there were direct correlations of depth, salinity stratification, degree of hypoxia, and observed liquid mud conditions (Livingston, 1997a).

The Choctawhatchee Bay system was characterized by relatively low FII indices (Figure 6.18). Deeper areas of Choctawhatchee Bay (stations C25, C31, C34) had low secondary production. This is consistent with observed hypoxia at depth and existing sediment quality. The imbalances of

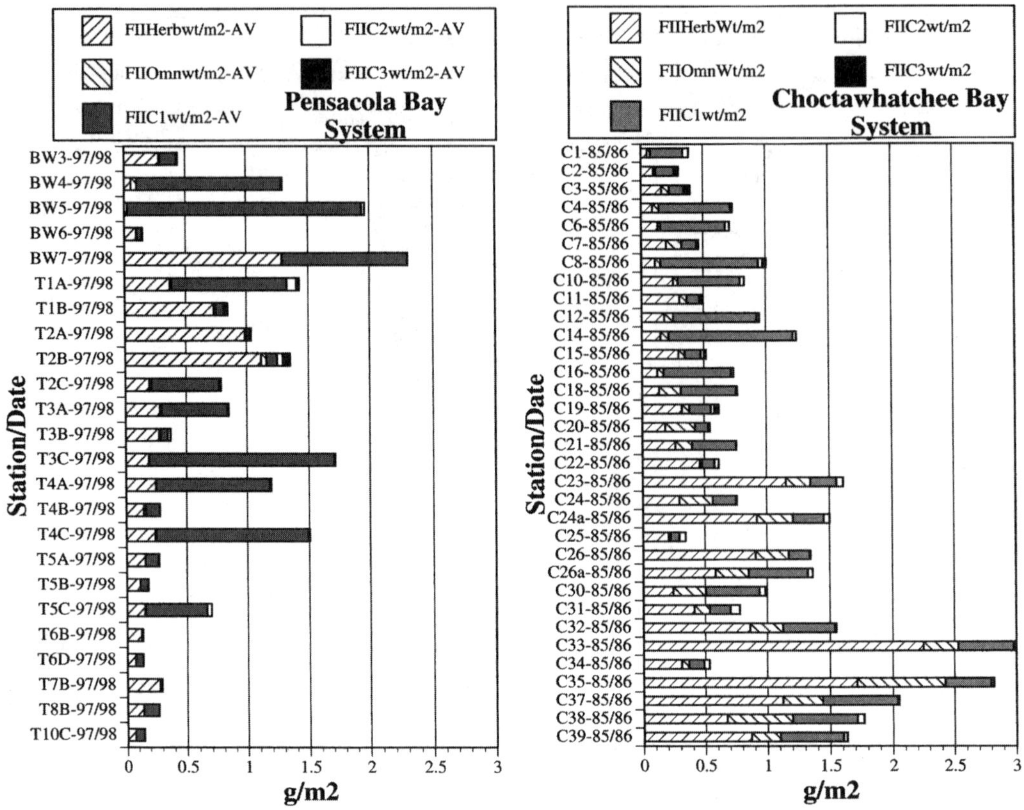

Figure 6.18 Total biomass (g m^{-2}) of FII indices by station in the Pensacola Bay system as annual averages from 1997 to 1998 and in the Choctawhatchee Bay system as annual averages from 1985 to 1986. The FII indices include herbivores, omnivores, C1 carnivores, C2 carnivores, and C3 carnivores.

secondary production (relatively low levels of herbivores in eastern sections of the bay where liquid muds predominate) resembled the disjunct food webs of salinity-stratified areas of Perdido Bay. This was probably a trophic reaction to the reduced quality of the sedimentary habitat due to the anthropogenous alteration of the salinity regime in Choctawhatchee Bay. In the better-mixed areas of the western section of the bay, where sandy bottoms and seagrass beds predominated, there was some increase in the herbivore groups although secondary production remained low. It is possible that generally low phytoplankton production (Livingston, 1987b; 1996a, b) as a result of sequestered nutrients in the sediments contributed to the observed low numbers of invertebrates and fishes in deeper parts of the bay. The current condition of the Choctawhatchee system has not been scientifically documented.

When compared with other river-dominated estuaries, the Pensacola Bay system had a relatively low level of FII biomass (Figure 6.18). The lowest FII biomass in the Blackwater/East Bay system was at station BW6, a relatively deep part of the bay that was associated with salinity stratification and the development of liquid mud. The Blackwater Bay biomass was highest in sandy substrates (BW7); it was in such areas that the benthic herbivore populations appeared to be more abundant. The lowest FII biomass in the Blackwater/East Bay system was thus associated with salinity stratification and the development of liquid mud as the principal habitat type. In Escambia Bay, herbivores were concentrated in the upper Bay (Figure 6.18). The primary carnivore biomass was concentrated in shallow parts of upper Escambia Bay and the Blackwater/East Bay system, whereas the biomass of secondary and tertiary carnivores was highest in deeper parts of upper Escambia

Bay. These data suggested that food web biomass was concentrated in the areas of highest chlorophyll *a* concentrations and primary productivity in upper Escambia Bay. During periods of high-salinity stratification, areas in upper Escambia Bay were associated with reduced DO at depth. This hypoxia was evident along the north-south axis of Escambia Bay. Chlorophyll *a* concentrations were moderately high in upper Escambia Bay with a general decreasing trend along the north-south axis. Relatively low FII indices were noted at stations T2C and T3B in upper Escambia Bay and at stations T4B, T5A, and T5B in middle and lower Escambia Bay. The FII index levels were extremely low in Pensacola Bay. The lowest FII biomass levels in Escambia Bay and Pensacola Bay were noted in areas known to be subject to point and non-point sources of pollution. These areas were also characterized by high concentrations of toxic substances in the sediments.

There were some parallels between the habitat features and FII indices in the Escambia Bay and Apalachicola Bay systems. In both bays, the highest FII biomass occurred in areas that (1) received direct river flow, (2) were high in primary production and chlorophyll *a*, and (3) featured a silty benthic habitat. In upper Escambia Bay, the dominant populations of fishes and invertebrates were concentrated in areas characterized by low benthic DO and silty sediments with high concentrations of liquid mud. However, the Apalachicola system was, at the time of sampling, relatively unpolluted with virtually no plankton blooms and little development of liquid mud. In part, this was due to the fact that Apalachicola Bay is relatively shallow. However, the primary difference between the two estuaries was that the Apalachicola system had not been subjected to anthropogenously enhanced nutrient loading and remained free of plankton blooms, whereas the Escambia system has had a long history of hypereutrophication due to nutrient loading. The relatively high dominance of plankton bloom species and corresponding deterioration of the sediments in Escambia Bay resembles conditions in Perdido Bay. It is thus highly likely that the impoverished level of secondary production in both upper Escambia Bay and Perdido Bay is a response to changes in the plankton assemblages (and possibly associated sediment conditions) created by anthropogenous nutrient loading.

A comparison of the general distribution of FII biomass in the various river-dominated estuaries of the northeast Gulf Coast (Figure 6.19) illustrates the various spatial relationships of secondary production in addition to showing the temporal changes that took place in Perdido Bay before and after the initiation of prolonged plankton blooms. In each case, the distributions of animal biomass were related to a combination of local habitat conditions and more widespread effects based on alterations of the benthic habitat through dredging and/or changes in the plankton associations. The effects of isolated concentrations of toxic substances in sediments such as in Escambia Bay are not as widespread as the baywide problems associated with physical alterations and nutrient loading.

A statistical analysis (PCA regression; Appendix II) was run with the FII biomass data totaled across all five components for all six river-dominated bays (Figure 6.20). The secondary production in the Choctawhatchee and Apalachicola systems was positively associated with river flow. Apalachicola Bay secondary production was also positively associated with bottom DO and silt composition of the sediments. In a bay that was relatively free of salinity stratification, debilitating plankton blooms, and low bottom DO, silty sediments were equated with increased secondary production. In Perdido Bay and Choctawhatchee Bay, the total FII biomass was negatively associated with depth, bottom salinity, salinity stratification (bottom salinity – surface salinity), mean grain size, percent organics and silty sediments; positive associations in both systems included bottom DO anomalies and sandy sediments. In the Choctawhatchee system, chlorophyll *a* was inversely related to secondary production. Perdido Bay and Choctawhatchee Bay were dominated by salinity stratification and corresponding reductions of dissolved oxygen below the halocline. Stabilization of the water column was also responsible for the occurrence of liquid mud (as represented by high percentages of organic matter and silt–clay particles) that was inversely related to the FII biomass.

The situation was more complex when Perdido Bay and Escambia Bay were compared. Upper Escambia Bay was characterized by high percentages of bloom species, hypoxia at depth, liquid

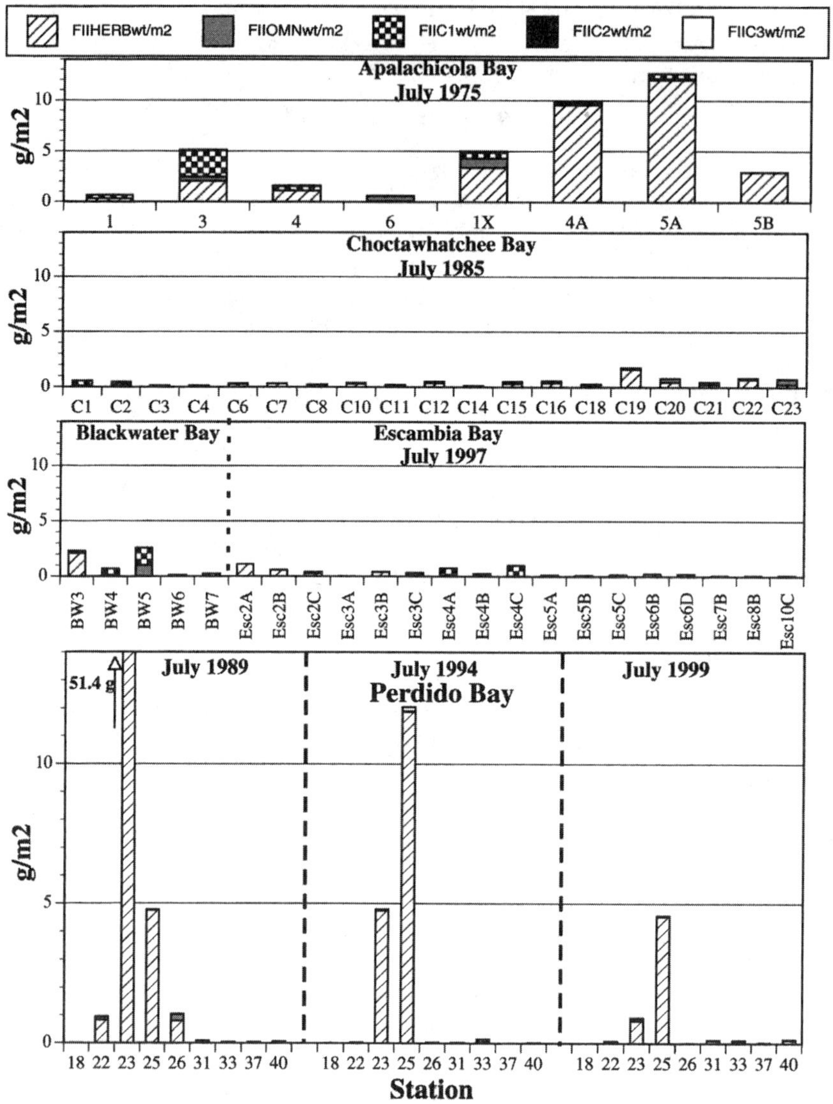

Figure 6.19 Comparison of FII indices (biomass m^{-2}) in Apalachicola Bay, Choctawhatchee Bay, Blackwater Bay, Escambia Bay, and Perdido Bay during the month of July. The Perdido Bay data are shown for periods before and after the phytoplankton blooms.

mud, and the general deterioration of the sediments. Upper Perdido Bay, during the period 1988 to 1990, had relatively healthy phytoplankton assemblages, high FII indices, and little liquid mud development due to the shallowness of this part of the bay. The situation in Perdido Bay deteriorated with time, with precipitous declines of FII biomass after the development of the plankton blooms in upper Perdido Bay. The relatively low FII production in the Escambia Bay–Pensacola Bay system was probably associated with previously described deterioration of the benthic habitat due to a combination of hypereutrophication and toxic substances in sediments of the lower bay. A regression analysis indicated that most of the FII production that did exist in the Pensacola Bay system was found in upper parts of Escambia Bay. However, FII biomass was statistically associated with phosphorus loading, high chlorophyll *a*, highly organic, silty sediments, and low bottom DO. The FII biomass was negatively associated with sandy sediments in Escambia Bay. The Blackwater Bay system, dominated by low primary and secondary production (compared to the alluvial

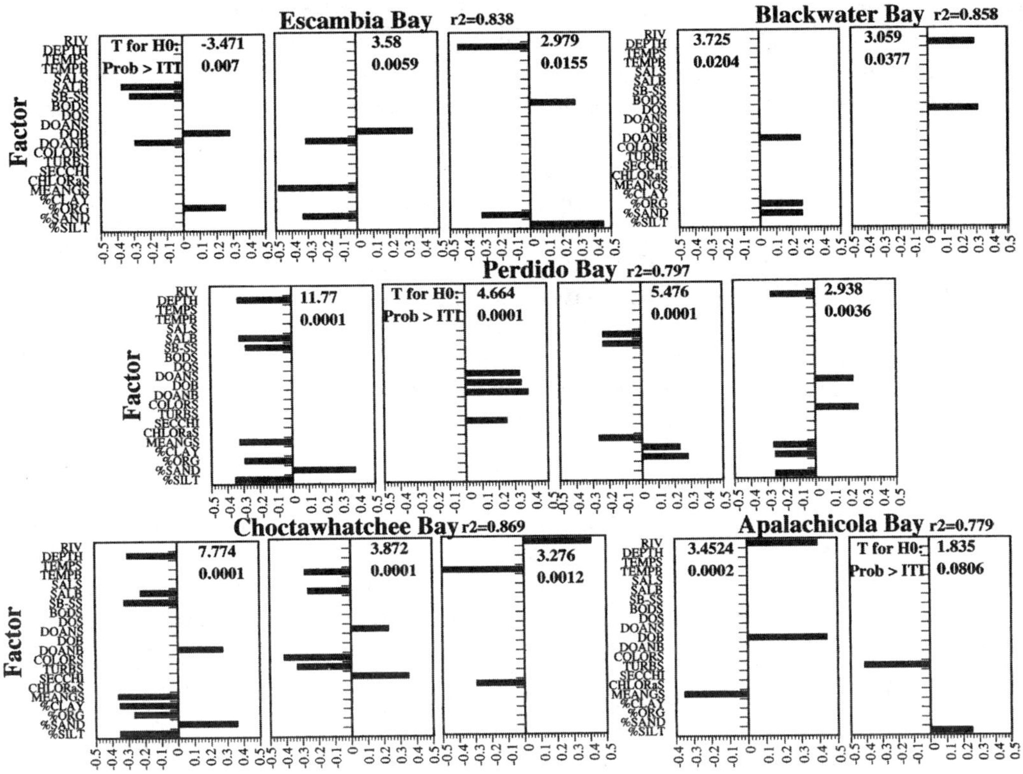

Figure 6.20 Results of PCA-regression analysis (Appendix II) of totals for all FII indices in a series of estuaries: Escambia Bay, Blackwater Bay, Perdido Bay, Choctawhatchee Bay, and Apalachicola Bay. Habitat variables include river flow rates, water depth, surface and bottom water quality features (temperature, salinity, biochemical oxygen demand, dissolved oxygen, color, turbidity, Secchi depths, chlorophyll *a*) and sediment features (percent organics, sand–silt–clay percentages). Only statistically significant ($p = 0.05$) eigenvectors are presented and only those factors with values exceeding 0.25 are shown.

systems), had positive FII associations with sandy, well-oxygenated habitats. Previous analyses of the Pensacola system (Livingston, 1999) indicated that organisms in more productive areas of the system (i.e., upper Escambia Bay) had very different affinities for sediment type and bottom DO concentrations than organisms in less productive areas of the system (i.e., Blackwater Bay and Pensacola Bay). In other words, organisms are present in highest numbers in areas of where their food is located (i.e., the upper parts of Escambia Bay where nutrient loading from the alluvial rivers provides the basis for phytoplankton production). Primary production and food web interactions are thus of higher priority than habitat type in upper Escambia Bay. Such areas are also the most vulnerable parts of the estuary when it comes to the initiation of phytoplankton blooms resulting from anthropogenous nutrient input.

The relationship between habitat and productivity is important to the understanding of the distribution of commercially important estuarine species such as penaeid shrimp and blue crabs in the Pensacola system. The data are consistent with the hypothesis that nutrient (i.e., inorganic phosphorus) loading into upper Escambia Bay from the Escambia River, together with associated habitat changes (low, highly variable salinity, moderately high chlorophyll *a*), were the primary determinants of secondary production in the Pensacola Bay system. Trophic organization of the system determined the distribution of dominant populations at all levels of biological organization. The relative influence of habitat factors such as sediment type, DO, and salinity, while important in determining the general distribution of organisms in the bay system as a whole, can only be understood within the context of the distribution of primary production in the system. In other words, the organisms go where the food is located; only major changes in the food web due to plankton blooms and associated deterioration of the benthic habitat can alter such conditions that can then lead to reduced secondary production. This highlights the relatively increased importance of the upper parts of the respective estuaries in terms of both primary and secondary production. However, the same features that contribute to the high productivity of such areas also contribute to increased latent vulnerability of such areas to phytoplankton blooms and the impacts of hypereutrophication.

The results of this study show that both the overall biomass per square meter and the spatial distributions of herbivore/omnivore biomass per square meter in the Perdido and Choctawhatchee systems were strongly influenced by salinity stratification and sediment quality. Silt-dominated sediments of small grain size were usually found in areas exceeding 2 m in depth; depth restrictions of liquid mud occurrence were caused by reductions of vertical transport by the well-established haloclines in the respective systems. Periodic hypoxia was secondarily associated with reductions of overall biomass and the herbivore/omnivore groups. Hypoxic effects were largely seasonal whereas the establishment of liquid mud occurred throughout the year, which explains the paucity of biota in areas dominated by liquid mud during periods not characterized by subhaloclinal hypoxia. Thus, sediment effects were probably more effective in reducing the biota than the periodic hypoxic events. The connection of salinity stratification and phytoplankton blooms with sediment quality is thus an important determinant of overall secondary production in estuaries along the northeastern Gulf.

In the shallow, well-oxygenated Apalachicola estuary, the invertebrate/fish biomass exceeded that of the salinity-stratified systems and was marked by enhanced representation of herbivores with the occurrence of silty sediments. There was little evidence of plankton blooms and there was no extended development of liquid mud in the shallow Apalachicola estuary. The areas of lowest overall biological productivity in the Apalachicola system had been disturbed by dredging effects (Livingston, unpublished data). Sediment type played a secondary role in the occurrence of the herbivore and omnivore groups in the shallow Apalachicola system. Overall, sediments were most important where liquid mud was prevalent. However, phytoplankton production, as a function of interannual variations in river flow and nutrient/microalgal distributions in space and time, has a major effect on the interannual cycles of the trophic organization of the bay.

There are obvious species-specific adaptations that are critical to the distribution of organisms in a given estuary, and both sediment type and water quality are important determinants of such assemblages (Rosenberg, 1977; Gaston, 1985). Thus, the development of liquid muds together with the status of the phytoplankton assemblages in river-dominated estuaries have had profound effects on the basic productivity and food web characteristics of the different bay systems in this study. Where such sediments were not dominated by liquid mud and plankton blooms and were subject to the direct effects of nutrient loading from upland sources (i.e., East Bay of the Apalachicola system, upper Perdido Bay during nonbloom years), silty conditions actually enhanced a relatively high level of biological productivity. Seasonal changes in hypoxic conditions had only secondary effects on such factors so that the emphasis on covariates such as DO in various studies (Dauer et al., 2000) may be spurious.

CHAPTER 7

Restoration, Research, and Regulation

I. RESTORATION OF THE PERDIDO SYSTEM

Phytoplankton blooms undergo regular short- and long-term cycles with distinct interannual trends of bloom species replacement after prolonged anthropogenous nutrient loading. Nutrient control that addresses specific phytoplankton requirements should take into consideration both the seasonal aspects of nutrient loading and long-term temporal changes of interactions among the various plankton species (Bernhard and Peele, 1997). There is evidence that nutrient removal from some estuaries has resulted in water quality improvements, for example, phosphorus removal in the upper tidal freshwater Potomac (Jaworski et al., 1972; Jaworski, 1981); the upper portion of the Thames River estuary (Gameson et al., 1973); Kaneohe Bay, and Hawaii (Smith, 1981). In most cases, it was not ascertained if the phytoplankton community and/or associated food webs had actually been either affected by the blooms or restored.

Under natural conditions, nutrient loading from upland sources drives the high productivity of river-dominated estuaries (Livingston et al., 1997). The relatively important heterotrophic elements of the estuarine food webs do not preclude the significance of autotrophic processes in such systems. However, such processes are not temporally uniform, and seasonal/interannual cycles of river flow, nutrient loading, and nutrient concentrations must be taken into consideration when plankton blooms are evaluated. Even under natural conditions, periodic droughts (seasonal and interannual) alter the relationships of nutrient loading to secondary production (Livingston et al., 1997). The time factor is critical to eutrophication studies both in terms of the natural cycling of bay conditions and the various stages of hypereutrophication. Drought periods represent maximal vulnerability to bloom generation due to anthropogenous nutrient loading because of habitat response to reduced freshwater inflow. Various factors such as turnover rates of the water, nutrient gradients (which may be increased during droughts without increases in the loading levels), and the effects of increased temperature, salinity, and light penetration should be taken into consideration when evaluating a given situation. Baseline (natural) nutrient loading sources of alluvial rivers should also be part of the evaluation along with the usual anthropogenous point and non-point sources of nutrients. In any given situation, nutrient loading from all primary sources, natural and anthropogenous, should be part of the background analysis.

Based on the long-term interdisciplinary database for the Perdido drainage system, an empirical approach was used to determine nutrient loading targets for the Cantonment paper mill. The preliminary biological data indicated that nutrient loading from Elevenmile Creek during the first 2 to 3 years of sampling (1988 to 1991) was not detrimental to the plankton assemblages and overall secondary production in upper Perdido Bay. Although it was evident that subsequent increases of orthophosphate and ammonia loading from the mill led to plankton blooms in the bay, it was not clear how such loading was associated with the initiation of the blooms. It took years of analysis before the interannual bloom successions could be determined and the complex interactions of the various controlling factors could be quantified with regard to the development of a realistic restoration program. There was little in the scientific literature to guide the choice of target nutrient loading requirements.

Determination of target nutrient loadings for the pulp mill was carried out with models and calculations based on data taken over the 11-year sampling period (T. Gallagher, S. Keolsch, personal communication). It was assumed that nutrient sources other than the pulp mill in upper Perdido Bay would not change, at least in the near future. Based on the rate of urbanization in the Elevenmile Creek subbasin, this assumption should be periodically reviewed although model results indicated that non-point sources of nutrients to upper Perdido Bay were minimal compared with the mill contributions. It was also assumed that the nutrient loading levels during the early years of sampling in the Perdido system were not accompanied by adverse impacts to the immediate receiving areas of the bay. The period 1988 to 1991 constituted a period of relatively natural conditions and a balanced food web in upper Perdido Bay (Livingston, 1992). It was assumed that nutrient loads during a 7Q10 event in 1990 could be safely discharged in the future and that these loads would be protective of the biological integrity of the bay. Uncertainty existed relative to the increased sensitivity of various bloom species to nutrient loading after exposure to long-term increases of orthophosphate and ammonia in the upper bay. This unknown factor supported a conservative approach in the determination of nutrient loading targets for the mill.

A review was made concerning ammonia loading to upper Perdido Bay and ammonia concentrations in Elevenmile Creek (see Figure 4.5). It was estimated that the level of ammonia loading by the mill should also ensure that ammonia concentrations at the mouth of Elevenmile Creek would remain below concentrations that could be toxic to phytoplankton species (Livingston, unpublished data). A target of maximal ammonia loading from the mill was determined that would approximate loading levels at the beginning of the study when phytoplankton community structure was not adversely affected by ammonia loading from Elevenmile Creek. With data concerning orthophosphate loading from upper Perdido Bay and orthophosphate concentrations in Elevenmile Creek over the past 11 years (Figure 4.5), the same process was used to determine targets for orthophosphate loading from the mill. A two-level method based on high and low flow rates in Elevenmile Creek was used as a conservative approach. The lowest target loads and concentrations of orthophosphate were imposed during the droughts at which times the bay was considered to be at greater risk for bloom generation. Because phosphorus monitoring at the mill was carried out in terms of total phosphorus, the TP loading and concentration data were used to estimate target loadings of TP from the mill (Figure 4.4b). It was estimated that TP levels approximating those observed during the first few years of sampling should not be exceeded. Based on the loading rates of these nutrients from Elevenmile Creek during the period of reduced mill phosphorus loading, these targets were thought to represent realistic caps on ammonia, orthophosphate, and TP loading from the mill.

Based on the empirical data and the modeling effort, it was recommended that a limit of the quarterly average of 361 kg d^{-1} ammonia be utilized for future mill discharges. At 361 kg d^{-1} and a mill flow of 23 mg d^{-1}, the concentration of mill effluent would be 4.1 mg l^{-1}, which, due to dilution with creek water, would not adversely impact the biological integrity of Perdido Bay even during drought periods. It was recommended that a limit of 50 kg d^{-1} as a quarterly average for final effluent orthophosphate be established for Elevenmile Creek flows > 65 cubic feet sec (cfs), and that a limit of 36.7 kg d^{-1} for final effluent orthophosphate be established for Elevenmile Creek flows <65 cfs. A loading cap of 83.5 kg d^{-1} was suggested for total phosphorus.

On a day-to-day basis, the calculation of loading rates often is delayed due to unavoidable lag times in the availability of river flow rates needed for the loading determinations. In addition, during drought periods, comparable levels of nutrient loading may lead to higher nutrient concentrations at the mouth of Elevenmile Creek than during periods of moderate or high river flows. Because of

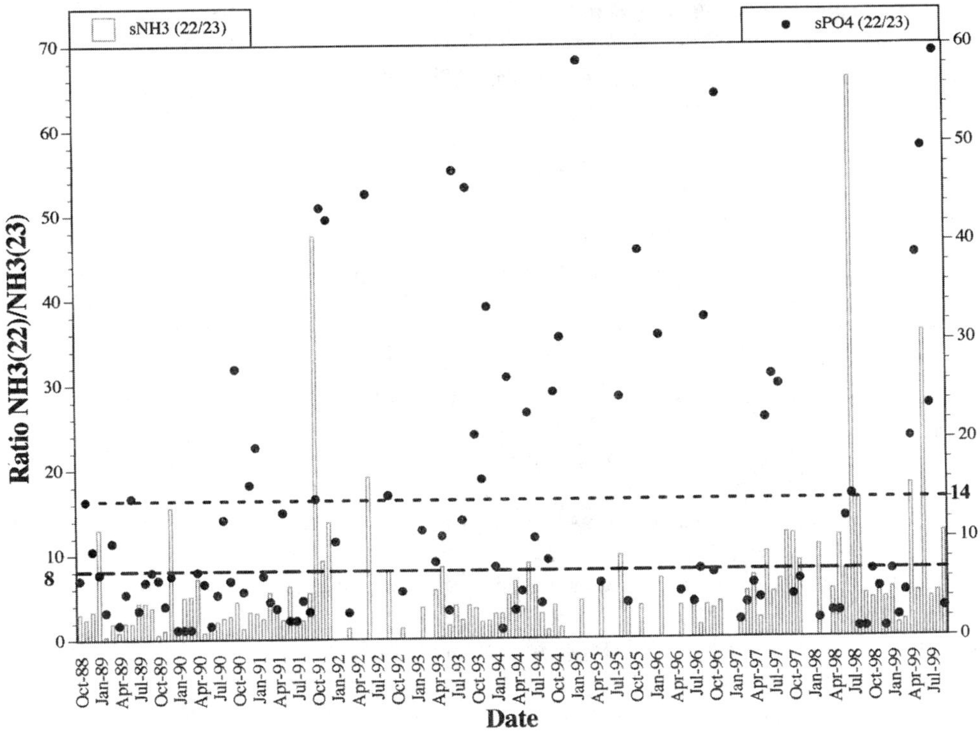

Figure 7.1 Orthophosphate and ammonia ratios (P22/P23) at the end of Elevenmile Creek. Data are plotted at monthly to quarterly intervals from October 1988 through September 1999.

the above limitations in the process of evaluation, an additional method of evaluating the efficacy of meeting the nutrient loading targets was initiated. This involved continuous monitoring of ammonia and orthophosphate gradients at the end of Elevenmile Creek. Based on the analysis of the 11-year database, it was estimated that ratios of 8 (ammonia) and 14 (orthophosphate) mark the upper boundary of safe concentration gradients (Figure 7.1) for these nutrients. Should monthly monitoring indicate exceedances of these ratios, an immediate review would be made of the nutrient loading from the mill with the aim of reducing such ratios as quickly as possible. As of September 1999, both orthophosphate and ammonia loading by the mill had been reduced to levels compatible with the above recommendations. The bay is currently being monitored so that an evaluation can be made regarding the effectiveness of the nutrient loading restrictions in eliminating the plankton blooms and restoring the Perdido Bay food webs and secondary production. The mechanism and timing of the postulated recovery will be addressed in this research.

The determination of a reasonable restoration program is possible with regard to point sources of nutrient loading such as the pulp mill. However, the question of restoration in areas affected by non-point nutrient sources (urban and agricultural runoff) has not been addressed much less answered. The restoration of Wolf Bay and lower Perdido Bay is highly problematical considering the fact that pollutant loading from agricultural activities and urban development is neither controlled nor regulated. In addition, the widespread, adverse effects due to the continuing practice of dredging through barrier islands to provide outlets for boaters has been completely ignored by state and federal regulatory agencies. The deleterious impacts of such dredging on Apalachicola Bay, Choctawhatchee Bay, and Perdido Bay have been well established here and elsewhere in the scientific literature. Restoration of these effects is not being contemplated at this time. The evidence indicates that dredging openings to the Gulf leads to severe salinity stratification in inshore waters with attendant adverse effects on benthic habitats in deeper parts of the affected areas. These effects enhance the impacts of eutrophication and therefore should be taken into consideration as important modifying factors when evaluating the impacts of nutrient loading and when determining the scope of restoration programs for hypereutrophicated bay systems.

II. "ECOSYSTEM" RESEARCH

The scientific determination of the impacts of nutrient loading is fundamental to regulatory processes and restoration efforts. Until now, the generally accepted methods used in ecological research in coastal systems has not been appropriate for determinations involving the initiation and impacts of plankton blooms on coastal systems (Bricker et al., 1999). In what has turned out to be a prophetic paper, Dayton (1979) gave an early warning to what he described as the fragmentation of the various disciplines of ecology into warring camps. This fragmentation has reduced the serious and complex problems of aquatic ecology into pedestrian positions rather than new and creative ways of using scientific inquiry to understand how these systems work. The compartmentalization of what is supposed to be an holistic science has led to a general lack of attention to the interacting components of complex coastal ecosystems. The inability of ecologists to frame appropriate questions at the ecosystem level has accompanied the continued success of influential faddists who currently control research funding. The Japanese term *mu* ("no thing"; Dayton, 1979) appropriately describes the current state of much of what passes for ecosystem ecology today. The *mu* concept means that the general context of many of the research questions being asked are simply not relevant; the results of such research are "not even wrong" (Dayton, 1979). Inadequate sampling over relatively short periods has resulted in a plethora of publications that define the unexplainable as "noise." Few papers in ecological journals even bother to give details concerning the database that was used for the research that underlies the written word. Consequently, interpretation of the validity of the results is often difficult if not impossible. False assumptions are then generated into easy generalizations that, with repetition, become scientific facts.

As for the funding process in aquatic ecology today, Dayton's "Matthew Effect" (i.e., "For unto every one that hath shall be given, and he shall have abundance; but from him that hath not shall be taken away even that which he hath") (Dayton, 1979) is still alive and well. The wars have continued between the theoretical ecologists in their air-conditioned offices spinning complex and totally untestable models on how nature should work, and the natural historians who too often adhere to the "Sorcerer's Apprentice" approach to data collection. Taxonomy increasingly has become a lost art, and natural history continues to be ignored (Peterson, 1992). Experimental elegance based on the *mu* principal is still preferred to the difficult, time-consuming interdisciplinary evaluation of ecosystem processes over periods that account for both seasonal and interannual variation. The "publish or perish" syndrome in most universities has contributed to the patch-quilt approach to ecosystem research that is so popular today. This minimalist approach continues to drive most of the ecological research concerning aquatic systems. The correlation of the average duration of an ecological project with the length of time it takes for an M.S. degree in ecology is probably not spurious. The current reductionist approach that dominates academic research in the environmental sciences also continues to produce crops of graduate students who know little with regard to the methods necessary for successful holistic, ecosystem-level research.

Ecosystem research is currently one of the least well understood disciplines of modern science. Ecosystem analyses encompass various disciplines that are coordinated to answer specific questions concerning how large, multidimentional systems work. This kind of research entails integration of diverse studies in a representative area over periods long enough to account for both seasonal and interannual variability of basic physical–chemical and biological factors. The entire research process should be built around a set of preliminary questions. The effort should include descriptive monitoring, field and laboratory experiments, and modeling that includes integration of the key components of the system in question (Livingston et al., 2000). Field sampling should be synoptic, with integration of all of the variables (physical, chemical, biological) as the basis for the analytical (statistical, modeling) framework of the project. Sampling methods should be quantified before project activities are begun. Field sampling should be carried out with regard to known periods of variation of the variables to be analyzed. The resulting database should be designed with regard to assumptions of the statistical models to be used for analysis. Data taken haphazardly in terms of station location and time of sampling cannot be used for such purposes. It is not possible to stitch together disparate databases taken in different areas over unrelated time periods. In short, the output needed for a given research question should be consistent with the methods used in collecting the data and the models necessary for analyses of the data. Fundamental changes in field sampling methodology, laboratory processing of the field collections, and taxonomic identification of biological samples during the course of a given project can lead to inconsistencies that make interpretation of the final database difficult if not impossible. In the case of long-term ecosystem studies, consistency in project methodology is extremely important. Although it is inevitable or even necessary that new research questions be developed as information is acquired, it is also necessary that the results of the project be continuously analyzed so that new program initiatives can be added without compromising the integrity of the database.

A. Gulf of Mexico

One example of a recent "ecosystem" approach to problems concerning coastal hypereutrophication involves a multiagency effort in the Gulf of Mexico that has been defined in the "Gulf of Mexico Symposium 2000" (U.S. Environmental Protection Agency). Unfortunately, despite a plethora of papers, scientific and otherwise, to be presented at the Symposium, there is not one ecosystem-level study in the entire program. Nowhere is there any mention of an integrated approach to meaningful coastal research concerning important problems in the Gulf of Mexico. Part of the problem lies in the increasing specialization of the various fields in environmental sciences. This trend of knowing more and more about less and less is particularly apparent at the university level, and is derived, in part, by the preoccupation of many ecological researchers with so-called experimental ecology. Although experimental work is essential in the determination of specific hypotheses in environmental science, the corollary (i.e., everything that is not experimental is worthless) has become the basis for the current preoccupation with reductionist approaches to research in aquatic systems. There is little attention to the development of inclusive scientific questions concerning ecosystem-level processes. Many so-called ecosystem analyses have been largely reduced to a series of unrelated studies with limited research goals and with little commonality of purpose or design. Holistic approaches to what should be an integrative research effort are therefore largely lacking in many of the current environmental analyses of coastal systems. In short, the so-called Gulf of Mexico initiatives for the evaluation of eutrophication problems remain in a primitive state that is saddled by an inadequate approach to the ecological definition of problems associated with nutrient loading. The lack of direction of the research effort translates directly into the absence of effective solutions to what has been identified by various research agencies as a major problem that will inevitably grow to even greater proportions in terms of affected coastal systems in the near future.

B. Chesapeake Bay

There are many examples in the recent scientific literature with regard to the use of ecosystem research in the evaluation of the past history and current status of coastal systems. The Chesapeake Bay system is one of the most studied estuaries in the world. However, despite decades of ample research funding and a plethora of projects concerning this system in recent years, there have been unprecedented declines of fisheries production during the past decade in what was once the most productive estuarine system in the United States. The history of the deterioration of the Chesapeake Bay system has been well documented. For example, according to MacKenzie (1996), oyster production from Maine to eastern Florida has declined in recent decades. In former times, the Chesapeake was one of the prime sources of oysters (*Crassostrea virginica*) in the northern hemisphere. The loss of most of the oyster production in Chesapeake Bay in the early 1990s has been attributed mainly to disease that has also "devastated" oyster production areas of Delaware Bay. It has been reported that the density of phytoplankton in Chesapeake Bay has been such that seagrass beds have been reduced or completely eliminated in major parts of the system, and hypoxia at depth has increased. According to Newell (1988), algal blooms have resulted, in part, from the loss of the oysters, which, in less-polluted times, cropped most of the phytoplankton in a relatively short period of time. The possibility that changes in the phytoplankton community could have contributed to the unprecedented loss of oysters in the bay was not contemplated.

Recent publications concerning Chesapeake Bay reflect the results of a series of well-funded research programs. Some of these studies have provided the basis for a review of the eutrophication question. Malone et al. (1996), in a review of nutrient limitation of phytoplankton productivity in Chesapeake Bay, found that phytoplankton growth rates were limited by dissolved inorganic phosphorus (DIP) during spring when biomass reaches an annual maximum. Such growth rates are limited by dissolved inorganic nitrogen (DIN) during summer when phytoplankton growth rates are maximal. Riverine DIN inputs are associated with seasonal accumulation of phytoplankton biomass in salt-intruded reaches of the bay, whereas the magnitude of the spring diatom bloom is limited by the supply of dissolved silica. Phytoplankton production was defined by chlorophyll concentrations in these studies.

Dauer et al. (2000) used associations of macrobenthic communities as indices of anthropogenous effects in Chesapeake Bay. However, there were certain qualifications for these results that are common to many ecosystem studies being published today. The Chesapeake analyses were carried out only during summer months with the assumption that it is only during warm periods that the primary impacts of cultural eutrophication occur. The analysis was based on samples taken from July through September from 1984 through 1991 due to "sampling limitations of some of the data sets." Statistical tests were run on averaged (not site-specific) data because "the Chesapeake Bay monitoring program elements were conducted independently, and the temporal and spatial distribution of sampling differs considerably among them." In other words, there was no effort to coordinate the various elements of the field sampling program, and disparate sampling collections were carried out independently and were later stitched together and averaged for statistical purposes. In addition, as is the rule in many publications today, no detailed (i.e., quantitative, species-specific) phytoplankton data were reported; chlorophyll *a* was used as a substitute for information concerning phytoplankton activity. Thus, the single most important group of organisms concerning the question of eutrophication effects was largely ignored. The determinations were also carried out in mesohaline areas of the bay with little regard to the multiple nutrient entry points to the system.

Dauer et al. (2000) found that the distribution of low DO was extensive and, based on regression analyses, explained 42% of the benthic index of biotic activity. Sediment contamination (i.e., toxic agents) accounted for 10% of the variation of this index. Residual variation of condition index was only weakly associated with eutrophication indices (total nitrogen, total phosphorus, chlorophyll *a*) after removal of the effects of DO. The benthic condition was negatively associated with urbanization, point source loadings, and total nitrogen loadings. These results were consistent with other studies that associated severely degraded benthic communities with very low oxygen concentrations in the Chesapeake system (Holland et al., 1977; Dauer et al., 1982; 1992; Diaz and Rosenberg, 1995). The naturally hypoxic condition due to stratification (Malone et al., 1988; Malone, 1992) was mentioned as a possible cause in addition to high rates of deposition of particulates to the benthos (Kemp and Boynton, 1992; Malone, 1992). These effects were attributed to organic matter passed to the stratified mesohaline parts of the bay from oligohaline areas (which were not part of the study). Benthic communities below the pycnocline in mesohaline waters were the most degraded in the bay (Holland et al., 1977; Dauer et al., 1992). Thus, based on relatively few months of an incomplete data set in only one part of the Chesapeake system, the main conclusion was that the primary problem in the bay was hypoxia at depth. Conditions in the bay during fall, winter, and spring periods were not considered relevant to the eutrophication question. It was not considered that hypoxia, while an important contributing factor to the deterioration of the benthos, might have been only the hands of the clock. It is possible that the actual clock mechanism was to be found elsewhere (i.e., in the connection between altered phytoplankton assemblages due to anthropogenous nutrient loading, deterioration of sediment quality, and associated alterations of the food webs of the bay).

It is possible that, using the same results, another interpretation could be made concerning the deteriorated state of mesohaline parts of the Chesapeake system. The situation in Chesapeake Bay could be analogous to that in Perdido Bay. In addition to ignoring the bay during cooler months of the year, Dauer et al. (2000) did not take into account the potential effects of long-term changes of meteorological (i.e., drought–flood) cycles on bay habitats. According to the authors, there was a minimal effect of eutrophication on the benthos in the absence of low DO. However, it is possible that the effects of eutrophication could have been temporally disconnected. Nutrient loading into oligohaline portions of the estuary during winter–spring periods could have resulted in a series of plankton blooms in succeeding months with associated discontinuities of the phytoplankton community that could have been translated into disjunct food web relationships that eventually resulted in the observed deterioration of the benthic communities. It is quite possible that hypoxia was a contributing factor during warm periods, but that basic alterations of the bay continued throughout the entire year, with the observed results reflecting years of accumulated impacts on the sediments. Thus, DO, while an important contributing factor, could have been only part of the problem relative to the condition of the bay.

It has been generally acknowledged that non-point sources are among the leading cause of water quality problems in the Chesapeake system (Dauer et al., 2000). Currently, there are commendable efforts to start restoration efforts regarding non-point sources of nutrients. However, although agricultural and urban land uses were considered responsible for the degraded benthic community condition, the relationship of the phytoplankton communities to these connections was not explained since phytoplankton have not been studied in a systematic and long-term fashion. At intermediate levels of eutrophication, macrobenthic communities responded with increased abundance and/or biomass whereas, at more advanced stages of the eutrophication processes, there could have been an associated deterioration of the benthos that could have been explained as a direct result of adverse effects on the food web. The observed deterioration of seagrass beds and bivalve mollusks would be consistent with this hypothesis. The relationship between nitrogen loadings and benthic condition, and the absence of this relationship with phosphorus loadings was consistent with the conclusion that nitrogen was more important to eutrophication in the Chesapeake system, at least during summer months. However, phosphorus loading to the upper bay during winter–spring periods could have produced blooms that had either delayed direct effects or indirect effects on the subject (mesohaline) areas of the bay that would have gone unnoticed, based on the assumptions made concerning the overall sampling program. Without adequate phytoplankton data, these questions were not even asked much less unanswered.

The results of the studies in the Chesapeake system (Dauer et al., 2000) are thus consistent with an alternate hypothesis that the correlations of benthic deterioration with nutrient loading could be explained by long-term direct and indirect effects of phytoplankton blooms on benthic associations in the upper bay. It is possible that such events could have followed processes not unlike those observed in some of the hypereutrophicated Gulf estuaries. It could also be argued that periods other than summer months (i.e., cold months when DO is not a problem, spring periods of recruitment, etc.) could play an important role in the observed effects on mesohaline parts of the bay. Seasonally directed nutrient loading into oligohaline areas and the resulting plankton blooms could contribute to the observed conditions in deeper, salinity-stratified mesohaline areas during summer months.

Sediment deterioration and food web alterations due to bloom-induced changes in the phytoplankton communities could be a primary reason for the observed changes in the benthic invertebrates in many coastal systems affected by anthropogenous nutrient loading. Dissolved oxygen, while seasonally important as a contributing factor to the observed effects, could be a secondary factor relative to the year-round deterioration of sediment quality. The strong correlation of hypoxia with reduced biotic quality could be related to the fact that many published studies are restricted to summer periods. However, hypoxia is seasonal whereas sediment deterioration and disruption of associated food webs can occur throughout the year. There has been a severe reduction of SAV in Chesapeake Bay (250,000 ha to current areas of 25,000 to 35,000 ha; Moore et al., 2000). Regrowth in mesohaline areas of the bay has been defined as "poor." These changes are also consistent with an impact that could be related to phytoplankton responses to nutrient loading. It is highly unlikely that hypoxia is the primary cause of the loss of seagrasses in the Chesapeake system. The deterioration of the oyster industry in the Chesapeake system is also consistent with a possible effect of altered phytoplankton communities due to anthropogenous nutrient loading. In short, correlations based on inadequate data could lead to misleading conclusions regarding cause-and-effect relationships of nutrient loading and bay response.

C. Florida Bay

There are virtually no detailed studies of phytoplankton community dynamics that have been coupled with long-term, detailed analyses of associated benthic and epibenthic coastal communities. The lack of such information has led to the publication of studies that purport to determine ecosystem-level processes that are related to nutrient loading and eutrophication effects when, in fact, such processes remain largely undetermined. An example of this trend is the published scientific literature concerns the deterioration of Florida Bay in recent years.

Over the past decade, seagrass beds in Florida Bay have undergone extensive deterioration (Fourqurean et al., 1999). The Florida Bay situation, however, can only be understood within the context of the long-term, systematic destruction of the Kissimmee River–Lake Okeechobee–Florida Everglades–Florida Bay–Florida Keys/Reefs system over the past half century. The Kissimmee River has been virtually destroyed by the U.S. Army Corps of Engineers as part of an extensive and largely successful operation to channelize the river. Currently, Lake Okeechobee is undergoing extensive bluegreen algae blooms due to cultural eutrophication based, in part, on erratic control of lake water levels and the loading of agricultural wastes (high in nutrients) into the lake. Major parts of the Florida Everglades have been destroyed by hydrological alterations by the U.S. Army Corps of Engineers and nutrient (i.e., phosphorus) loading by agricultural interests. The agricultural interests have benefited from the draining of the swamps and redirection of water flows in one of the most extensive plumbing jobs in history. Florida Bay, which has been studied for years by various state and federal agencies, recently experienced algal blooms and corresponding losses of seagrass beds in major parts of the bay. There has been a general deterioration of Florida Bay between 1987 and 1991 (Fourqurean and Robblee, 1999). For the past three to four decades, the Florida Keys have undergone massive urbanization with very little in the way of regulation regarding water or sediment quality relative to urban runoff. The associated hermatypic reefs of south Florida are in a state of decline with recent reports of extensive algal growths. The entire system, from the Kissimmee River to the Florida Keys, is thus in a state of deterioration. A common factor in this vast system relates to interconnected freshwater flows that, during earlier periods before the massive assault by state and federal agencies, urban development, and agricultural activities, contributed to an ecological system that was unique in the northern hemisphere.

Despite a series of relatively well-funded studies that have been undertaken in Florida Bay in recent years, the factors that are responsible for the observed habitat deterioration of this system remain "poorly known" (Fourqurean et al., 1999). The deterioration of Florida Bay was, at one time, considered to be associated with the anthropogenous destruction of the natural flow of fresh water from the Everglades into Florida Bay and the rampant urbanization of the Florida Keys. However, the almost complete lack of relevant nutrient loading information and water quality data prior to the algal blooms contributed to confusion regarding the causes and effects of the observed losses of seagrass beds. Boesch et al. (1993), in a review of the Florida Bay research, indicated that algal blooms predated the seagrass die-offs, and that such deterioration may have been initiated by long-term increases of land-based nutrient loading somewhere in the system. This possible explanation was in direct conflict with the findings of Fourqurean and Robblee (1999) that associated the blooms with preceding deterioration of the seagrass beds. The Boesch panel found that "virtually nothing was published on the Florida Bay algal blooms" and "surprisingly little quantitative historic information exists on the Bay's water quality."

The results of the most recent studies of Florida Bay can be found in an entire issue of the journal *Estuaries* (Fourqurean et al., 1999). This volume consists of the following: three papers on historic reconstructions of the history of Florida Bay (mollusk shell isotope records, paleoecological analyses, proxy chemical records in coral skeletons), seagrass distribution analyses, recruitment records of pink shrimp (*Penaeus duorarum*), a series of descriptive fish studies (based on very limited data), and an analysis of the American crocodile in Florida Bay. Analyses of water clarity and bottom albedo in Florida Bay (Stumpf et al., 1999) shed little light on processes associated with the decline of the seagrass beds. Rudnick et al. (1999) evaluated the importance of the Everglades watershed as a source of nitrogen and phosphorus to Florida Bay. It was determined that <3% of all phosphorus inputs and <23% of all nitrogen inputs were from freshwater runoff from the Everglades. The Gulf of Mexico was viewed as major source even though nutrient loading from the south was not really studied. Nutrient data used for the loading determinations were derived mainly from reports by federal agencies (i.e., the U.S. Environmental Protection Agency, 1993). No comprehensive nutrient loading analyses were reported in the Florida Bay papers.

Boyer et al. (1999) gave a review of water quality in Florida Bay from 1989 through 1997. Phytoplankton and zooplankton data were not taken; the phytoplankton component was represented by turbidity, TP, and chlorophyll *a*. Conclusions from what amounted to 6 years of water quality data were tentative at best. It was concluded that "the death and decomposition of large amounts of seagrass biomass can at least partially explain some of the changes in water quality of Florida Bay, but the connections are temporally disjoint and the processes indirect and not well understood." Tomas et al. (1999) found that the blooms were mixed populations of cyanobacteria and diatoms although no consistent community-level data concerning long-term changes of the phytoplankton were given. The blooms appeared to be seasonal and varied in different parts of the bay. The authors quoted Fourqurean et al. (1993) who hypothesized that offshore water was a source of phosphorus for Florida Bay. Nutrient ratios were used to estimate nutrient limitation for phytoplankton (Tomas et al., 1999); however, the authors qualified their results by the limitations involved in using static nutrient concentrations to describe a dynamic process. The source(s) of the nutrients loaded to Florida Bay remained undocumented.

Many of the recommendations for research in Florida Bay by Boesch et al. (1993) were not followed by researchers in the Florida Bay system (Fourqurean et al., 1999). Overall, very little oceanographic information was presented. The current structure, turnover rates, and most of the physical processes that might be involved in bloom generation in Florida Bay were not reported. There was little in the way of detailed phytoplankton analyses. There were no trophic analyses concerning how the food webs of the bay were affected by observed changes in the seagrasses. The primary nutrient sources to Florida Bay remain unknown. The question of whether the nutrients came from the seagrass deterioration or whether the seagrass beds were actually killed by plankton blooms caused by allochthonous nutrient loading has not been resolved. Nutrient loading, nutrient fluxes to the bay, nutrient concentrations in the bay, and the time-based dynamics of phytoplankton assemblages of Florida Bay were not studied in a comprehensive, long-term fashion. In short, the situation in Florida Bay remains nebulous.

Based on the published studies of Florida Bay, it will be difficult to evaluate how a proposed $8 billion restoration program for the Florida Everglades will affect this system. There has been little evaluation of the dynamic interactions between Florida Bay and the Florida Keys. Porter et al.

(1999) indicated that "very little is known regarding the quality, frequency of occurrence, distribution, and transport of Florida Bay water masses into and within the Florida Keys National Marine Sanctuary." The possible adverse effects of Florida Bay water masses on the coral reefs of south Florida remain unknown. The potential effects of urban development in the Florida Keys on Florida Bay have not been systematically evaluated. Porter et al. (1999) pointed out that the pattern of measured coral decline in Florida Bay and the Florida Keys is consistent with adverse effects of Florida Bay water. However, without adequate oceanographic data, nutrient loading analyses, and phytoplankton information, we are left almost completely in the dark with respect to the processes involved in the possible connections between Florida Bay and the south Florida coral reefs.

The citations of studies in the Chesapeake and Florida areas are merely representative of worldwide research efforts that are routinely published in scientific journals. The individual authors of these studies are competent scientists who are well respected in the scientific community. The fault lies in the systematic assumption by many aquatic ecologists today that an ecosystem analysis can be successfully carried out based on a series of disconnected, short-term studies by diverse research interests that operate more or less independently of one another. The losses of ecologically unique ecosystems of south Florida have occurred over a period of decades. The lack of consistent, long-term ecological data has contributed to the current state of confusion and controversy regarding restoration. This situation, involving proposed restoration projects that can be measured in $billions, has been due, in part, to an almost complete lack of a systematic ecosystem approach leading to a comprehensive, well-analyzed base of information. How did this happen? The situation in Florida Bay is indicative of various ecosystem-level projects in coastal areas of the United States. The almost blind adherence to a failed patch-quilt approach to ecosystem research by funding agencies, scientific journals, and the reductionist researchers who dominate the funding agencies has resulted in massive and continuing destruction of aquatic resources in Florida. This situation can be extended throughout the country. As stated by Hackney and Adams (1992), it is difficult to "avoid pessimism" in the contemplation of the future of aquatic ecosystems in the southeastern United States or, for that matter, in the country as a whole.

D. Scientific Overview of the Eutrophication Question

As is the case in most scientific research, funding is central to the successful completion of a given project. A long-term, reliable source of funding is essential to ecosystem studies. There have been successful efforts in the past such as the Great Lakes Program of the U.S. Environmental Protection Agency. However, this model has not been followed in most ongoing coastal programs. There has been a scarcity of consistent, long-term funding for system-level coastal research for decades. This problem has been exacerbated by political interference at various levels of government concerning research projects that are viewed as a threat to land development schemes and the operations of influential industries. The general lack of information concerning the long-term effects of non-point pollution (i.e., agricultural activities and urban development) is one example of how political pressure at the local, state, and federal levels has precluded a meaningful evaluation of important ecosystem issues such as cultural eutrophication. Although there has been a recent effort by the U.S. Environmental Protection Agency to address non-point source problems in aquatic systems, the outcome is in doubt in the face of open hostility by members of the U.S. Congress.

The lack of funding for ecosystem research is not the only stumbling block. Often the scientific community gets in the way of good research even when funding is available. In some cases, the research funds are divided up by influential scientists along political lines rather than scientific needs so that different research institutions get their "cut" regardless of their contribution. There is usually no central control of the project with respect to integration of research efforts, the production of a uniform and consistent database, and comprehensive analysis of the results. At the end of a given project, isolated pieces of information are usually patched together, and are analyzed using statistical analyses that are often not appropriate for the data taken. These results are then published as an ecosystem study regardless of the inadequacy of the information generated. In many publications, the adequacy of the database is often ignored by reviewers and journal editors with the result that problems associated with incomplete and uncoordinated field data are papered over by elegant statistical and modeling efforts. Often, the results consist of analyses of a series of unrelated databases with no possibility of an integrated approach that can then be used to answer specific (ecosystem-level) questions. This misuse of funds is protected by in-house control of the review process, a scientific version of the "good-old-boy/good-old-girl" political process that runs many of the governmental and regulatory activities in the United States today. The final result is scientific uncertainty and the continued destruction of coastal resources.

In their comprehensive review of southeastern estuaries, Dardeau et al. (1992) listed the areas of estuarine and coastal research "in which information is lacking or poorly understood." In the area of nutrient loading/eutrophication, the list included the following:

1. Nutrient sources and control technology.
2. Role in stimulating primary producers: water column and benthic coupling of primary production and secondary production.
3. Exchange rates and recycling of nutrients and organic material: geochemical and biological.
4. Carrying capacity.

The central issues of research on eutrophication processes remain essentially the same today. The above definition should include a realistic appraisal of spatial and temporal variation of the many factors that contribute to the proliferation of plankton blooms in coastal areas. Due to recent publicity generated concerning plankton blooms in East Coast estuaries, efforts in this direction have begun. The recent publication of a "national strategy for the development of regional nutrient criteria" (U.S. Environmental Protection Agency, 1998) lists the priorities for assessments of estuaries and coastal marine areas. The overall organization of the strategy was broken into three categories (seagrass-dominated systems, plankton-dominated systems, and macroalgae-dominated systems) even though these systems are often inextricably linked in any given coastal system. The emphasis was on chlorophyll *a*, cyanophytes, dinoflagellates, and diatoms (including HABs), documentation of the incidence and location of blooms, DO, the role of silica relative to the nitrogen and phosphorus in phytoplankton blooms, and changes in consumer communities. No criteria were outlined for analysis of phytoplankton community dynamics, the linkage of nutrients with alterations of plankton communities, the possible relationships of the water column and sediment microphyte assemblages, sediment quality as a major criterion for processes related to hypereutrophication, and the effects of nutrient-induced changes on associated food webs.

In a recent review of eutrophication by the National Academy of Sciences, Howarth et al. (2000) defined the principal issues concerning the eutrophication question in coastal systems. In the development of quantitative measures of estuarine condition, the authors concluded the following:

> The primary quantitative indicators may include variations in algal composition, elevated concentration of chlorophyll *a*, and an increase in the extinction coefficient for a given waterbody....*Because algal composition changes are typically determined with expensive techniques such as microscopy or pigment analyses, this indicator is less useful for large-scale screening analyses* (emphasis added). Secondary indicators of eutrophication include changes in the dissolved oxygen regime of the system; changes in the areal extent of seagrass beds; incidence of harmful algal blooms or nuisance blooms and changes in their frequency, duration, and areal extent.... For mesotrophic (less impacted) systems, annual measures in July or August when symptoms could be expected to be more severe, should suffice to characterize worse-case conditions. For eutrophic or hypereutrophic systems, more frequent measurements are needed, at least twice per year.

Howarth et al. (2000) pointed out that there are few "well-documented examples" of the effects of nutrients on phytoplankton community structure. Nevertheless, the authors then gave relatively little consideration to seasonal successions among bloom species or the individual nutrient requirements of planktonic bloom forms that can be described as noxious to other species. Interannual progressions of bloom species composition were not considered. The dismissal of "algal composition" as an indicator of hypereutrophication sidesteps the fact that phytoplankton community composition is central to questions concerning nutrient loading problems. The food web connections are critical to the eutrophication question. As for the suggested sampling frequency, what possible use is annual sampling when blooms can and do occur throughout the year? In addition, the connection of blooms to sediment quality was largely overlooked, and the possible connection between benthic microalgae and the sediment quality was not addressed. The authors made only tangential reference to the effects of altered phytoplankton composition on associated food webs. The emphasis on nitrogen as the primary limiting nutrient by Howarth et al. (2000) overlooks a growing body of research in the Gulf of Mexico that indicates the importance of phosphorus as a limiting and/or co-limiting nutrient. This is due largely to overgeneralizations that accompany the preoccupation with relatively few estuaries along the East Coast by politically influential scientists. The possibility of nitrogen + phosphorus interactions in the eutrophication process, even in East Coast estuaries, should be considered. Given the current state of knowledge concerning the effects of nutrient loading on diverse coastal systems, the emphasis on nitrogen is premature at best.

Once again, the Howarth committee represents highly competent and well-respected scientists; the problem lies in assumptions made on admittedly poorly understood processes associated with phytoplankton bloom response to anthropogenous nutrient loading and the effects of such changes on associated food webs. The lack of attention to these issues, in addition to problems with sampling inadequacy in space and time, indicates a misunderstanding of the basic processes involved with phytoplankton blooms in coastal systems.

The increased frequency of plankton blooms (Smayda, 1990) has been broadly associated with increases in the areal extent of bottom-water hypoxia in various marine systems (Officer et al., 1984; Justic, 1987; Anderson and Rydberg, 1988; Rabalais et al., 1996). Rabalais et al. (1996) suggest that increases in nitrogen loading may lead to nondiatom production and suggest that increased nitrogen and phosphorus concentrations will favor flagellate growth. Rabalais et al. (1996) related increased freshwater phosphorus loading to increased silica burial in freshwater sediments in the Mississippi River system. These authors found that increased loading of nitrate and phosphorus and decreased loading of silica led to shifts of phytoplankton species in Gulf areas offshore of the Mississippi River. These shifts of phytoplankton community structure in response to increased nutrient loading were based on observed changes in the biological uptake of silicate and the accumulation of biologically bound silica in sediments, which increased in response to increased nitrogen loading in the Mississippi system. Hypoxia in offshore areas followed the increased primary production.

Using published reports of phytoplankton species composition for 1955 to 1957 near the delta (Simmons and Thomas, 1962) and other data from west of the delta in 1972 to 1973 (Fucik, 1974; Ward et al., 1979), Rabalais et al. (1996) made a comparison of this information with more recent phytoplankton data taken near the delta and in other areas (Dortch et al., unpublished data). The comparison was considered "qualitative since there were differences of locations, seasons sampled, and methodology." Heavily silicified diatom species found during earlier periods were absent in the most recent samples. Other species such as *Trichodesmium* sp., which were present in later (but not earlier) periods, were used as possible indicators of decreased availability of silica. The species *Nitzschia pungens* (considered to be a toxic form that is now called *Pseudonitzschia pseudodelicatissima*) found in recent blooms in the area of interest is a heavily silicified diatom which has not responded to the reduced SI:N ratios. However, other purported shifts included the recent discovery of the dinoflagellate *Dinophysis caudata* that was not found in earlier studies. The authors concluded that the shift from heavily to lightly silicified forms may have altered carbon fluxes with subsequent carbon loading to the bottom causing oxygen depletion. Another hypothesis put forward by Rabalais et al. (1996) was that less preferentially grazed plankton increased carbon fluxes to the bottom with the same result. These shifts, together with the benthic foraminifera data were thus directly related by Rabalais et al. (1996) to increased organic loading to the bottom which was accompanied by increased bottom water hypoxia. The authors suggested that decreased phosphorus loading to freshwater areas and reduced offshore nitrogen loading (together with increased silica availability offshore) would lead to a return to former phytoplankton communities with associated decreases in benthic hypoxia.

Howarth et al. (2000) correctly point out that national documentation of nutrient overenrichment "is limited by the inconsistency of data collection." This criticism could be made for the overwhelming majority of monitoring efforts, past and present. Such efforts can be characterized by a lack of consistent monitoring schedules, relatively little in the way of uniform protocol development, a lack of coordination among contributing researchers (e.g., station placement, synoptic sampling efforts, etc.), and a general paucity of consistent, interannual biological data in representative coastal systems. The use of indicators such as nutrients, DO, light conditions, primary production, and respiration taken once or twice a year would not have indicated the problems experienced with nutrient loading in the Perdido system. It is questionable that 1 or 2 years of data would have been anything but misleading in terms of causes and effects of nutrient loading on that system. By the time the seagrass beds in upper Perdido Bay showed adverse effects due to the blooms, a great deal of damage had already been done to the most productive parts of the estuary. As for modeling efforts, Howarth et al. (2000) correctly point out that "in spite of the effort and resources expended on the Chesapeake Bay model …," it has been suggested that there are caveats in interpretations of these watershed-water quality models. These include (1) model predictions based on several uncertain assumptions, (2) models that calculate "average" conditions "in a variable world," and (3) models that assume immediate benefits of source reductions (Boesch et al., 1997). There is an even more important problem with the current preoccupation with models by various government agencies. There is so little quantitative data concerning changes that occur in phytoplankton assemblages with even moderate anthropogenous additions of nutrient loading, that no model has been developed that can predict the initiation and succession of plankton blooms even in well-studied areas such as Chesapeake Bay.

Many processes associated with eutrophication in coastal areas are poorly understood, and therefore modeling efforts should be qualified by their assumptions according to Howarth et al. (2000). Models are only as good as the data on which they are based. Model calibrations should be field-verified with actual data. Currently, there is little in the way of a rigorous examination of the assumptions of eutrophication modeling. The commonly held belief that chlorophyll *a* is a good indicator of phytoplankton communities should be reviewed and possibly altered. Although chlorophyll *a* was significantly associated with phytoplankton biomass and numbers in the Perdido system, the relatively low r^2 values indicate that direct representation of the phytoplankton by this index should be qualified. The fact that chlorophyll *a* was not closely associated with bloom successions in Perdido Bay could indicate that reliance on this index for the characterization of phytoplankton blooms is not correct. In addition, the lack of consistent, long-term scientific data that has accompanied the use of linear statistical models may be problematic in attempts to define processes that could be better described using nonlinear statistical models. False assumptions and oversimplification of the complexity involved in the interactions between phytoplankton assemblages and associated food webs could lead to inadequate restoration efforts that would be costly in terms of lost coastal productivity.

It has become popular to refer to "ecosystem ecology" when more than one scientific discipline is applied to a given environmental question. Within this context, studies of 1 to 2 years are commonly considered "long term." According to Dauer et al. (2000) and Howarth et al. (2000), one or two samples per year are adequate to monitor coastal systems. The assumptions that underlie this generalization bring up the issue of just which 2-year period sampled once or twice each year would have been representative of processes in Perdido Bay or Apalachicola Bay? Unfortunately, integrated, long-term, interdisciplinary research remains a scientific rarity today. The first step in any ecosystem-level study should be the definition of the research questions and a detailed review of the proposed field program to define and qualify the context of the research questions within the limits of the study area and the temporal sampling regime. Ecosystem analyses require considerable planning prior to initiation of the study. An understanding of the scale of the effort is fundamental to the successful completion of system-level operations (Livingston, 1987b). A preliminary field screening program is therefore necessary to establish the geographic extent of the study area, and to determine the specific hypotheses that form the basis of the study program. Habitat stratification follows, with a resulting plan regarding the establishment of station locations and the timed sequence(s) of sampling at various physical, chemical, and biological levels of organization. Sampling schedules should take into account the specific life-history cycles of the populations in question. The sampling program should then carried out with field data analyzed on a descriptive and quantitative basis leading to the establishment of preliminary models that allow summarization of the data and a redefinition of the research questions. Continuity of the synoptic sampling effort is critical to the successful completion of multiyear projects. Consistent quality control of the taxonomic aspects of the research over the years is essential.

The single most important factor in the execution of successful ecosystem-level research is a strong central organization for coordination of the project. There should be someone who is responsible for the development of an integrated, uniform database that is consistent with the assumptions of the statistical models that are to be used in the analysis. Field analyses should be carried out in a way that provides the context for experimental tests of hypotheses both in the field and the laboratory. These experimental tests can be carried out in many ways. The results of such tests can then be related back to the long-term field studies in the form of time series analyses, statistical determinations of relationships of field factors, and the resolution of specific experimental hypotheses relative to real and projected impacts of both natural and anthropogenous interventions.

Specific fields of research necessary to the eutrophication question have been simply left out of many of the current programs concerned with HABs. The connection between plankton bloom sequences and sediment quality remains undetermined. There is virtually no scientific literature concerning long-term changes of benthic microphyte communities. Fjerdingstad (1965) organized benthic microphytes relative to their ability to survive in polluted conditions. This early listing of benthic microphytes presented a detailed taxonomic framework within which the overall organization of this group of plants could be evaluated. These studies have not been followed up with regard to the effects of nutrient loading on coastal areas. The response of many bloom species is directly related to both increased loading of limiting nutrients (phosphorus, nitrogen, urea, iron, etc.) and the gradients/interactions of such nutrients as they enter the estuaries. The generation of blooms is governed by complex combinations of environmental factors that are subject to seasonal nutrient limitation requirements and interannual fluctuations in habitat variables. These variables have not even been identified much less quantified with respect to impact. Bloom species in the eastern Gulf are associated with environmental factors that follow seasonal changes in state variables and interannual drought–flood cycles. Temperature, salinity, and light appear to be integral to the limiting factors that direct the bloom sequences. Blooms initiated at the head of the estuary may eventually take over major parts of the bay and may even enter the Gulf, feeding on nutrients provided by both natural (i.e., rivers) and anthropogenous (point and non-point) sources. Blooms generated in the lower parts of a given estuary may have a completely different *modus operandi* with regard to both origin and effects. There is therefore a need to identify the location of bloom origin. The result of changes in the seasonal and interannual sequences of plankton blooms can lead to the reorganization of an entire food web in a given estuary. However, without integrated sampling programs, such changes can go unnoticed.

III. INFORMATION, REGULATORY "ACTION," AND POLITICAL CONTROL

The problems with the effects of nutrient loading on coastal systems are not solely located in limitations of the scientific research effort. There are other controlling factors. Florida represents one of the fastest growing states in the country so it is not surprising that urban development has been a dominant economic force in the state for decades. As part of the prevailing economic hegemony, urban development interests have been relatively unhindered by environmental considerations. This is the result of a number of causes that include prevailing regulatory policies at the local, state, and federal level, inadequate enforcement of existing environmental laws, and the general lack of adequate ecosystem-level scientific data concerning eutrophication issues. Currently, the adverse effects of nutrient loading on aquatic systems throughout Florida are extensive. Important parts of the entire ecosystem from the Kissimmee River to the coral reefs of the Florida Keys have been seriously damaged by nutrient pollution. At one point, the situation in the Florida Everglades deteriorated to the degree that a federal attorney and the Department of Justice had to file suit against the Florida Department of Regulation for pollution of Everglades National Park. Kushlan (1991) noted that 65% of the primitive Everglades has been drained and major parts of the entire system have been adversely affected by altered hydroperiods and changes in the flood–drought sequence. The inhibition of freshwater flows to many of Florida's coastal areas because of urbanization and agricultural activities has complemented the nutrient loading problems (McPherson and Nammett, 1991; Estevez et al., 1991).

Inadequate sewage treatment in coastal areas is a continuing problem in a number of Florida's coastal systems. According to Markley et al. (1990), high concentrations of coliform bacteria indicated that "serious sewage contamination problems occur in Biscayne Bay." Coliform bacteria densities exceeded Class III (recreational) water quality criteria at least 10% of the time at stations located in the tributaries or at bay stations near the mouth of the Miami River. Fecal coliform bacteria exceeded state water quality criteria at least 10% of the time at 28% of the stations monitored; all of these stations were located in tributaries to the bay. According to a comprehensive review of Biscayne Bay by the South Florida Water Management District (1994), this system has been historically characterized by serious problems related to urbanization of upland areas. The Metropolitan Dade County Wastewater Collection System was subject to thousands of spills of raw sewage per year with resultant impacts on freshwater and estuarine drainage systems associated with Biscayne Bay. The most obvious water quality problems in the bay were high ammonia and nitrate/nitrite concentrations (South Florida Water Management District, 1994). Various canals and the Miami River were considered to be the most degraded parts of the system. Sewage contamination was viewed as a "persistent water quality problem." The South Florida Water Management District (1994) indicated that "extensive contamination of the bay and tributaries by raw sewage continues to occur." Sources included boats, interconnected sewage and storm water systems, and emergency overflows from sanitary systems. Phosphorus was probably the chief limiting nutrient in the bay that could explain the typically low concentrations of this nutrient in the bay. Abnormalities in fishes as indicators of pollution were most often found in northern Biscayne Bay; 28% of the blue crabs collected from the mouth of the Miami River had external abnormalities.

As land development has moved to north Florida, formerly undeveloped areas have come under increasing population pressure. The St. Johns River system in northeastern Florida has been subject to extensive damage due to urbanization and industrialization (DeMort, 1991). In recent years, nutrient pollution has been associated with continuing plankton blooms and adverse effects on fishes in the St. Johns system. Sewage discharges have destroyed parts of wetlands creeks in the lower basin of the Apalachicola drainage system. The Escambia Bay and Perdido Bay systems have both been adversely affected by anthropogenous nutrient loading from point and non-point sources. The combination of sewage spills and storm water runoff has also led to serious problems in nearby Choctawhatchee Bay. Anecdotal information (K. Spencer, personal communication) indicates that there have been increased incidences of sewage spills into Choctawhatchee Bay in recent years with reports of recent damage to wildlife and fish kills in the bay and in associated offshore Gulf areas. Although extensive red tide (*Gymnodinium breve*) events have been associated with the deaths of hundreds of bottlenose dolphin, there has been little scientific verification of the source and extent of the hypereutrophic effects of anthropogenous nutrient loading on the bay. There is a general perception of serious deterioration of bay secondary production in the Choctawhatchee system, although there have been virtually no recent ecosystem-level studies to determine the current status of the bay. As usual, there has been virtually no regulatory activity associated with urban development in any of these north Florida systems.

The results of ecosystem research in coastal areas often conflict with local and regional development plans and, of course, the political forces that are aligned with such development. After carrying out 14 years of research in the Apalachicola system with funding from the Florida Sea Grant (National Oceanic and Atmospheric Administration) and local matching funds from the Franklin County Commission, the Florida State University research program (Livingston, 1984a) was eliminated as a result of pressure from local politicians and bureaucrats in the Florida Department of Natural Resources. Local politicians, with support from real estate development interests on St. George Island in Apalachicola Bay, were able to cancel an oyster project with the Florida Sea Grant even though the project received good outside reviews and was fundamental to the development of adequate planning for the region. There were other, associated problems. Research boats were sabotaged. There were anonymous death threats. Local and state developers were concerned that scientific data would help to perpetuate the Apalachicola oyster industry that could be then used to enforce water quality laws in the Apalachicola system. This was viewed as a threat to real estate development in sensitive parts of the system. Later, these same political forces got the Franklin County Commission to cancel the local matching funds for the oyster project. All of this was aimed at getting rid of research that would have evaluated the state of Apalachicola Bay with respect to pollution problems associated with urban development. At the same time, an effort to carry on ecological research in the bay through the established Apalachicola National Estuarine Sanctuary (now Reserve) was canceled by officials of the Florida Department of Natural Resources. To this day, such research has not been carried out in this important area, and the current ecological state of the bay remains in question.

Another form of political pressure concerning the generation of scientific data in coastal areas destined for urban development has developed through the legal system. As part of continuing pressure on those concerned with the protection of the Apalachicola resource during the 1980s, a major real estate developer filed a $60 million law suit against the Franklin County Commission, state and local planners, and the author of this book. In the largest civil rights suit to that time, these officials were charged with illegally interfering with the developers' rights to build condominiums on St. George Island. The case was dismissed 4 years later by a federal judge who took the unusual step of issuing awards of legal fees to the defendants with the following admonition:

> From the beginning of this case, the litigation was conducted in a manner that indicates the plaintiff's intent was to use the judicial system to harass those who opposed his development plans.…He brought suit against Dr. Robert Livingston … who advised officials in Franklin County on ways to protect Apalachicola Bay.… The court can only conclude that this litigation is being pursued in bad faith. (Judge William Stafford, Tallahassee, Florida)

The lack of adequate funding for long-term ecosystem work is thus the result of a number of contributing factors. Inadequate or corrupt management of research funding at various levels of government can be involved in the lack of useful information generated in coastal systems. Political control of the processes involved with ecosystem research has a long history. In the early 1980s, the newly elected Reagan administration established a "hit list" of scientists who served on the

Science Advisory Board of the U.S. Environmental Protection Agency. Various scientists (including this author) were listed with accompanying derogatory comments concerning their expertise and attitudes. The Reagan administration decimated the Science Advisory Board of the U.S. Environmental Protection Agency and filled the void with more compliant advisors.

The extensive and ongoing deterioration of aquatic resources in Florida over the past three to four decades is thus the result of complex combinations of many contributing factors. The results of such actions can be summarized as follows:

> The pattern has been to allow various interests to destroy public resources and then, when there is sufficient destruction to warrant a public outcry, to spend considerable amounts of money in restoration. This approach is both costly and ineffective in terms of any lasting solutions. In many cases, the actual value of the restoration effort is questionable. Unless preventive environmental action is taken through realistic and effective management of the rivers of Florida, the result will be the continued degradation of one of the most productive series of habitats in the country. (Livingston, 1991a).

A prime example of this process concerns the current (controversial) $8 billion effort to "restore" the Florida Everglades. The proposed restoration program represents the end product of decades of environmental destruction fostered by suppression of scientific inquiry, the proliferation of misinformation, and official indifference to a unique aquatic system. Much of the destruction of this system was carried out at public expense. The history of water diversion and adverse effects of anthropogenous nutrient loading in the Everglades illustrate how such impacts can be subtle early in the eutrophication process. With time, the impacts eventually become more extensive until it becomes difficult to ignore them. Restoration then becomes the new politically correct order of the day. Often the very agencies responsible for the original destruction are rewarded with extensive funds for the restoration effort. The Everglades model can be applied to the situation in many urbanized coastal areas of Florida and the country as a whole. The same mistakes are repeated with the same dreary results.

Little has been done to correct current problems associated with hypereutrophication. Based on the connection of nutrient-altered phytoplankton community functions and deterioration of associated coastal food webs, the projection of future impacts on coastal productivity due to anthropogenous nutrient loading (Bricker et al., 1999) should be viewed as extremely serious. Regulation of water quality in coastal areas of the United States is both selective and sporadic. Without enforcement of water quality laws, the likelihood of increasing deterioration of coastal habitat along the Gulf Coast becomes inevitable. However, without ecosystem-level data, enlightened regulation of continuing destruction of estuarine and coastal resources is not possible. The effects of cultural eutrophication are not always easy to define, especially when the most important organisms in the process (i.e., the phytoplankton) are largely ignored. There are definite, localized problems with various toxic substances such as PAHs, pesticides, certain metals, and PCBs in various coastal areas. However, the current emphasis on toxic substances in regulatory matters is misplaced. Such substances usually affect only parts of coastal areas (Dauer et al., 2000) whereas the initiation of blooms by nutrient loading can cause extensive habitat deterioration in the most important parts of river-dominated estuaries. Most of the outstanding unregulated sources of nutrients involve inadequate or malfunctioning sewage treatment plants and non-point sources due to urban and agricultural activities. There is a need for integrated and enlightened regulatory action concerning nutrient loading into coastal systems.

The basis of environmental regulation is adequate knowledge of the source of the concern and the extent of the problem. An example of misinformation currently being promulgated as fact can be found in a report by Summers (1999) and published under the auspices of the U.S. Environmental Protection Agency. Using monitoring data for "nitrogen, chlorophyll, and dissolved oxygen as indicators of eutrophication," Summers indicated that the percentage of the Gulf areas affected by nutrient loading was relatively low.

> The Gulf estuaries had moderate conditions overall for nutrients and dissolved oxygen. Definite nutrient problems were observed in >25% of the estuarine area in Louisiana and Texas and definite dissolved oxygen problems were observed in Alabama. (Summers, 1999).

This review differs considerably from the more comprehensive report by Bricker et al. (1999) who accurately reported that the Gulf of Mexico was significantly affected by increased eutrophication with almost half the estuaries characterized by high levels of eutrophic conditions. According to Bricker et al. (1999),

> Of all regions, the Gulf of Mexico has the greatest percentage of estuaries with high eutrophic conditions, despite low to moderate nutrient inputs. In addition to the prevalence of moderate to high susceptibility, the generally long growing season and warm waters result in higher levels of expression of eutrophic conditions.

This review (Bricker et al., 1999) included references to Perdido Bay and Choctawhatchee Bay even though there were relatively few such problems in the northeastern parts of the Gulf according to Summers (1999). Our data from Perdido Bay indicated that few of the usual indicators that have been used for monitoring eutrophication are effective indicators of the impacts of hypereutrophication. For example, dissolved oxygen may actually increase during certain phases of the hypereutrophication process. The utility of chlorophyll *a* as an indicator of plankton blooms is limited. Sumners (1999) did not even mention deterioration of sediment habitat due to baywide effects of hypereutrophication. His treatment of harmful algal blooms was simplistic and erroneous. It is especially disconcerting that such misinformation was promulgated by an agency whose Gulf Breeze Laboratory has been located on the Pensacola Bay system for decades. Some of the most extensive fish kills ever recorded due to hypereutrophication occurred in Escambia Bay. It is significant that the bay is currently being polluted from various point and non-point sources, most of which are associated with urban runoff that is virtually free of state or federal regulation. Summers (1999) stated that, in Escambia Bay, a sediment problem was attributed to poor flushing and high suspended sediment input from tributaries. However, neither of these factors constitutes the primary problems of the bay (Livingston, 1999).

These basic misunderstandings on the part of the Environmental Protection Agency concerning causes and effects of cultural eutrophication (exemplified by but not restricted to Summers [1999]) are the result of the lack of comprehensive scientific analyses of estuaries as ecosystems. For example, despite the long-term presence of the EPA Gulf Breeze Laboratory on the Pensacola Bay system, virtually no sustained ecosystem studies have been carried out in this estuary by U.S. EPA researchers. No phytoplankton studies have been reported. Although the past hypereutrophication of the bay was responsible for the destruction of seagrass beds, oyster bars, and unvegetated benthic habitat, no sustained, ecosystem-level studies of this condition have been carried out. The continued presence of algal bloom species as dominants of the plankton community have been associated with sediment deterioration and the present low levels of secondary production in upper Escambia Bay (Livingston, 1999). However, state and federal regulatory authorities have not addressed these problems. Consequently, little or no control of nutrient discharges has been carried out, a common occurrence of Gulf coastal systems.

There are other factors that have been overlooked by state and federal regulatory agencies. The adverse impacts of dredging and anthropogenous physical changes to the Gulf coastal systems have been almost completely ignored by Summers (1999) in his overview of the current state of Gulf estuaries. The connection between physical changes of habitat through dredging and connections of such effects with hypereutrophication, associated phytoplankton community changes, and impacts on coastal food webs has not even been considered. The lack of attention to the sedimentary environment of highly productive river–estuaries is a major problem with current regulatory practices. State and federal regulatory agencies do not regulate sediment quality. Again according to Summers (1999),

> Based on the percent area of each estuary that was contaminated, however, the majority of estuarine systems in all Gulf states were identified by EMAP as having fair to good sediment quality.

This statement was based on the incidence of toxic substances and not on sediment deterioration due to hypereutrophication. Considering the documented changes in many of the northeastern Gulf estuaries in recent times, the statements in this U.S. EPA report concerning the present state of these systems cannot be taken seriously.

Based, in part, on the lack of consistent ecosystem-level data, there has been little in the way of enlightened regulation of water and sediment quality in Gulf estuarine and coastal systems. Nutrient loading is not regulated. Dredging operations leading to openings through barrier islands on the northeastern Gulf Coast and maintenance dredging of such artificial passes are not regulated. Sediment quality is not regulated. The discharge of most non-point sources of pollution, including agricultural wastes and urban runoff, is not effectively regulated. The use of taxonomically difficult groups of organisms such as phytoplankton is almost completely ignored in regulatory decisions. Because changes associated with nutrient loading are often subtle, and usually involve the disappearance of organisms due to habitat deterioration rather than dramatic fish kills, they often go unnoticed until the damage leads to the collapse of sports and commercial fisheries. By this time, the changes in a given coastal ecosystem are widespread and long-lasting, requiring extremely expensive programs that are often beyond the economic capacity or the political will of agencies responsible for restoration. The destruction of considerable areas of the Gulf of Mexico due to nutrient loading from the Mississippi River system (Rabalais, 1992) is an example of how this process works.

The basis for all scientific research in aquatic ecology rests on an enlightened review of past studies, assimilation of such facts into a comprehensive base of knowledge, and the encouragement of new studies designed to answer important questions that are derived from such knowledge. Unfortunately, this ideal has not been achieved with regard to coastal ecosystem research. The funding of such studies in the United States has, in fact, been subject to political manipulation, both within the scientific community, and from outside political forces. Political manipulation of aquatic research efforts is not uncommon. What is uncommon is an honest discussion of such control. Most scientists are loath to talk about funding problems or to make public findings that may go against existing political establishments at various levels of government because they are afraid of losing what funding they already have. This situation is not restricted to the United States. For example, Meinesz (1999) described the events leading up to the proliferation of the invasive alga *Caulerpa taxifolia* in the Mediterranean Sea. The cumulative effects of negligence in the political evaluation of the problem distorted the scientific assessment and favored inaction. This alga was introduced into the Mediterranean by the Oceanographic Museum in Monaco, and has now grown to such proportions that it has seriously affected various systems along the coasts of Spain, France, and Italy. Meinesz outlined some of the problems regarding how ecological science is conducted in France:

> The progression of this species was aided by both scientific incompetence and bureaucratic ineptitude and dishonesty.... Productive organizers of research teams are rare and increasingly marginalized by their colleagues in biology.... The current generation of biological researchers is composed essentially of laboratory scientists. Their field of expertise has been reduced to a single level, the molecular. This approach to the study of life is termed "reductionist"...here and there, ecologists undertake studies concerning habitat conversion or pollution. Their professional opinions are not always well received. It is well known that certain teams have been accused of thwarting some huge project or branch of industry. Blacklisted by the promoters and their political allies, they have seen their funding decline drastically. ... Passivity is the most widespread response (of ecologists): don't say anything. Don't criticize an attack on the environment in one's area of competence, in order to preserve the means to keep functioning....
>
> Ecology deals with very broad concepts; it is said to be "holistic." ...Specialists on faunistic and floristic groups are not replaced after they leave or retire. They end their service before transmitting

> to younger colleagues their experience and their knowledge, acquired over several decades.... Scientific journals are increasingly specialized, reflecting the atomization of knowledge into ever more specialized micro-disciplines.... To continue to work, research teams are constrained to find external funding from other institutions or from industry.... Some scientists are emboldened and warn the relevant authorities. But the authorities do not like to hear from Cassandras, because it is difficult for elected officials to enhance their standing by managing a controversial problem.

Meinesz's statements can be applied to the state of ecosystem science in the United States today with the added observation that many of the reductionists are, in fact, ecologists. These ecological minimalists have dominated the funding and publication of environmental research for decades in this country. They have effectively blocked efforts to carry out ecosystem-level studies. There is little doubt that the ecological sciences represent the neglected stepchild of American science with the molecular types dominating both funding and publication activities. The center of reductionist thinking remains within the vaunted National Science Foundation. Major funding agencies such as the National Science Foundation continue to restrict ecosystem-level funding for long-term, interdisciplinary projects. When limited funding is available, it is usually divided among a group of scientific institutions based not on ability or need but as a political sop to influential scientist-politicians. Universities have become increasingly devoted to minimalist "experiments" without the least idea of how the results apply to the bigger (i.e., ecosystem) picture. And all this is carried out with a sanctimonious nod to the glorious peer-review system which often functions as a good-old-boy/good-old-girl shell game that has managed to discourage innovation and creativity in aquatic research for decades.

The "atomization" of Meinesz is even more frustrating in the United States because of the relatively high level of funding available for various areas of science. The favored funding model is to pass out large numbers of relatively small grants. When an ecosystem project is funded, it is partitioned among the favored few, with little regard to an integration of the final results. The history of funding in the Chesapeake and Hudson River–estuaries represents a good example of how relatively large sums of money have been expended on a long line of unrelated 1 to 2 year studies. It is as if the murals on the ceiling of the Sistine Chapel were commissioned to dozens of isolated portrait artists who then contributed to a series of patch-quilt representations instead of the integrated final fresco by Michelangelo. What is worse, the universities have produced generations of graduate students who remain almost completely ignorant of the goals and methods of ecosystem science. Taxonomy has been virtually eliminated from many curricula in favor of politically correct "experimental ecology."

The eutrophication question in coastal systems should be placed within a broad perspective. Bricker et al. (1999) remain realisitic regarding the continuing destruction of coastal resources due to nutrient loading. In many cases, a dependence on patch-quilt data sets instead of integrated ecosystem analyses has led to seriously flawed estimates of the actual status of various riverine and coastal systems. The lack of scientific determination of causative factors associated with plankton blooms in coastal areas (Bricker et al., 1999) has contributed to the deterioration of useful production throughout the country. The continued minimization of the dimensions of research efforts to understand interannual aspects of system response has led to the misrepresentation of the long-term effects of human activities on coastal systems. In his review of processes involved in the shaping of soft-sediment systems, Peterson (1992) decried the lack of application of information from such areas to broader ecological concepts. He cited the chief problem as the "regionally parochial" approach taken in many such studies. The lack of adequate studies has long been defended by funding agencies as too difficult and/or expensive to carry out. The same excuses have been used recently to simply ignore changes in the phytoplankton community as an index of cultural eutrophication because phytoplankton are too complex and therefore too expensive to study (Howarth et al., 2000). Consequently, despite considerable and very expensive (albeit piecemeal) research efforts, we are still without even the most basic understanding of how estuarine and coastal systems function relative to increasing discharges of nutrients to coastal systems. In addition, conjecture in a series of ancient papers continues to be elevated to fact by repetition and overgeneralization in the scientific literature.

The result of all this is the almost continuous elimination of aquatic resources that quietly disappear. Thanks to inadequate ecosystem evaluations of coastal systems, such effects remain largely hidden. The same federal funding agencies that denied generation of long-term, interdisciplinary research programs 30 years ago are now making excuses for the lack of information as many of our major coastal systems undergo systematic damage that in some cases the damage is exacerbated by the very agencies that are supposed to be protecting the environment. The destruction of the Kissimmee–Okeechobee–Everglades–Florida Bay–Keys Coral Reef system is just one of many examples of the actions (or inaction) of state and federal agencies that have carried out political mandates that have resulted in serious harm to important aquatic systems.

The current situation regarding coastal resources is the practical result of what has amounted to universal indifference to Dayton's (1979) definition of ecology as a religious philosophy rather than a scientific discipline. The environmental costs of the actions of the scientists and administrators who have politically controlled not only the research but also how the research results have been applied to practical problems have made this an ethical issue. According to J. B. Battle and M. I. Lipeles (*Water Pollution*, 1998; Anderson Publishing Co.), problems concerning the loading of nutrients into coastal systems are universal:

> America has made essentially no progress in addressing polluted runoff over the past quarter-century,... there is compelling data indicating that non-point pollution has gotten worse rather than better over the past 25 years. ... polluted runoff is the largest reason that roughly 40% of America's surveyed waters are too polluted for basic uses....

The reasons for such a problem are complex but are closely connected to the political control of research in this country at all levels of govenment. The real authors of the lack of enforcement of the water quality laws are elected officials who keep a tight grip on the budgets of state and federal regulatory agencies. It seems that lawmakers do not believe in enforcement of the law if it inconveniences the powerful interests with whom they are so closely connected. Their actions are largely kept quiet by collusion of many elements of the news media that contribute to the proliferation of ignorance and confusion regarding ecological processes in coastal systems. Consequently, the American public remains largely ignorant of many of our most damaging environmental problems.

Although protected climate changes due to global warming cannot be made with any precision, there is a likelihood that the southeastern United States will be affected by increased temperatures during the 21st century (Smith and Torpak, 1989). Projected rises in sea level would be accompanied by increased salinity in coastal areas. It is also possible, according to one modeling scenario, that the Southeast will also be affected by reduced rainfall in the next 50 to 100 years (Smith and Torpak, 1989). Should there be changes in these factors, with increased coastal temperature and salinity and possible reductions of river flow due to reduced precipitation, the projected changes in coastal habitats would favor enhanced bloom generation. When viewed within the projected increases of anthropogenous nutrient loading along the Gulf Coast (Bricker et al., 1999), the results of the Perdido study linking plankton blooms with impaired food webs and reduced secondary production could be translated into major losses of Gulf fisheries, should the projected effects of global warming become reality.

With respect to the current problems concerning the effects of anthropogenous nutrient loading on coastal systems, Anderson and Garrison (1997) have defined the problem succinctly:

> Toxin monitoring programs and other practical activities are obviously needed from a management standpoint, but these are only treating the symptoms of an HAB problem that is steadily worsening.

> If we are ever to reverse this trend, we must first unravel the complexities of these phenomena and identify the factors that regulate population dynamics of the HAB species and the other components of the ecosystem in which they occur. This requires a thorough understanding of the ecology and oceanography of harmful algal blooms.

To this should be added that the understanding of HAB ecology and phytoplankton community associations should be undertaken with reference to associated food webs that form the foundation of useful productivity of estuarine and coastal systems. It is high time that the journeymen scientists, the field-oriented "grunts" who are actually knowledgeable concerning what is happening, stand up and confront the handful of politically powerful scientist-politicians and their bureaucratic stooges who currently run ecosystem science in the United States today. Unless these people do something to change the current situation, the rate of coastal deterioration will only accelerate.

CHAPTER 8

Conclusions

1. Estuaries associated with alluvial rivers are highly productive because of nutrient loading and associated high phytoplankton productivity. In the northeastern Gulf of Mexico, phosphorus is limiting in such areas during winter–spring periods whereas phosphorus and nitrogen are often co-limiting during warmer months of the year. However, secondary productivity, in the form of infauna (animals living in the sediments), epibenthic macroinvertebrates, and fishes is not evenly distributed. The highest levels of such productivity are located in the immediate receiving areas at the heads of alluvial estuaries (Figures 8.1 and 8.2). These areas represent primary nursery grounds for shrimp, crabs, and sciaenid fishes. Freshwater influxes thus determine the distribution of primary and secondary production in alluvial estuaries in addition to controlling important habitat variables. However, trophic linkages take precedence over habitat distribution, and estuarine organisms are concentrated in areas receiving freshwater influxes as a trophic response to enhanced phytoplankton productivity. When anthropogenous nutrient loading causes an imbalance in the phytoplankton communities, there are resultant alterations in the associated food webs, many of which are benthic. Such changes can eventually result in lost secondary production in the most productive parts of the system. Altered phytoplankton associations are thus translated into losses of benthic productivity. Preliminary analyses link plankton blooms with losses of benthic microalgae. Although the exact mechanisms of such effects on the sediments remain largely unknown, the association of the quality of planktonic communities with effects on benthic productivity could be an important factor in the observed reduction of secondary production in coastal systems affected by cultural eutrophication. These effects can be subtle and, without detailed information concerning phytoplankton community response to seasonal and interannual trends of nutrient loading, significant changes in the overall productivity of coastal systems can occur without detection.

2. In coastal waters, the causes and impacts of blooms must be understood with respect to background conditions that include regular interannual cycles of drought–flood conditions. In addition, habitat alteration through human activities such as dredging can have major effects on estuarine food web structure. Our long-term studies of alluvial estuaries and associated coastal systems indicated that periodic droughts were associated with increased water column stability, reduced turnover rates, and increased temperature, salinity, and light penetration. These habitat changes encouraged plankton blooms even under natural conditions of nutrient loading. By dredging openings into barrier islands of the Apalachicola, Perdido, and Choctawhatchee Bays, influxes of high-salinity water from the Gulf increased salinity at depth and caused widespread salinity stratification. Vertical stabilization of the water column caused serious losses of secondary production. Apalachicola Bay lost major oyster reefs due to predation by stenohaline organisms that were allowed into the bay due to the saltwater influxes through Sike's Cut. The Perdido and Choctawhatchee estuaries were modified by dredged openings (Perdido Pass, East Pass) that led to the development of liquid mud at depth and seasonal benthic hypoxia in salinity-stratified areas. Liquid

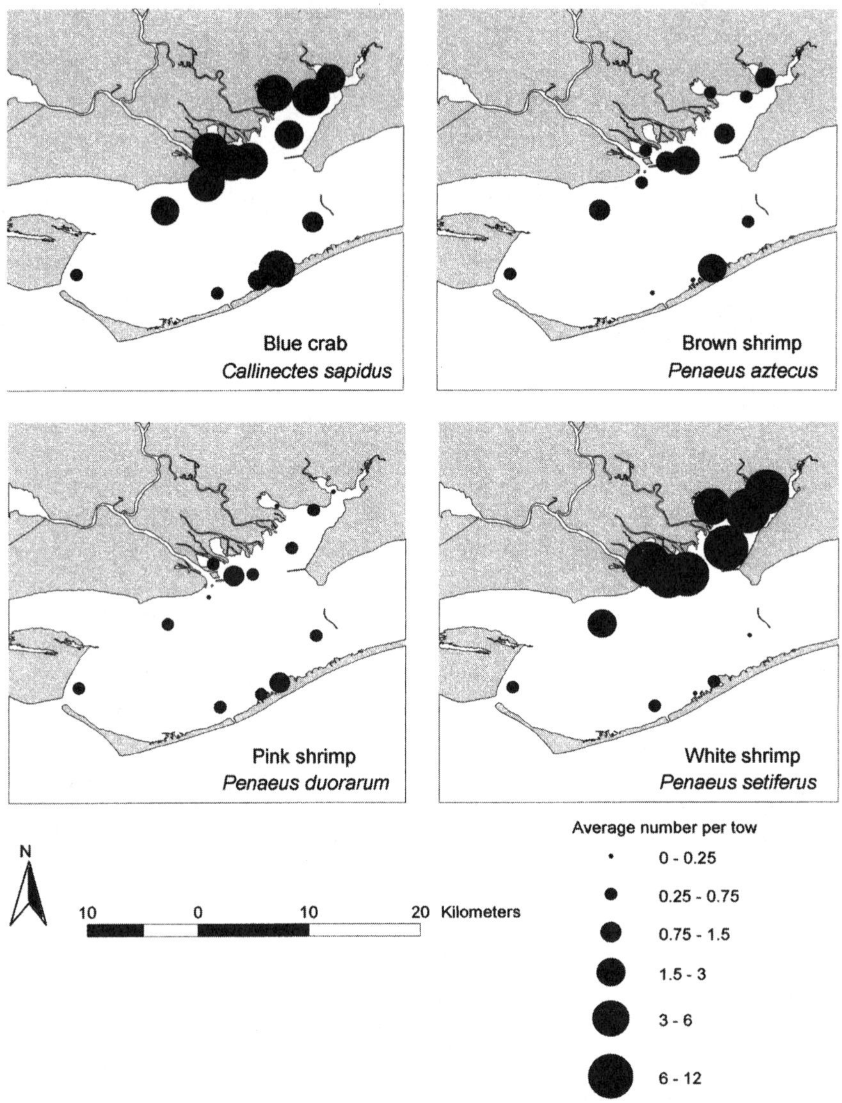

Figure 8.1 Numbers of dominant fishes and invertebrates in the Apalachicola Bay system. Data are averaged over a 12-year period of monthly, baywide sampling.

mud or nephlos (fine particles suspended above the sediment/water interface due to stabilization of the water column and reduced turbulence) was a poor habitat for infaunal macroinvertebrates, which form the base of many estuarine food webs. Generation of liquid mud through dredged passes interfered with the basic structural components of associated food webs leading to losses of secondary production throughout major parts of these estuaries. Such effects also exacerbated the impacts of cultural eutrophication.

3. This 11-year interdisciplinary study of the Perdido system has shed a good deal of light regarding nutrient utilization in this estuary. Nutrient limitation experiments indicated that covariates P+N were stimulatory during all months of the year in the Upper Bay. During winter–spring periods, phosphorus was limiting to phytoplankton production in the Upper Bay and was the most important single limiting factor there. In the presence of nitrogen, production surpassed the effect of either of the other nutrients by themselves, except in the winter. The combination of N+P was

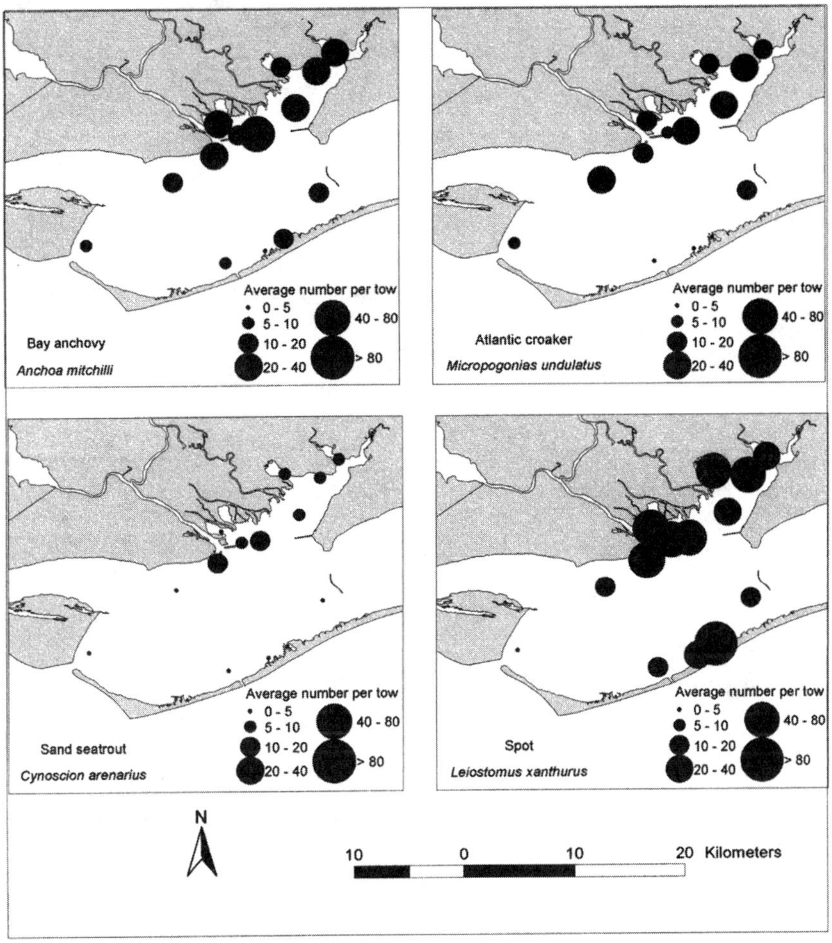

Figure 8.1 (continued)

most stimulatory during late summer–autumn periods. Other parts of the bay had somewhat different responses to nutrient inputs.

4. In Upper Perdido Bay, during the period 1988 to 1991, nutrients from the mill (orthophosphate, ammonia) stimulated the growth of naturally occurring phytoplankton and secondary production. However, increased nutrient discharges from the Champion Pensacola Mill into upper Perdido Bay led to a series of phytoplankton blooms from 1994 to 1999. These blooms affected major parts of the bay. Since October 1999, there have been reductions of mill discharge of ammonia and orthophosphate. The response of Perdido Bay to these reductions is currently being evaluated.

5. As water from Elevenmile Creek, containing effluent from the mill, entered the bay, concentration gradients of orthophosphate and ammonia appeared to initiate many of the plankton blooms. These gradients, together with overall nutrient loading rates, tended to define the nature and extent of the phytoplankton blooms.

6. The phytoplankton blooms from 1994 to 1999 were dominated by species that had distinctive seasonal and interannual successions (Figures 8.3–8.6). The initial blooms were mainly small-celled diatoms (*Leptocylindrus danicus*, *Cyclotella choctawhatcheeana*) that were eventually replaced by larger-celled raphidophytes (summer; *Heterosigma akashiwo*) and dinoflagellates (winter; *Prorocentrum minimum*). Of the diatom bloom species, *L. danicus* was a winter bloom species whereas *C. choctawhatcheeana* and *Synedropsis* sp. were spring–summer bloom species. In time, it took

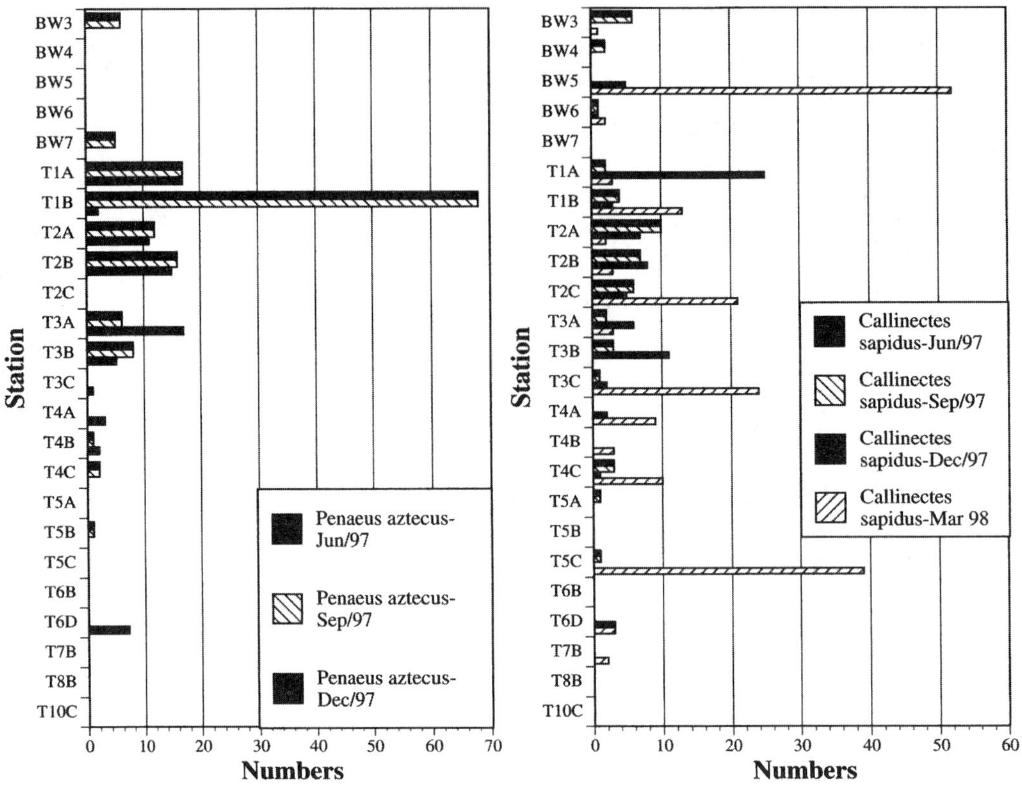

Figure 8.2 Numbers of dominant fishes and invertebrates in the Pensacola Bay system. Data are averaged over a 12-month period of baywide sampling.

less nutrient loading to cause a bloom in the bay, indicating possible sensitization after prolonged anthropogenous nutrient loading.

7. Chlorophyll *a* was significantly associated with phytoplankton numbers and biomass but accounted for a low percentage of the variation. Chlorophyll *a* was not significantly associated with the plankton blooms. Benthic dissolved oxygen increased during the initial blooms, and then decreased after dominance was established by the larger-celled bloom species and after associated increases in plankton biomass. Nutrient concentrations in the bay were altered by the phytoplankton in complex ways that varied through time. Current assumptions regarding the use of these factors as indicators of phytoplankton communities should be reviewed.

8. Sediment quality was an important variable in the overall condition of the estuarine food webs and resulting secondary production. The water column blooms were associated with decreased abundance and species richness of microphytes in the sediments. The decrease of sedimentary microalgae could be a factor in the observed deterioration of benthic food webs that accompanied the plankton blooms. The exact process associated with the declines of the sediment microalgae was not determined and the limitation of only 17 months of sediment microphyte data precluded broad generalization of the results.

9. The plankton blooms were associated with major reductions of infaunal and epibenthic macroinvertebrates and fishes. There was a serious disruption of the estuarine food webs with the loss of planktivorous infauna. Similar impacts associated with altered phytoplankton assemblages were found in other estuaries along the Gulf Coast. The comparative results supported the connection between the plankton and the benthic infauna as the gateway to the primary estuarine food webs.

10. The Pensacola Mill was responsible for only a small proportion of the dissolved oxygen loss at depth in Perdido Bay during the first 3 years of study. However, during later bloom periods,

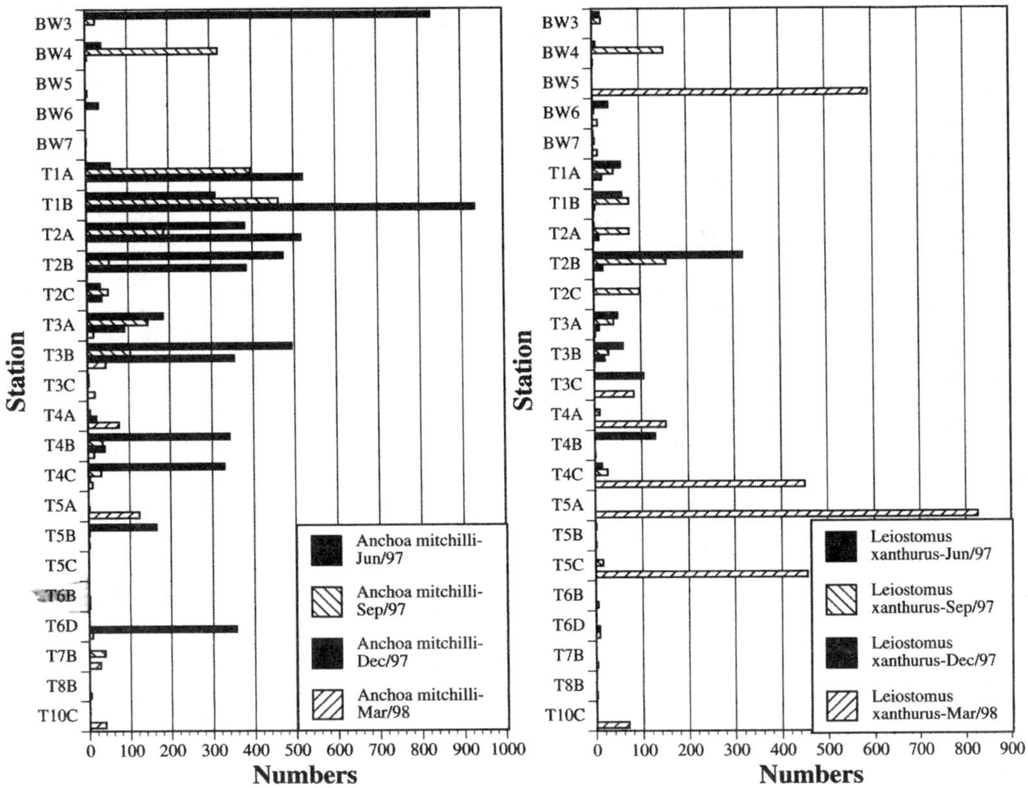

Figure 8.2 (continued)

there were associated reductions of dissolved oxygen at depth. Salinity stratification resulting from the dredged opening at Perdido Pass was still the principal cause for the low dissolved oxygen at depth although the blooms appeared to exacerbate conditions of hypoxia at depth in deeper parts of the bay.

11. The general good health of upper Perdido Bay during 1988 to 1991 period suggests that limits on discharges of nutrients from sources into the Upper Bay should be maintained within the limits observed during that time. Although complicated by a number of factors, improvements in the bay following ongoing nutrient management implementation at the mill further support this contention. A target of maximal ammonia discharge from the mill (or into Upper Perdido Bay) would be a quarterly average of 361 kg d^{-1}. It was estimated that total phosphorus discharges to Elevenmile Creek (or the Upper Bay) should not exceed 83.5 kg d^{-1} on a quarterly basis. Limits of total soluble phosphorous (orthophosphate) discharges were estimated at 50 kg d^{-1} (quarterly averages) during high-flow periods (greater than 65 cfs at station P22) and 36.7 kg d^{-1} (quarterly averages) during low-flow periods (less than 65 cfs at station P22). These limits should apply to all discharges to the Upper Bay as well as Elevenmile Creek because of the limited uptake of nutrients in the run of the creek, and the fact that the mill was responsible for most of the nutrient loading to the bay (non-point-source loading was found to be insignificant). With these levels, reasonable gradients of nutrients in the Upper Bay will be maintained.

12. In the United States, nutrient loading and sediment quality are not regulated by federal and state regulatory agencies. Community analyses of phytoplankton assemblages are rarely carried out. Information concerning nutrient-induced changes of phytoplankton communities has not been an integral part of coastal research. This information will be needed if the connections among nutrient loading, plankton blooms, habitat deterioration, and reduced secondary production are to

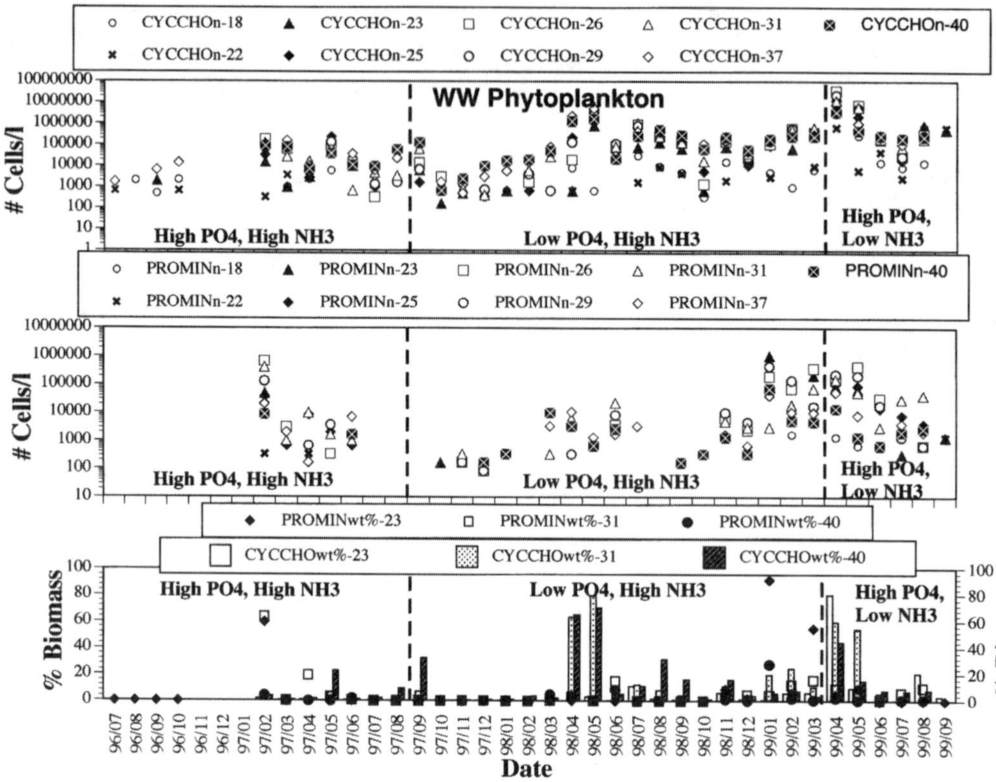

Figure 8.3 The dinoflagellate *Prorocentrum minimum* (= Promin) and the diatom *Cyclotella choctawhatcheeana* (Cyccho) taken as whole-water samples from Perdido Bay monthly during the period July 1996 to September 1999. Data are presented as station-specific numbers of cells l^{-1} and as percent biomass of the total whole-water plankton sample.

be evaluated. The relationship of increased anthropogenous nutrient loading and increased incidence of plankton blooms could involve the interaction of water column microalgae and benthic microphytes. This connection should be further studied. Without adequate information concerning the role of planktonic and sedimentary microalgae in coastal systems, appropriate management of productive river-dominated estuaries will be difficult if not impossible. The result of the lack of enlightened nutrient management will be the continued loss of important coastal resources.

13. The ongoing system of environmental research in the United States continues to use a combination of political control and inadequate, reductionist approaches to perpetuate the continuing destruction of aquatic resources. The dependence on patch-quilt data sets instead of integrated ecosystem analyses has led to seriously flawed estimates of the actual status of various riverine and coastal systems. The lack of scientific determination of causative factors associated with cultural eutrophication in coastal areas has contributed to the deterioration of useful production in many of the most important estuaries of the United States. Nutrient loading due to urbanization, industrialization, and agricultural activities is a worldwide problem. The continued minimization of the dimensions of research efforts to understand interannual aspects of system response has led to the serious misrepresentation of the long-term effects of human activities on coastal systems.

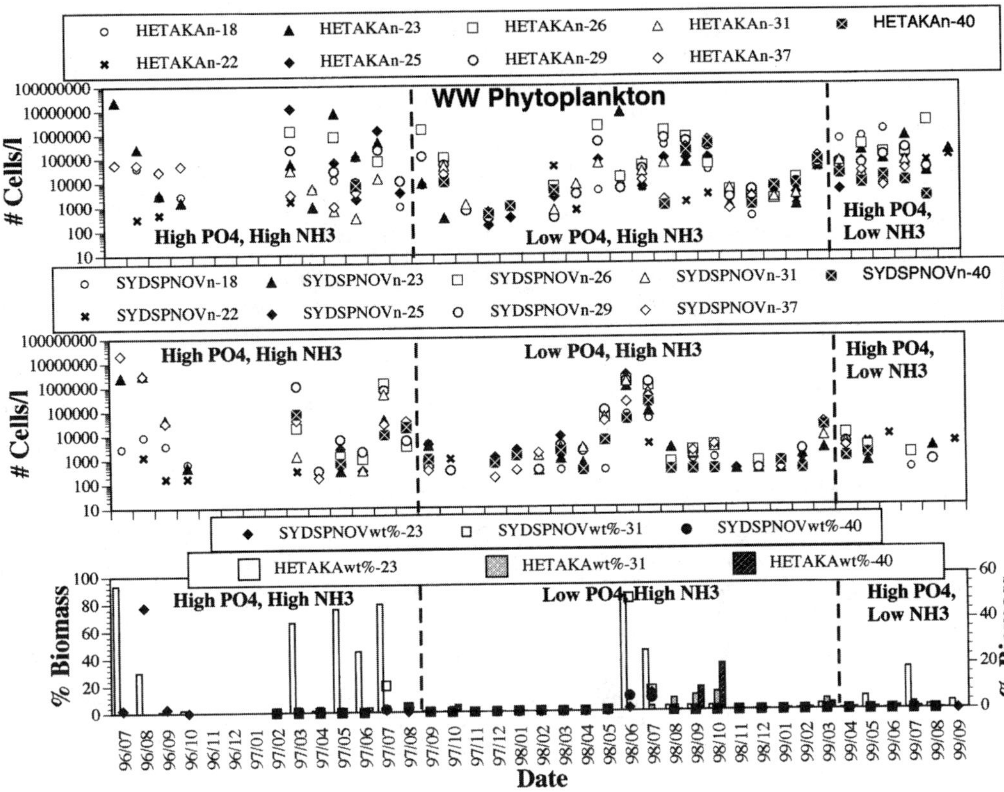

Figure 8.4 The raphidophyte *Heterosigma akashivo* (= Hetaka)) and the diatom *Synedropsis* sp. nov. (= Synsonov) taken as whole-water samples from Perdido Bay monthly during the period July 1996 to September 1999. Data are presented as station-specific numbers of cells l^{-1} and as percent biomass of the total whole-water plankton sample.

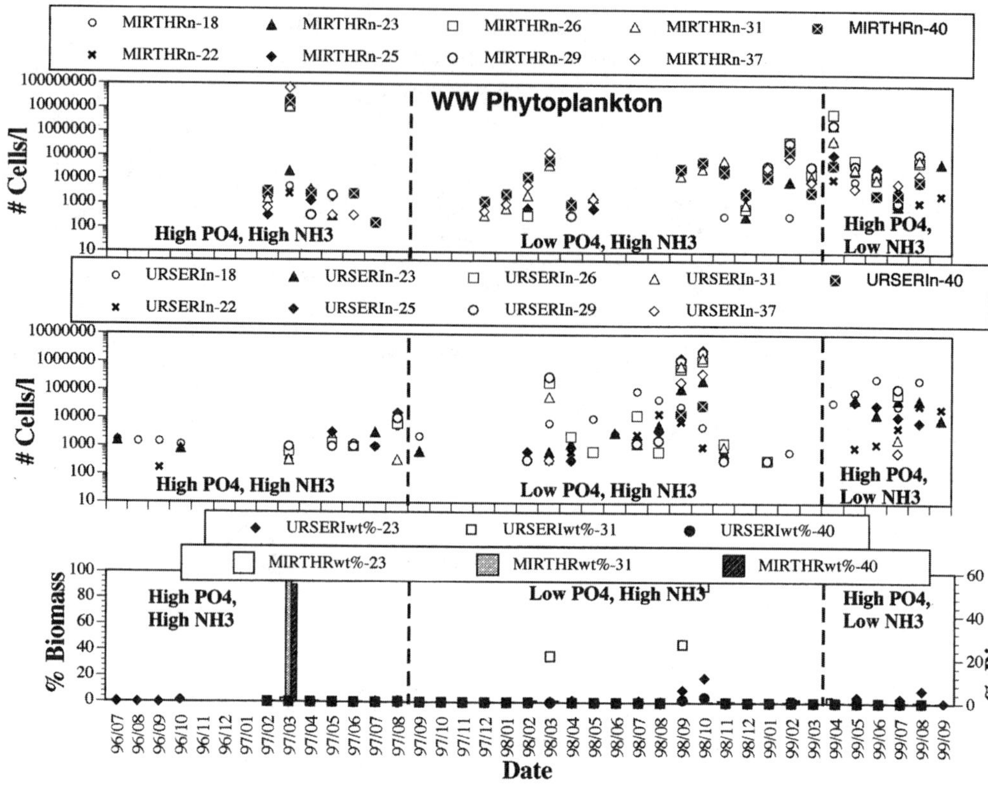

Figure 8.5 The diatom *Miraltia throndsenii* (= Mirthr) and the diatom *Urosolenia eriensis* (= Urser) taken as whole-water samples from Perdido Bay monthly during the period July 1996 to September 1999. Data are presented as station-specific numbers of cells l^{-1} and as percent biomass of the total whole-water plankton sample.

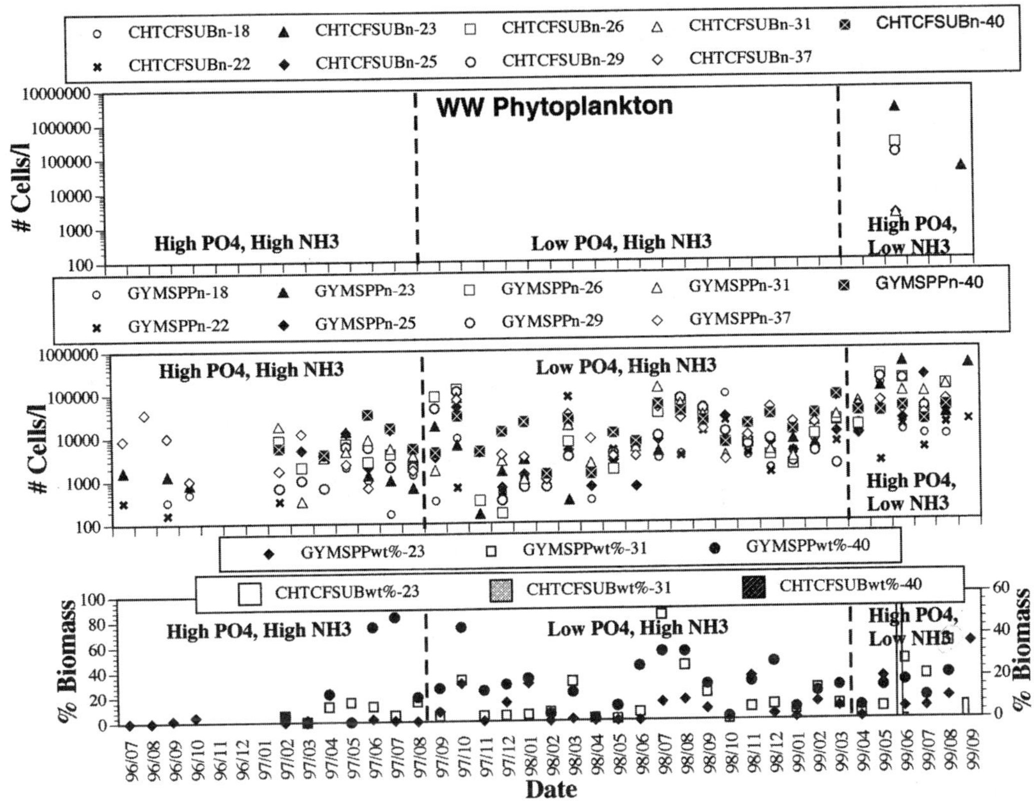

Figure 8.6 The raphidophyte *Chattonella subsalsa* (= Chtcfsub) and the dinoflagellate genus *Gymnodinium* (= Gymspp) taken as whole-water samples from Perdido Bay monthly during the period July 1996 to September 1999. Data are presented as station-specific numbers of cells l^{-1} and as % biomass of the total whole-water plankton sample.

CHAPTER 9

Summary of Results

A continuous, multidisciplinary study (field and laboratory) study was carried out in seven river–estuarine and coastal systems in the northeastern Gulf of Mexico from 1970 to the present. Among the central research issues investigated were the analysis of eutrophication processes, the effects of anthropogenous nutrient loading on the initiation and succession of plankton blooms, and food web responses to seasonal and interannual changes of phytoplankton community structure that were associated with the blooms. The use of long-term, interdisciplinary information along with species-specific phytoplankton data and detailed analyses of associated food web organization in a series of estuaries supported a comprehensive analysis of the effects of natural and anthropogenous nutrient loading on Gulf coastal systems.

One of the primary target areas for the comprehensive nutrient loading analyses was the Perdido Bay system in northwestern Florida. Perdido Bay is somewhat unique in that it is a relatively small system with only one primary anthropogenous source of nutrification to the upper bay, a pulp mill. Kraft pulp mill effluents tend to have relatively high concentrations of orthophosphate and ammonia. Most of the land in the upper Perdido drainage basin is lightly populated. The most concentrated agricultural and urban runoff is located in the lower bay. This combination of factors established upper Perdido Bay as an appropriate subject for research concerning nutrient loading and eutrophication. The results of these analyses were then compared to similar studies in six other estuarine systems in the northeastern Gulf: Apalachee Bay (Econfina estuary, Fenholloway estuary), Apalachicola Bay, Choctawhatchee Bay, Blackwater Bay, and Escambia–Pensacola Bay.

Long-term (11-year) studies were carried out in the Perdido drainage system to determine the effects of pulp mill discharges of ammonia and orthophosphate on Perdido Bay. The Perdido River, an alluvial stream, also loads nutrients into upper Perdido Bay. A detailed analysis of the nutrient loading characteristics of the Perdido drainage system indicated that, during early years of analysis (1988 to 1991), orthophosphate and ammonia loading from the pulp mill enhanced secondary production in the immediate receiving area of the upper estuary. There were no serious phytoplankton blooms during this period and the phytoplankton community structure remained highly diverse during this period. The chrysophytes, and to a lesser degree, the chlorophytes, tended to be abundant and were associated with a balanced food web and relatively high secondary production in upper bay areas that received nutrient loading from the paper mill. However, during a winter–spring drought in 1993–1994, the pulp mill increased orthophosphate loading to the bay. The combination of drought conditions and increased orthophosphate loading was associated with the initiation of a series of phytoplankton blooms that were dominated by various diatom species. There was an orderly seasonal succession of these bloom species in the bay. Continued high orthophosphate loading over the next 3 years led to increased phytoplankton bloom frequency and intensity throughout the bay.

During the period 1995 to 1998, there were periodic increases of ammonia loading to the upper bay by the pulp mill. From late summer 1997 through spring 1999, the pulp mill reduced its orthophosphate loading, thus establishing an ecosystem experiment whereby the chief source of orthophosphate to the upper bay was lowered to levels that were common during prebloom years. Concurrent with the reduced orthophosphate loading, there were reductions of the high relative dominance of certain bloom species, especially during winter–spring periods. By the winter–spring of 1999, phytoplankton species richness increased to levels comparable to those during prebloom periods (1988 to 1991). These changes were complicated by increased ammonia loading from the mill during summer 1998 that was associated with temporary returns of dominance by summer bloom species and associated reductions of phytoplankton species richness. The evidence indicated a partial recovery of the phytoplankton associations during the period of low orthophosphate loading to upper Perdido Bay. However, by spring 1999, the mill again resumed high loading of orthophosphate to the bay. This loading was accompanied by baywide spring and summer blooms, increased dominance of bloom species, and associated reductions of phytoplankton species richness. The long-term trends of phytoplankton community indices indicated that species richness and relative dominance of the bloom species were reasonable indicators of the overall condition of the phytoplankton assemblages in Perdido Bay, and that such indicators followed generalized nutrient loading patterns. There was considerable background noise in these patterns, and it is possible that other nutrients could have played a role in the noted changes. However, the postulated effects of increased orthophosphate and ammonia loading were usually correlated with general sequences of bloom species and associated changes in the phytoplankton community structure that were consistent with observed natural history characteristics of the diatoms, raphidophytes, and dinoflagellates that comprised the bloom types. The bloom sequences were also consistent with nutrient limitation data that indicated orthophosphate limitation in the upper bay during winter–spring periods and nitrogen and/or phosphorus + nitrogen limitation during warmer months.

The relatively simple organization of Perdido Bay and the controlled changes of nutrient loading over the 11-year study period permitted detailed analyses of the relationships of various levels of nutrient loading with associated changes in the estuarine phytoplankton community structure and species-specific responses to the nutrient loading. There was a distinct seasonal pattern to the bloom events, with individual species forming blooms in specific months or seasons. These temporal progressions were attributed to seasonal differences in the nutrient requirements and the distinctive life histories of the bloom species. There were also qualitative and quantitative interannual dominance shifts within the bloom sequences over the period from 1993 through 1999 whereby larger-celled raphidophyte and dinoflagellate species replaced diatom species that had been predominant during the initial blooms. Thus, the overall weight of field evidence indicated that increased orthophosphate and ammonia loading to upper Perdido Bay by the paper mill led to seasonal and interannual successions of plankton blooms. There was evidence that interspecific competition and predation within the plankton community could have affected the interannual successions. The high nutrient loading tended to favor cryptophytes, cyanophytes, dinoflagellates, and raphidophytes. These groups are well represented by prominent bloom species in coastal and marine waters. During periods of high winter orthophosphate loading, the dinoflagellates were numerically dominant, whereas summer ammonia loading was associated with increased dominance by raphidophytes. The chrysophytes, and to a lesser degree, the chlorophytes, tended to be more abundant during periods of low nutrient loading. The diatoms were good indicators of the long-term nutrient loading patterns into upper Perdido Bay with reduced percent abundance during periods of high nutrient loading and associated blooms.

There was a long-term increase of plankton numbers and biomass during the years of increased nutrient loading that was related, at least in part, to the replacement of the smaller-celled diatoms with larger dinoflagellate and raphidophyte species. Nutrient loading from the mill into Elevenmile Creek was associated with increased concentrations of orthophosphorus and ammonia in areas where the creek enters upper Perdido bay. There were distinct concentration gradients of these nutrients in areas where many of the plankton blooms originated. These concentration gradients appeared to be a stimulus for the initiation of many of the plankton blooms. The concentration gradients of orthophosphorus and ammonia, in addition to the nutrient loading, appeared to be reliable predictors of bloom initiation in the bay. It is likely that changes of the relative concentrations of these compounds at the entry point between Elevenmile Creek and upper Perdido Bay provided the spark that ignited at least some of the plankton blooms. Additional nutrient loading from the Perdido River system and other sources in the upper bay provided supplementary fuel for the blooms as they expanded through the bay.

During the initial plankton blooms in 1993–1994, a survey was conducted of sediment-associated microalgae. Sedimentary microalgae form the basis of primary production in bay sediments. The benthic microalgae were subject to decreases that were concurrent with the initiation of the phytoplankton blooms in 1993–1994. Numbers, biomass, and species richness of the sedimentary microalgae were significantly reduced during and after the primary phytoplankton blooms of winter–spring 1994. There was evidence that bloom-induced changes of water column phytoplankton assemblages were somehow associated with reductions of sedimentary microalgae. The exact process associated with the declines of the sediment microalgae was not determined, and the limitation of only 17 months of data precluded broad generalization of the results. However, these microalgae represent an important component of the benthic food webs of the bay as they are fed upon by planktivorous and deposit-feeding infaunal macroinvertebrates. The infauna, in turn, play an important role in overall structure of the estuarine food web since such organisms are gateway species to the various carnivore groups. During the initial plankton blooms in 1993–1994, there were associated declines of abundance and species richness of the infaunal and epibenthic macroinvertebrates and fishes in Perdido Bay. These reductions were most apparent in organisms associated with sediments. The reduced secondary production followed the bloom sequence in succeeding years both in terms of the spatial distribution and temporal sequencing of bloom occurrence. Upper bay seagrass areas that originally were not affected by the blooms eventually showed similar reductions of secondary production during the last 2 to 3 years of the study. By spring 1999, at the end of the period of reduced orthophosphate loading, there were indications of a partial recovery of epibenthic macroinvertebrates throughout the bay. However, after resumption of orthophosphate loading by the mill during spring 1999, secondary production continued to deteriorate. During the period 1997 to 1999, there was an almost complete collapse of the benthic biota that forms the basis of the food webs and secondary production in the Perdido system.

The changes in the numbers, biomass, and species richness of the infaunal and epibenthic macroinvertebrates and fishes in Perdido Bay appeared to be a trophic response of the estuarine food web to postulated effects of the phytoplankton blooms on the benthic environment. Analyses of heavy metals and various organic compounds in Perdido Bay sediments established that the observed long-term changes of secondary production were not due to anthropogenous loading of toxic agents into Perdido Bay. The descriptive biological data (quantitative determinations of infaunal and epibenthic macroinvertebrates and fishes) were transformed into trophic equivalents and a detailed food web structure. The sharp declines of sediment-associated infaunal and epifaunal macroinvertebrates appeared to be associated with food web processes. After initiation of the blooms in 1993–1994, there was an immediate reduction of planktivorous infaunal macroinvertebrates. This group of herbivorous worms and bivalve mollusks, as primary biomass dominants, forms a major link between the microalgal producers in the sediments and the upper (carnivorous) trophic levels of the bay. As the loss of the herbivorous infauna continued, ending in the almost complete collapse of this important group during the last 3 years of the study, there were associated reductions of the carnivore groups that represent the dominant epibenthic macroinvertebrates and fishes in the Perdido system (i.e., penaeid shrimp, blue crabs, sciaenid fishes). The most severe reductions of the overall secondary production in the bay occurred during a period of increasing numerical and biomass dominance by raphidophytes and dinoflagellates. Thus, the trophic response of Perdido Bay in space and time was consistent with a change in the sediments that was closely related to the plankton blooms, associated reductions of sedimentary microflora, and the loss of planktivorous infaunal macroinvertebrates.

There are many possible ways that the plankton blooms could have affected the benthic organization of the bay. The association between the blooms and the loss of sedimentary organisms via reductions of benthic microphytes represents the simplest and most likely explanation of reduced secondary production in areas affected by the blooms. However, evidence for such a connection remains largely correlative, and a significant linkage of the water column and sedimentary microalgae awaits experimental verification. Most of the dominant epibenthic invertebrates and fishes in the bay are benthic feeders. At most stations, the collapse of planktivorous infaunal macroinvertebrates during the blooms was associated in space and time with general increases of bloom activity and serious reductions of the primary and secondary carnivores. Areas in the upper bay that were not initially affected by the plankton blooms (i.e., seagrass beds) were not characterized by reduced secondary production during the early bloom years; however, in later years, even these areas experienced reduced secondary production. By the end of the 11-year study period, food web components in major portions of Perdido Bay collapsed with a severely disjunct food web response to the long-term plankton bloom sequences.

A comparison of these results with other bay systems in the northeastern Gulf of Mexico confirmed the probable association of planktonic bloom generation, altered phytoplankton community structure, and associated changes of food webs in estuaries affected by high anthropogenous nutrient loading. In Wolf Bay (lower Perdido Bay), there was increased nutrient loading due to postulated agricultural runoff. A similar pattern of habitat deterioration was noted whereby high (anthropogenous) nutrient loading was associated with increased relative abundance of plankton bloom species, reduced phytoplankton species richness, deterioration of the sedimentary habitat, and the collapse of secondary production through inhibition of planktivorous infaunal macroinvertebrates. Lower Perdido Bay, subject to urban runoff, showed a similar pattern of deterioration of the infaunal assemblages. Thus, non-point-source nutrient loading in lower Perdido Bay led to similar reductions of secondary production as that noted in the upper bay. The general absence of well-defined nutrient gradients in Wolf Bay indicated that loading of phosphorus and nitrogen compounds per se could lead to damaging plankton blooms. The relationship of reduced phytoplankton species richness and increased dominance by bloom species was also noted in Escambia Bay which has been subject for decades to high anthropogenous nutrient loading, phytoplankton blooms, losses of oyster beds and submerged aquatic vegetation, deterioration of unvegetated soft-sediment habitat, and fish kills. In a recent survey, it was found that upper Escambia Bay is still characterized by increased dominance of phytoplankton bloom species, deterioration of the sedimentary environment, food web alteration through elimination of planktivorous infaunal macroinvertebrates, and relatively low secondary production. This represents a similar situation to that in Perdido Bay. In the Escambia system, point source pollution and storm water runoff continue to play a major role in nutrient loading trends and associated deterioration of the bay. Sediments in upper Escambia Bay still show signs of deterioration due to nutrient loading. Likewise, anecdotal information indicates that urban storm water runoff and periodic sewage overflows into Choctawhatchee Bay have been associated with increased frequency of extensive red tide events, fish kills, reduced secondary production, and the deaths of hundreds of marine mammals. Bays such as Apalachicola Bay and Apalachee Bay, where there has been no hypereutrophication, had relatively high secondary production with functional food webs compared with the bloom-infested Escambia and Perdido estuaries.

The response of coastal systems to anthropogenous nutrient loading should be placed within the context of regional habitat conditions and regular interannual sequences of drought and flood conditions. Periodic droughts in alluvial systems along the northeastern Gulf are correlated with habitat changes in the receiving estuaries that favor bloom conditions, even in unpolluted systems such as the Apalachicola. Increased bay temperatures, reduced turnover rates, increased light penetration, and higher nutrient gradients during prolonged droughts favored planktonic bloom activity. Long-term trends of natural freshwater runoff in alluvial river–bay systems should thus be taken into consideration when evaluating the impact of anthropogenous nutrient loading. In addition, physical alterations of the coastal systems by dredging have exacerbated the effects of anthropogenous nutrient loading. Dredged openings of lower Choctawhatchee Bay and Perdido Bay have allowed influxes of high salinity Gulf water into deeper waters, forming highly stable haloclines throughout major parts of these estuaries. Stable salinity stratification usually occurs at depths exceeding 2 m. Water column stratification has led to subhaloclinal hypoxia in both systems during warm periods.

The lack of vertical mixing has also stabilized the sediments in these systems, leading to the formation of liquid mud, an amorphous suspension of fine particles above the sediment/water interface. The liquid mud or nephlos, maintained by the lack of turbulence engendered when the water column is stabilized, represents a poor habitat for bottom-dwelling organisms. In both the Perdido and Choctawhatchee Bay systems, the liquid mud has been generally associated with disjunct food webs and reduced secondary production. Thus, the combination of periodic, warm-season hypoxia and liquid mud formation has resulted in serious reductions of benthic production in major areas of both the Choctawhatchee and Perdido estuaries. In Perdido Bay, the deeper, depositional areas were already damaged by anthropogenous salinity stratification. The plankton blooms only added to the deterioration of secondary production in these depositional parts of the bay. Shallow seagrass habitats were least affected by stratification and the plankton blooms although, after years of bloom activity, even these areas showed deterioration of secondary production. Increased plankton biomass due to the bloom successions was also associated with increased benthic biochemical oxygen demand and reduced subhaloclinal dissolved oxygen during warm months. These changes could have added to the postulated effects of the blooms on the sedimentary habitats in salinity-stratified parts of the bay. In this way, background habitat condition is another important consideration when evaluating the effects of plankton blooms in coastal systems.

The detailed characterization of the long-term changes in plankton assemblages in Perdido Bay led to an examination of water quality factors as indicators of phytoplankton activity. For example, chlorophyll *a* is widely used as an indicator of phytoplankton assemblages in both descriptive analyses and modeling studies. However, chlorophyll *a* trends were not representative of the incidence and succession of phytoplankton blooms in Perdido Bay. Chlorophyll *a* was significantly associated with phytoplankton numbers and biomass, but accounted for relatively little of the variation. The current widespread use of chlorophyll *a* as a substitute for phytoplankton communities should be reviewed. The widespread use of this factor could lead to misleading conclusions concerning cause-and-effect processes associated with plankton blooms. The same can be said for other water quality indicators such as dissolved oxygen, nutrients, and organic carbon that undergo complicated, long-term changes due to sequential changes of species-specific bloom occurrences. Although dissolved oxygen is commonly used as an indicator of hypereutrophication, this factor undergoes complex changes during the long-term plankton successions. Also, hypoxia is seasonal whereas the impacts on the benthos occur year-round thus implicating sediment change (physical and/or biological) as a primary factor in the deterioration of the benthic food webs. The complex response of these water quality factors to both seasonal and interannual bloom successions and phytoplankton community responses thus complicates the indiscriminate use of water quality as the sole indicator of the overall health of coastal ecosystems. The importance of sediments in this process also precludes the simplistic use of water quality indices in regulatory matters. The microalgal community structure, both in the sediments and the water, could be a more reliable indicator of the health of estuarine and coastal food webs than the usual water quality indices that are in common use today.

The long-term changes in Perdido Bay that followed increased nutrient loading and the associated plankton blooms are summarized in Figures 9.1 and 9.2. Trends of orthophosphate and ammonia loading by the pulp mill were associated with the origin and proliferation of plankton blooms in Perdido Bay. There were both seasonal progressions and interannual successions of

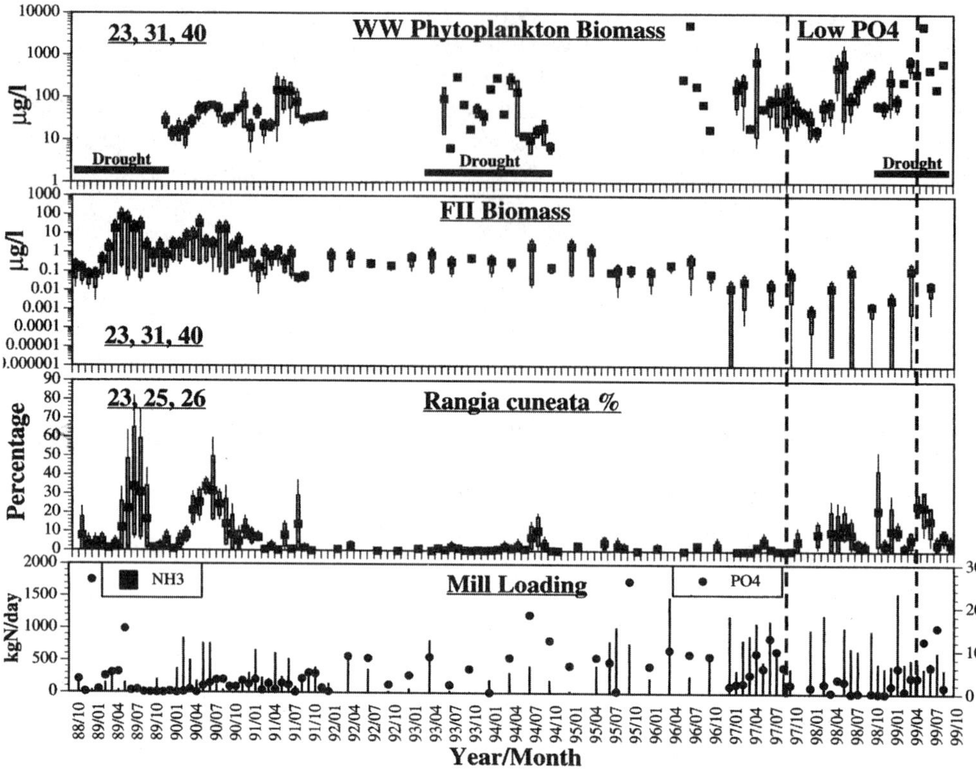

Figure 9.1 Total biomass (μg l^{-1}) of whole-water phytoplankton taken in Perdido Bay are compared with total biomass (μg l^{-1}) of infauna, epibenthic macroinvertebrates, and fishes (FII) and percentage numerical abundance of the bivalve *Rangia cuneata* taken in Perdido Bay from October 1988 through September 1999. Data are presented as means across stations (P23, P31, P40) with the 10th, 25th, 75th, and 90th percentiles. The biological trends are compared with ammonia (kg N d^{-1}) and orthophosphate (kg P d^{-1}) loading rates by the Cantonment paper mill over the study period. The period of reduced orthophosphate loading by the mill is also shown (dashed lines).

bloom species. Increased plankton biomass following dominance by raphidophytes and dinoflagellates were also correlated with major declines of the bay food webs and secondary production. Indicator species such as *Rangia cuneata*, an herbivorous bivalve mollusk, declined after the initiation of the blooms. Following a period of reduced orthophosphate loadings by the mill from late 1997 through spring 1999, there was a partial recovery of the plankton assemblages as measured by indices such as species richness. Continued loading of ammonia during this period could have been at least partially responsible for the incomplete recovery. The loss of secondary production was related to food web deterioration and the reduction of planktivorous herbivores during the bloom successions. Models of the long-term data trends (Figures 9.3 and 9.4) outline the differences between the responses of unstratified upper bay areas and salinity-stratified lower bay areas to the long-term successions of the plankton blooms. The already damaged benthos in areas of high-salinity stratification and liquid mud was almost completely eliminated during the period of plankton blooms in Perdido Bay.

Overall, long-term cycles of river flows and associated estuarine habitat variables, seasonal temperature successions, nutrient loading and associated nutrient concentration gradients, and anthropogenous changes in the physiography of the bay through dredged openings to the Gulf of Mexico affected the initiation and temporal successions of plankton blooms in Perdido Bay. Similar blooms in other bays along the northeastern Gulf Coast were associated with habitat changes that eventually led to the loss of secondary production. The response of the trophic organization of

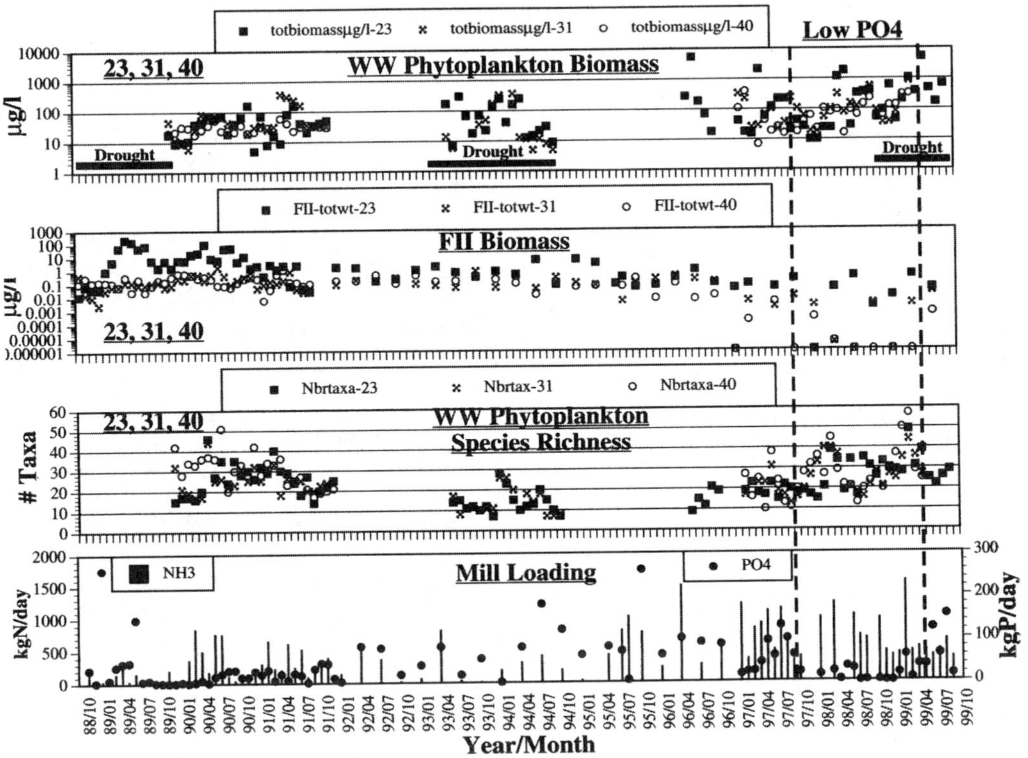

Figure 9.2 Total biomass (μg l^{-1}) by station of whole-water phytoplankton taken in Perdido Bay over the 11-year study period and total biomass (μg l^{-1}) by station of the infauna, epibenthic macroinvertebrates, and fishes (FII) taken in Perdido Bay from October 1988 through September 1999. Also shown are the station-specific whole-water phytoplankton species richness taken over the study period and compared with ammonia (kg N d^{-1}) and orthophosphate (kg P d^{-1}) loading rates by the Cantonment paper mill. The period of reduced orthophosphate loading by the mill is also shown (dashed lines).

these bays was the most important process in the observed temporal changes of estuarine invertebrates and fishes. The changes in the estuarine food webs were found in various bays (i.e., Wolf Bay, Escambia Bay) where nutrient loading, plankton blooms, and altered phytoplankton community structure occurred. Thus, the apparent association of plankton blooms with sediment/water quality and reductions of secondary production via food web response in Perdido Bay was strengthened by the interbay comparisons. The exact mechanism that connects the plankton blooms and associated deterioration of the benthos remains undetermined.

The results of the comparative studies of the effects of pulp mill effluents on estuaries along the northeastern Gulf Coast indicated that essentially similar industrial effluents in terms of chemical composition and loading characteristics of nutrients and dissolved organic carbon have very different effects in different systems. The effluents from the Fenholloway drainage seriously damaged the submerged aquatic vegetation (SAV) in the relatively clear, oligotrophic Apalachee Bay system. Nutrient and dissolved organic carbon (DOC) loading resulted in changes of water quality factors and light transmission characteristics in the receiving Gulf area. Sediments in affected offshore areas had increased silt–clay fractions and increased nutrient concentrations relative to reference sites. Based on the field data, the best (statistically significant) predictors of SAV distribution were photic depths, qualitative aspects of wavelength distributions, water quality factors (color, DOC, and chlorophyll *a*), and sediment characteristics. Due to high inshore color levels, there were no plankton blooms in the light-limited receiving areas of Apalachee Bay. The differences in the responses of Apalachee Bay and Perdido Bay to nutrient loading thus indicate that regulatory action concerning

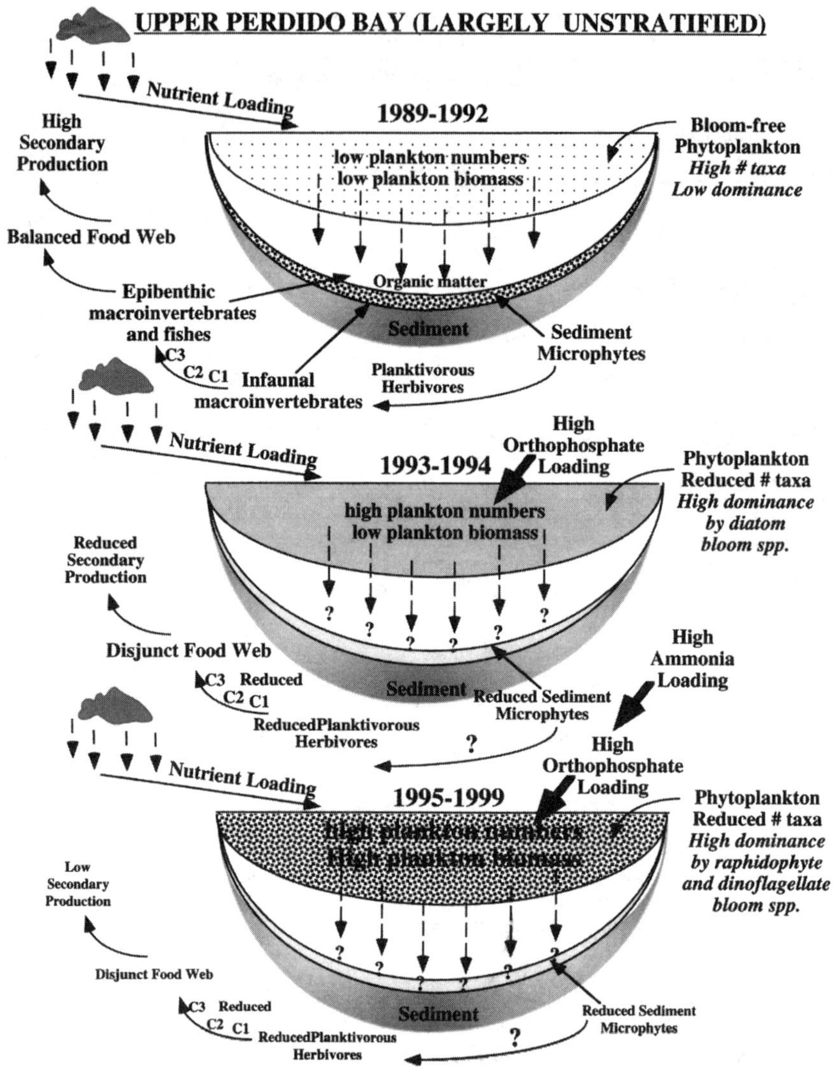

Figure 9.3 Models showing changes in phytoplankton community indices, trophic organization, and secondary production in upper Perdido (unstratified) before (1989 to 1992) and after (1993 to 1994, 1995 to 1999) increased nutrient loading and phytoplankton blooms. Question marks represent unknown changes in the sediments based on observed reductions of benthic microphytes after initiation of the plankton blooms.

releases of anthropogenous nutrients should be based on the assimilative capacity of the receiving system. This means that ecosystem-level knowledge is fundamental to the regulation of nutrient loading into aquatic systems.

The findings in Perdido Bay reveal major problems both with past and current ecosystem research initiatives and regulatory practices. There has been a historical rejection of interdisciplinary, long-term studies of coastal systems by funding sources and regulatory agencies. This lack of relevant data is exacerbated by the absence of detailed, species-specific changes of the phytoplankton communities in response to increased anthropogenous nutrient loading throughout the country. The use of taxonomically difficult groups of organisms such as phytoplankton is largely ignored in regulatory decisions. Based, in part, on the lack of consistent ecosystem-level data, there has been virtually no state or federal regulation of nutrient loading to estuarine and coastal systems. In addition, dredging operations leading to openings through barrier islands and maintenance

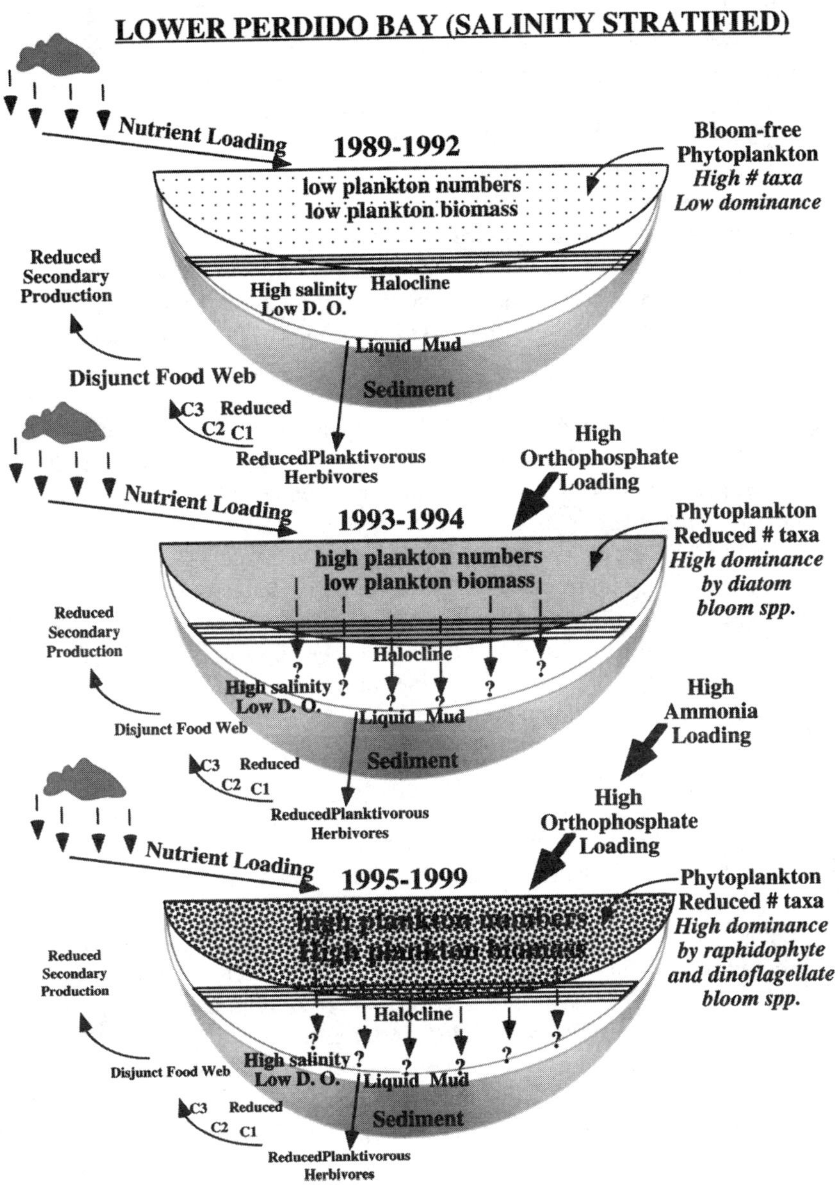

Figure 9.4 Models showing changes in phytoplankton community indices, trophic organization, and secondary production in lower Perdido (salinity stratified) before (1989 to 1992) and after (1993 to 1994, 1995 to 1999) increased nutrient loading and phytoplankton blooms. Question marks represent unknown changes in the sediments based on observed reductions of benthic microphytes after initiation of the plankton blooms.

dredging of existing artificial passes are not regulated. Sediment quality is not regulated. The discharge of most non-point sources of pollution, including agricultural wastes and urban runoff, is not regulated. The emphasis on toxic substances in regulatory matters is misplaced as such substances usually affect only parts of coastal areas, whereas changes in plankton communities due to the stimulation of blooms by nutrient loading can cause baywide habitat deterioration and the continuing loss of useful secondary production. The potential effects of nutrient-based changes of plankton on sediment quality cast doubt on studies that are restricted to the effects of toxic substances on sediments. Because adverse ecological effects due to hypereutrophication are subtle, and often involve the disappearance of organisms as a result of habitat deterioration rather than dramatic fish kills, such changes often go unnoticed until associated damage leads to the collapse of sports and commercial fisheries. By this time, the changes in a given coastal ecosystem are widespread and long-lasting, requiring extremely expensive restoration programs that are often beyond the economic capacity or the political will of agencies responsible for such restoration. In many cases, the efficacy of such restoration remains in doubt because of the uncertainties engendered by the lack of real ecosystem research.

The initial restoration of systems affected by anthropogenous nutrient loading can be relatively simple in terms of scientifically based restrictions on such loading. A restoration program of proposed reductions of orthophosphate and ammonia loading by the pulp mill to Elevenmile Creek was based on the long-term Perdido Bay database is now under way. Orthophosphate and ammonia loading by the pulp mill has already been reduced to levels that approximate loadings during periods when the bay was free of blooms and secondary production in immediate receiving areas was relatively high. An evaluation is being conducted regarding the effectiveness of the nutrient loading restrictions. The mechanism and timing of the postulated recovery will be addressed in this research. This restoration program is possible with regard to point sources of nutrient loading. However, the question of such restoration in areas affected by multiple point and non-point nutrient sources (urban and agricultural runoff) has not even been addressed, much less answered. The lack of regulatory attention to non-point nutrient pollution complicates restoration efforts in areas such as lower Perdido Bay, Escambia Bay, and Choctawhatchee Bay. This conundrum is general throughout the United States and is currently a growing problem worldwide. Recent analyses indicate that 44 estuaries, representing 40% of the total estuarine surface area in the United States, suffer extreme eutrophic conditions with an additional 40 estuaries having moderate eutrophic conditions. By the year 2020, it has been estimated that eutrophic conditions will have worsened in 86 estuaries. Based on our analyses of the effects of plankton blooms on secondary production in coastal systems of the northeastern Gulf, the adverse impacts of anthropogenous nutrient loading on estuarine productivity in the country as a whole are already significant, and such effects will be catastrophic within a relatively short period if anthropogenous nutrient loading is not controlled.

The current system of environmental research in aquatic ecology continues to be driven by a combination of political control and inadequate, reductionist approaches to phytoplankton research. The lack of funding for adequate, long-term studies has long been defended by research funding agencies with the excuse that such studies are too time-consuming and too expensive to carry out. The resulting dearth of long-term, interdisciplinary studies in coastal systems suffers from the same minimalist approach that dominates both the agencies that fund coastal research and the scientific journals where the results of such research are published. The dependence on patch-quilt data sets instead of integrated ecosystem analyses has led to seriously flawed estimates of the current status of many river–estuarine and coastal systems. The general acceptance of false and even misleading assumptions concerning the causes and effects of phytoplankton blooms has contributed to the ongoing lack of attention to nutrient-driven impacts on coastal systems. The somewhat limited effectiveness of scientific determinations of the causative factors associated with continuously increasing levels of anthropogenous nutrient loading has led to generally pessimistic projections of coastal deterioration due to cultural eutrophication. Despite considerable and very expensive (albeit piecemeal) research efforts, we are still without even the most basic understanding of how estuarine and coastal systems have responded to fundamental alterations of the eutrophication process. Efforts should be made to increase our understanding of phytoplankton ecology and the effects of blooms on phytoplankton community structure. This research should also be undertaken with reference to impacts of hypereutrophication on associated food webs that form the foundation of useful productivity of estuarine and coastal systems.

Issues concerning phytoplankton community response to anthropogenous nutrient loading are fundamental to the food web ecology in coastal systems. Until now, the complexity of phytoplankton community structure has precluded research concerning the long-term aspects of how changes in phytoplankton organization affect associated coastal food webs. The trend away from taxonomic concerns has contributed to the current emphasis on individual HABs. However, important issues concerning initiation of plankton blooms and the effects of such blooms on coastal systems will not be addressed by autecological analyses of blooms. Until there is more emphasis on how the phytoplankton community as a whole is being changed by anthropogenous nutrient loading, coastal hypereutrophication will remain a phantom issue. Based on long-term projections concerning increased water temperature and reduced precipitation in the southeastern United States due to global warming, the predicted increases of anthropogenous nutrient loading in Gulf estuaries, together with our findings regarding the effects of hypereutrophication on coastal food webs, could have serious implications concerning future systemwide losses of secondary production in some of the most productive coastal systems in the world.

References

Admiraal, W. 1984. The ecology of estuarine sediment-inhabiting diatoms. In: E. Round and G. Chapman (Eds.), Progress in Phycological Research 3. Biopress LD, pp. 269–321.

Alexander, C., R. Smith, B. Loganathan, J. Ertel, H. L. Windom, and R. F. Lee. 1999. Pollution history of the Savannah River Estuary and comparisons with Baltic Sea pollution history. Limnologica 29: 267–273.

Anderson, D. M. 1996. Control and mitigation of red tides. December 4. Manuscript prepared for Project Start (Solutions to Avoid Red Tide), Sarasota, FL, 59 pp.

Anderson, D. M. 1997a. Bloom dynamics of toxic *Alexandrium* species in the northeastern U.S. Limnol. Oceanogr. 42: 1009–1022.

Anderson, D. M. 1997b. Physiology and bloom dynamics of toxic *Alexandrium* species, with emphasis on life cycle transitions. In: D. M. Anderson, A. D. Cembella, and G. M. Hallegreaff, Eds., Physiological Ecology of Harmful Algal Blooms. Springer-Verlag, Berlin, pp. 13–28.

Anderson, D. A. and D. J. Garrison (Eds.). 1997. The ecology and oceanography of harmful algal blooms. Limnol. Oceanogr. 42: 1305.

Andersson, L. and L. Rydberg. 1988. Trends in nutrient and oxygen conditions within the Kattegar: Effects on local nutrient supply. Estuarine Coastal Shelf Sci. 26: 559–579.

APHA. 1989. Standard Methods for the Examination of Water and Wastewater, 17th ed. American Public Health Association, Washington, D.C.

Baird, D. and R. E. Ulanowicz. 1989. The seasonal dynamics of the Chesapeake Bay ecosystem. Ecol. Monogr. 59: 329–364.

Barnett, E., J. Lewis, J. Marx, and D. Trimble. 1995. Ecosystem Management Implementation Strategy. Florida Department of Environmental Protection, Tallahassee.

Bass, D. G., Jr. 1990. Stability and persistence of fish assemblages in the Escambia River, Florida. Rivers 1, 4 (October 1990): 296–306.

Bass, D. G., Jr. 1991. Riverine fishes of Florida. In: R. J. Livingston (Ed.), The Rivers of Florida. Springer-Verlag, New York, pp. 65–84.

Becker, M. E. 1996. The effect of nutrient loading on benthic microalgal biomass and taxonomic composition. M.S. thesis, University of North Carolina at Wilmington.

Bernhard, A. E. and E. R. Peele. 1997. Nitrogen limitation of phytoplankton in a shallow embayment in northern Puget Sound. Estuaries 20: 759–769.

Bidagare, R. R. and M. E. Ondrusek. 1996. Spatial and temporal variability of phytoplankton pigment distributions in the central equatorial Pacific Ocean. Deep-Sea Res. 43: 809–833.

Bloesch, J. and H. Burgi. 1989. Changes in phytoplankton and zooplankton biomass and taxonomic composition reflected by sedimentation. Limnol. Oceanogr. 34: 1048–1061.

Boesch, D. F., D. M. Anderson, R. A. Horner, S. E. Shumway, P. A. Tester, and T.E. Whitledge. 1997. Harmful Algal Blooms in Coastal Waters: Options for Prevention, Control, and Mitigation. NOAA Coastal Ocean Program Decision Analysis Series No. 10. NOAA Coastal Ocean Office, Silver Spring, MD. 46 pp. + Appendix.

Boesch, D. F., N. E. Armstrong, C. F. D'Elia, N. G. Maynard, H. W. Paerl, and S. L. Williams. 1993. Deterioration of the Florida Bay ecosystem: an evaluation of the scientific evidence. National Fish and Wildlife Foundation, National Park Service, South Florida Water Management District. Unpublished report.

Bonin, D. J., M. R. Droop, S. Y. Maestrini, and M. C. Bonin. 1986. Physiological features of six micro-algae to be used as indicators of seawater quality. Cryptogamie Algol. 7: 23–83.

Bortone, S. 1991. Seagrass mapping of Perdido Bay. Unpublished report.

Botton, M. L. 1979. Effects of sewage sludge on the benthic invertebrate community of the inshore New York Bight. Estuarine Coastal Mar. Sci. 8: 169–180.

Boyer, J. N., J. W. Fourqurean, and R. D. Jones. 1999. Seasonal and long-term trends in the water quality of Florida Bay. In: Fourqurean, J. W., M. B. Robblee, and L. A. Deegan (Eds.), Florida Bay: A Dynamic Subtropical Estuary. Estuaries 22: 417–430.

Boynton, W. R., W. M. Kemp, and C. W. Keefe. 1982. A comparative analysis of nutrients and other factors influencing estuarine phytoplankton production. In: V. S. Kennedy (Ed.), Estuarine Comparisons. Academic Press, New York, pp. 69–90.

Breitburg, D. L. 1990. Near-shore hypoxia in the Chesapeake Bay: patterns and relationships among physical factors. Estuarine Coastal Shelf Sci. 30: 593–609.

Breitburg, D. L. 1992. Episodic hypoxia in Chesapeake Bay: interacting effects of recruitment, behavior, and physical disturbance. Ecol. Monogr. 62: 525–546.

Bricker, S. B., C. G. Clement, D. E. Pirhalla, S. P. Orlando, and D. R. G. Farrow. 1999. National Estuarine Eutrophication Assessment; Effects of Nutrient Enrichment in the Nation's Estuaries. National Oceanic and Atmospheric Administration, National Ocean Service, Special Projects Office and the National Centers for Coastal Ocean Science, Silver Spring, MD, 71 pp.

Brim, M. S. 1993. Toxics characterization report for Perdido Bay, Alabama, and Florida. U.S. Fish and Wildlife Service Publ. No. PCFO-EC-93-94, 137 pp.

Brockman, U. H., R. W. P. M. Lane, and H. Postma. 1990. Cycling of nutrient elements in the North Sea. Neth. J. Sea Res. 26: 239–264.

Brown, M. B. and A. B. Forsythe. 1974. Robust tests for the equality of variances. J. Amer. Stat. Assoc. 69: 364–367.

Brush, G. S. 1984. Stratigraphic evidence of eutrophication in an estuary. Water Resour. Res. 20: 531–541.

Brush, G. S. 1991. Long-term trends in Perdido Bay: a stratigraphic study. Florida Department of Environmental Regulation. Unpublished report.

Burkholder, J. A. and H. B. Glasgow, Jr. 1997. *Pfiesteria piscicidia* and other *Pfiesteria*–dinoflagellates: behaviors, impacts, and environmental controls. Limnol. Oceanogr. 42: 1052–1075.

Burkholder, J. M., E. J. Noga, C. H. Hobbs, and H. B. Glasgow, Jr. 1992. New "phantom" dinoflagellate is the causative agent of major estuarine fish kills. Nature 358: 407–410.

Buskey, E. J. and D. A. Stockwell. 1993. Effects of a persistent "brown tide" on zooplankton populations in the Laguna Madre of South Texas. In: Toxic Phytoplankton Blooms in the Sea. Proc. 5th Int. Conf. on Toxic Marine Phytoplankton. Elsevier, New York, pp. 659–665.

Buskey, E. J., P. A. Montagna, A. F. Amos, and T. E. Whitledge. 1997. Disruption of grazer populations as a contributing factor to the initiation of the Texas brown tide algal bloom. Limol. Oceano. 42: 1215–1222.

Canuel, E. A. and A. R. Zimmerman. 1999. Composition of particulate organic matter in the southern Chesapeake Bay: sources and reactivity. Estuaries 22: 980–994.

Caperon, J., S. A. Cattell, and G. Krasnick. 1971. Phytoplankton kinetics in a subtropical estuary: eutrophication. Limnol. Oceanogr. 26: 599–607.

Caraco, N., A. Tamse, O. Boutros, and I. Valiela. 1987. Nutrient limitation of phytoplankton growth in brackish coastal ponds. Can. J. Fish. Aquat. Sci. 4: 473–476.

Cardwell, R. D., S. Olsen, M. I. Carr, and E. W. Sanborn. 1979. Causes of oyster mortality in South Puget Sound NOAA Tech. Mem. ERL MESA-39. Washington Department of Fisheries, Salmon Research and Development, Brinnan, WA.

Carvalho, L. R., E. J. Cox, S. C. Fritz, S. Juggins, P. A. Sims, F. Gasse, and R. W. Battarbee. 1995. Standardizing the taxonomy of saline lake *Cyclotella* spp. Diatom Res. 10: 229–240.

Cerco, C. F. and S. P. Seitzinger. 1997. Measured and modeled effects of benthic algae on eutrophication in Indian River-Rehoboth Bay, Delaware. Estuaries 20: 231–248.

Chang, E. H. 1988. Distribution, abundance and size composition of phytoplankton off Westland, New Zealand, February 1982. N.Z. J. Mar. Freshwater Res. 22: 345–367.

Chang, F. J., C. Anderson, and N. C. Boustead, 1990. First record of *Heterosigma* (Raphidophyceae) bloom with associated mortality of cage-reared salmon in Big Glory Bay, New Zealand. N.Z. J. Mar. Freshwater Res. 24: 461–469.

Chanton, J. and G. Lewis. 1999. Plankton and dissolved inorganic carbon isotopic composition in a river-dominated estuary: Apalachicola Bay, Florida. Estuaries 22: 575–583.

Christensen, H. and E. Kanneworff. 1985. Sedimenting phytoplankton as a major food source for suspension and deposit feeders in the Oresund. Ophelia 24: 223–244.

Clement, A. and G. Lembeye. 1993. Phytoplankton monitoring program in the fish farming of south Chile. In: T. J. Smayda and Y. Shimizu (Eds.), Toxic Phytoplankton Blooms in the Sea. Elsevier, New York, pp. 223–228.

Clements, W. H. and R. J. Livingston. 1983. Overlap and pollution-induced variability in the feeding habits of filefish (Pisces: Monacanthidae) from Apalachee Bay, Florida. Copeia (1983): 331–338.

Clements, W. H. and R. J. Livingston. 1984. Prey selectivity of the fringed filefish *Monacanthus ciliatus* (Pisces: Monacanthidae): role of prey accessibility. Mar. Ecol. Prog. Ser 16: 291–295.

Cloern, J. E. 1996. Phytoplankton bloom dynamics in coastal ecosystems: a review with some general lessons from sustained investigations of San Francisco Bay, California. Rev. Geophys. 34: 127–168.

Coffin, R. B. and L. A. Cifuentes. 1992. A stable isotope study of carbon sources and microbial transformations in the Perdido Estuary, FL. In: R. J. Livingston, Ecological Study of the Perdido Drainage System. Florida Department of Environmental Protection. Unpublished report.

Coffin, R. B. and L. A. Cifuentes. 1999. A stable isotope analysis of carbon cycling in the Perdido Estuary, Florida. Estuaries 22: 917–926.

Collard, S. B. 1991a. The Pensacola Bay System: Biological Trends and Current Status. Water Resources Special Report 91-3. Northwest Florida Water Management District, Havana.

Collard, S. B. 1991b. Management Options for the Pensacola Bay System: The Potential Value of Seagrass Transplanting and Oyster Bed Refurbishment Programs. Water Resources Special Report 91-4. Northwest Florida Water Management District, Havana.

Cooper, S. R. 1995a. Diatoms in sediment cores from the mesohaline Chesapeake Bay, U.S.A. Diatom Res. 10: 39–89.

Cooper, S. R. 1995b. An abundant, small brackish water *Cyclotella* species in Chesapeake Bay, U.S.A. In: J. P. Kociolek and M. J. Sullivan (Eds.), A Century of Diatom Research in North America: A Tribute to the Distinguished Careers of Charles W. Reimer and Ruth Patrick. Koeltz Scientific Books, Champaigne, pp. 133–140.

Corredor, J. E., R. W. Howarth, R. R, Twilley, and J. M. Morrell. 1999. Nitrogen cycling and anthropogenic impact in the tropical interamerican seas. Biogeochemistry 46: 163–178.

Cross, R. D. and D. L. Williams. 1981. Proceedings of the National Symposium on Freshwater Inflow to Estuaries. U.S. Fish and Wildlife Service, Office of Biological Services. FWS/OBS-81/04. 2 Vol.

Cullen, J. J., A. M. Ciotti, R. F. Davis, and M. R. Lewis. 1997. Optical detection and assessment of algal blooms. Limol. Oceanogr. 42: 1223–1239.

Culliton, T. J., M. A. Warren, T. R. Goodspeed, D. G. Remer, C. M. Blackwell, and J. J. McDough, III. 1990. 50 years of population growth along the Nation's coasts, 1960–2010. Strategic Assessment Branch, Ocean Assessments Division, Office of Oceanography and Marine Assessment, National Ocean Survey, National Oceanic and Atmospheric Administration.

Curl, H. 1962. Analysis of carbon in marine plankton organisms. J. Mar. Res. 20: 181–188.

Dale, B., T. A. Thorsen, and A. Fjellsa. 1999. Dinoflagellate cysts as indicators of cultural eutrophication in the Oslofjord, Norway. Est. Coast. Shelf Sci. 48: 371–382.

Dardeau, M. R., R. F. Modlin, W. W. Schroeder, and J. P. Stout. 1992. Estuaries. In: C. T. Hackney, S. M. Adams, and W. H. Martin (Eds.), Biodiversity of the Southeastern United States. John Wiley & Sons, New York, 779 pp.

Dauer, D. M., R. W. Ewing, G. H. Tourtellotte, W. T. Harlan, J. W. Sourbeer, and H. R. Baker. 1982. Predation, resource limitation and the structure of lower Chesapeake Bay. Int. Rev. Gesamt. Hydrobiol. 67: 477–489.

Dauer, D. M., A. J. Rodi, and J. A. Ranasinghe. 1992. Effects of low dissolved oxygen events on the macrobenthos of the lower Chesapeake Bay. Estuaries 15: 384–391.

Dauer, D. M., J. A. Ranasinghe, and S. B. Weisberg, 2000. Relationships between benthic community condition, water quality, sediment quality, nutrient loads, and land use patterns in Chesapeake Bay. Estuaries 23: 80–96.

Davis, W. P., M. R. Davis, and D. A. Flemer. 1999. Observations on the regrowth of subaquatic vegetation following transplantation: a potential method to assess environmental health of coastal habitats. In: S. Bortone (Ed.), Seagrasses; Monitoring, Ecology, Physiology, and Management. CRC Press, Boca Raton, FL, pp. 231–238.

Dawson, C. E. 1955. A contribution to the hydrograph of Apalachicola Bay. Publ. Tex. Inst. Mar. Sci. 4: 15–35.

Dayton, P. K., 1979. Ecology: a science and a religion. In: R. J. Livingston (Ed.), Ecological Processes in Coastal and Marine Systems. Plenum Press, New York, pp. 3–18.

Dayton, P. K., M. J. Tegner, P. E. Parnell, and P. B. Edwards. 1992. Temporal and spatial patterns of disturbance and recovery in a kelp forest community. Ecol. Monogr. 62: 421–445.

De Jong, V. N. 1995. The Ems Estuary, the Netherlands. In: A. J. McComb (Ed.), Eutrophic Shallow Estuaries and Lagoons. CRC Press, Boca Raton, FL, pp. 81–108.

De Jong, V. N. and W. van Raaphorst, 1995. Eutrophication of the Dutch Wadden Sea (Western Europe), an estuarine area controlled by the River Rhine. In: A. J. McComb (Ed.), Eutrophic Shallow Estuaries and Lagoons. CRC Press, Boca Raton, FL, pp. 81–108.

D'Elia, C. F., J. G. Sanders, and W. R. Boynton. 1986. Nutrient enrichment studies in a coastal plain estuary — Phytoplankton growth in large-scale, continuous cultures. Can. J. Fish. Aquat. Sci. 43: 97–406.

DeMort, C. L. 1991. The St. Johns river system. In: R. J. Livingston (Ed.), The Rivers of Florida. Springer-Verlag, New York, pp. 97–120.

Diaz, R. J. and R. Rosenberg. 1995. Marine benthic hypoxia: a review of its ecological effects and the behavioral responses of benthic macrofauna. Oceanogr. Mar. Biol.: Annu. Rev. 33: 245–305.

Dolan, D. M, A. K. Yui, and R. D. Geist. 1981. Evaluation of river load estimation methods for total phosphorus. Great Lakes Res. 7: 207–214.

Dortch, Q. 1990. The interaction between ammonium and nitrate uptake in phytoplankton. Mar. Ecol. Prog. Ser. 61: 183–201.

Doudoroff, P. and D. L. Shumway. 1967. Dissolved oxygen criteria for the protection of fish. American Fisheries Society, Special Publication No. 4, pp. 13–19.

Duarte, C. M. 1995. Submerged aquatic vegetation in relation to different nutrient regimes. Ophelia 41: 87–112.

Dugan, P. J. and R. J. Livingston. 1982. Long-term variation in macroinvertebrate communities in Apalachee Bay, Florida. Estuarine Coastal Shelf Sci. 14: 391–403.

Duxbury, A. C. 1975. Orthophosphate and dissolved oxygen in Pudget Sound. Limnol. Oceanogr. 20: 270–274.

Edmonson, W. T. 1970. Phosphorus, nitrogen, and algae in Lake Washington after diversion of sewage. Science 169: 690–691.

Elmgren, R. and U. Larsson. 1997. Himmelfjorden foranderingen i ett naringsbelastat kustekybsystem in Ostersjon. Rapport 4565. Naturvardsuerket Forlag.

Estabrook, R. H. 1973. Phytoplankton Ecology and Hydrography of Apalachicola Bay. M.S. thesis, Florida State University, Tallahassee, 166 pp.

Estevez, E. D., L. K. Dixon, and M. F. Flannery. 1991. West-coastal rivers of peninsular Florida. In: R. J. Livingston (Ed.), The Rivers of Florida. Springer-Verlag, New York, pp. 187–222.

Federle, T. W., M. A. Hullar, R. J. Livingston, D. A. Meeter, and D. C. White. 1983a. Spatial distribution of biochemical parameters indicating biomass and community composition of microbial assemblies in estuarine mud flat sediments. Appl. Environ. Micro. 45: 58–63.

Federle, T. W., R. J. Livingston, D. A. Meeter, and D. C. White. 1983b. Modifications of estuarine sedimentary microbiota by exclusion of epibenthic predators. J. Exp. Mar. Biol. Ecol. 73: 81–94.

Federle, T. W., R. J. Livingston, L. E. Wolfe, and D. C. White. 1986. A quantitative comparison of microbial community structure of estuarine sediments for microcosms in the field. Can. J. Microbiol. 32: 319–325.

Fernald, E. A. and D. J. Patton (Eds). 1984. Water Resources Atlas of Florida. Institute of Science and Public Affairs, Florida State University, Tallahassee, 291 pp.

Fernald, E. A. and E. D. Purdum (Eds.). 1992. Atlas of Florida. University of Florida Press, Gainesville, 280 pp.

Fisher, T. R., E. R. Peele, J. W. Ammerman, and L. W. Harding, Jr. 1992. Nutrient limitation of phytoplankton in Chesapeake Bay. Mar. Ecol. Prog. Ser. 82: 51–63.

Fjerdingstad, E. 1965. Taxonomy and saprobic valency of benthic phytomicro-organisms. Int. Rev. Ges. Hydrobiol. 50: 475–604.

Flemer, D. A., R. J. Livingston, and S. E. McGlynn. 1997. Phytoplankton nutrient limitation: seasonal growth responses of sub-temperate estuarine phytoplankton to nitrogen and phosphorus — an outdoor microcosm experiment. Estuaries 21: 145–159.

Foehrenbach, J. 1971. Eutrophication. J. Water Pollut. Control Fed. 43: 1191–1202.

Fourqurean, J. W. and M. B. Robblee. 1999. Florida Bay: a history of recent ecological changes. Estuaries 22: 345–357.

Fourqurean, J. W., I. R. D. Jones, and J. C. Zieman. 1993. Processes influencing water column nutrient characteristics and phosphorus limitation of phytoplankton biomass in Florida Bay, FL, USA: inferences from spatial distributions. Estuarine Coastal Shelf Sci. 36: 295–314.

Fourqurean, J. W., M. B. Robblee, and L. A. Deegan. 1999. Florida Bay: a dynamic subtropical estuary. Estuaries 22.

Freeman, D. B. 1989. The Distribution and Trophic Significance of Benthic Microalgae in Masonboro Sound, North Carolina. M.S. thesis, University of North Carolina at Wilmington.

Frost, B. and W. Grazing. 1980. In: I. Morris (Ed.), The Physiological Ecology of Phytoplankton. University of California Press, Berkeley, pp. 465–492.

Fryxell, G. A. and M. C. Villac. 1999. Toxic and harmful marine diatoms. In: E. F. Stoermer and J. P. Smol (Eds.), The Diatoms: Applications for the Environmental and Earth Sciences. Cambridge Press, New York, pp. 419–428.

Fucik, K. W. 1974. The Effect of Petroleum Operations on the Phytoplankton Ecology of the Louisiana Coastal Waters. M.S. thesis, Texas A&M University, College Station.

Fuhs, G. W., S. D. Demmerle, E. Canelli, and M. Chen. 1972. Characterization of phosphorus-limited plankton algae. Limnol. Oceanogr., Special Symposium, 1: 113–133.

Fukuyo, Y., H. Takano, M. Chihara, and K. Matsuoka. 1990. Red Tide Organisms in Japan — An Illustrated Taxonomic Guide. Uchida Rokakuho, Tokyo, 407 pp.

Fulmer, J. M. 1997. Nutrient Enrichment and Nutrient Input to Apalachicola Bay, Florida. M.S. thesis, Florida State University, Tallahassee, 59 pp.

Galehouse, J. S. 1971. Sedimentation analysis. In: R. E. Carver (Ed.), Procedures in Sedimentary Petrology. Wiley-Interscience, New York, pp. 69–94.

Gameson, A. L. H., N. J. Barrett, and J. S. Strawbridge. 1973. The aerobic Thames estuary. In: S. H. Jenkins (Ed.), Advances in Water Pollution Research, Sixth Information Congress, Jerusalem. Pergamon Press, New York, pp. 843–850.

Gaston, G. R. 1985. Effects of hypoxia on macrobenthos of the inner shelf off Cameron, Louisiana. Estuarine Coastal Shelf Sci. 20: 600–613.

Geary, R. C. 1936. Moments of the ratio of the mean deviation to the standard deviation for normal samples. Biometrika 28: 295–307.

Gerhart, D. Z. and G. E. Likens. 1975. Enrichment experiments for determining nutrient limitation: four methods compared. Limnol. Oceanography 20: 649–653.

Gilbert, P. M., D. J. Conley, T. R. Fisher, L. W. Harding, and T. C. Malone. 1995. Dynamics of the 1990 winter/spring bloom in Chesapeake Bay. Mar. Ecol. Prog. Ser. 122: 27–43.

Gorsline, D. S. 1963. Oceanography of Apalachicola Bay, Florida. In: Essays in Marine Geology in Honor of K.O. Emery. University of Southern California Press, Los Angeles, pp. 145–176.

Graneli, E. 1987. Nutrient limitation of phytoplankton in a brackish water bay highly influenced by river discharge. Estuarine Coastal Shelf Sci. 25: 555–565.

Graneli, E. and P. Carlsson. 1997. The ecological significance of phagotrophy in photosynthetic flagellates. In: D. M. Anderson, A. D. Cembella, and G. M. Hallegreaff (Eds.), Physiological Ecology of Harmful Algal Blooms. Springer-Verlag, Berlin, pp. 540–557.

Graneli, E., K. Wallstrom, U. Larsson, W. Graneli, and R. Elmgren. 1990. Nutrient limitation of primary production in the Baltic Sea area. Ambio 19: 142–151.

Greening, H. S. and R. J. Livingston. 1982. Diel variations in the structure of epibenthic macroinvertebrate communities of seagrass beds (Apalachee Bay, Florida). Mar. Ecol. Prog. Ser. 7: 147–156.

Greve, W. and T. R. Parsons. 1977. Photosynthesis and fish production Hypothetical effects of climate change and pollution. Helgol. Wiss. Meeresunters. 30: 666–672.

Hackney, C. T. and S. M. Adams. 1992. Aquatic communities of the southeastern United States: past, present, and future. In: C. T. Hackney, S. M. Adams, and W. J. Martin (Eds.), Biodiversity of the Southeastern United States. John Wiley & Sons, New York, pp. 747–760.

Hagy, J. D. 1996. Residence Times and Net Ecosystem Processes in Patuxent River Estuary. M.S. thesis, University of Maryland, College Park, 205 pp.

Haines, E. B. 1979. Interactions between Georgia salt marshes and coastal waters: a changing paradigm. In: R. J. Livingston (Ed.), Ecological Processes in Coastal and Marine Systems. Plenum Press, New York, pp. 35–46.

Hallegreaff, G. M. 1995. Harmful algal blooms: a global overview. In: Hallegreaff, G. M., D. M. Anderson, and A. D. Cembella (Eds.), Manual on Harmful Marine Microalgae. IOC Manuals and Guides No. 33. UNESCO, pp. 1–22.

Hallegreaff, G. M. and S. Fraga. 1997. Bloom dynamics of the toxic dinoflagellate *Gymnodinium catenatum*, with emphasis on Tasmanian and Spanish coastal waters. In: D. M. Anderson, A. D. Cembella, and G. M. Hallegreaff (Eds.), Physiological Ecology of Harmful Algal Blooms. Springer-Verlag, Berlin, pp. 59–80.

Hallegreaff, G. M., D. M. Anderson, and A. D. Cembella. 1995. Manual on Harmful Marine Microalgae. IOC Manuals and Guides No. 33. UNESCO.

Hand, J., J. Col, and L. Lord. 1996. 1996 Florida Water Quality Assessment, 305(b) Technical Appendix. Florida Department of Environmental Protection, Tallahassee.

Hannah, R. P., A. T. Simmons, and G. A. Moshiri. 1973. Some aspects of nutrient–primary productivity relationships in a bayou estuary. J. Water Pollut. Control Fed. 45: 2508–2520.

Hansen, P. J. 1997. Phagotrophic mechanisms and prey selection in mixotrophic phytoflagellates. In: D. M. Anderson, A. D. Cembella, and G. M. Hallegreaff (Eds.), Physiological Ecology of Harmful Algal Blooms. Springer-Verlag, Berlin, pp. 523–537.

Hara, Y. and M. Chihara, 1987. Morphology, ultrastructure, and taxonomy of the Raphidophycean alga *Heterosigma akashiwo*. Bot. Mag. Tokyo 100: 151–163.

Harding, L. W. 1994. Long-term trends in the distribution of phytoplankton in the Chesapeake Bay: role of light, nutrients and stream-flow. Mar. Ecol. Prog. Ser. 104: 267–291.

Hargraves, P. H. 1990. Studies on marine planktonic diatoms: V. Morphology and distribution of *Leptocylindrus minimus* Gran. Beih. Nova Hedw. 100: 47–60.

Hasle, G. R. and G. A. Fryxell, 1995. Taxonomy. Diatoms. In: G. M. Hallegreaff (Ed.), Manual on Harmful Marine Microalgae. IOC Manuals and Guides No. 33. UNESCO, pp. 339–364.

Hecky, P. E. 1998. Low nitrogen:phosphorus ratios and the nitrogen fix: why watershed nitrogen removal will not improve the Baltic. In: T. Hellstrom (Ed.), Effects of Nitrogen in the Aquatic Environment. Swedish National Committee for IAWQ, Royal Swedish Academy of Sciences, Stockholm.

Hein, M., M. F. Pedersen, and K. Sand-Jensen. 1995. Size-dependent nitrogen uptake in micro- and macroalgae. Mar. Ecol. Prog. Ser. 118: 247–253.

Hellstrom, T. 1996. An empirical study of nitrogen dynamics in lakes. Water Environ. Res. 68: 55–65.

Hitchcock, G. L. 1982. A comparative study of the size-dependent organic composition of marine diatoms and dinoflagellates. J. Plankton Res. 4: 363–377.

Hodgkiss, I. J. and W. S Yim. 1995. A case study of Tolo Harbour, Hong Kong. In: A. J. McComb (Ed.), Eutrophic Shallow Estuaries and Lagoons. CRC Press, Boca Raton, FL, pp. 41–58.

Holland, A. F., N. K. Mountford, and J. A. Mihursky. 1977. Temporal variation in upper bay mesohaline benthic communities: 1. The 9-m mud habitat. Chesapeake Sci. 18: 370–378.

Honjo, T. 1993. Overview on bloom dynamics and physiological ecology of *Heterosigma akashiwo*. In: T. J. Smayda, and Y. Shimizu (Eds.), Toxic Phytoplankton Blooms in the Sea. Proceedings of the Fifth International Conference on Toxic Marine Phytoplankton. Elsevier, New York, pp. 33–41.

Hooks, T. A., K. L. Heck, and R. J. Livingston. 1975. An inshore marine invertebrate community: structure and habitat associations in the N.E. Gulf of Mexico. Bull. Mar. Sci. 26: 99–109.

Hopkinson, C. S. and R. Wetzel. 1982. *In situ* measurements of nutrient and oxygen fluxes in a coastal marine benthic community. Mar. Ecol. Prog. Ser. 10: 29–35.

Hopkinson, C. S., Jr., I. Buffam, J. Hobbie, J. Vallino, M. Perdue, B. Eversmeyer, F. Prahl, J. Covert, R. Hodson, M. Moran, E. Smith, J. Baross, B. Crump, S. Findleay, and K. Forman. 1998. Terrestrial inputs of organic matter to coastal ecosystems: an intercomparison of chemical characteristics and bioavailability. Biogeochemistry 43: 211–234.

Hopkinson, C. S., Jr., A. E. Giblin, J. Tucker, and R. H. Garritt. 1999. Benthic metabolism and nutrient cycling along an estuarine salinity gradient. Estuaries 22: 863–881.

Horner, R. A., D. L. Garrison, and E. G. Plumley. 1997. Harmful algal blooms and red tide problems on the U.S. west coast. Limnol. Oceanogr. 42: 1076–1088.

Howarth, R. W. 1988. Nutrient limitation of net primary production in marine ecosystems. Annu. Rev. Ecol. Syst. 19: 89–110.

Howarth, R. W. and R. Marino. 1998. A mechanistic approach to understanding why so many estuaries and brackish waters are nitrogen limited. In: T. Hellstrom (Ed.), Effects of Nitrogen in the Aquatic Environment. KVA Report 1, Royal Swedish Academy of Sciences, Stockholm.

Howarth, R. W., H. S. Jensen, R. Marino, and H. Postma. 1995. Transport to and processing of phosphorus in near-shore and oceanic waters. In: H. Tiessen (Ed.), Phosphorus in the Global Environment. Wiley, New York.

Howarth, R. W., D. M. Anderson, T. M. Church, H. Greening, C. S. Hopkinson, W. C. Huber, N. Marcus, R. J. Nainman, K. Segerson, A. N. Sharpley, and W. J. Wiseman. 2000. Clean Coastal Waters: Understanding and Reducing the Effects of Nutrient Pollution. Ocean Studies Board and Water Science and Technology Board, National Academy Press, Washington, D.C., 391 pp.

Hydroqual, Inc. 1992. Elevenmile Creek and Perdido Bay Modeling Analysis. Hydroqual, Inc., 1 Lethbridge Plaza, Mahwah, NJ. Unpublished report.

Hudson, T. J. and D. R. Wiggins. 1996. Comprehensive Shellfish Harvesting Area Survey of Pensacola Bay System Escambia and Santa Rosa Counties, Florida. Florida Department of Environmental Protection, Tallahassee.

Imai, I., M. Yamaguchi, and M. Watanabe. 1997. Ecophysiology, life cycle, and bloom dynamics of *Chattonella* in the Seto inland sea, Japan. In: D. M. Anderson, A. D. Cembella, and G. M. Hallegreaff (Eds.), Physiological Ecology of Harmful Algal Blooms. Springer-Verlag, Berlin, pp. 93–112.

Inoue, Y. and K. Nozawa, 1989. Separation of toxin from harmful red tides occurring along the coast of Kagoshima Prefecture. In: Okaichi, T., D. M. Andersen, and T. Nemato (Eds.), Red Tides: Biology, Environ. Sci. Toxicol., pp. 371–374.

Isphording, W. C. and R. J. Livingston. 1989. Report on synoptic analyses of water and sediment quality in the Perdido Drainage System. Florida Department of Environmental Regulation, Tallahassee. Unpublished report.

Iverson, R. L. and H. F. Bittaker. 1986. Seagrass distribution and abundance in eastern Gulf of Mexico coastal waters. Estuarine Coastal Shelf Sci. 22: 577–602.

Iverson, R. L., W. Landing, B. Mortazawi, and J. Fulmer. 1997. Nutrient transport and primary productivity in the Apalachicola River and Bay. F. G. Lewis (Ed.), Apalachicola River and Bay Freshwater Needs Assessment. Report to the ACF/ACT Comprehensive Study. Northwest Florida Water Management District, Havana, FL. Unpublished report.

Jaworski, N. A. 1981. Sources of nutrients and the scale of eutrophication problems in estuaries. In: B. J. Neilson and L. E. Cronin (Eds.), Estuaries and Nutrients. Humana Press, Clifton, NJ, pp. 81–100.

Jaworski, N. A., D. W. Lear, Jr., and O. Villa. 1972. Nutrient management in the Potomac estuary. In: G. E. Likens (Ed.), Nutrients and Eutrophication. Am. Soc. Limnol. Oceanogr. Spec. Symp. Vol. 1. Lawrence, KS, pp. 246–272.

Jorgensen, B. B. and K. Richardson. 1996. Eutrophication in Coastal Marine Systems. American Geophysical Union, Washington, D.C.

Justic, D. 1987. Long-term eutrophication of the northern Adriatic Sea. Mar. Pollut. Bull. 89: 1–5.

Kaul, L. W. and P. N. Froelich, Jr. 1984. Modeling estuarine nutrient geochemistry in a simple system. Geochim. Cosmochem. Acta 48: 1417–1433.

Kemp, W. M. and W. R. Boynton. 1984. Spatial and temporal coupling of nutrient inputs to estuarine primary production: the role of particulate transport and decomposition. Bull. Mar. Sci. 35: 522–535.

Kemp, W. M. and W. R. Boynton. 1992. Benthic–pelagic interactions: nutrient and oxygen dynamics. In: D. E. Smith, M. Leffler, and G. Mackiernan (Eds.), Oxygen Dynamics in the Chesapeake Bay. A Synthesis of Recent Research. Maryland Sea Grant College, College Park.

Kennish, M. J. 1997. Estuarine and Marine Pollution. CRC Press, Boca Raton, FL, 524 pp.

King, R. J. and B. R. Hodgson. 1995. Tuggerah Lakes system, New South Wales, Australia. 1995. In: A. J. McComb (Ed.), Eutrophic Shallow Estuaries and Lagoons. CRC Press, Boca Raton, FL, pp. 19–30.

Kirk, J. T. O. 1983. Light and Photosynthesis in Aquatic Ecosystems. Cambridge University Press, New York.

Kirk, J. T. O. 1986. Optical limnology — a manifesto. Limnology in Australia 1: 33–62.

Kivi, K., S. Kaitala, H. Kuosa, J. Kuparinen, E. Leskinen, R. Lignell, B. Marcussen, and T. Tamminen. 1993. Nutrient limitation and grazing control of the Baltic plankton community during annual succession. Limnol. Oceanogr. 38: 893–905.

Klein, C. J. 1991. Hydrological modeling of Perdido Bay. Florida Department of Environmental Regulation. Unpublished report.

Komarek, J. and K. Anagnostidis 1998. Cyanoprokaryota. 1. Chroococcales. In: Susswasserflora von Mitteleuropa, Bd. 19. G. Fischer, Jena, 548 pp.

Kuo, A. Y. and B. J. Neilson. 1987. Hypoxia and salinity in Virginia estuaries. Estuaries 10: 277–283.

Kushlan, J. A. 1991. The Everglades. In: R. J. Livingston (Ed.), The Rivers of Florida. Springer-Verlag, New York, pp. 121–142.

Lam, C. W. Y. and K. C. Ho. 1989. Red tides in Tolo Harbor, Hong Kong. In: T. Okaichi, D. M. Anderson, and T. Nemoto (Eds.), Red Tides: Biology. Environ. Sci. Toxicol. Elsevier, New York, pp. 49–52.

Lassus, P. and J. P. Berthome. 1988. Status of 1987 algal blooms in IFREMER. ICES/annex III C: 5–13.

Latasa, M. and R. R. Bidagare. 1998. Comparison of phytoplankton populations of the Arabian Sea during the spring inter-monsoon and southwest monsoon of 1995 as described by HPLC-analyzed pigments. Deep-Sea Res. 11: 2133–2170.

Laughlin, R. A. 1979. Trophic Ecology and Population Distribution of the Blue Crab, *Callinectes sapidus* Rathbun, in the Apalachicola estuary (North Florida, U.S.A.). Ph.D. dissertation, Florida State University, Tallahassee.

Laughlin, R. A. and R. J. Livingston. 1982. Environmental and trophic determinants of the spatial/temporal distribution of the brief squid (*Lolliguncula brevis*) in the Apalachicola Estuary (North Florida, USA). Bull. Mar. Sci. 32: 489–497.

Leber, K. M. 1983. Feeding Ecology of Decapod Crustaceans and the Influence of Vegetation on Foraging Success in a Subtropical Seagrass Meadow. Ph.D. dissertation, Florida State University, Tallahassee.

Leber, K. M. 1985. The influence of predatory decapods, refuge, and microhabitat selection on seagrass communities. Ecology 66: 1951–1964.

Legendre, L. and S. Demers. 1985. Auxiliary energy, ergoclines, and aquatic biological production. Nat. Can. 112: 5–14.

Lehman, P. W. 1992. Environmental factors associated with long-term changes in chlorophyll concentration in the Sacramento–San Joaquin delta and Suisun Bay, California. Estuaries 15: 335–348.

Leitman, H. M., J. E. Sohm, and M. A. Franklin. 1983 Wetland hydrology and tree distribution of the Apalachicola River flood plain, Florida. U.S. Geological Survey Water-Supply Paper 2196-A, 52 pp.

Leitman, S. F., L. Ager, and C. Mesing. 1991. The Apalachicola experience: environmental effects of physical modifications to a river for navigational purposes. In: R. J. Livingston (Ed.), The Rivers of Florida. Springer-Verlag, New York, pp. 223–246.

Lenihan, H. S. and C. H. Peterson. 1998. How habitat degradation through fishery disturbance enhances impacts of hypoxia on oyster reefs. Ecol. Appl. 8: 128–140.

Levene, H. 1960. Robust tests for equality of variances. In: I. Olkin (Ed.), Contributions to Probability and Statistics. Stanford University Press, Palo Alto, CA, pp. 278–329.

Levinton, J. S. 1982. Marine Biology. Prentice-Hall, Englewood Cliffs, NJ.

Lewis, F. Graham, III. 1974. Avoidance Reactions of Two Species of Marine Fishes to Kraft Pulp Mill Effluent. M.S. thesis, Florida State University, Tallahassee.

Lewis, F. Graham, III and R. J. Livingston. 1977. Avoidance of bleached kraft pulpmill effluent by pinfish (*Lagodon rhomboides*) and gulf killifish (*Fundulus grandis*). J. Fish. Res. Bd. Can. 34: 568–570.

Lindeman, R. L. 1942. The trophic-dynamic aspect of ecology. Ecology 23: 399–418.

Lirdwitayaprasit, T., S. Nishio, S. Montani, and T. Okaichi. 1990. The biochemical processes during cyst formation in *Alexandrium catenella*. In. L. E. Graneli, B. Sundstorm, L. Edler, and D. M. Anderson (Eds.), Toxic Marine Phytoplankton. Elsevier, Amsterdam, pp. 294–299.

Livingston, R. J. 1975. Impact of kraft pulp-mill effluents on estuarine and coastal fishes in Apalachee Bay, Florida, USA. Mar. Biol. 32: 19–48.

Livingston, R. J. 1976. Diurnal and seasonal fluctuations of organisms in a north Florida estuary. Estuarine Coastal Mar. Sci. 4: 374–400.

Livingston, R. J. 1979. Multiple factor interactions and stress in coastal systems: a review of experimental approaches and field implications. In: F. J. Vernberg (Ed.), Marine Pollution: Functional Responses. Academic Press, New York, pp. 389–413.

Livingston, R. J. 1980. Ontogenetic trophic relationships and stress in a coastal seagrass system in Florida. In: V. S. Peterson (Ed.), Estuarine Perspectives. Academic Press, New York, pp. 423–435.

Livingston, R. J. 1982. Trophic organization in a coastal seagrass system. Mar. Ecol. Prog. Ser. 7: 1–12.

Livingston, R. J. 1984a. The ecology of the Apalachicola Bay system: an estuarine profile. U.S. Fish and Wildlife Service FWS/PBS 82/05, Gainesville, FL, 148 pp.

Livingston, R. J. 1984b. Trophic response of fishes to habitat variability in coastal seagrass systems. Ecology 65: 1258–1275.

Livingston, R. J. 1984c. River-derived input of detritus into the Apalachicola estuary. In: R. D. Cross and D. L. Williams (Eds.), Proceedings of the National Symposium on Freshwater Inflow to Estuaries. Fish and Wildlife Service, Washington, D.C., pp. 320–332.

Livingston, R. J. 1985a. The relationship of physical factors and biological response in coastal seagrass meadows. Proceedings of a Seagrass Symposium, Estuarine Research Foundation. Estuaries 7: 377–390.

Livingston, R. J. 1985b. Aquatic field monitoring and meaningful measures of stress. In: H. H. White (Ed.), Concepts in Marine Pollution Measurements, Washington, D.C., pp. 681–692.

Livingston, R. J. 1986a. The Choctawhatchee River–Bay System. Final Report. Northwest Florida Water Management District, Havana, FL. Unpublished report.

Livingston, R. J. 1986b. Analysis of Field Data Concerning Old Pass Lagoon (Choctawhatchee Bay, Florida: September, 1985–February, 1986). Final Report. Northwest Florida Water Management District, Havana, FL. Unpublished report.

Livingston, R. J. 1987a. Historic trends of human impacts on seagrass meadows in Florida (invited paper). Symposium Proceedings: Subtropical–Tropical Seagrasses of the Southeastern U.S. Fla. Mar. Res. Publ. 42: 139–151.

Livingston, R. J., 1987b. Distribution of Toxic Agents and Biological Response of Infaunal Macroinvertebrates in the Choctawhatchee Bay System. Final Report, Office of Coastal Management, Florida Department of Environmental Regulation, Tallahassee. Unpublished report.

Livingston, R. J. 1987c. Field sampling in estuaries: the relationship of scale to variability. Estuaries 10: 194–207.

Livingston, R. J. 1988. Inadequacy of species-level designations for ecological studies of coastal migratory fishes. Environ. Biol. Fish. 22: 225–234.

Livingston, R. J. 1989. The Ecology of the Choctawhatchee River System. Final Report. Northwest Florida Water Management District, Havana, FL. Unpublished report.

Livingston, R. J. 1990. Inshore marine habitats. In: R. L Myers and J. J. Ewel (Eds.), Ecosystems of Florida. University of Central Florida Press, Orlando, pp. 549–573.

Livingston, R. J. 1991a. The Rivers of Florida. Springer-Verlag, New York.

Livingston, R. J. 1991b. Medium sized rivers: gulf coastal plain. In: C. T. Hackney (Ed.), Biodiversity of the Southeastern U.S. Ecological Society of America, John Wiley & Sons, New York, 434 pp.

Livingston, R. J. 1991c. Historical relationships between research and resource management in the Apalachicola River-estuary. Ecol. Appl. 1: 361–382.

Livingston, R. J. 1991d. The Rivers of Florida. Ecology of the Choctawhatchee River System. Springer-Verlag, New York.

Livingston, 1992. Ecological study of the Perdido Drainage System. Florida Department of Environmental Regulation, Tallahassee. Unpublished report.

Livingston, R. J. 1993a. River-Gulf Study. Florida Department of Environmental Regulation, Tallahassee. Unpublished report.

Livingston, R. J. 1993b. Estuarine Wetlands. In: J. Berry and M. Dennison (Eds.), Wetlands, Noyes Publications, Park Ridge, NJ, pp. 128–153.

Livingston, R. J. 1994. Ecological Study of Jack's Branch and Soldier Creek. Florida Department of Environmental Regulation, Tallahassee. Unpublished report.

Livingston, R. J. 1995. Nutrient analysis of upper Perdido Bay. Florida Department of Environmental Regulation, Tallahassee. Unpublished report.

Livingston, R. J. 1997a. Eutrophication in estuaries and coastal systems: relationships of physical alterations, salinity stratification, and hypoxia. In: F. J. Vernberg, W. B. Vernberg, and T. Siewicki (Eds.), Sustainable Development in the Southeastern Coastal Zone. University of South Carolina Press, Columbus, pp. 285–318.

Livingston, R. J. 1997b. Update on the ecological status of Perdido Bay. Florida Department of Environmental Regulation, Tallahassee. Unpublished report.

Livingston, R. J. 1997c. Final analyses of Perdido data base: The Soldier Creek System. Florida. Florida Department of Environmental Regulation, Tallahassee. Unpublished report.

Livingston, R. J. 1997d. Trophic response of estuarine fishes to long-term changes of river runoff. Bull. Mar. Sci. 60: 984–1004.

Livingston, R. J. 1998. Perdido Bay Analysis: 10/88–9/98. Florida Department of Environmental Regulation, Tallahassee. Unpublished report.

Livingston, R. J. 1999. Pensacola Bay System Environmental Study; Ecology and Trophic Organization. Florida Department of Environmental Regulation, Tallahassee. Unpublished report.
Livingston, R. J. and E. A. Joyce, Jr. 1977. Proceedings of the conference on the Apalachicola drainage system. Florida DNR Marine Research Publication No. 26. St. Petersburg.
Livingston, R. J. and G. L. Ray. 1989. A simplified and rapid method for assessing the biological disturbances resulting from stormwater and marina discharges in estuaries. Final Report, Florida Institute of Government and Florida Department of Environmental Regulation, Tallahassee. Unpublished report.
Livingston, R. J., R. L. Iverson, R. H. Estabrook, V. E. Keys, and J. Taylor, Jr. 1974. Major features of the Apalachicola Bay system: physiography, biota, and resource management. Fla. Sci. 4: 245–271.
Livingston, R. J., R. S. Lloyd, and M. S. Zimmerman. 1976. Determination of adequate sample size for collections of benthic macrophytes in polluted and unpolluted coastal areas. Bull. Mar. Sci. 26: 569–575.
Livingston, R. J., X. Niu, F. G. Lewis, and G. C. Woodsum. 1997. Freshwater input to a Gulf estuary: long-term control of trophic organization. Ecol. Appl. 7: 277–299.
Livingston, R. J., A. W. Niedoroda, T. W. Gallagher, and A. Thurman. 1998a. Environmental Studies of Perdido Bay. Florida Department of Environmental Regulation, Tallahassee. Unpublished report.
Livingston, R. J., S. E. McGlynn, and X. Niu. 1998b. Factors controlling seagrass growth in a gulf coastal system: water and sediment quality and light. Aquat. Bot. 60: 135–159.
Livingston, R. J., R. L. Howell, X. Niu, F. G. Lewis, and G. C. Woodsum. 1999. Recovery of oyster reefs (*Crassostrea virginica*) in a Gulf estuary following disturbance by two hurricanes. Bull. Mar. Sci. Gulf Caribbean 64: 75–94.
Livingston, R. J., F. G. Lewis III, G. C. Woodsum, X. Niu, B. Galperin, W. Huang, J. D. Christensen, M. E. Monaco, T. A. Battista, C. J. Klein, R. L. Howell IV, and G. L. Ray. 2000. Use of coupled physical and biological models: response of oyster population dynamics to freshwater input. Estuarine Coastal Shelf Sci. 50: 655–672.
Ljung, G. M. and G. E. P. Box. 1978. On a measure of lack of fit in time series models. Biometrika 65: 297–303.
Loesch, H. 1960. Sporadic mass shoreward migrations of demersal fish and crustaceans in Mobile Bay, Alabama. Ecology 41: 292–298.
Long, E. R., G. M. Sloane, R. S. Carr, K. J. Scott, G. B. Thursby, E. Crecelius, C. Peven, and H. L. Windom. 1997. Magnitude and extent of sediment toxicity in four bays of the Florida Panhandle: Pensacola, Choctawhatchee, St. Andrew, and Apalachicola. NOAA Technical Memorandum NOS ORCA 117, Silver Spring, MD.
Lovegrove, T. 1960. An improved form of sedimentation apparatus for use with an inverted microscope. J. Cons. Chem. 25: 279–284.
McComb, A. J. 1995. Eutrophic Shallow Estuaries and Lagoons. CRC Press, Boca Raton, FL, 240 pp.
McComb, A. J. and R. J. Lukatelich, 1995. The Peel-Harvey estuarine system, Western Australia. In: A. J. McComb (Ed.), Eutrophic Shallow Estuaries and Lagoons. CRC Press, Boca Raton, FL, pp. 5–18.
McPherson, B. F. and K. M. Hammett. 1991. Tidal rivers of Florida. In: R. J. Livingston (Ed.), The Rivers of Florida. Springer-Verlag, New York, pp. 31–46.
Macauley, J. M., V. D. Engle, J. K. Summers, J. R. Clark, and D. A. Flemer. 1995. An assessment of water quality and primary productivity in Perdido Bay, a northern Gulf of Mexico estuary. Environ. Monitox. Assess. 78: 1–15.
MacKenzie, C. L., Jr. 1996. Management of natural populations. In: V. S. Kennedy, R. I. E. Newell, and A. F. Eble (Eds.), The Eastern Oyster, *Crassostrea virginica*. Maryland Sea Grant College, College Park, pp. 707–721.
Maestrini, S. Y., D. J. Bonin, and M. R. Droop. 1984a. Phytoplankton as indicators of seawater quality: bioassay approach and protocols. In: L. E. Shubert (Ed.), Algae as Ecological Indicators. Academic Press, New York, pp. 71–132.
Maestrini, S. Y., M. R. Droop, and D. J. Bonin. 1984b. Test algae as indicators of sea water quality: prospects. In: L. E. Shubert (Ed.), Algae as Ecological Indicators. Academic Press, New York, pp. 131–188.
Mahoney, B. M. S. and R. J. Livingston. 1982. Seasonal fluctuations of benthic macrofauna in the Apalachicola estuary, Florida, USA: the role of predation. Mar. Biol. 69: 207–213.
Mahoney, J. B., F. H. Midlige, and D. G. Deuel. 1973. A fin rot disease of marine and euryhaline fishes in the New York Bight. Trans. Am. Fish. Soc. 102: 596–605.
Mallin, M. A. 1994. Phytoplankton ecology in North Carolina estuaries. Estuaries 17: 561–574.
Mallin, M. S. and H. W. Paerl. 1994. Planktonic trophic transfer in an estuary; seasonal, diel, and community structure effects. Ecology 75: 2168–2184.

Malone, T. C. 1977. Environmental regulation of phytoplankton productivity in the lower Hudson estuary. Estuarine Coastal Mar. Sci. 13: 157–172.
Malone, T. C. 1992. Effects of water column processes on dissolved oxygen, nutrients, phytoplankton and zooplankton. In: D. E. Smith, M. Leffler, and G. Mackiernan (Eds.), Oxygen Dynamics in the Chesapeake Bay. A Synthesis of Recent Research. Maryland Sea Grant College, College Park.
Malone, T. C., L. H. Crocker, S. E. Pike, and B. A. Wendler. 1988. Influences of river flow on the dynamics of phytoplankton in a partially stratified estuary. Mar. Ecol. Prog. Ser. 32: 149–160.
Malone, T. C., D. J. Conley, T. R. Fisher, P. M. Gilbert, L. W. Harding, and K. G. Sellner. 1996. Scales of nutrient-limited phytoplankton productivity in Chesapeake Bay. Estuaries 19: 371–385.
Marcomini, A., A. Sfriso, B. Pavoni, and A. A. Orio. 1995. Eutrophication of the Lagoon of Venice: Nutrient loads and exchanges. In: A. J. McComb (Ed.), Eutrophic Shallow Estuaries and Lagoons. CRC Press, Boca Raton, FL, pp. 59–80.
Marino, D., M. Montresor, and A. Zingone. 1987. *Miraltia throndsenii* gen. et sp. nov. a planktonic diatom from the Gulf of Naples. Diatom Res. 2: 205–211.
Marino, D., G. Giuffre, M. Montresor, and A. Zingone. 1992. An Electron microscope investigation on Chaetoceros minimus (Levander) Comb. nov. and new observations on *Chaetoceros throndsenii* (Marineo, Montresor & Zingone) comb. nov. Diatom Res. 6: 317–326.
Markley, S. M., D. K. Valdes, and R. Menge. 1990. Sanitary sewer contamination of the Miami River. Metro Dade DERM Technical Report 90-9.
Marshall, H. G. 1982a. Meso-scale distribution patterns for diatoms over the northeastern continental shelf of the United States. 7th Diatom-Symposium, pp. 393–400.
Marshall, H. G. 1982b. Phytoplankton distribution along the eastern coast of the USA. Part IV. Shelf waters between Cape Lookout, North Carolina, and Cape Canaveral, Florida. Proc. Biol. Soc. Washington 95: 99–113.
Marshall, H. G. 1984. Phytoplankton distribution along the eastern coast of the USA. Part V. Seasonal density and cell volume patterns for the northeastern continental shelf. J. Plankton Res. 6: 169–193.
Marshall, H. G. 1988. Distribution and concentration patterns of ubiquitous diatoms for the northeastern continental shelf of the United States. Proc. 9th Int. Diatom Symp. 1986, Otto Koeltz, Stuttgart, pp. 75–85.
Marshall, H. G. and J. A. Ranasinghe. 1989. Phytoplankton distribution along the eastern coast of the U.S.A. Part VII. Mean cell concentrations and standing crop. Cont. Shelf Res. 9: 153–164.
Martin, D. F. and W. H. Taft. 1998. Management of the Florida red tide — revisited. Fla. Sci. 61: 10–16.
Martin, D. F, M. T. Doig III, and R. H. Pierce, Jr. 1971. Distribution of naturally occurring chelators (humic acids) and selected trace metals in some Florida streams, 1968–1969. Fla. Dept. Nat. Resources Prof. Pap. Ser. 12.
May, E. B. 1973. Extensive oxygen depletion in Mobile Bay, Ala. Limnol. Oceanogr. 18: 353–366.
Meeter, D. A. and R. J. Livingston. 1978. Statistical methods applied to a four-year multivariate study of a Florida estuarine system. In: J. Cairns, Jr., K. Dickson, and R. J. Livingston (Eds.), Biological Data in Water Pollution Assessment: Quantitative and Statistical Analyses. American Society for Testing and Materials, Special Technical Publication 652, pp. 53–67.
Meeter, D. A., R. J. Livingston, and G. C. Woodsum. 1979. Short and long-term hydrological cycles of the Apalachicola drainage system with application to Gulf coastal populations. In: R. J. Livingston (Ed.), Ecological Processes in Coastal and Marine Systems. Plenum Press, New York, pp. 315–338.
Meinesz, A. 1999. Killer Algae. University of Chicago Press, Chicago, 360 pp.
Micheli, F. 1999. Eutrophication, fisheries, and consumer-resource dynamics in marine pelagic ecosystems. Science 285: 1396–1398.
Montgomery, R. T., B. F. McPherson, and E. T. Emmons. 1991. Effects of nitrogen and phosphorus additions on phytoplankton productivity and chlorophyll *a* in a subtropical estuary, Charlotte Harbor, Florida. U.S. Geological Survey, Water-Resources Investigations Report, 91-4077, Denver, CO.
Moore, K. A., D. J. Wilcox, and R. J. Orth. 2000. Analysis of abundance of submersed aquatic vegetation communities in Chesapeake Bay. Estuaries 23: 115–127.
Moshiri, G. A. 1976. Interrelationships between Certain Microorganisms and Some Aspects of Sediment-Water Nutrient Exchange in Two Bayou Estuaries. Phases I and II. University of Florida Water Resources Research Center Res. Ctr. Publ. No. 37, 45 pp.
Moshiri, G. A. 1978. Certain mechanisms affecting water column-to-sediment phosphate exchange in a bayou estuary. J. Water Pollut. Cont. Fed. 50: 392–394.

Moshiri, G. A. 1981. Study of Selected Water Quality Parameters in Bayou Texar. Final Report on Contract No. DACW01-80-0252, University of West Florida, Pensacola.

Moshiri, G. A. and W. G. Crumpton. 1978. Some aspects of redox trends in the bottom muds of a mesotrophic bayou estuary. Hydrobiologia 57: 155–158.

Moshiri, G. A., W. G. Crumpton, and D. A. Blaylock. 1978. Algal metabolites and fish kills in a bayou estuary: an alternative explanation to the low dissolved oxygen controversy. J. Water Pollut. Cont. Fed. 50: 2043–2046.

Moshiri, G. A., W. G. Crumpton, and N. G. Aumen. 1979. Dissolved glucose in a bayou estuary, possible sources and utilization by bacteria. Hydrobiologia 62: 71–74.

Moshiri, G. A., N. G. Aumen, and W. G. Swann III. 1980. Water Quality Studies in Santa Rosa Sound, Pensacola, Florida. U.S. Environmental Protection Agency, Gulf Breeze Environmental Research Lab, Pensacola.

Moshiri, G. H., N. G. Aumen, and W. G. Crumpton. 1987. Reversal of the eutrophication process: a case study. In: B. G. Neilson and L. E. Cronin (Eds.), Estuaries and Nutrients. Humana Press, Clifton, NJ, pp. 370–390.

Myers, V. B. and R. I. Iverson. 1981. Phosphorus and nitrogen limited phytoplankton productivity in northeastern Gulf of Mexico coastal waters. In: B. J. Neilson and L. E. Cronin (Eds.), Estuaries and Nutrients. Humana Press, Clifton, NJ, pp. 569–582.

Nakazima, M. 1965. Studies on the source of shellfish poison in Lake Hamana. III. Poisonous effects on shellfishes feeding on *Prorocentrum* sp. Bull. Jpn. Soc. Sci. Fish. 31: 281–285.

National Research Council. 1999. From Monsoons to Microbes: Understanding the Ocean's Role in Human Health. National Academy Press, Washington, D.C.

Netherlands Institute of Ecology. 1999. Progress Report 1998. A. Biere, F. Rienks, L. S. Galdon, and K. Soetaert (Eds.), Nieuwersluis, the Netherlands.

Newell, R. I. E. 1988. Ecological changes in Chesapeake Bay: are they the result of over-harvesting the American oyster, *Crassostrea virginica*? Understanding the estuary: Advances in Chesapeake Bay Research. Proceedings of a Conference, 29–31 March 1988. Baltimore, MD. Chesapeake Research Consortium Publication 129, CBP/TRS 24/88, pp. 536–546.

Niedoroda, A. W. 1992. Hydrography and oceanography of Perdido Bay. In: R. J. Livingston (Ed.), Ecological Study of the Perdido Drainage System. Unpublished report, Florida Department of Environmental Protection, Tallahassee.

Niedoroda, A. W. 1999. Pensacola Bay System Environmental Study; Physical Processes and Oceanography. Unpublished report, Florida Department of Environmental Regulation, Tallahassee.

Nihoul, J. C. 1986. Marine Interfaces — ecohydrodynamics. Elsevier Oceanography Series 42. Elsevier, New York.

Nilsson, P., B. Jonsson, I. Swanberg, and K. Sundback. 1991. Response of a marine shallow-water sediment system to an increased load of inorganic nutrients. Mar. Ecol. Prog. Ser. 71: 275–290.

Nixon, S. W. 1980. Between coastal marshes and coastal waters — a review of twenty years of speculation and research on the role of salt marshes in estuarine productivity and water chemistry. In: P. Hamilton and K. B. MacDonald (Eds.), Estuarine and Wetland Processes. Plenum Press, New York, pp. 438–525.

Nixon, S. W. 1981. Remineralization and nutrient cycling in coastal marine ecosystems. In: B. G. Neilson and L. E. Cronin (Eds.), Estuaries and Nutrients. Humana Press, Clifton, NJ, pp. 111–138.

Nixon, S. W. 1988. Chesapeake Bay nutrient budgets: A reassessment. Biogeochem. 4: 77–90.

Nixon, S. W. 1995. Coastal marine eutrophication: a definition, social causes, and future concerns. Ophelia 41: 199–219.

Nixon, S. W., M. E. Q. Pilson, C. A. Oviatt, P. Donaghy, B. Sullivan, S. Seitzinger, D. Rudnick, and J. Frithsen. 1984. Eutrophication of a coastal marine ecosystem — an experimental study using the MERL microcosms. In: M. J.-R. Fasham (Ed.), Flows of Energy and Materials in Marine Ecosystems: Theory and Practice. Plenum Press, New York, pp. 105–135.

Nordlie, F. G. 1990. Rivers and springs. In: R. L. Myers and J. J. Ewel (Eds.), Ecosystems of Florida. University of Central Florida Press, Orlando, pp. 392–428.

Norin, L. L. 1977. Bioassays with the natural phytoplankton in the Stockholm Archipelago. Ambio Special Report No. 5: pp. 15–21.

Officer, C. B., R. B. Biggs, J. L. Taft, L. E. Cronin, M. Tyler, and W. R. Boynton. 1984. Chesapeake Bay hypoxia: origin, development, and significance. Science 223: 222–227.

O'Connor, D. J. 1981. Modeling of eutrophication in estuaries. In: B. G. Neilson and L. E. Cronin (Eds.), Estuaries and Nutrients. Humana Press, Clifton, NJ, pp. 182–223.

O'Connor, D. J., R. V. Thomann, and D. M. DiToro. 1977. Water quality analysis of estuarine systems. In: Estuaries, Geophysics and the Environment. National Academy of Sciences, Washington, D.C., pp. 71–83.

Odum, H. T. and R. F. Wilson. 1962. Further studies on reaeration and metabolism of Texas bays, 1958–1960. Publ. Inst. Mar. Sci. U. Texas 8: 23–55.

Odum, W. E. and E. J. Heald. 1972. Trophic analyses of an estuarine mangrove community. Bull. Mar. Sci. 22: 671–738.

Odum, W. E., J. S. Fishes, and J. C. Pickral. 1979. Factors controlling the flux of particulate organic carbon from estuarine wetlands. In: R. J. Livingston (Ed.), Ecological Processes in Coastal and Marine Systems. Plenum Press, New York, pp. 69–82.

Odum, W. E., C. C. McIvor, and T. G. Smith. 1982. The Florida mangrove zone: a community profile. FWS/OBS-82/24. U.S. Fish and Wildlife Service, Office of Biological Services, Washington, D.C., 144 pp.

Officer, C. B., R. B. Biggs, J. L. Taft, L. E. Cronin, M. A. Tyler, and W. R. Boynton. 1984. Chesapeake Bay anoxia: origin, development, and significance. Science 223: 22–27.

Okaichi, T. 1997. Red tides in the Seto Inland Sea. In: T. Okaichi and T. Yanagi (Eds.), Sustainable Development in the Seto Inland Sea, Japan: From the Viewpoint of Fisheries. Terra Scientific Publishing Company, Tokyo, pp. 251–304.

Okaichi, T. and Y. Imatomi. 1979. Toxicity of *Prorocentrum minimum* var. Mariae-Lebouriae assumed to be a causative agent of short-necked clam poisoning. In: Toxic Dinoflagellate Blooms. Elsevier North-Holland, New York, pp. 385–388.

Olsen, L. A. 1973. Food and Feeding in Relation to the Ecology of Two Estuarine Clams, *Rangia cuneata* (Gray) and *Polymesoda caroliniana* (Bosc). M.S. thesis, Florida State University, Tallahassee, 103 pp.

Oshima, Y., C. J. Bolch, and G. M. Hallegraeff. 1992. Acute effects of the cyanobacterium cysts of *Alexandrium tamarense* (Dinophyceae). Toxicon 30: 1539–1544.

Oviatt, C. A., P. Doering, B. Nowicki, I. Reed, J. Cole, and J. Frithsen. 1995. An ecosystem level experiment on nutrient limitation in temperate coastal marine environments. Mar. Ecol. Prog. Ser. 116: 171–179.

Paasche, E. E. and S. R. Erga. 1988. Phosphorus and nitrogen limitation of phytoplankton in the Inner Oslofjord (Norway). Sarsia 73: 229–243.

Paerl, H. W. 1997. Coastal eutrophication and harmful algal blooms: the importance of atmospheric and groundwater as "new" nitrogen and other nutrient sources. Limnol. Oceanogr. 42: 1154–1165.

Parsons, T. R., Y. Maita, and C. M. Lalli, 1984. A Manual of Chemical and Biological Methods for Seawater Analysis. Pergamon Press, New York, 173 pp.

Patrick, L. 1991. Land use characterization report: Perdido River and Bay. U.S. Fish and Wildlife Service, Panama City, FL.

Patrick, R. M. 1967. Diatom communities in estuaries. Estuaries 83: 311–315.

Patrick, R. M. 1971. Diatom communities. Am. Microsc. Soc. Symp. Monogr. 3: 151–164.

Patrick, R. M. and C. W. Reimer. 1966. The diatoms of the United States. Vol. 1. Monograph 13, Monographs of the Academy of Natural Sciences of Philadelphia, 688 pp.

Paulic, M. and J. Hand. 1994. Florida water quality assessment 1994, 305(b) report. Florida Department of Environmental Protection, Tallahassee, 261 pp.

Pearson, T. H. 1980. Marine pollution effects of pulp and paper industry wastes. Helgol. Wiss. Meeresunters. 33: 340–365.

Pearson, T. H. and R. Rosenberg. 1978. Macrobenthic succession in relation to organic enrichment and pollution of the marine environment. Oceanogr. Mar. Biol. A. Rev. 16: 229–311.

Pennock, J. R. 1987. Temporal and spatial variability in plankton, ammonium, and nitrate uptake in the Delaware River. Estuarine Coastal Shelf Sci. 24: 841–857.

Pennock, J. R. and J. H. Sharp. 1994. Temporal alteration between light- and nutrient limitation of phytoplankton production in a coastal plain estuary. Mar. Ecol. Prog. Ser. 111: 275–288.

Peterson, B. J. and R. W. Howarth. 1987. Sulfur, carbon, and nitrogen isotopes used to trace organic matter flow in the salt-marsh estuaries of Sapelo Island, GA. Limnol. Oceanogr. 32: 1195–1213.

Peterson, C. H. 1979. Predation, competitive exclusion, and diversity in the soft-sediment benthic communities of estuaries and lagoons. In: R. J. Livingston (Ed.), Ecological Processes in Coastal and Marine Systems. Plenum Press, New York, pp. 233–264.

Peterson, C. H. 1982. The importance of predation and intra- and interspecific competition in the population biology of two infaunal suspension feeding bivalves, *Protothaca staminea* and *Chione undatella*. Ecol. Monogr. 52: 437–475.

Peterson, C. H. 1992. Competition for food and its community-level implications. Benth. Res. 42: 1–11.

Philippart, C. J. M., G. C. Cadee, W. van Raaphorst, and R. Riegman. 2000. Long-term phytoplankton-nutrient interactions in a shallow coastal sea: algal community structure, nutrient budgets, and denitrification potential. Limnol. Oceanogr. 45: 131–144.

Platt, T. L., M. Dickie, and R. W. Trites. 1970. Spatial heterogeneity of phytoplankton in a near-shore environment. J. Fish. Res. Bd. Can. 27: 1453–1473.

Porter, J. S., S. K. Lewis, and K. G. Porter. 1999. The effect of multiple stressors on the Florida Keys coral reef ecosystem. A landscape hypothesis and a physiological test. Limnol. Oceanogr. 44: 941–939.

Posey, M. H., T. D. Alphin, L. Cahoon, D. Lindquist, and M. E. Becker. 1999. Interactive effects of nutrient additions and predation on infaunal communities. Estuaries 22: 785–792.

Postma, H. 1985. Eutrophication of Dutch coastal waters. Neth. J. Zool. 35: 348–359.

Potts, M. 1980. Blue-green algae (Cyanophyta) in marine coastal environments of the Sinai Peninsula; distribution, zonation, stratification and taxonomic diversity. Phycologia 19: 60–73.

Powers, C. F., D. W. Schults, K. W. Malueg, R. M. Brice, and M. D. Schuldt. 1972. Algal responses to nutrient additions in natural waters. II. Field experiments. Limnol. Oceanogr. Spec. Symp. 1: 141–156.

Prasad, A. K. S. K. and G. A. Fryxell. 1991. Habit, frustule morphology and distribution of the Antarctic marine benthic diatom *Entopyla australis var. gigantea* (Greville) Fricke (Entopylaceae). Br. Phyc. J. 26: 101–122.

Prasad, A. K. S. K. and R. J. Livingston. 1995. A microbiological and systematic study of *Coscinodiscus jonesianus* (Bacillariophyta) from Florida coastal waters. Nova Hedw. Beih. 112: 247–263.

Prasad, A. K. S. K., and R. J. Livingston. 1998. Fine structure and taxonomy of *Fryxelliella* — a new genus of centric diatom (Triceratiaceae: Bacillariophyta) with a new valve feature, a circumferential marginal tube, and descriptions of *F. floridana* sp. nov. from the Atlantic coast of Florida and *F. inconspicua* (Rattray) comb. nov. from the Miocene. Phycologia 36: 305–323.

Prasad, A. K. S. K. J. A. Nienow, and R. J. Livingston. 1990. The genus *Cyclotella* (Bacillariophyta) in Choctawhatchee Bay, Florida, with special reference to *C. striata* and *C. choctawhatcheeana* sp. nov. Phycologia 29: 418–436.

Premila, V. E. and M. Umamaheswara Rao. 1977. Distribution and seasonal abundance of *Oscillatoria nigroviridis* Thwaites ex. Gomant in the waters of Visakhapatnam Harbour. Ind. J. Mar. Sci. 3: 79–91.

Rabalais, N. N. 1992. An updated summary of status and trends in indicators of nutrient enrichment in the Gulf of Mexico. Report to Gulf of Mexico Program, Nutrient Enrichment Subcommittee. Publication No. EPA/800-R-92-004, U.S. Environmental Protection Agency, Office of Water, Gulf of Mexico Program, Stennis Space Center, Mississippi, 421 pp.

Rabalais, N. N., J. Berg, and E. Hagmeier. 1990. Long-term changes of the annual cycles of meteorological, hydrographic, nutrient, and phytoplankton time series at Helgoland and at LV. ELBE 1 in the German Bight. Cont. Shelf Res. 10: 305–328.

Rabalais, N. N., R. E. Turner, D. Justic, Q. Dortch, W. J. Wiseman, Jr., B. K. Sen Grupta. 1996. Nutrient changes in the Mississippi River and system responses on the adjacent continental shelf. Estuaries 19: 386–407.

Rabalais, N. N., R. E. Turner, D. Justic, Q. Dortch, and W. J. Wiseman, Jr. 1999. Characterization of Hypoxia: Topic 1 Report for the Integrated Assessment of Hypoxia in the Gulf of Mexico. NOAA Coastal Ocean Program, Silver Spring, MD, 167 pp.

Radach, G., J. Berg, and E. Hagmeier. 1990. Long-term changes of the annaul cycles of meteorological, hydrographic, nutrient, and phytoplankton time series at Helgoland and at LV Elbe I in the German Bight. Cont. Shelf Res. 10: 305–328.

Reddy, P. M. and V. Venkateswarlu. 1986. Ecology of algae in the paper mill effluents and their impact on the River Tungabhadra. J. Environ. Biol. 7: 215–223.

Reyer, A. J., D. W. Field, J. E. Cassells, C. E. Alexander, and C. L. Holland. 1988. The distribution and areal extent of coastal wetlands in estuaries of the Gulf of Mexico. National Coastal Wetlands Inventory. National Oceanic and Atmospheric Administration, Rockville, MD, 18 pp.

Riegman, R. 1998. Species composition of harmful algal blooms in relation to macronutrient dynamics. In: D. M. Anderson, A. D. Cembella, and G. M. Hallegreaff (Eds.), Physiological Ecology of Harmful Algal Blooms. Springer-Verlag, Berlin, pp. 475–488.

Riegman, R., A. A. M. Noordeloos, and G. C. Cadee. 1992. *Phaeocystis* blooms and eutrophication of the continental coastal zones of the North Sea. Mar. Biol. 112: 479–484.

Rizzo, W. 1990. Nutrient exchanges between the water column and a subtidal benthic microalgal community. Estuaries 13: 219–226.

Rizzo, W. M., G. L. Lackey, and R. R. Christian. 1992. Significance of eutrophic, subtidal sediments to oxygen and nutrient cycling in a temperate estuary. Mar. Ecol. Prog. Ser. 86: 51–61.

Roelke, D. L., L. A. Cifuentes, and P. M. Eldridge. 1997. Nutrient and phytoplankton dynamics in a sewage-impacted Gulf Coast estuary: a field test of the PEG-model and equilibrium resource competition theory. Estuaries 20: 725–742.

Rosenberg, R. 1977. Benthic macrofaunal dynamics, production, and dispersion in an oxygen-deficient estuary on west Sweden. J. Exp. Mar. Biol. Ecol. 26: 107–133.

Rosenberg, R., R. Elmgren, S. Fleischer, P. Jonsson, G. Persson, and H. Dahhm. 1990. Marine eutrophication case studies in Sweden. Ambio 19: 102–108.

Rowe, G. T. and G. S. Boland. 1991. Benthic Oxygen Demand and Nutrient Regeneration on the Continental Shelf of Louisiana and Texas, Abstract for LUMCON meeting.

Rudnick, D. T., Z. Chen, D. L. Childers, J. N. Boyer, and T. D. Fontaine III. 1999. Phosphorus and nitrogen inputs to Florida Bay. The importance of the Everglades watershed. In: J. W. Fourqurean, M. B. Robblee, and L. A. Deegan (Eds.), Florida Bay: A Dynamic Subtropical Estuary. Estuaries 22: 398–416.

Ryther, J. H. 1954. The ecology of phytoplankton blooms in Moriches Bay and Great South Bay, Long Island, New York. Biol. Bull. 106: 198–209.

Ryther, J. H. and W. M. Dunstan. 1971. Nitrogen, phosphorus, and eutrophicationin the coastal marine environment. Science 171: 1008–1013.

Sakshaug, E. and Y. Olsen. 1986. Nutrient status of phytoplankton blooms in Norwegian waters and algal strategies for nutrient competition. Can. J. Fish. Aquat. Sci. 43: 389–396.

Santos, S. H. and S. A. Bloom. 1980. Stability in an annually defaunated estuarine soft-bottom community. Oecologia 46: 290–294.

Santos, S. L. and J. L. Simon. 1980a. Marine soft-bottom community-establishment following annual defaunation: larval or adult recruitment? Mar. Ecol. 2: 235–241.

Santos, S. L. and J. L. Simon. 1980b. Response of soft-bottom benthos to annual catastrophic disturbance in a south Florida estuary. Mar. Ecol. 3: 347–355.

Schnable, J. E. 1966. The Evolution and Development of Part of the Northwest Florida Coast. Ph.D. dissertation, Florida State University, Tallahassee, 231 pp.

Schroeder, W. W. 1978. Riverine influence on estuaries: a case study. In: V. S. Kennedy (Ed.), Estuarine Interactions. Academic Press, New York, pp. 347–364.

Schroeder, W. W. and W. J. Wiseman, Jr. 1985. An analysis of the winds (1974–1984) and sea level elevations (1973–1983) in coastal Alabama. Mississippi-Alabama Sea Grant Consortium. Publ. No. MASGP-84-024. Jackson, MS.

Schroeder, W. W. and W. J. Wiseman. 1986. Low-frequency shelf-estuarine exchange processes in Mobile Bay and other estuarine systems on the northern Gulf of Mexico. In: D. A. Wolfe (Ed.), Estuarine Variability. Academic Press, New York.

Schropp, S. J., F. D. Calder, G. M. Sloane, J. C. Carlton, G. L. Holcomb, H. L. Windom, F. Huan, and R. B. Taylor. 1991. A report on physical and chemical processes affecting the management of Perdido Bay: results of Perdido Bay Interstate Project, Alabama Department of Environmental Management and Florida Department of Environmental Regulation, Tallahassee.

Seal, T. L., F. D. Calder, G. M. Sloane, S. J. Schropp, and H. L. Windom. 1994. Florida Coastal Sediment Contaminants Atlas: A Summary of Coastal Sediment Quality Surveys. Florida Department of Environmental Protection, Tallahassee.

Seliger, H. H. and J. A. Boggs. 1988. Long term pattern of anoxia in the Chesapeake Bay. In: M. P. Lynch and E. C. Krome (Eds.), Understanding the Estuary: Advances in Chesapeake Bay Research. Chesapeake Research Consortium, Solomons, MD, pp. 570–583.

Sellner, K. G., S. E. Shymway, M. W. Luckenbach, and T. L. Cucci. 1995. The effects of dinoflagellate blooms on the oyster *Crassostrea virginica* in Chesapeake Bay. In: P. Lassus, G. Arzul, E. Erard, P. Gentien, and C. Marcalliou (Eds.), Harmful Marine Algal Blooms. Intercept Ltd., Lavoisier, pp. 505–511.

Sheridan, P. F. 1978. Trophic Relationships of Dominant Fishes in the Apalachicola Bay System (Florida). Dissertation, Florida State University, Tallahassee.

Sheridan, P. F. 1979. Trophic resource utilization by three species of sciaenid fishes in a northwest Florida estuary. Northeast Gulf Sci. 3: 1–15.

Sheridan, P. F. and R. J. Livingston. 1979. Cyclic trophic relationships of fishes in an unpolluted, river-dominated estuary in North Florida. In: R. J. Livingston (Ed.), Ecological Processes in Coastal and Marine Systems. Plenum Press, New York, pp. 143–161.

Sheridan, P. F. and R. J. Livingston. 1983. Abundance and seasonality of infauna and epifauna inhabiting a *Halodule wrightii* meadow in Apalachicola Bay, Florida. Estuaries 6: 407–419.

Shimada, H., T. Hayashi, and T. Mizushima. 1996. Spatial distribution of *Alexandrium tamarencse* "species complex" (Dinophyceae): dispersal in the North American and West Pacific regions. Phycologia 34: 472–485.

Shimada, M., T. H. Murakami, T. Imahayashi, H. S. Ozaki, T. Toyashima, and T. Okaichi. 1983. Effects of sea bloom, *Chattonella antiqua*, on grill primary lamellae of the young yellowtail, *Seriola quinqueradiata*. Acta Histochem. Cytochem. 16: 232–244.

Simmons, E. G. and W. H. Thomas. 1962. Phytoplankton of the eastern Mississippi delta. Publ. Inst. Mar. Sci. Univ. Texas 8: 269–298.

Sin, Y., R. L. Wetzel, and I. C. Anderson. 1999. Spatial and temporal characteristics of nutrient and phytoplankton dynamics in the York River estuary. Estuaries 22: 260–275.

Smayda, T. J. 1965. A quantitative analysis of the phytoplankton of the Gulf of Panama. II. On the relationship between ^{14}C assimilation and the diatom standing crop. Bull. Inter-Am. Trop. Commun. 9: 46–531.

Smayda, T. J. 1978. From phytoplankters to biomass. In: A. Sournia (Ed.), Phytoplankton Manual. UNESCO, Paris, pp. 273–279.

Smayda, T. J. 1980. Phytoplankton species succession. In: I. Morris (Ed.), The Physiological Ecology of Phytoplankton. University of California Press, Berkeley, pp. 493–570.

Smayda, T. J. 1989. Primary production and the global epidemic of phytoplankton blooms in the sea: a linkage? In: E. M. Cosper, J. Carpenter, and V. M. Bricelj (Eds.), Novel Phytoplankton Blooms: Causes and Impacts of Recurrent Brown Tides and Other Unusual Blooms. Springer-Verlag, Berlin.

Smayda, T. J. 1990. Novel and nuisance phytoplankton blooms in the sea. Evidence for a global epidemic. In: E. Granneli, B. Sundstrom, R. Edler, and D. M. Anderson (Eds.), Toxic Marine Phytoplankton: Proceedings of the Fourth International Conference. Elsevier, New York, pp. 29–40.

Smayda, T. J. 1997a. What is a bloom? A commentary. Limol. Oceanogr. 42: 1132–1136.

Smayda, T. J. 1997b. Harmful algal blooms. Their ecophysiology and general relevance to phytoplankton blooms in the sea. Limnol. Oceanogr. 42: 1137–1153.

Smayda, T. J. 1997c. Ecophysiology and bloom dynamics of *Heterosigma akashiwo* (Raphidophyceae). In: D. M. Anderson, A. D. Cembella, and G. M. Hallegreaff (Eds.), Physiological Ecology of Harmful Algal Blooms. Springer-Verlag, Berlin, pp. 111–131.

Smayda, T. J. and Y. Shimizu. 1993. Toxic Phytoplankton Blooms in the Sea. Proceedings of the Fifth International Conference on Toxic Marine Phytoplankton. Elsevier, New York, 952 pp.

Smith, J. B. and D. Torpak. 1989. The Potential Effects of Global Climate Change on the United States. Report to Congress. U.S. Environmental Protection Agency, Office of Research and Development, Washington, D.C., 413 pp.

Smith, D. E., M. Laffler, and G. Mackiernan. 1992. Oxygen dynamics in the Chesapeake Bay. Maryland Sea Grant College, College Park, MD.

Smith, S. M. and G. L. Hitchcock. 1994. Nutrient enrichment and phytoplankton growth in the surface waters of the Louisiana Bight. Estuaries 17: 740–753.

Smith, S. V. 1981. Response of Kaneohe Bay, Hawaii to relaxation of sewage stress. In: B. G. Neilson and L. E. Cronin (Eds.), Estuaries and Nutrients. Humana Press, Clifton, NJ, pp. 391–410.

Smith, S. V. 1991. Stoichiometry of C:N:P fluxes in shallow-water marine ecosystems. In: J. Cole, G. Lovett, and T. Finley (Eds.), Comparative Analyses of Ecosystems — Patterns, Mechanisms, and Theories. Springer-Verlag, New York, pp. 259–286.

Snedaker, S., D. de Sylva, and D. Cottrell. 1977. A review of the role of freshwater in estuarine ecosystems. Final Report, Southwest Florida Water Management District, Brooksville, FL, 126 pp.

Sorokin, Y. I., F. Dallocchio, F. Gelli, and L. Pregnollato. 1996. Phosphorus metabolism in anthropogenically transformed lagoon ecosystems: the Comacchio Lagoons (Ferrar, Italy). J. Sea Res. 35: 243–250.

Sournia, A. 1978. Phytoplankton Manual. Monographs on Oceanic Methodology 6. UNESCO, Paris, 337 pp.

Sournia, A., M. J. Chretiennot-Dinet, and M. Ricard. 1991. Marine phytoplankton: how many species in the world ocean? J. Plankton Res. 13: 1093–1099.

South Florida Water Management District. 1994. An update of the surface water improvement and management plan for Biscayne Bay. Draft Final Report, West Palm Beach, FL.

Squires, L. E. and N. A. Sinnu. 1982. Seasonal changes in the diatom flora in the estuary of the Damour River, Lebanon. 7th Diatom Symposium, pp. 359–372.

Steidinger, K. A., J. T. Davis, and J. Williams. 1966. Observations of *Gymnodinium breve* Davis and other dinoflagellates. In: Observations of an Unusual Red Tide. A Symposium. Prof. Pap. Ser. 8. Florida Board of Conservation, Marine Laboratory, St. Petersburg, pp. 9–15.

Steidinger, K. A., J. H. Landsberg, D. W. Truby, and B. S. Roberts. 1998a. First report of *Gymnodinium pulchellum* Larsen 1994 (Dinophyceae) in North America and associated fish kills in the Indian River, Florida. J. Phycol. (in press).

Steidinger, K. A., P. Carlson, D. Baden, C. D. Rodriguez, and J. Seagle. 1998b. Neurotoxic shellfish poisoning due to toxin retention in the clam *Chione cancellata*. In: B. Reguera, J. Blanco, M. L. Fernandez, and T. Wyatt (Eds.), Harmful Algae, Xunta de Galicia and IOC (in press).

Stoner, A. W. 1976. Growth and Food Conversion Efficiency of Pin-fish (*Lagodon rhomboides*) Exposed to Sublethal Concentrations of Bleached Kraft Mill Effluents. M.S. thesis, Florida State University, Tallahassee.

Stoner, A. W. 1982. The influence of benthic macrophytes on the foraging behavior of pinfish *Lagodon rhomboides* (Linnaeus). J. Exp. Mar. Biol. Ecol. 58: 271–284.

Stoner, A. W. and R. J. Livingston. 1980. Distributional ecology and food habits of the banded blenny *Paraclinus fasciatus* (Clinidae): a resident in a mobile habitat. Mar. Biol. 56: 239–246.

Stoner, A. W. and R. J. Livingston. 1984. Ontogenetic patterns in diet and feeding morphology in sympatric sparid fishes from seagrass meadows. Copeia (1984): 174–187.

Stoner, A. W., H. S. Greening, J. D. Ryan, and R. J. Livingston. 1982. Comparison of macrobenthos collected with cores and suction dredge. Estuaries 6: 76–82.

Strathmann, R. R. 1967. Estimating the organic carbon content of phytoplankton from cell volume or plasma volume. Limnol. Oceanogr. 12: 411–418.

Streckis, R. A., I. C. Potter, and R. C. J. Lenanton. 1995. The commercial fisheries in three southwestern Australian estuaries exposed to different degrees of eutrophication. In: A. J. McComb (Ed.), Eutrophic Shallow Estuaries and Lagoons. CRC Press, Boca Raton, FL, pp. 189–204.

Strickland, J. D. H. 1960. Measuring the production of marine phytoplankton. Bull. No. 112. Fish Res. Bd. Can., Ottawa.

Stumpf, H. P., M. L. Frayer, M. J. Durako, and J. C. Brock. 1999. Variations in water clarity and bottom albedo in Florida Bay from 1985 to 1997. In: J. W. Fourqurean, M. B. Robblee, and L. A. Deegan (Eds.), Florida Bay: A Dynamic Subtropical Estuary. Estuaries: 431–444.

Summers, J. K. 1999. The ecological condition of estuaries in the Gulf of Mexico. U.S. Environmental Protection Agency, Washington, D.C.

Sundback, K., V. Enoksson, W. Graneli, and K. Pettersson. 1991. Influence of sublittoral microphytobenthos on the oxygen and nutrient flux between sediment and water: a laboratory continuous flow study. Mar. Ecol. Prog. Ser. 74: 263–279.

Suttle, C. A. and P. J. Harrison. 1988. Ammonium and phosphate uptake rates, N:P supply ratios, and evidence for N and P limitation in some oligotrophic lakes. Limnol. Oceanogr. 33: 186–202.

Tanner, W. F. 1960. Florida Coastal Classification. Trans. Gulf Coast Assoc. Geol. Soc. 10: 259–266.

ter Braak, C. J. F. 1996. Unimodal Models to Relate Species to Environment. DLO-Agricultural Mathematics Group, Wageningen, the Netherlands, 266 pp.

ter Braak, C. J. F. and P. Smilauer. 1998. Canoco 4. Center for Biometry, Wageningen, the Netherlands, 349 pp.

Thornton, J. A., J. McComb, and S. O. Ryding. 1995a. The role of sediments. In: A. J. McComb (Ed.), Eutrophic Shallow Estuaries and Lagoons. CRC Press, Boca Raton, FL, pp. 205–224.

Thornton, J. A., H. Beekman, G. Boddington, R. Dick, W. R. Harding, M. Lief, I. R. Morrison, and A. J. R. Quick. 1995b. The ecology and management of Zandvlei (Cape Province, South Africa), an enriched shallow African estuary. In: A. J. McComb (Ed.), Eutrophic Shallow Estuaries and Lagoons. CRC Press, Boca Raton, FL, pp. 109–128.

Thorpe, P., R. Bartel, P. Ryan, K. Albertson, T. Pratt, and D. Cairns. 1997. The Pensacola Bay system surface water improvement and management plan. Northwest Florida Water Management District, Havana, FL. Unpublished report.

Tomas, C. R., B. Bendis, and D. K. Johns. 1999. Role of nutrients in regulating plankton blooms in Florida Bay. In: H. Kumpf, K. Steidinger, and K. Sherman (Eds.), The Gulf of Mexico: Large Marine Ecosystem. Blackwell Science, New York, pp. 323–337.

Turner, R. E., W. W. Schroeder, and W. J. Wiseman, Jr. 1987. The role of stratification in the deoxygenation of Mobile Bay and adjacent shelf bottom waters. Estuaries 10: 13–20.

Tyrrell, T. 1999. The relative influences of nitrogen and phosphorus on oceanic primary production. Nature 368: 619–621.

U.S. Army Corps of Engineers. 1976. Statement of Findings: Perdido Pass Channel (maintenance dredging), Baldwin County, Alabama.

U.S. Army Corps of Engineers. 1980. Water Resources Study: Escambia-Yellow Rivers Basins. Mobile District, Mobile, Alabama.

U.S. Environmental Protection Agency. 1983. Methods for the chemical analysis of water and wastes. EPA-600/4-79-020.

U.S. Environmental Protection Agency. 1993. Water-quality protection program for the Florida Keys National Marine Sanctuary. Phase II report. U.S. Environmental Protection Agency, Washington, D.C.

U.S. Environmental Protection Agency. 1998. National strategy for the development of regional nutrient criteria. U.S. Environmental Protection Agency, Washington, D.C.

U.S. Fish and Wildlife Service. 1990. Aerial photography of the seagrass beds of Perdido Bay. Pensacola, FL. Unpublished report.

Valiela, L., J. McClelland, J. Hauxwell, P. J. Behr, D. Hersh, and K. Foreman. 1997. Macroalgal blooms in shallow estuaries: controls and ecophysiological and ecosystem consequences. Limnol. Oceanogr. 42: 1105–1118.

van den Hoek, C., D. G. Mann and H. M. Jahns. 1995. Algae: An Introduction to Phycology. Cambridge University Press, Cambridge, U.K.

Van Es, F. B., M. A. Van Arkel, L. A. Bowman, and H. G. J. Schroder. 1980. Influence of organic pollution on bacterial, macrobenthic and meiobenthic populations in intertidal flats of the Dollard. Neth. J. Sea Res. 14: 288–304.

Van Hoose, M. S. 1987. Biological assessment of the Perdido Bay system. Alabama Marine Resources Division, Mobile. Unpublished report.

Virnstein, R. W. 1977. The importance of predation by crabs and fishes on benthic infauna in Chesapeake Bay. Ecology 58: 1199–1217.

Vollenweider, R. A. 1985. Elemental and biochemical composition of plankton biomass; some comments and explorations. Arch. Hydrobiol. 105: 11–29.

Wald, A. and J. Wolfowitz. 1940. On a test whether two samples are from the same population. Ann. Math. Stat. 11: 147–162.

Ward, C. H., M. E. Bender, and D. J. Reish (Eds.). 1979. The Offshore Ecology Investigation: Effects of oil drilling and production in a coastal environment. Rice Univ. Stud. 65: 1–589.

Welch, B. L., R. B. Whitlatch, and W. F. Bohlen. 1982. Relationship between physical characteristics and organic carbon sources as a basis for comparing estuaries in southern New England. In: V. S. Kennedy (Ed.), Estuarine Comparisons. Academic Press, New York, pp. 53–67.

Wetzel, R. G. and G. E. Likens. 1990. Limnological Analyses, 2nd ed. Springer-Verlag, Berlin, 212 pp.

Whitfield, W. K., Jr., and D. S. Beaumariage. 1977. Shellfish management in Apalachicola Bay: Past, present, and future. In: R. J. Livingston and E. A. Joyce, Jr. (Eds.), Proceedings of the Conference on the Apalachicola Drainage System. Florida Department of Natural Resources Mar. Res. Publ. 26: 130–140.

Whitlatch, R. B. and R. N. Zajac. 1985. Biotic interactions among estuarine infaunal opportunistic species. Mar. Ecol. Prog. Ser. 21: 299–311.

Wikfors, G. H. and R. M. Smolowitz. 1993. Detrimental effects of a *Prorocentrum* isolate upon hard clams and bay scallops in laboratory feeding studies. In: T. J. Smayda and Y. Shimizu (Eds.), Toxic Phytoplankton Blooms in the Sea. Proceedings of the Fifth International Conference on Toxic Marine Phytoplankton. Elsevier, New York, pp. 447–452.

Wilcoxon, F. 1945. Individual comparisons by ranking methods. Biometrics 1: 50–83.

Wilcoxon, F. 1949. Some Rapid Approximate Statistical Procedures. American Cyanamid Co., Stamford, CT.

Woelke, C. E. 1961. Pacific oyster *Crassostrea gigas* mortalities with notes on common oyster predators in Washington waters. Proc. Nat. Shellfish Assoc. 50: 53–66.

Wolfe, S. H., J. A. Reidenauer, and D. B. Means. 1988. An ecological characterization of the Florida Panhandle. U.S. Fish Wild. Serv. Biol. Rep 88 (12), 277 pp.

Young, D. L. and R. T. Barber. 1973. Effects of waste dumping in New York Bight on the growth of natural populations of phytoplankton. Environ. Pollut. 5: 237–252.

Zimmerman, M. S. and R. J. Livingston. 1976a. The effects of kraft mill effluents on benthic macrophyte assemblages in a shallow bay system (Apalachee Bay, North Florida, U.S.A.). Mar. Biol. 34: 297–312.

Zimmerman, M. S. and R. J. Livingston. 1976b. Seasonality and physico-chemical ranges of benthic macrophytes from a north Florida estuary (Apalachee Bay). Cont. Mar. Sci., Univ. Texas 20: 34–45.

Zimmerman, M. S. and R. J. Livingston. 1979. Dominance in benthic macrophyte assemblages from a north Florida estuary (Apalachee Bay). Bull. Mar. Sci. 29: 27–40.

APPENDIX I

Methods Used for Data Collection in the Field and Laboratory

The research effort in the northern Gulf of Mexico is based on written, peer-reviewed protocols for all field and laboratory operations. Water quality and biological analyses have been taken by personnel from the Center for Aquatic Research and Resource Management (CARRMA, Florida State University, Tallahassee, Florida) and Environmental Planning & Analysis, Inc. (Tallahassee, Florida) with additional work carried out at a series of laboratories in various countries. Water quality methods and analyses and specific biological methods have been continuously certified through the Quality Assurance Section of the Florida Department of Environmental Protection (Comprehensive QAP #940128 and QAP#920101). Various methods that have been used to take field information and run laboratory analyses have been published in the reviewed literature. This includes the following: Flemer et al., 1997; Livingston, 1975; 1976;1980; 1982; 1984a, b; 1985a; 1987c; 1988; 1992; 1997a; Livingston et al., 1974; 1976; 1997; 1998a,b).

A. PHYSICAL–CHEMICAL SAMPLING AND ANALYSIS

Designated, fixed stations in the Perdido Bay system were sampled (surface and bottom) for water quality factors at monthly to quarterly intervals from October 1988 through September 1999. Vertical collections of water temperature, salinity, conductivity, dissolved oxygen, depth, Secchi depth, and pH were taken at the surface (0.1 m) and at 0.5 m intervals below the surface to about 0.5 m above the bottom. Surface and bottom water samples were taken with Niscon bottles for further chemical analyses in the laboratory.

A complete list of variables is given below:

Physical–chemical Factors

1. Temperature
2. Salinity/conductivity
3. Dissolved oxygen
4. pH
5. Secchi depth
6. Depth
7. Light transmission
8. True color, NCASI colorimetric method
9. True color, spectroscopic method
10. Turbidity, nephelometric method
11. Total dissolved solids
12. Total suspended solids

13. Dissolved Oxygen, YSI DO meter
14. Particulate organic carbon
15. Dissolved organic carbon
16. Total inorganic carbon
17. Total organic carbon
18. Total carbon
19 Chlorophyll *a*
20 Chlorophyll *b*
21 Chlorophyll *c*

Nutrients

1. Nitrogen nitrate and nitrite
2. Ammonia nitrogen
3. Organic nitrogen and organic phosphate
   a. Total organic nitrogen and total organic phosphate
   b. Dissolved organic nitrogen and dissolved organic phosphate
   c. Particulate organic nitrogen and particulate organic phosphate
4. Total nitrogen
5. Total phosphorus
6. Orthophosphorus
7. Dissolved reactive silicate

Physical–chemical data and water samples for chemical analysis were taken according to methods defined by the U.S. Environmental Protection Agency (1983). Temperature and salinity were measured with YSI Model 33 S-C-T meters calibrated in the laboratory with commercial standards. Dissolved oxygen was taken with YSI Model 57 oxygen meters calibrated by the azide-modified Winkler technique. The pH was measured with an Orion Model 250A pH meter equipped with a Calomel electrode. Aquatic light readings were taken with a Li-Cor LI-1800UW Underwater Spectroradiometer, which measures the spectral composition of photon flux density at 1 to 2 nm intervals from 300 to 850 nm.

A Beckman DU-64 spectrophotometer was used to analyze true color (American Public Health Association, or APHA, 1989). Both ratio and nephelometric turbidity analyses were carried out. A ratio turbidimeter and a Hach model 2100A turbidimeter were used for turbidity analyses. Quantification of dissolved inorganic nitrogen (nitrite, nitrate) and dissolved inorganic phosphorus (orthophosphate) followed methods outlined by Parsons et al. (1984). Ammonium was measured with an ion electrode (U.S. EPA 1983, method 350.3) after raising the pH to 11. This method had some essential features (e.g., minimal interference from waters highly stained with humic materials and paper mill effluents) that required special attention to the results. Consequently, the level of detection was relatively high (e.g., 2.0 μm ammonium-N) but adequate for this study. Particulate organic nitrogen (U.S. EPA 1983, method 351.4) and particulate organic phosphorus (U.S. EPA 1983, method 364.4) were collected on Gelman glass fiber filters combusted at 500°C, oxidized to inorganic fractions, and organic N measured as ammonia and organic P measured as particulate organic phosphorus. Total dissolved solids (APHA Method 209-B), total suspended solids (APHA Method 209-C), dissolved organic carbon (APHA Method 415.1), chlorophyll *a* (APHA Method 1002-G), and biochemical oxygen demand (APHA Method 405.1) were carried out according to the methods outlined by APHA (1989). Particulate organic carbon was analyzed according to Parsons et al. (1984) (method 3.1). Nutrient analyses were carried out according to APHA (1989).

B. NUTRIENT LOADING MODELS

The U.S. Geological Survey (Tallahassee, Florida; M. Franklin, personal communication) provided flow data for the Perdido River system, Bayou Marcus Creek, and Eleven-mile Creek. Nutrient sampling was carried out in the various rivers leading into the bay (see Figure 4.3). Monthly nutrient concentrations from these rivers were used in the models. Overland runoff from unmonitored areas was estimated using models devised by A. Niedoroda (personal communication). Loading models for the 11-year database were run on monthly, quarterly, and annual intervals.

The nutrient loading models were based on a ratio estimator developed by Dolan et al. (1981). This model corrects for the biases due to sparse temporal sampling. It uses auto- and cross-covariance values of flows and nutrients for correction of the loading calculations. Nutrient data for the models was usually taken monthly where river data were taken daily. Program modifications were made by A. Niedoroda and G. Han (personal communication) whereby (1) due to discontinued river monitoring at Bayou Marcus Creek and the Blackwater River in September 1991, river flows for these areas were estimated from regressions (log/log, Elevenmile Creek and Bayou Marcus Creek; Styx and Blackwater Rivers); (2) stations 9 and 21 were sampled monthly from March 1993 to October 1994 and flows were then estimated by the following, St. 09 — (2.52 * St. 05) + (1.01 * St. 04) + (1.10 * St. 07); St. 21 — (2.64 * [St. 13-32]) + 32. These data are considered questionable at best and should not be used for any serious conclusions.

Dr. Alan Niedoroda and Mr. Gregory Han of Woodward-Clyde designed the loading estimator program. Data were grouped according to methods described in previous reports. Model applications were carried out according to a formula given by Dolan et al. (1981).

$$\bar{\mu}_y = \mu_x \frac{m_y}{m_x} \left(\frac{1 + \frac{1}{n} \frac{S_{xy}}{m_x m_y}}{1 + \frac{1}{n} \frac{S_x^2}{m_x^2}} \right)$$

where μ_y = estimated load
μ_x = mean daily flow for the year
m_y = mean daily loading for the days on which concentrations were determined
m_x = mean daily flow for the days on which concentrations were determined

$$S_{xy} = \frac{1}{(n-1)} \sum_{i=1}^{n} x_i y_i - n m_x m_y$$

$$S_x^2 = \frac{1}{(n-1)} \sum_{i=1}^{n} x_i^2 - n m_x^2$$

n = number of days on which concentrations were determined
x_i = individual measured flows
y_i – daily loading for each day on which concentrations were determined

This formulation corrects for biases introduced by sparse (temporal) sampling by using auto- and cross-covariance values of the flows and nutrients to correct the loading calculations for variability in the river flow which influences the nutrient concentrations. This applies where flow data are usually taken regularly (usually daily), whereas nutrient concentrations are usually taken at monthly or quarterly intervals.

C. SEDIMENT ANALYSES

Sediments were taken with coring devices. Sediment analyses were run from samples taken at the fixed stations throughout the field descriptive and experimental activities. Variables determined included phi unit distribution, percent organics, sand/silt/clay fraction percentages, total organic carbon, total nitrogen, and total phosphorus. The sediments were analyzed for particle size distribution and organic composition according to methods described by Mahoney and Livingston (1982) as defined by Galehouse (1971). A diver-deployed Benthic Lander was used in the determinations of sediment oxygen demand and nutrient transfer rates (nitrogen and phosphorus compounds) across the sediment/water interface. The Benthic Lander was designed and built by a research team at Texas A& M University under the direction of Dr. Gilbert T. Rowe. A description of the instrument is given in a paper given by Gilbert T. Rowe and Gregory S. Boland ("Benthic Oxygen Demand and Nutrient Regeneration on the Continental Shelf of Louisiana and Texas," Abstract for LUMCON meeting, 1991). Duplicate samples were taken after the instrument was grounded in sediments at a given station. Dissolved oxygen probes (one for each of the duplicate units) were attached to YSI dissolved oxygen meters which were monitored at 2 to 5 min intervals. *In situ* determinations of sediment oxygen demand were also taken. The battery-operated Lander had dual stirrers that maintained a regular rate of stirring in each 7-l (v) chamber. Each chamber covered 0.09 m^2 of sediment. Syringe ports were available for removal of water within each chamber for chemical analysis. Water was taken at the onset of the Lander placement and at the end of a 2 to 3 hour period of deployment. Nutrient analyses were carried out on the water samples as described in the protocols and the standard operating procedures (SOP) as part of our COMQUAP.

Analyses were carried out during the summer and fall of 1994 to determine the nutrient distribution in the sediments at chosen stations for the experimental work (stations 23, 29, 33). This work included vertical sediment nutrient analyses. Sediments were pumped by divers at regular (2 to 5 cm) intervals measured from the sediment/water interface. A device was used that allowed the sampling of discrete vertical sections of the sediments. Two 25 cm (d) PVC disks with a separation of 1 cm were attached to an outlet in the middle of the top disk. This outlet was attached to a diaphragm pump and water was channeled at discrete depths into labeled 2-l containers for full sediment chemistry analyses. Descriptions of the methods used are given in our QA/QC and SOP documents. *In situ* pore water nutrient samplers were constructed. These "peepers" were placed in the sediments at chosen bay stations and left for 6 to 7 days of equilibration. Methods used for nutrient analyses followed those described by Adams (Chapter 7, *CRC Handbook of Technique for Aquatic Sediments Sampling*).

D. LIGHT MEASUREMENTS

Spectroradiometry includes the measurement and analysis of the flux (e.g., energy) emitted by a radiation source (in this case, the sun) at different wavelengths. Such flux can be influenced by various processes such as reflection, transmission, fluorescence, or scattering of intermediary mediums; in this case, the water is the medium and the process of light penetration is dependent on water quality (turbidity and color) as function of various physical–chemical factors and biological processes (Kirk, 1983).

Aquatic light readings were taken with a Li-Cor LI-1800UW Underwater Spectroradiometer, which measures the spectral composition of photon flux density at 1 to 2 nm intervals from 300 to 850 nm. The basis for the instrumentation used in this study is given by Publication No. 8405-0037 of Li-Cor, Inc. (Revision 2; August, 1989), the instruction manual for the LI-1800UW Underwater Spectroradiometer. This instrument is designed to measure the spectral composition of photon flux density at various wavelengths (at 1, 2, 5, or 10 nm intervals). Underwater scan limits are 300 to 850 nm. A software package with the system allows data reduction routines for conversion to photon flux units, photosynthetically active radiation, illuminance, linear combinations of files, and ratios/products of files. The instrument has been calibrated by Li-Cor with calibration information stored as data files in the memory of the machine. Calibration is by use of a standard lamp of known spectral power distribution and this process establishes the response function of the instrument. The working standard is prepared by calibrating against the primary standard supplied by the U.S. National Institute of Standards and Technology.

In the field, three to five scans (replicates) were taken and averaged for each reading. Spectral radiometric light readings were taken only on cloudless days with winds less than 10 to 12 kts and within 2.5 hours of solar noon. Data were taken at 1 to 2 nm intervals in the air and water. The sensor was lowered from the sunny side of the boat at an angle that kept the sensor out of the shadow of the boat. Readings were taken only under full sunlight conditions (no clouds) with the exception of some experiments where readings were taken under varying levels of cloud cover. Some experimentation with cloud cover indicated that this factor was extremely important in the amount and type of light noted. For any series of collections, light conditions (taken mainly between 10 A.M. and 2:30 P.M.), wave action and other variables related to water conditions were standardized as best as possible. The data were taken within 2.5 hours of solar noon to minimize the possible effects of a changing solar angle.

The spectral irradiance was given in terms of units of energy (micro-mole photons) nm^{-1}/m^{-2}. The photosynthetically active radiation (PAR) represented the radiation in the 400 to 700 nm wave band. The output of the quantum sensor was an integration of the spectral flux that was incident on the sensor within the spectral range above. Just after each field deployment, a water sample was taken for measurement of the various water quality factors such as chlorophyll *a*, color, turbidity, and Seston at each depth. By dividing each aqueous irradiance reading by the corresponding atmospheric irradiance, the spectral radiometric readings were "normalized." The K_d (downwelling) attenuation values were calculated with reference to different wavelengths (Kirk, 1983; 1986). K_d (PAR) was calculated as the attenuation coefficient of the photosynthetically active radiation (400 to 700 nm). The lower limit of the photic zone was considered the PAR at 10% of the surface light (expressed as depth; Kirk, 1983; 1986). The lower limit of the euphotic zone was considered the PAR at 1% of the surface light (expressed as depth; Kirk, 1983; 1986).

E. BIOLOGICAL SAMPLING AND TROPHIC ANALYSES

1. Phytoplankton

Baywide phytoplankton collections were taken using nets (monthly, 10/88–9/91; 6/93–5/94) and whole-water samples (monthly, 11/89–12/91; 6/93–5/94; 2/97–9/99). There are basic strengths and weaknesses in the use of both methods. Net plankton collections take the larger species and are thus not comprehensive in the detection of smaller plankton. However, most of the larger species of phytoplankton are relatively well known taxonomically and therefore such collections can usually be relatively complete in terms of species identifications. Whole-water samples take the entire spectrum of plankton. However, the amount of water sampled is considerably less than that taken in the net collections. Species identification from whole-water samples is more difficult. Also, phytoplankton collections should be made synoptically (i.e., all samples taken within a given tidal cycle) in the field together with metered determinations of water quality characteristics and light characteristics. This should include collection of water for laboratory determination of nutrients, organic carbon compounds, chlorophyll, and toxicants. The use of synoptic surveys allows a more complete determination of associations among the biological and physical–chemical factors.

The transient nature of phytoplankton population and community trends increases the difficulty of documenting cause-and-effect relationships, especially when such collections are made at weekly to monthly intervals. The need for specific methods of preservation and the use of sophisticated microscopes and scanning electron microscopy (SEM) for species identifications add to the cost and time of analysis of the plankton samples. The generally high level of taxonomic complexity associated with phytoplankton identifications requires a high level of expertise. In the current analysis of the coastal phytoplankton in the Gulf study areas, we have maintained high standards of taxonomic effort so that the species-specific components of the phytoplankton communities have been determined. Net samples — 25 and 64 μm nets (multiple nets, bongo configuration; duplicate runs) — were taken over periods of 1 to 2 min. Three 1-l water samples, treated as separate replicates, were also collected. Samples were immediately fixed in Lugol's solution in its acid version (Lovegrove, 1960) to give the sample a weak brown color. Fixed samples were transported back to the laboratory for processing. The unconcentrated samples were settled and concentrated by decanting or by pipetting to 100 or 200 ml depending on the density of the samples.

Phytoplankton cells were counted with a standard research microscope (see Semina in Sournia, 1978) with phase contrast illumination at 400× magnification. Three aliquots of 0.05 ml each were transferred separately using a disposable pipette onto a glass slide and were covered by a 24 × 30 mm cover glass. All phytoplankton cells were counted in each aliquot and identified to species or to the lowest taxonomic category possible (genus/order/class). The location (coordinates on the stage of the microscope) of specimens difficult to identify during the counting procedure was noted; these were examined immediately after the counting was over with oil-immersion objectives. Diatoms, dinoflagellates, and cryptophytes, whose orientation was unsuitable for taxonomic determination, were examined by slightly touching the cover glass with a preparation needle to make such cells turn around. Abundance was expressed in number of cells per liter. Permanent slides of diatoms were prepared using standard procedures of acid-cleaning described in Prasad et al. (1990). Material for scanning electron microscopy was prepared as described by Prasad and Fryxell (1991). After preparation, electron micrographs were taken on a Polaroid 4x5 Land film type 55/positive-negative using the JEOL-840 scanning electron microscope at Florida State University operating at an accelerating voltage of 10 or 20 KV.

Cell volume for this study was based on data collected from Perdido Bay from October 1988 to Sepember 1999. Phytoplankton were preserved in Lugol's solution. Abundance and size estimates of dominant species were based on species-specific determinations. As in cell size, cell shape also varies to a great extent. To determine cell volume, the linear measurements per cell and geometric solids corresponding to the shapes for the species were used. As recommended by Smayda (1978), the appropriate dimensions of at least 25 randomly selected cells in preserved materials were measured in 1000× or 400× magnification, depending on the size of the cells. The volume of each of the measured cells was calculated after determining the approximate geometric shapes (as given in Wetzel and Likens, 1990) from which, the mean cell volume was then calculated. Because it was difficult to measure *in situ* levels of living algal carbon, the relationship between cell carbon and cell volume was provided using methods outlined by Strathmann (1967) and Smayda (1965; 1978). The carbon weight was converted to ash-free dry weight (AFDW) using separate equations for diatoms and other phytoplankton groups.

$\text{Log}_{10} C = 0.758 (\log_{10} V) - 0.422$ for diatoms (Equation 5 of Smayda 1978, p. 278)
$\text{Log}_{10} C = 0.0866 (\log_{10} V) - 0.460$ for other phytoplankton groups (Equation 6 of Smayda, 1978)

with V representing total cell volume (μm^3) and C representing the amount of carbon in pg cell^{-1}. Since the majority of the diatoms had a lower organic carbon per unit cell volume, they were treated separately from other groups of phytoplankton. Strathmann's (1967) observations of high cellular carbon concentrations in flagellates, in comparison to diatoms, suggested that the dinoflagellates may possess higher cellular caloric values than diatoms. The carbon weight was converted to AFDW using separate equations for diatoms and other phytoplankton groups:

$C = 38.2\%$ AFDW for diatoms (Curl, 1962)

with $C = 50\%$ AFDW for other phytoplankton groups (Vollenweider, 1985).

2. Zooplankton

Zooplankton were taken in various ways. Multiple tows were made with 202-µm nets run over varying time periods that depended on zooplankton density. Sample volumes were recorded so that a quantitative estimate could be made of zooplankton numbers. Another approach, used under certain conditions, involved pumping water (a volume that was determined in the field) through 202-µm nets. Zooplankton preservation was made using 10% formalin. All counts were made to species.

3. Macroinvertebrates and Fishes

Infaunal macroinvertebrates were taken using coring devices (7.6 cm diameter, 10 cm depth). The number of cores taken was determined using large initial samples (40 cores) and a species accumulation analysis based on rarefaction of cumulative biological indices (Livingston et al., 1976). Based on this analysis, multiple (10) core samples were taken randomly at each station with the assured representation of at least 80% of the species taken in the initial sampling. All infaunal samples were preserved in 10% buffered formalin in the field, sieved through 500 µm screens and identified to species wherever possible. Epibenthic fishes and invertebrates were collected with 5 m otter trawls (1.9 cm mesh wing and body, 0.6 cm mesh liner) towed at speeds of about 3.5 to 4 km/h for 2 min resulting in a sampling area of about 600 m^2 per tow. Repetitive samples (2 or 7) were taken at each site; sampling adequacy of such nets had been determined by Livingston (1976). All organisms were preserved in 10% buffered formalin, sorted and identified to species, counted and measured (standard length for fishes, total length for penaeid shrimp, carapace width for crabs). Representative samples of fishes and invertebrates were dried and weighed and regressions were run so that data from the biological collections could be converted into dry and ash-free dry mass. The numbers of organisms were then converted to biomass m^{-2}.

4. Trophic Determinations

All biological data (as biomass$\cdot m^{-2} \cdot mo^{-1}$ of the infauna, epibenthic macroinvertebrates, and fishes) were transformed from species-specific data into a new data matrix based on trophic organization as a function of ontogenetic feeding stages of the species found in Perdido Bay over the multiyear sampling program. Ontogenetic feeding units were determined from a series of detailed stomach-content analyses carried out with the various epibenthic invertebrates and fishes in the nearshore Gulf region (Sheridan, 1978; 1979; Laughlin, 1979; Sheridan and Livingston, 1979; 1983; Livingston, 1980; 1982; 1984b, unpublished data; Stoner and Livingston, 1980; 1984; Laughlin and Livingston, 1982; Stoner, 1982; Clements and Livingston, 1983; 1984; Leber, 1983; 1985). Based on the long-term stomach-content data for each size class, the fishes and invertebrates were reorganized first into their trophic ontogenetic units using cluster analyses. Infaunal macroinvertebrates were also organized by feeding preference based on a review of the scientific literature (Livingston, unpublished data). The field data in Perdido Bay were then reordered into trophic levels so that temporal changes in the overall trophic organization of this system could be determined over the 8-year study period. We assumed that feeding habits (at this level of detail) did not change over the period of observation; this assumption is based on previous analyses of species-specific fish feeding habits that remained stable in a given area over a 6- to 7-year period (Livingston, 1980).

The long-term field data were used to establish a database based on the total ash-free dry biomass$\cdot m^{-2} \cdot mo^{-1}$ as a function of the individual feeding units of the infauna, epibenthic macroinvertebrates and fishes. Data from the various stations in Perdido Bay were used (g m^{-2}) for all statistical analyses. The data were summed across all taxonomic lines and translated into the various trophic levels that included herbivores (feeding on phytoplankton and benthic algae), omnivores (feeding on detritus and various combinations of plant and animal matter), primary carnivores (feeding on herbivores and detritivorous animals), secondary carnivores (feeding on primary carnivores and omnivores), and tertiary carnivores (feeding on primary and secondary carnivores and omnivores). All data are given as ash-free dry mass$\cdot m^{-2}\cdot mo^{-1}$ or as percent ash-free dry mass$\cdot m^{-2}\cdot mo^{-1}$. In this way, the long-term database of the collections of infauna, epibenthic macroinvertebrates, and fishes was reorganized into a quantitative and detailed trophic matrix based not solely on species (Livingston, 1988) but on the complex ontogenetic feeding stages of the various organisms. A detailed review of this process and the resulting feeding categories is given by Livingston et al. (1977).

F. EXPERIMENTAL METHODOLOGY

1. Nutrient Limitation Experiments

We constructed 18-1, all-glass microcosms (width: 20; length: 21.5; and depth: 48 cm) using nontoxic sealant. A wooden rack assembly held microcosms at the surface of large temperature-controlled reservoirs. Bay ambient water temperatures were maintained within ±2′C through use of heating and cooling coils placed in the thermal reservoir. Large volumes of bay water were gently pumped (Ruhl model 1500 rubber diaphragm pump) from approximately 0.5 m depth at each station into polyethylene containers by first passing through a set of nested Nitex® plankton nets held in a standpipe system to equilibrate water pressure (i.e., a 202-µm net placed inside a 64-m net; 0.5 × 2.0 m) and to remove larger zooplankton.

Sampling was conducted during late afternoon and samples were transported to an experimental "microcosm tank facility" near Tallahassee, during early evening to reduce heating from daytime peak temperatures. Experiments began the next day. Because a continuous supply of estuarine phytoplankton was logistically infeasible, we used a modified static renewal approach with daily removal of 10% of the volume of each microcosm. A supply of the field sample for each station was maintained as renewal water. This water was filtered through a 25-µm Nitex net and held at 4°C in the dark. Thus, early experimental artifacts such as production of allelopathic substances would be less likely in such a system compared with a static experimental design. Before use, all collecting vessels, culture tanks, and Tygon air lines were acid-washed (10% HCl), rinsed twice with fresh water, rinsed once with deionized water, and air-dried. Microcosms were and sealed with nontoxic silicon sealant. Air was delivered to each microcosm to maintain mixing via Tygon tubing, plastic control valves, and "air stones" from an oil-free aquaculture air pump.

Water collected from a station the previous day was placed in a fiberglass mixing tank during early morning to minimize light shock and was gently mixed with a rubber paddle to ensure homogeneous filling requirements. Water was poured sequentially in small volumes into glass microcosms to ensure homogeneous filling requirements. Reference samples were taken from the mixing tank for chlorophyll *g*, dissolved inorganic phosphate (PO_4), ammonium (NH_3), nitrite (NO_2), and nitrate (NO_2) particulate organic nitrogen (PON) and phosphorus (POP) to characterize initial nutrient conditions and standing stock of phytoplankton. Chlorophyll *a* was collected on Gelman A/E glass fiber filters and extracted with 90% acetone buffered with MgCO, and measured according to Parsons et al. (1984). Quantification of dissolved inorganic N (i.e., NO_2 and NO_3) and P (PO_4) followed Parsons et al. (1984). Ammonium was measured as NH_3 with an ion electrode (U.S. EPA, 1983, method 350.3) after converting NH_4 to NH_3 by raising the pH to 11. This method has some essential features (e.g., minimal interference from waters highly stained with humic materials and paper mill effluents. However, the level of detection typically was relatively high (e.g., 2.0 µ*M* NH_3-N) but adequate for this study. Particulate organic N (U.S. EPA, 1983, method 351.4) and POP (U.S. EPA, 1983, method 364.4) were collected on Gelman glass fiber filters combusted at 500°C oxidized to inorganic fractions and organic N measured as NH_3 and organic P measured as PO_4.

Single samples were collected daily at approximately 0800 hours from each microcosm for chlorophyll analysis to measure changes in phytoplankton biomass. Nutrient enrichment experiments included the following treatments triplicated for each station: three control tanks, three P-enriched tanks at 10 μM PO_4 above ambient, three N-enriched tanks at 50 µm NH_3N above ambient NH_3, and three combined $NH_3 + PO_4$ (referred to as N+P) above ambient as described for single additions. We used NH_3 enrichments because of interest in potential increased riverine transport into the bay. Although aware of possible suppression of NO_3 uptake by NH_3 this process was judged to be relatively unimportant in bioassays lasting over 5 to 12 days (D'Elia et al., 1986) and at high NO_3 concentrations (Pennock, 1987). Dortch (1990) concluded from a comprehensive literature review that NH_3 suppression of NO_3 uptake under field conditions is quite variable and often undetectable. Nutrient additions were made at time 0 of the experiment. On day 2 and daily thereafter, 18 1 of water were collected from each microcosm for chemical and biological analyses. This volume was replaced by an equal volume of renewal water adjusted to ambient temperature before introduction into the microcosms. Hydrographic variables (e.g., surface water temperature, °C) and salinity (8b) (YSI model 33 conductivity meter), and 30-cm all-white Secchi disk readings (m) were collected generally within 1 week of samples for nutrient enrichment experiments.

A randomized block design was used with statistical significance of ($P < 0.05$) within and among stations. The four treatments in each row were randomly selected so that each of the three station-specific experiments had 12 treatments (randomized four treatment by three replicates for three stations). Treatments were oriented in a north/south direction, to minimize shading from changes in the annual sun angle, and to allow detection of a possible blocking effect. A repeated-measures, randomized analysis of variance (ANOVA) was run for each experiment. ANOVA assumptions were checked using residual boxplots, the Lilliefors test on the standard residuals, and Bartletts test for equality of variances. A log transformation was necessary to achieve equality of variances. The two-way interaction P-values for treatment by day and for block by day were calculated. To determine how treatments differed by day, a randomized block analysis of variance was run for each station, month, and day. As above, a log transformation was necessary to satisfy assumptions. Scheffe's S-test was used as a *post hoc* treatment.

2. Diver-Deployed (*in situ*) Benthic Lander

a. Introduction

The various problems and sources of errors with respect to sediment oxygen demand (SOD) analyses have been reviewed by J. F. Davis et al. (1987: Delaware Estuary use attainability project, Delaware River Basin Commission, West Trenton, New Jersey). We have analyzed the relatively broad literature on this subject and the method chosen reflects our interpretation of results from the studies of others in addition to our own experience with SOD analyses in our South Carolina work. Replicate samples were used as an indication of experimental error. *In situ* tests have the most potential for error. The laboratory approach, while suffering from the problem of not being carried out in the field, has proved to be the most reliable in terms of variability of results. We followed techniques that have a record of being the most reliable on the basis of intersystem comparisons.

We worked with *in situ* samplers developed and constructed at Texas A&M University (Dr. G. Rowe). The diver-deployed Benthic Lander used in the SOD measurements was designed and built by a research team at Texas A&M University under the direction of Dr. Gilbert T. Rowe. A description of the instrument is given in a paper presented by Gilbert T. Rowe and Gregory S. Boland ("Benthic Oxygen Demand and Nutrient Regeneration on the Continental Shelf of Louisiana and Texas," Abstract for LUMCON meeting, 1991).

b. Instrument Specifications

Duplicate samples are taken after the instrument is grounded in sediments at a given station. Dissolved oxygen probes (one for each of the duplicate units) are attached to YSI dissolved oxygen meters which are monitored 2 to 5 min intervals at the surface. The battery-operated machine has dual stirrers that maintain a regular rate of stirring in each 7-l (v) chamber. Each chamber covers 0.09 m^2 of sediment. Syringe ports are available for removal of water within each chamber for chemical analysis. The Lander is housed in a basic ¾ in. aluminum frame to which is attached a battery housing with two power Sonic batteries. Ballast weights are used to keep the device in the sediments. Total negative weight is –28.23 kg and the total positive lift (largely from the 13 in. instrument spheres) is 13 kg.

c. Operation of the Lander

1. Laboratory Preparations

1. The YSI meters should be prepared. This includes installation of probe membranes, redlining the instrument at salinity 0, adjustment of temperature to ambient for calibration, calibrating probes for ambient temperature, setting meter scales to desired dissolved oxygen range, checking calibration with 5 to 10% sodium sulfite solution (0 ppm oxygen). The temperature should be reset to the expected bottom temperature (based on previous experience). Salinity should also be set in a similar fashion. Glass and gasket should be cleaned with a solvent (toluene).
2. Make sure that the silicon oil fills the impeller housing.
3. Charge batteries to +13 V and place in housing.
4. Check o-ring on battery and use light treatment of silicon grease on the battery chamber o-rings Close chamber.
5. Check for chamber volume by measuring distance from the bottom of the rim to the flange (40 mm = 7.0 l).
6. Grease and clean the pins for attachment to the DO batteries (large pin is +). Attach battery wires and start/check stirring assemblies.
7. Secure device in frame and set/secure for travel.

2. Boat Operations (Deployment)

1. Turn on stirrers by connecting the battery leads. Calibrate the DO meters and place probes in housing.
2. Check syringes for the removal of water.
3. Slide apparatus off boat and let down gently with marker buoy attached to frame.
4. Divers should remove the chamber unit and secure the bungee cords. One diver on each side should lift the apparatus out and away from the frame. Then the divers should flip the chamber unit over to release all air bubbles. Make sure that stirrers are operating.
5. Look for flat area to place chambers and set chambers down gently on sediment, pressing down until the gray flange is sitting on the sediment surface. Again check the stirrers. DO chamber wires should be bundled together.
6. Take water samples (200 ml) with syringes. Filter samples through 0.2-μm filter.
7. Take DO and temperature readings at 5-min intervals, recording data on the sample sheet. Such sheets should be part of the overall data sheet (see BIFS field data sheet summary).
8. After a significant drop in DO has been recorded (1 to 3 hours), divers should take a final water sample with the syringes (same as above). The chamber should be picked up and turned upside down to check for the impeller movement (sediment may interfere with such mechanism). Chamber unit should be placed in lender housing and attached with bungee cords. Dive lifts should be inflated and divers should escort apparatus to surface. The apparatus should be carefully placed on the boat and the battery leads detached.

3. Determination of Oxygen Consumption Rates

$$\text{Rate } R = \frac{(\text{Final } O_2, \text{ ml}^{-1} - \text{Initial } O_2, \text{ ml}^{-1})(\text{Volume of chamber} = 7 \text{ l})}{(\text{area of chamber, } 0.091 \text{ m}^2)(\text{Total time, h, } T_{final} - T_{initial})}$$

Conversion of mg l^{-1} to ml^{-1}: mg = 0.7 ml

3. Laboratory (*in vivo*) SOD Determinations

Four replicate samples of undisturbed core samples were taken with the SOD corer apparatus at each of the three stations. Sediment filled the corer to at least half the height of the PVC coring tube. Water filled the rest of the corer. Great care was taken so that there was minimal disturbance to the sample. Detachable cores were capped and placed in racks. The samples were sealed with PVC end caps and placed in an upright position in the rack in a Gott cooler. The seawater bath in the Gott cooler was maintained at the level of water in the cores to lessen pressure on the seals. The samples were taken immediately to the laboratory and equilibrated for 10 hours with gentle aeration without disturbing the sediment. Laboratory water temperature was controlled at field levels.

We followed the batch method of analysis. The laboratory SOD measurements were made on undisturbed sediments in the original core tubes. Water circulation was maintained with an internal mixing device. Two replicates from each station set were poisoned with HCN; the remaining two samples were left without any additions or changes. Dissolved oxygen probes were pre- and postcalibrated using the micro-Winkler technique. Temperature and depletion of dissolved oxygen were monitored (without stirring) every 30 min for 4 to 6 hours or until 1.5 to 3.0 mg l^{-1} DO depletion was obtained. If the initial DO readings in the cores was below 4.0 mg l^{-1}, the core water was aerated without sediment disturbance to have an initial DO reading of at least 4.0 mg l^{-1}.

DO depletion curves were constructed with the cumulative DO depletion data plotted against time. The slope was calculated from the linear portion of the DO depletion curve using the first degree regression function: the results are to be represented in units of g O_2 m^{-2} h^{-1}.

4. Primary Productivity (*in situ* light/dark bottle study)

Primary productivity is the rate at which inorganic carbon is converted to organic carbon via chlorophyll-bearing organisms such as phytoplankton. The plankton respiration term was an important part of evaluating the oxygen-uptake mechanism in the bay. Therefore, we evaluated the phytoplankton net and gross photosynthesis using the classical light and dark bottle oxygen technique (Strickland, 1960; Standard Methods, 1989). Some unpublished data using the C_{14} uptake method were obtained by personnel of the Environmental Research Laboratory, Gulf Breeze for Perdido Bay. Data taken using this method were comparable to our own.

At present, there is no available method of measuring primary productivity that is free of artifacts. Various authors have recently emphasized the importance of production processes occurring at strong physical gradients (Legendre and Dimers, 1985) and an emerging area of emphasis has concentrated on ecohydrodynamics. The approach in this study examined bay primary productivity in terms of vertical stratification and its relationship to the euphotic zone, the longitudinal salinity gradient in the bays, and temporal patterns over the warm season when hypoxic conditions become prevalent. The scaling of measurements was made mostly at the meso- to macroscale; fine-grain scaling of observations was based on the present design. Integral water column rates of photosynthesis allowed an estimate of system production minus the relatively small contribution from submerged aquatic vegetation and a somewhat more important contribution of phytomicrobenthos.

The approach here is developed from that described in Standard Methods (1989). All procedures involving bottle cleaning, checking for supersaturation, the handling of water samples, and calculations are given in the above citation. The oxygen method will be used whereby clear (light) and darkened (dark) bottles will be filled with water samples and suspended at regular depth intervals for appropriate intervals. The duration of exposure will depend on the rate of net photosynthesis and respiration with a need to avoid bubble formation in light bottles or hypoxia in dark bottles except at depths where hypoxia is present. We will start with sunrise to sunset deployment with possible adjustment to 0900 to 1500 hours or some equal distribution of time around noon (CST). Any given bubble development (>0.3 mm diameter) at a single depth will not result in a change in deployment schedule. Bubble development at different depths will be reason to change such deployment based on a initial screening by periodically checking the bottle strings.

a. End Points

Changes in the concentration of DO will be measured with a YSI Model 58 DO meter equipped with a BOD probe. A magnetic stirrer and stirring bar will be required (a 12 V system will be used). The specifications of this DO unit is ±0.01 mg^{-1} with an accuracy of ±0.03 mg l^{-1}. These specifications are minimally satisfactory for the warm season in Perdido Bay. More sensitive measurement equipment is not seen to be seaworthy in a small research vessel. If necessary, a more sensitive system will be available at a shore-based system, which will require transport of samples to this laboratory setup. The present method will allow a spatial coverage that is important to the project.

b. Methods

Two numbered light bottles and two dark bottles will be used for each depth at each station. Nonbrass ("marine") wire will be used to attach a strong (plastic or stainless steel) snap-catch to the neck of each 300-ml BOD bottle. Wheaton "400" brand borosilicate (high-purity) glass bottles are to be used. Plastic bottle caps will be used to protect the stoppers. The bottles are to be acid-cleaned (warm 10% HCl) and rinsed with distilled water prior to use. Just before the filling of each bottle, the container will be rinsed with the water to be tested. Phosphorus-containing cleaning agents will not be used. The experimental setup has been constructed so that the supporting lines or racks do not shade the suspended bottles.

c. Calculations

The increase in oxygen concentration in the light bottle during incubation is a measure of the net production (somewhat less than gross production due to respiration). The loss of oxygen in the dark bottle is an estimate of the total plankton respiration.

$$\begin{aligned} \text{Net photosynthesis} &= \text{light bottle DO} - \text{initial DO} \\ \text{Respiration} &= \text{initial DO} - \text{dark bottle DO} \\ \text{Gross photosynthesis} &= \text{light bottle DO} - \text{dark bottle DO} \end{aligned}$$

Results from duplicates are to be averaged. The calculations of the gross and net production for each incubation depth should be plotted.

$$\text{mg carbon fixed/m}^2 = \text{mg oxygen released/l} \times 12/32 \times 1000 \times K$$

where K = photosynthetic quotient ranging from 1 to 2, depending on the N supply. One mole of oxygen (32 g) is released for each mole of carbon (12 g) fixed. The productivity of a vertical column of water 1 m square is determined by plotting productivity for each exposure depth and graphically integrating the area under the curve.

G. VERTICAL SAMPLING PROGRAM

Phytoplankton, zooplankton, and water quality conditions (temperature, salinity, DO) were sampled at the five stations. Samples were taken from integrated samples above and below the halocline. The intake hoses were moved up and down continuously. At each sampling (depth by station), three 1-l samples were taken for phytoplankton analysis with preservation with Lugol's solution as noted in previous samplings. Zooplankton were taken by pumping water (a volume determined in the field) through a 202-μm net with preservation with formalin as outlined in the previous sampling protocols for zooplankton. The zooplankton pumping data was supplemented with net samples taken according to prescribed procedures. A single water sample was taken during the second day of the field operations at each depth at each station.

H. SALINITY AND DISSOLVED OXYGEN CLASSIFICATION

A salinity stratification classification was carried out based on the cooperative efforts of our research group and scientists from the National Oceanic and Atmospheric Administration (John Klein and Paul Orlando, personal communication). This transformation was carried out with surface and bottom salinity determinations in the various estuarine systems. Data were organized on an annual basis (by month, by station) for each of the subject estuaries according to the following criteria:

Highly stratified (HS): Salinity difference (bottom – surface) ≥10 ppt
Partially stratified, strong (PSS): Salinity difference ≥ 5 and < 10 ppt
Partially stratified, weak (PSW): Salinity difference ≥ 2 and < 5 ppt
Vertically homogeneous (VH): Salinity difference < 2 ppt

A review was carried out using our long-term database of selected scientific papers concerning the relationships of low DO to estuarine populations. Based on these analyses, we reordered the DO data taken in each system according to potential impact on the biological components.

Anoxic to severely hypoxic (ASH): DO ≤ 2 mg l^{-1}
Hypoxic (HYP): DO > 2 and ≤ 4 mg l^{-1}
Biologically "neutral" (BN): DO > 4 mg l^{-1}

The ASH classification includes DO levels that are actively detrimental to most estuarine populations. The HYP classification includes DO levels that may be stressful to more sensitive organisms and thus can be associated with loss of certain populations. The exact meaning of this transitional level of DO is, as yet, not well understood and probably has a stress value that remains relative to the stratigraphic characteristics of the area and the time of year.

APPENDIX II

Statistical Methods

A series of statistical methods were used to analyze the long-term data. Scattergrams of the long-term field data were examined and either logarithmic or square root transformations were made, where necessary, to approximate the best fit for a normalized distribution. These transformations were used in all statistical tests of significance.

To examine spatial differences in mean values of various field characteristics, the nonparametric Wilcoxon Signed Rank test for paired differences (i.e., applied to the differences between paired observations through time at the sites being compared) was used. The Wilcoxon test (Wilcoxon, 1949) is well suited for these types of field data comparisons because the differencing used tends to remove much of the serial correlation in the data sets, and the test does not require the normality assumption, which is often difficult to meet with field data. In all such comparisons, the null hypothesis ($\mu_1 = \mu_2$) was tested against the alternative ($\mu_1 \neq \mu_2$) using a 2-tailed test with $\alpha = 0.05$. Variances of the field characteristics were tested for significant differences using a modified Levene's statistic (Levene, 1960) when the underlying populations were non-normal. Both W_{10} and W_{50} statistics were examined using the recommendations of Brown and Forsythe (1974). Nonparametric testing, while avoiding problems with normality, is still subject to the assumptions of independence. Dependence was examined in both the Wilcoxon and Levene tests with an overall Q statistic (Ljung and Box, 1978).

We used a number of basic statistical tests regarding comparisons of two or more data sets. These included the following:

1. Independence Tests

Random, independent observations are integral to the comparison of two samples. The analysis can check to see if there are certain forms of numeric dependence in the data; if there is dependence due to problems of design or sampling methodology, for example, it cannot be detected. The independence tests used for preliminary analyses examine each sample for autocorrelation at a number of lags. For each sample, autocorrelation is computed for the lesser of 24 or $n/4$ lags, where n is the number of observations. The program then computes the Q statistic of Ljung and Box (1978) for an overall test of the autocorrelations as a group where the null hypothesis is that correlation at each lag equals zero. This statistic is approximately chi-square distributed under the null hypothesis with K-m degrees of freedom, where K is the number of lags examined and m is the number of estimated parameters (1 if an AR1 filter has been applied to the data, 0 otherwise). The p-value of the chi-square test is checked and, if it remains less than 0.05, the null hypothesis is rejected.

The output sheet also contains plots of the autocorrelation functions for the two samples. These graphs are bar-charts showing the values of autocorrelation coefficients at all computed lags, and are overlain with dashed lines at the critical value (value for which any individual correlation may be considered significant) for each series. Critical values are equal to $2/\sqrt{n}$ where n is the number of observations.

2. Normality Tests

The parametric F-test for comparison of variances and t-test for comparison of means assume that the data come from normally distributed populations. In support of those tests the two-sample spreadsheet provides a (chi-square) goodness-of-fit test which compares distributions of the samples to normal distributions having the same means and variances as the samples. Because the number of observations is generally fairly small, the program is designed to divide the data into seven equally sized intervals (frequency classes) for comparison with expected (normally distributed) frequencies. The program also examines the expected frequencies for the leftmost and rightmost classes and expands (doubles) the width of these classes if the initial expected frequency is less than one. These procedures both help to minimize the problem of having extremely low expected frequencies in the denominator of the chi-square computation. However, very low sample sizes and sample distributions with extreme departures from normality can still interfere with the test. For this reason, the program prints a warning when sample size is below 25. The chi-square test should not be considered reliable for such low sample sizes.

The spreadsheet contains a graph of the frequency distribution (with a normal curve overlain for visual comparison purposes) for each of the two samples. Additionally, the sheet contains calculated values for skewness, kurtosis, and Geary's g statistic (Geary, 1936), which is another measure of kurtosis and which may be compared with the value 0.7979 (g value for a normal distribution). All of these values may be used by investigators to assist in their consideration of the normality issue.

3. Variance Testing

For independent, random samples from normally distributed populations, the F-test can be used to compare the variances of two samples. F is computed as the ratio of the variances of the samples:

$$F = \frac{s_1^2}{s_2^2}, \text{ where the null hypothesis is } H_0\text{: } \frac{\sigma_1^2}{\sigma_2^2} = 1,$$

and where s_1^2 is the larger of the two sample variances. The program prints the value of F, the associated p-value from an F-table with $n_1 - 1$ numerator degrees of freedom and $n_2 - 1$ denominator degrees of freedom. It also prints a decision to reject or not to reject the null hypothesis (using an alpha level of 0.05), and a warning if it seems that one or more test assumptions have been violated. If one or both of the data series violate the independence assumption due to serial dependence at lag 1, one can use the AR(1) filter option in the two-sample menu. Running the filter will prompt the program to use a modified formula for the calculation of F, which computes F in the presence of lag 1 dependence.

For those cases where the samples do not satisfy the assumption of normality, two modifications to Levene's statistic (Levene, 1960) for variance comparisons are included in the spreadsheet. The statistics are W_{10} and W_{50}, which are both robust to departures from normality in the underlying populations (Brown and Forsythe, 1974). They use robust measures of central location (W_{10}) uses the mean of the middle 80% of the observations; W_{50} uses the median observation) and thereby decrease or eliminate the influence of extreme values in testing variance. Brown and Forsythe (1974) recommend using W_{50} for testing the variances of asymmetric distributions and using W_{10} for testing long-tailed distributions. Both formulas involve the generation of a new data series z_{ij}

which also must be free of serial dependence. If a warning message is printed with the robust tests, it applies to the new series not the original series, and it is possible to have dependence in the z_{ij} series even if the original data were independent. Dependence is examined with an overall Q statistic (Ljung and Box, 1978) and a chi-square test developed in a fashion similar to that described above in the independence testing section. The output of the variance tests should not be used if a transformation (log, square root, arcsin, normalization) has been applied to the data. Also, performing a sequential average tends to "smooth" the data (i.e., variation is lowered in both data sets) and may have an effect on the variance test.

4. Means Testing

For independent, random samples from normally distributed populations, the parametric *t*-test can be used to compare the sample means. For unpaired samples the spreadsheet provides either a *t*-test with equal variance or a *t*-test with unequal variance (the results of the variance test above are used as the criterion for deciding which *t*-test to output). For paired samples, the sheet provides the results of a paired *t*-test. For all tests the output contains the computed value of the t statistic, a probability value (two-sided test, alpha = 0.05) from the *t*-distribution, a reject or do not reject decision regarding the null hypothesis, and a warning if it seems that one or more test assumptions have been violated. If one or both of the data series violate the independence assumption due to serial dependence at lag 1, one can use the AR(1) filter option from the two-sample menu. Running the filter will prompt the program to use a modified formula for the calculation of t, which computes t in the presence of lag 1 dependence.

For cases where one or both of the data sets violate the assumption of normality, a data transformation can be made to bring the data into normality (although variance testing is affected), or use one of the nonparametric tests which have been included in the spreadsheet. The tests are the rank sum test for unpaired comparisons (Wilcoxon, 1945) and the signed rank test for paired differences (Wilcoxon, 1949). Both tests are subject to the assumption of random independent samples. For the rank sum test both of the data series should be independent, while for the signed rank test only the series of differences needs to be independent. This can make the signed rank test a valuable tool for two-sample testing because, in practice, serial correlation often disappears when dependent data sets are differenced. Wilcoxon test results should not be used if an AR1 filter has been applied to the data.

Output for the Wilcoxon tests includes a description of the rejection region, the computed value for the test statistic T for small samples and z for larger samples ($n > 25$ for signed rank test, $n > 10$ for rank sum), a reject or do not reject decision regarding the null hypothesis (with a probability value based on a two-tailed test and alpha level of 0.05), and a warning if the independence assumption has been violated. Output for the signed rank Wilcoxon test also includes for the differenced data series an overall Q statistic and chi-square probability value, along with a graph of the autocorrelation function.

5. Distributional Shape Testing

The two-sample spreadsheet also includes a chi-square test, which compares the shapes of the distributions for the two samples. Although the samples may individually exhibit dependence, the assumption is made that the two samples are independent of each other. There are no assumptions regarding the form of the underlying distributions. To examine shape only, the program adjusts the two data sets by subtracting the respective mean value from each observation such that both series then have a common mean value of zero. Then the overall range of data in the two series is divided into the lesser of 9 or $n/4$ classes and a frequency distribution is produced for each sample. The frequency distributions are compared with a chi-square test.

Output for the shape test includes the computed values for the chi-square statistic and its associated probability value, a reject or do not reject decision regarding the null hypothesis (based on a one-tailed test and alpha level of 0.01), and a warning if the sample size is too small for a reliable test ($n < 25$) or if the independence assumption has been violated.

6. Principal Components Analysis and Regressions

Principal Component Analysis (PCA) and associated correlation matrices were determined using monthly data and the SAS™ statistical software package. A PCA was carried out as a preliminary review of the data using river flow information, sediment data, and water quality variables. The PCA was used to reduce the physical–chemical variables into a smaller set of linear combinations that could account for most of the total variation of the original set. For the physical–chemical variables in this study, the series were stationary after appropriate transformations; thus the sample correlation matrix of the variables was a good estimate of the population correlation matrix. Therefore, the standard PCA could be carried out based on the sample correlation matrix.

A matrix of the water and sediment quality data associated with the sampling stations in the Perdido system was prepared. The data were then grouped by station by year. Values for the dependent variables (total biomass m^{-2}, herbivore biomass m^{-2}, omnivore biomass m^{-2}, primary carnivore biomass m^{-2}, secondary carnivore biomass m^{-2}, tertiary carnivore m^{-2}) were then paired with the water/sediment quality data (independent variables) taken for stations within each sector. Unless otherwise defined, these statistics were run using SAS™, Systat™, and SuperAnova™. Data analyses were run on the three independent data sets. Significant principal components were then used to run a series of regression models with the biological factors as dependent variables. Residuals were tested for independence using serial correlation (time series) analyses and the Wald–Wolfowitz (Wald and Wolfowitz, 1940) runs test. A chi-square test was run to evaluate normality.

Index

A

Acartia tonsa, 107
Algae blooms, *see* Bloom effects on secondary production; Phytoplankton blooms
Ammonia
  coastal systems loading study
    nutrient loading, 183–184
    water quality, 188
  concentration gradient, Perdido Bay, 49–51
  concentrations, Elevenmile Creek, 139–140
  loading analysis, Perdido Bay, 45–49, 76–80, 118
  loading from Elevenmile Creek, 155
  water quality and, 251
Anabaena, 7
Anacystis, 7
Anchoa mitchilli (bay anchovies), 229
Animals and food web contamination dangers, 31
Anthropogenic stresses, *see also* Nutrient loading; Pollution
  in Escambia Bay, 203–204
  eutrophication process and, 2
  evaluating for phytoplankton, 4
  HABs in Perdido Bay and, 81–82
  human activities in Perdido Bay, 34–35, 44
  nutrient loading, 3–4
  nutrient loading effects, 9–10, 35
Apalachee Bay
  phytoplankton numbers and dominance table, 208–224
  phytoplankton suppression in, 212
  phytoplankton trends, 210, 212
  site description and characteristics, 26–27
Apalachicola Bay
  delta geography, 15
  drainage system, 13
  embayment geography, 14
  FII indices, 229–230, 233
  flow rates and nutrients study, *see* Gulf coastal systems comparison
  site description and characteristics, 27–29
  trophic organization interactions, 164–165
Aphanizomenon, 7
Atlantic croakers (*Micropogonias undulatus*), 229
Atlantic salmon (*Salmo salar*), 93

B

Bacillaria paxillifer, 124
Bacillariophyceae (diatoms), 100
Bairdiella chrysoura (silver perch), 229
Bay anchovies (*Anchoa mitchilli*), 229
Benthic lander
  experimental methodology, 307–308
  instrument specifications, 308
  operation, 308–309
Benthic microalgae in Perdido Bay
  cell numbers table, 126–135
  community role, 2
  food web impacted by, 122
  infauna changes, 125, 136
  inter-connections hypothesis, 136
  long-term trophic trends analysis, 167–168
  nutrient fluxes and sediments, 121–122
  salinity factors, 123
  secondary production impacted by, 160–162
  species richness, 124–125
  water column phytoplankton and, 122–123
  whole-water phytoplankton biomass and, 124
Biochemical oxygen demand (BOD), 75
Biscayne Bay, 251
Blackwater Bay and River, *see* Gulf coastal systems comparison; Pensacola Bay System
Bloom effects on secondary production, *see also* Phytoplankton blooms
  fishes, 29, 97–99
  food web and, 9
  overview, 153
  poisonous effects, 100
  red tide outbreaks, 96–97
  study results summary, 173
  trends analysis
    ammonia loading, 155
    epibenthic macroinvertebrates, 160–162
    fishery trends, 162
    infaunal numbers/species richness, 158–160, 163–164
    long-term changes, 155
    physical-chemical habitat changes, 155, 156–157
  trophic organization, *see* Trophic relationships and organization

Blue crabs (*Callinectes sapidus*), 25, 29, 228–229
Blue-green aglae structural components, 91–93
BOD (Biochemical oxygen demand), 75
Brief squid in Choctawhatchee Bay, 29

C

Callinectes sapidus (blue crabs), 25, 29, 228–229
Caulerpa taxifolia, 255
Chaetoceros spp., 10, 94
Chattahoochee Anticline, 14
Chattonella subsalsa, 6
  association with chlorophyll *a*, 143
  fish-killing mechanism, 97–99
  in Perdido Bay, 91, 115, 116
  structural components, 91
Chesapeake Bay
  bloom effects on oysters, 100
  complexity of system, 43–44
  inadequacies of research programs
    absence of needed data, 243
    consequences of study deficiencies, 244
    inappropriate assumptions, 242–243
    lack of long-term observation, 243
    lack of sampling elements coordination, 242
Chlorophyll *a*
  coastal systems loading study, 187–188, 189–191
  as indicator of water quality, 137, 261
  long-term water quality trends and, 138
  summary of effects, 261
  water quality in Perdido Bay and, 80, 83, 140–143
Chlorophyta (green algae), 103
Choctawhatchee Bay
  FII indices, 231–232
  fishery species, 29
  flow rates and nutrients study, *see* Gulf coastal systems comparison
  habitats, 29–30
  nutrient loading, 30–31
  physiography and location, 29
  phytoplankton numbers and dominance table, 208–224
  phytoplankton organization study, 212, 217
  seagrass deterioration, 30
Chrysophyceae (golden brown algae), 102–103
Clay percentages, 145*; see also* Sediment analysis
Coho salmon (*Oncothuynchus kisutch*), 93
Crassostrea virginica, *see* Oysters
Cryptophytes, 104
Cyanophytes, 104
Cyclotella choctawhatcheeana
  in Apalachee Bay, 210
  association with chlorophyll *a*, 143
  cell numbers in Perdido Bay, 112
  cell size, 141
  dominance in Perdido Bay, 114
  in Escambia Bay, 207
  nutrient loading effects, 117
  occurrences in Perdido Bay, 86, 90, 91, 217
  postbloom abundance, 119–120
  progression in abundance, 106
  salinity factors effecting, 95–96
  silica levels and, 104
  structural components, 91
  in Wolf Bay, 197
  zooplankton association, 107
Cynoscion
  arenarius (sand seatrout), 229
  nebulosus (spotted seatrout), 25
  regalis (sea trout), 93

D

Data collection methods, Perdido Bay study
  biological sampling and trophic analyses
    macroinvertebrates and fishes, 305
    phytoplankton, 303–304
    trophic determinations, 305–306
    zooplankton, 305
  experimental methodology
    benthic lander, diver-deployed, 307–309
    laboratory SOD determinations, 309
    nutrient limitation, 306–306
    primary productivity, 309–311
  light measurements, 302–303
  nutrient loading models, 300–301
  physical-chemical sampling and analysis, 299–300
  salinity and dissolved oxygen classification, 311
  sediment analyses, 301–302
  vertical sampling, 311
Diatoms, 4
  description and trends, 100
  silica abundance and, 104–106
  structural components, 91–93
DIN (dissolved inorganic nitrogen), 242
Dinoflagellates
  description and trends, 100–102
  life history, 6–7
  nutrient loading effects, 104
  structural components, 91–93
Dinophysis caudata, 248
Dissolved inorganic nitrogen (DIN), 242
Dissolved inorganic phosphorus (DIP), 242
Dissolved oxygen (DO)
  anomalies in, 138–139
  distribution in Gulf coastal systems, 180
  in Elevenmile Creek, 73–75
  as indicator of water quality, 137
  water quality in Perdido Bay and, 66, 67, 72–73
Distributional shape testing, 315–316
DO, *see* Dissolved oxygen
Dolphins, 31

E

Econfina River, *see* Apalachee Bay
Elevenmile Creek, *see also* Perdido Bay
  ammonia source, Perdido Bay, 155
  dissolved oxygen in, 73–75

flow rates and nutrients study, *see* Gulf coastal systems comparison
light distribution and productivity, 70–72
long-term water quality trends, 137–140
nutrient concentration gradient, 49–51
nutrient loading analysis, 45–49, 118–121
nutrients in, 76–80
silica distribution, 104–106
toxic agents in, 64–65
water color effects from, 69–70
Environmental Protection Agency, U.S. (EPA), 244, 253
EOCL (extractable, organically bound chlorine or halogen), 64
Escambia Bay, *see also* Pensacola Bay System
anthropogenic stresses, 203–204
FII indices, 232–233
flow rates and nutrients study, *see* Gulf coastal systems comparison
outlook, 208, 210
phytoplankton numbers and dominance table, 208–224
phytoplankton organization, 203–210
species richness, 207–208, 209–210
water flow, 208
Everglades National Park, 251
Extractable, organically bound chlorine or halogen (EOCL), 64

F

Falcula hyalina, 106
Fecal coliform bacteria, 251
Fenholloway River, *see* Apalachee Bay
Fish and Wildlife Service, U.S., 35
Fishes, *see also* Bloom effects on secondary production
bloom fish-killing mechanism, 97–99
long-term trophic trends analysis, 167
phytoplankton blooms effects on, 29
red tide outbreaks and, 96–97
species in Choctawhatchee Bay, 29
Fish-invertebrate-infauna (FII) indices
biomass distribution and, 233–235
habitat features and, 233
long-term trophic trends analysis, 166–167, 169
in Perdido Bay, 230–231
throughout Gulf coast, 229–233
Florida Bay
consequences of poor regulation, 244
inadequacies of research programs
absence of whole-system data, 245–246
lack of cause-and-effect analysis, 244–245
lack of systematic approach, 246
study papers incompleteness, 245
Florida Department of Natural Resources, 252
Florida Department of Regulation, 251
Florida Keys, 251
Floridan Aquifer, 21
Florida Panhandle, *see* Northeast Gulf of Mexico
Florida Peninsula Sedimentary Province, 15
Florida Sea Grant, 252
Florida State University, 252
Flounder (*Paralichthys*), 229
Flushing analysis, Perdido Bay, 42–43, 181–182
Food web
benthic algae impact on, 122
blooms affect on processes, 9
coastal systems loading study, 196
complexity of, 1
importance of processes, 2–3
long-term trophic trends analysis
deterioration, 169
structure, 168–169
productivity determination, 1–2
pulp mill effluent responses, 27
responses to plankton in Gulf coast
existing conditions, 228
FII biomass distribution, 233–235
FII indices and habitat features, 233
FII indices throughout, 229–233
habitat/productivity relationship, 235–236
nutrient/food web interactions, 226–227
species in estuaries, 228–229
summary observations, 236
trophic relationships, 224, 226, 227–228
response to nutrient loading, 9–10
structure in Perdido Bay, 168
Franklin County Commission, 252

G

Global warming, 257–258
Golden brown algae (Chrysophyceae), 102–103
Grandidierella bonnieroides, 228
Great Lakes Program, 244
Green algae (Chlorophyta), 103
Gulf coastal systems comparison
flow rates and nutrient loading/concentrations
ammonia loading, 183–184
light and temperature limitations, 187
orthophosphate loading, 182–183
overall loading factor, 185
silica limitation effects, 186
summary observations, 188
systems studied, 182
food web responses to plankton organization
existing conditions, 228
FII biomass distribution, 233–235
FII indices and habitat features, 233
FII indices throughout, 229–233
habitat/productivity relationship, 235–236
nutrient/food web interactions, 226–227
species in estuaries, 228–229
summary observations, 236
trophic relationships, 224, 226, 227–228
habitat conditions, 175–176
nutrient loading effects, 4
phytoplankton indices, 217–218, 224
phytoplankton organization
Apalachee Bay, 210–212
Choctawhatchee Bay, 212, 217
community structure study results, 196

Escambia Bay, 203–210
food web impacted by, 196
Perdido Bay, 217
table of numbers and dominance, 208–223
Wolf Bay, 197–203
salinity stratification
analysis, 176–179
circulation and mixing, 181–182
dissolved oxygen distribution, 180
factors involved, 181
indices definition, 176
salinity gradients, 179–180
sediment quality, 191–196
study summary of results
bloom occurrence due to nutrient loading, 269
coordination need, 278
ecosystem-level knowledge importance, 276
geographic and temporal scope of study, 269
impact of inadequate regulations, 276–278
long-term changes summary, 273–275
long-term plankton trends, 270–271
nutrient loading, 270
restoration program elements, 278
sedimentary microalgae trophic response, 271–272
sediment quality, 273
species-specific responses, 270
water flow effects, 272–273
water quality, 188–191
Gulf flounder (*Paralichthys albigutta*), 229
Gulf of Mexico research programs, 241
Gulf of Mexico Sedimentary Basin, 14
Gymnodinium spp., 10
breve, *see* Red tide
occurrences in Perdido Bay, 91
splendens
in Perdido Bay, 116, 217
poisonous effects, 100
in Wolf Bay, 197

H

HAB, *see* Harmful algal blooms
Halodule wrightii (shoalgrass)
in Choctawhatchee Bay, 30
pulp mill effluent responses, 26
Harmful algal blooms (HAB), *see also* Phytoplankton blooms
cause-and-effect determination, 6
extent of, 5
general forms, 6
life history, 6–7
phytoplankton trends, Perdido Bay, 81–82
Health and safety issues from red tide effects, 6
Heterosigma spp.
akashiwo
chronological trends in Perdido Bay, 86, 90, 91
in Escambia Bay, 207
long-term changes in Perdido Bay, 115, 116, 217
red tide outbreaks and, 96–97
structural components, 91
in Wolf Bay, 197
Hobsonia florida, 228
Hogchokers (*Trinectes maculatus*), 229
Hypereutrophication impact in Pensacola Bay, 254–255

I

Independence tests, 313–314
Invertebrates
long-term trophic trends analysis, 167
phytoplankton blooms and, *see* Bloom effects on secondary production

K

Kissimmee River, 251

L

Leiostomus xanthurus (spot), 229
Leptocylindricus danicus, 10
distribution, 93
dominance in Perdido Bay, 114
occurrences in Perdido Bay, 86, 115, 217
progression in abundance, 106
structural components, 91
Light and temperature
distribution in Elevenmile Creek, 70–72
limitations found in Gulf coastal systems, 187
measurement collection methods, 302–303
Lingulodinium machaerophorum, 196
Liquid mud, 57–58

M

Mactra fragilis, 228
Means testing, 315
Mediomastus ambiseta
in Gulf coast food web, 228
in Perdido Bay, 125
Merismopedia tenuissima
cell size, 141
dominance in Perdido Bay, 114
occurrences in Perdido Bay, 86, 90, 91, 116
progression in abundance, 106
structural components, 91, 93
zooplankton association, 107
Metal contamination in Pensacola Bay system, 33
Microalgae, *see* Benthic microalgae in Perdido Bay
Micropogonias undulatus (Atlantic croakers), 229
Miraltia throndsenii
anthropogenic factors, 93–94
association with chlorophyll *a*, 143
cell size, 141

occurrences in Perdido Bay, 86, 91, 115, 217
structural components, 91
Mixotrophy, 100

N

National Science Foundation, 256
Navicula, 124
Nitrate/nitrite concentrations
gradient, Perdido Bay, 49–51
loading analysis, Perdido Bay, 45–49
water quality and, 251
Nitzschia spp., 124
palea, 4
pungens, 248
reversa, 91
Non-point-source pollution, *see also* Anthropogenic stresses; Pollution
nutrient loading from, 3–4, 9–10, 239–240
in Pensacola Bay System, 32
Normality tests, 314
Northeast Gulf of Mexico, *see also specific sites*
aquatic habitats, 22–23
climate
precipitation, 18–20
temperature, 17–18
wind, 20–21
geographic description, 11
physiography, 11–13
primary study sites
Apalachee Bay, 26–27
Apalachicola Bay, 27–29
Choctawhatchee Bay, 29–31
general characteristics, 25
Pensacola Bay, 31–33
Perdido Bay, 33–36
regional geology, 14–15, 17
river flow, 21–22
tides, 22
Nutrient loading, *see also* Pollution
ammonia
coastal systems loading study, 183–184, 188
concentration gradient, Perdido Bay, 49–51
concentrations, Elevenmile Creek, 139–140
loading analysis, Perdido Bay, 45–49, 76–80, 118
loading from Elevenmile Creek, 155
water quality and, 251
analyses models, 300–301
anthropogenic stresses
effects of, 3–4, 9–10, 35
eutrophication process and, 2
evaluating for phytoplankton, 4
HABs in Perdido Bay and, 81–82
human activities, 34–35, 44
chlorophyll *a*
coastal systems loading study, 187–188, 189–191
as indicator of water quality, 137
long-term water quality trends and, 138
water quality in Perdido Bay and, 83, 140–143
in Choctawhatchee Bay, 30–31
coastal systems effected by, 4
Elevenmile Creek
analysis, 45–49, 118–121
gradients, 49–51
extent of, 3
food web response to, 9–10
orthophosphates
coastal systems loading study, 182–183, 188
concentration gradient, Perdido Bay, 49–51
loading analysis, Perdido Bay, 45–49, 118
occurrence in Perdido Bay, 76–80
plankton response to, 4–5
primary effects, 3–4
silica concentrations
coastal systems loading study, 186, 188
distribution in Elevenmile Creek, 104–106
Elevenmile Creek, 140
in Perdido Bay, 78–80
study analysis details, Perdido Bay, 44–49
water flow rates and
ammonia loading, 183–184
light and temperature limitations, 187
orthophosphate loading, 182–183
overall loading factor, 185
silica limitation effects, 186
summary observations, 188
systems studied, 182

O

Oncothuynchus kisutch (coho salmon), 93
Organic matter quality variation, 64
Orthophosphates
coastal systems loading study
nutrient loading, 182–183
water quality, 188
concentration gradient, Perdido Bay, 49–51
loading analysis, Perdido Bay, 45–49, 118
in Perdido Bay, 76–80
Oscillatoria spp., 4, 8
Ostrea iurida (oyster), 100
Oysters (*Crassostrea virginica*), 25
bloom effects on, 100
decline in Chesapeake Bay, 242
mortality due to stenohaline predators, 28

P

PAHs (polynucleated aromatic hydrocarbons)
in Pensacola Bay System, 33
in Perdido Bay, 64
Palaemonetes spp. (shrimp), 228
Panaeus (penaeid shrimp), 228–229
Paralichthys (flounder), 229
Parandalia americana, 228
PCBs (polychlorinated biphenyls), 33
Pedinophora sp.

occurrences in Perdido Bay, 86
in Perdido Bay, 116
Penaeid shrimp (*Penaeus* spp.), 25, 29, 229
Pensacola Bay System
Escambia Bay
anthropogenic stresses, 203–204
FII indices, 232–233
outlook, 208, 210
phytoplankton numbers and dominance table, 208–224
phytoplankton organization, 203–210
species richness, 207–208, 209–210
water flow, 208
FII indices, 232–233
flow rates and nutrients study, *see* Gulf coastal systems comparison
fresh water sources, 32
long-term nutrient loading effects, 32–33
physiography, 31–32
Perdido Bay
data collection methods, *see* Data collection methods, Perdido Bay study
drainage system, 12
FII indices, 230–231
flow rates and nutrients study, *see* Gulf coastal systems comparison
fresh water sources, 33–34
geographic regions, 34
nutrient concentration gradients, 49–51
nutrient limitation
experimental techniques, 51–52, 54–55
phosphorus or nitrogen question, 52–53, 55–56
nutrient loading
human activities, 34–35, 44
study analysis details, 44–49
phytoplankton blooms
bloom species, *see* Phytoplankton blooms, species status in Perdido Bay
effects of, *see* Bloom effects on secondary production
status, 39–40
trends in benthic microalgae, *see* Benthic microalgae in Perdido Bay
phytoplankton numbers and dominance table, 208–224
phytoplankton organization, 217
phytoplankton trends
chronology, 85–86, 90–91
chronology summary table, 87–89
HABs effects, 81–82
long-term changes, 100–104
research shortcomings, 82–83
silica and diatom abundance, 104–106
study methodology, 83–85
whole-water phytoplankton, *see* Whole-water phytoplankton, Perdido Bay
zooplankton and, 106–109
pollution sources, 34
restoration study
non-point nutrient sources and, 239–240
nutrient loading targets determination, 237–238
nutrient monitoring, 239
river flow trends
flushing, 42–43
productivity and, 40–41
seasonal patterns, 41–42
water column stratification, 43–44
salinity fluxes, 34, 37
sampling program, 39
SAV changes, 35
sediment analysis, 36
sediment quality
isotope analyses, 63–64
liquid mud, 57–58
nutrient analysis, 58–63
physical characteristics, 57–58
productivity effected by, 56
toxic agents, 64–65
water depth factors, 56–57
sediment quality long-term trends
bloom frequency and, 145–146
clay percentages, 145
phosphorus levels, 143–144
statistical analysis of trends
bloom period, 1993–94, 146–147
comparisons of long-term, 148–151
study conclusions
bloom occurrence due to nutrient loading, 269
chlorophyll *a* association, 261
coordination need, 278
discharge limits recommendation, 262–263
ecosystem-level knowledge importance, 276
impact of inadequate regulations, 263, 276–278
impact of inadequate studies, 264
long-term changes summary, 273–275
long-term plankton trends, 270–271
nutrient loading, 260–261, 270
nutrient utilization, 260
phytoplankton bloom history, 261
phytoplankton blooms association, 262
productivity levels, 259
restoration program elements, 278
sedimentary microalgae trophic response, 271–272
sediment quality, 261–262, 273
species-specific responses, 270
statistical analysis, *see* Statistical analyses methods
understanding of blooms, 259–260
water flow effects, 272–273
study elements, 38
study overview, 37
study results summary, 173
water quality
biochemical oxygen demand, 75
chlorophyll *a*, 80
color, 69–70
dissolved oxygen and, 66, 67, 72–73
hypoxia/anoxia impact, 66–67
light distribution and productivity, 70–72
nutrient limitation, 67
nutrients, 76–80
overview of factors affecting, 65–66
pH, 75
phytoplankton blooms and, 137–143

salinity, 68
salinity stratification, 66
Secchi depth, 70
temperature, 68
turbidity, 70
Periclimenid shrimp, 29
Pfiesteria, 6–7
Phaeocystis, 196
pH and water quality in Perdido Bay, 75
Phosphorus
long-term sediment quality trends and, 143–144
water quality and, 251
Phytoplankton
nutrient loading response, 4–5
organization in Gulf coastal systems, 8–9
Apalachee Bay, 210–212
Choctawhatchee Bay, 212, 217
community structure study results, 196
Escambia Bay, 203–210
food web impacted by, 196
Perdido Bay, 217
table of numbers and dominance, 208–223
Wolf Bay, 197–203
water quality indicated by
dissolved oxygen and, 8–9
historic survey use, 8
lack of research data, 9
pollution presence, 7–8
usefulness of, 9
Phytoplankton blooms
benthic microalgae in Perdido Bay
cell numbers table, 126–135
community role, 2
food web impacted by, 122
infauna changes, 125, 136
inter-connections hypothesis, 136
nutrient fluxes and sediments, 121–122
salinity factors, 123
secondary production impacted by, 160–162
species richness, 124–125
water column phytoplankton and, 122–123
whole-water phytoplankton biomass and, 124
in Choctawhatchee Bay, 31
classes and divisions
Bacillariophyceae, 100
Chlorophyta, 103
Chrysophyceae, 102–103
Dinophyta, 100–102
definition, 40
effects on secondary production, *see* Bloom effects on secondary production
harmful algal blooms concept
cause-and-effect determination, 6
extent of, 5
general forms, 6
life history, 6–7
phytoplankton trends, Perdido Bay, 81–82
lack of understanding of, 4
species status in Perdido Bay
anthropogenic stresses, 93–94
environmental factors, 95–96
fish-killing mechanism, 97–99
long-term loading pattern, 99–100
poisonous effects, 100
red tide outbreaks, 96–97
structural components, 91–93
status in Perdido Bay, 39–40
toxins associated with, 5
understanding of, 259–260
Pleurosigma, 124
Political influences and pressures on research, 252, 255–256
Pollution, *see also* Anthropogenic stresses; Nutrient loading
in Florida, 251
non-point-source
nutrient loading, 3–4, 9–10, 239–240
in Pensacola Bay System, 32
point source nutrient loading, 9–10
pulp mill effluent responses
ammonia source, Perdido Bay, 45
effects on food web, 27
sources in Perdido Bay, 34, 45
Polychlorinated biphenyls (PCBs), 33
Polynucleated aromatic hydrocarbons (PAHs)
in Pensacola Bay System, 33
in Perdido Bay, 64
Prorocentrum micans, 4
Prorocentrum minimum
biomass changes in Perdido Bay, 113
dominance in Perdido Bay, 114–115
in Escambia Bay, 207
nutrient loading effects, 117
occurrences in Perdido Bay, 86, 90, 217
poisonous effects, 100
postbloom abundance, 120
species richness in Perdido Bay, 111
structural components, 91
in Wolf Bay, 197
Prorocentrum spp., 10
Pseudonitzschia pseudodelicatissima, 248
Pulp mill
ammonia source, Perdido Bay, 45
effluent effects on food web, 26, 27

R

Rangia cuneata
in Gulf coast food web, 228
in Perdido Bay, 108, 125
Raphidophytes
fish-killing mechanism, 97–99
nutrient loading effects, 104
structural components, 91–93
Red tide (*Gymnodinium breve*)
in Choctawhatchee Bay, 31
HABs and, 6
outbreaks worldwide, 96–97
Research on ecosystem ecology
inadequacies of programs examples
absence of needed data, 243
absence of whole-system data, 245–246

consequences of study deficiencies, 244
inappropriate assumptions, 242–243
lack of cause-and-effect analysis, 244–245
lack of direction in efforts, 241
lack of long-term observation, 243
lack of sampling elements coordination, 242
lack of systematic approach, 246
study papers incompleteness, 245
information and regulation
damages caused by development, 251–252
environmental costs of poor research, 256–257
impediments to funding, 255
inadequacies of studies, 253–254
lack of preventative action, 253
lack of regulation, 255
projected effects of global warming, 257–258
lack of funding, 244
lack of scientific approach, 240
political influences and inaction, 255–256
political pressure effects on research, 252
research process needed, 240–241
scientific overview of eutrophication
centralized control need, 246–247, 250
inadequacies of studies and models, 249–250
lack of understanding of blooms, 248
principal issues, 247–248
topics requiring study, 247, 250
Restoration study of Perdido Bay
non-point nutrient sources and, 239–240
nutrient loading targets determination, 237–238
nutrient monitoring, 239
Rhithropanopeus harrisi (xanthid crabs), 228
Rhizosolenia spp., 141
Rhopalodia gibberula, 4

S

Salinity factors
Apalachicola Bay, 28
fluxes, 34, 37
stratification in Perdido Bay, 66, 68
Salinity stratification
benthic microalgae and, 123
Gulf coastal systems comparison
analysis, 176–179
circulation and mixing, 181–182
dissolved oxygen distribution, 180, 311
factors involved, 181
indices definition, 176
salinity gradients, 179–180
Salmo salar (Atlantic salmon), 93
Science Advisory Board, EPA, 253
Seagrasses
deterioration in Choctawhatchee Bay, 30
pulp mill effluent responses, 26
Seatrout (*Cynoscion*), 25, 93, 229
Secchi depth, 70, 188
Secondary production, *see* Bloom effects on secondary production
Sediment analysis in Perdido Bay
clay percentages and quality, 145
data collection methods, 301–302
Gulf coastal systems comparison, 191–196
long-term trends
bloom frequency and, 145–146
clay percentages, 145
phosphorus levels, 143–144
nutrient fluxes and, 121–122
quality, 261–262, 273
bloom frequency and, 145–146
clay percentages, 145
isostope analyses, 63–64
liquid mud, 57–58
nutrient analysis, 58–63
phosphorus levels, 143–144
physical characteristics, 57–58
productivity effected by, 56
toxic agents, 64–65
water depth factors, 56–57
scope of, 36
Sediment oxygen demand (SOD)
coastal systems loading study, 195
laboratory determinations, 309
Shellfish, *see also* Bloom effects on secondary production
oysters, 25
bloom effects on, 100
decline in Chesapeake Bay, 242
mortality due to stenohaline predators, 28
poisonous effects of dinoflagellates, 100
shrimp, 25, 29, 228–229
Shoalgrass (*Halodule wrightii*)
in Choctawhatchee Bay, 30
pulp mill effluent responses, 26
Shrimp (*Palaemonetes* spp.), 228
Silica concentrations
coastal systems loading study
nutrient loading, 186
water quality, 188
distribution in Elevenmile Creek, 104–106, 140
in Perdido Bay, 78–80
Silver perch (*Bairdiella chrysoura*), 229
Skeletonema costatum, 8, 10
SOD (sediment oxygen demand)
coastal systems loading study, 195
laboratory determinations, 309
Soil clay percentages and sediment quality, 145
South Florida Water Management District, 251
Species richness
benthic microalgae in Perdido Bay, 124–125
bloom effects on secondary production, 158–160, 163–164
in Escambia Bay, 207–208, 209–210
Spot (*Leiostomus xanthurus*), 229
Spotted seatrout (*Cynoscion nebulosus*), 25
Statistical analyses methods
distributional shape testing, 315–316
independence tests, 313–314
means testing, 315
normality tests, 314
principal components analysis and regressions, 316
variance testing, 314–315

Streblospio benedicti
in Gulf coast food web, 228
in Perdido Bay, 125
Submerged aquatic vegetation (SAV), 35
Surficial Aquifer, 21
Suwannee Straits, 14–15
Synedropsis sp.
occurrences in Perdido Bay, 86, 90, 91, 115, 116, 217
structural components, 91
Syringodium filiforme (seagrass), 26

T

Temperature of Perdido Bay, 68
Thalassia testudinum (seagrass), 26
Toxic agents in blooms, 5, 64–65, 100*; see also* Red tide
Trichodesmium sp., 248
Trinectes maculatus (hogchokers), 229
Trophic relationships and organization
biological sampling and analyses
macroinvertebrates and fishes, 305
phytoplankton, 303–304
trophic determinations, 305–306
zooplankton, 305
food web and plankton, 224, 226, 227–228
long-term trends
benthic herbivores, 167–168
FII biomass analysis, 166–167, 169
fishes, 167
food web deterioration, 169
food web structure, 168–169
invertebrates, 167
trophic changes, 169–173
research approach
field data collection and treatment, 165–166
interactions in Apalachicola Bay, 164–165
space and time considerations, 165
trophic hierarchy use, 164
trophodynamic concept, 162, 164
sedimentary microalgae in Gulf coast, 271–272
Turbidity in Perdido Bay, 70

U

U.S. Army Corps of Engineers, 244
U.S. Environmental Protection Agency (EPA), 244, 253
Urosolenia eriensis, 86, 116

V

Vallisneria americana, 35
Variance testing, 314–315

W

Water column stratification, 43–44
Water quality in systems
biochemical oxygen demand, 75
dissolved oxygen, 72–73
pH, 75
phytoplankton as indicator
dissolved oxygen and, 8–9
historic survey use, 8
lack of research data, 9
pollution presence, 7–8
usefulness of, 9
phytoplankton blooms and
chlorophyll *a* and, 140–143
long-term trends, 137–140
overview, 137
Whole-water phytoplankton, Perdido Bay
long-term changes in distribution
biomass, 113–114
bloom species, 114–116
cell numbers, 112–113
nutrient limitation response, 116–118
seasonal trends, 110–111
species richness, 111–112
nutrient loading response
factors involved, 119
state of the bay, 120–121
summary, 121
variable nature of, 118
Wolf Bay, *see also* Perdido Bay
agricultural activities impact, 197, 203
infauna distribution, 199, 203
sediment quality, 198–199
species richness, 197–198, 200–202
water quality analysis, 197

X

Xanthid crabs (*Rhithropanopeus harrisi*), 228

Y

Yellow Bay, *see* Gulf coastal systems comparison; Pensacola Bay System

Z

Zooplankton vs. phytoplankton trends, 107–110